ANIMAL COLONIES
Development and
Function Through Time

Edited by
Richard S. Boardman, Alan H. Cheetham,
and **William A. Oliver, Jr.**
Associated Editors: **A. G. Coates** and **F. M. Bayer**

Dowden, Hutchinson
& Ross, Inc.
Stroudsburg, Pennsylvania

Library of Congress Cataloging in Publication Data
Main entry under title:

Animal colonies; their development and function through time.

Papers presented at a symposium held Nov. 1971 during the annual meet-
ings of the Geological Society of America and the Paleontological Society in
Washington.
  Includes bibliographies.
  1. Animal colonies—Congresses. 2. Invertebrates—Congresses. I. Board-
man, Richard S., ed. II. Cheetham, Alan H., ed. III. Oliver, William
Albert, 1926–   , ed. IV. Geological Society of America. V. Paleontol-
ogical Society.
QL364.5.A54   592'.05'24   73–14960
ISBN   0–87933–035–X

Copyright © 1973 by **Dowden, Hutchinson & Ross, Inc.**
Library of Congress Catalog Card Number: 73–14960
ISBN: 0–87933–035–X

Manufactured in the United States of America.
74 75 76   5 4 3 2
Exclusive distributor outside the United States and Canada:
JOHN WILEY & SONS, INC.

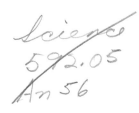
Contributors

Marie B. Abbott *Gray Museum, Marine Biological Laboratory, Woods Hole, Massachusetts 02543*

Leonard M. Bahr *Department of Biology, University of Georgia, Athens, Georgia 30601*

William C. Banta *Department of Biology, The American University, Washington, D.C. 20016*

Frederick M. Bayer *Rosenstiel School of Marine and Atmospheric Sciences, University of Miami, Miami, Florida 33149*

William B. N. Berry *Department of Paleontology, University of California, Berkeley, California 94720*

John W. Bishop *Department of Biology, University of Richmond, Richmond, Virginia 23173*

Daniel B. Blake *Department of Geology, University of Illinois, Urbana, Illinois 61801*

Richard S. Boardman *Department of Paleobiology, National Museum of Natural History, Smithsonian Institution, Washington, D.C. 20560*

Alan H. Cheetham *Department of Paleobiology, National Museum of Natural History, Smithsonian Institution, Washington, D.C. 20560*

A. G. Coates *Department of Geology, The George Washington University, Washington, D.C. 20006*

Bernd-Dietrich Erdtmann *Department of Earth and Space Sciences, Indiana University, Fort Wayne, Indiana 46805*

Jack D. Farmer *Department of Geology, University of California, Davis, California 95616*

Robert M. Finks *Department of Geology, Queens College, The City University of New York, Flushing, New York 11367*

Willard D. Hartman *Department of Biology and Peabody Museum of Natural History, Yale University, New Haven, Connecticut 06520*

J. A. E. B. Hubbard   *Department of Geology, King's College, Strand, London, England*

Valdar Jaanusson   *Naturhistoriska Riksmuseet, Paleozoologiska sektionen, Stockholm 50, Sweden*

R. K. Jull   *Department of Geology, University of Windsor, Windsor, Ontario, Canada*

Karl W. Kaufmann   *Department of Geophysical Sciences, University of Chicago, Chicago, Illinois 60637*

Don L. Kissling   *Department of Geology, State University of New York, Binghamton, New York 13901*

G. O. Mackie   *Department of Biology, University of Victoria, Victoria, B.C., Canada*

William A. Oliver, Jr.   *Paleontology & Stratigraphy Branch, United States Geological Survey, Washington, D.C. 20244*

Philip J. Phillips   *100 Belmont Place, Staten Island, New York 10301*

Henry M. Reiswig   *Redpath Museum, McGill University, Montreal, Canada*

Albert J. Rowell   *Department of Geology, University of Kansas, Lawrence, Kansas 66044*

Philip A. Sandberg   *Department of Geology, University of Illinois, Urbana, Illinois 61801*

Thomas J. M. Schopf   *Department of the Geophysical Sciences, and Committee on Evolutionary Biology, University of Chicago, Chicago, Illinois 60637*

Tracy L. Simpson   *Department of Biology, University of Hartford, West Hartford, Connecticut 06117*

Robert S. Takagi   *Department of Paleontology, University of California, Berkeley, California 94720*

Adam Urbanek   *Katedra Paleontologii, Uniwersytet Warszawski, al. Zwirki i Wigury 93, Warszawa 22, Poland*

John Utgaard   *Department of Geology, Southern Illinois University, Carbondale, Illinois 62901*

J. W. Wells   *Department of Geological Sciences, Cornell University, Ithaca, New York 14850*

Timothy S. Wood   *Department of Biological Sciences, Wright State University, Dayton, Ohio 45431*

Introducing Coloniality

This collection of papers is an attempt to analyze the basic nature of metazoan colonies and to provide new information on the development and function of key groups of colonial animals. New descriptive data permit evaluation of the concept of coloniality as applied to the Porifera, Coelenterata, Bryozoa, and Hemichordata (Class Graptolithina). These phyla have significant fossil records which document the evolution of the colonial habit.

During the past 20 years there has been a significant increase in the study of colonial animals. New techniques have allowed a major accumulation of new morphologic data on living and fossil colonies and physiologic data on living colonies. Interpretation of modes of growth and function, however, have been made in relatively few groups within the phyla that include colonial animals. The concept of the colony, which is based on these interpretations, has been critically evaluated in even fewer groups.

A fundamental feature of the colony, as used in this volume, is the apparent genetic constancy of its members (zooids, individuals of a colony). Morphologic and physiologic variation in populations of solitary animals can be ascribed to genetic, ontogenetic, and ecologic factors. Common ancestry through asexual reproduction ensures that differences among members of a colony are nongenetic, except possibly for somatic mutations and extrachromosomal factors, probably of negligible phenotypic expression in most groups. Fusion and intergrowth of zooids with different genotypes have been demonstrated in colonial tunicates (see Burnet, 1971, for review), but there is no genetic evidence for this in the groups discussed in this volume.

Assumption of genetic constancy permits direct evaluation of the other factors controlling morphologic and physiologic differences between the members of a colony. Contributors to this volume discuss such factors as ontogeny, astogeny, polymorphism, and microenvironment. None of these is unique to colonies as they may be sources of variation in any clone, but clones of solitary animals cannot be recognized as fossils and, if living, can generally be studied only in the laboratory. Polymorphic differences between individuals in certain nonclonal animals, such as social insects, can also be determined by factors other than genetic differences (Wilson, 1953, 1968, 1971), and in that respect are similar to polymorphism and possibly to astogenetic (generational) differences in clones.

Definitions of colony have ranged from the pragmatic to the philosophical and from exclusive to all inclusive. Perrier (1880, 1881, 1898) represented a nineteenth-century extreme that considered all multicellular animals to be colonies of cells that to some extent retained their individuality, and most higher animals to be colonies of higher-level individuals. He grouped "colonial" metazoans as (1) irregular colonies (sponges, coelenterates, bryozoans, tunicates); (2) linear colonies (segmented "worms," arthropods, and by extension vertebrates); and (3) coalesced linear colonies (echinoderms). Other Metazoa (e.g., mollusks, brachiopods) were considered to be more highly integrated colony derivatives. Perrier outlined a sequence of levels of coloniality which, although not the earliest, serves to indicate that this concept of progression is not new. Perrier's succession was "the formation of a colony, the division of physiological labor, the appearance of polymorphism, and the concentration [i.e., integration] of the parts so elaborated" (Perrier 1880, p. 634). In 1881 (p. 403) he modified the fourth level to "the transformation of the colony into an individual." Perrier reviewed earlier concepts relating to individuality and coloniality, but these and his own interpretations are now largely of historical interest.

Mackie (1963, p. 329–331) outlined the absurdities that can result from too philosophical an approach to the question of individuality in relation to colonies. Logic can carry one to the conclusion that all colonies are individuals or that all individuals are colonies: "It immediately becomes clear that individuality is too elusive a thing to serve as a basis for classifications . . ." (p. 330). The concept of the colony is too useful to be discarded and a more practical approach is necessary and has been adopted by Mackie and most other contemporary workers.

Thompson and Geddes (1931) included a wide-ranging discussion of coloniality in their general biology textbook. They outlined a seven-stage progression of colonies "from aggregates to integrates," with sponges and siphonophores illustrating the first and last of these (p. 115). Except possibly for the sponge stage, their discussion is compatible with most of the definitions in this symposium. Thompson and Geddes (p. 117) referred to the colony progression as one line of integrative evolution and drew a parallel between this and levels of increasing interdependence in social insects (p. 117).

The most comprehensive modern reviews and analyses of coloniality are those of Beklemishev (1944–1970), to which the reader is referred for discussions of the comparative anatomy of the living colonial Metazoa. Beklemishev suggests eight stages in "the weakening of the individuality of zooids" and eight more in the "intensification of the individuality of the colony" (1970, p. 483–489). These processes, according to Beklemishev, lead toward degrees of colonial organization approaching "individuals of a higher order" (p. 490). Beklemishev's text, in several editions, is a modern foundation for work on colonies and is cited in several of the following papers. Other work on coloniality is widely scattered in the literature and comprehensive review is beyond the scope of this preface.

Papers in this volume demonstrate the complexity of, and differences in, the concept of colony as applied to various groups of animals. Two common themes run through most of the expressed or implied definitions: members of the colony must be physically connected and they must have common ancestry through asexual reproduction. These factors exclude social insects, clones of physically discontinuous members, and certain "colonial worms" that are united by an apparently common skeleton but are not clones (e.g., serpulid "reefs"). Physical connection and asexual reproduction in the groups covered in this volume permit degrees of integration of zooids within a colony ranging from complete zooid autonomy at one extreme to almost complete integration (colony control) at the other.

Graptolites are considered by the authors in this volume to be clear-cut colonies having zooids varying morphologically in complex colony-wide patterns. These patterns are interpreted as indicating high levels of colony control, but with zooids retaining a distinct morphologic identity.

Bryozoans are all colonial as interpreted by the authors in this volume. The concept of colony in bryozoans is based on comparison of member zooids with individuals in solitary groups. Zooids range from feeding members capable of sexual reproduction to polymorphs of restricted function and divergent morphology. In addition, many bryozoan colonies have extrazooidal parts that support or connect the member zooids. The member zooids and extrazooidal parts of a bryozoan colony are developed from the expansion and partitioning of the body cavity and bounding walls of a primary zooid or group of zooids. The flexibility of this mode of development is indicated by the wide variety of arrangements of interior and exterior walls bounding zooids, of extrazooidal parts, and of patterns of morphologic variation of zooids in colonies, all suggesting varying degrees of colony control. Many of these arrangements and their interpretation are discussed by the authors in this volume.

The phylum Coelenterata includes solitary as well as colonial forms and much has been learned about colony biology through comparison with related solitary individuals. Solitary as well as colonial coelenterates can reproduce asexually, and simple colonies are little more than incompletely separated solitary animals. In some corals the connections are skeletal only and these represent the simplest type of colony in any of the groups discussed. Other coral colonies are more highly developed systems with polymorphs and varying degrees of morphologic and physiologic interdependence of members. At the upper end of the scale of coloniality, siphonophores are colonies of zooids so specialized and integrated that the zooids can be compared to organ systems in other animals:

> No one would suggest that the siphonophores are "higher animals," but they are the most complex coelenterates and the only ones to have explored fully the possibilities of colonialism. They have developed colonialism to the point where it has provided them with a means of escaping from the limitations of the diploblastic body-plan. The higher animals escaped these limitations by becoming triploblastic and using the new layer, the mesoderm, to form organs. The siphonophores have reached the organ grade of construction by a different method—that of converting whole individuals into organs. It is interesting to speculate that, had it not been for the invention of the mesoderm in some remote, diploblastic era, the highest animals on earth might now be, if not the Siphonophora, something similar to them in principle (Mackie, 1963, p. 336).

The interpretation of sponges as solitary or colonial animals has been controversial, and the controversy continues among the authors in this volume. The flexibility in arrangement of different cell types makes member units difficult to compare with zooids in more obviously colonial groups, but some functional units are asexually duplicated and retained in physical connection by the sponge.

Many of the interpretations in this symposium, as well as the concept of colony itself, are based on the assumption of genetic constancy within a colony. There is a large mass of empirical data supporting this assumption, but little experimental work has been done. The work on colonial tunicates recently reviewed by Burnet (1971) indicates that physiologic connection is possible between zooids of different genotypes having one "recognition" allele in common. It is important to know whether this kind of compatibility is possible and/or common in other colonial groups and how this compatibility can be recognized. This volume is intended to focus attention on the need for additional interpretive studies in the genetics and physiology of colonies as well as in their morphology.

This book is based on a Paleontological Society symposium, entitled Development and Function of Animal Colonies, held at the November 1971 Annual Meetings of the Geological Society of America and the Paleontological Society in Washington, D.C. It includes expanded versions of most of the papers presented there plus papers, primarily of a review nature, prepared after the symposium in an attempt to form an organized synthesis of the topic. The editors thank the participants in the symposium and the Paleontological Society for providing the forum for two half-days of papers and an evening discussion which gave rise to this volume.

Participants in the symposium were volunteers from a canvass of the sponge, coelenterate, bryozoan, and graptolite workers in North America. A few specialists from overseas who were in North America were also contacted. The authors of papers prepared subsequently include the editors and some of the "volunteers" who were invited to expand their contributions in specific ways so as to provide more comprehensive coverage of the four phyla.

The editors are grateful to John H. Bushnell, John Pojeta, Jr., Klaus Ruetzler, and Kenneth M. Towe for technical review of parts of the volume and to Donald A. Dean and JoAnn Sanner for assistance in assembling the manuscript.

The need for a symposium on coloniality was realized during a year-long seminar on colonial animals at the National Museum of Natural History, Washington, D.C., in 1970–1971. Discussions in this seminar revealed the differing views of "what a colony is" among specialists in different animal groups. The editors of this volume are grateful to the other members of that seminar: William C. Banta, David R. Budge, Susan Cummings, T. Gary Gautier, Eckart Håkansson, Olgerts Karklins, William J. Sando, and Adam Urbanek, and to Willard Hartman, guest participant; some members later contributed papers to the symposium.

REFERENCES

Beklemishev, W. N., 1944–1970. Principles of Comparative Anatomy of Invertebrates; vol. 1, Promorphology; vol. 2, Organology; English edition, 1970, Oliver & Boyd Ltd., Edinburgh, and Univ. Chicago Press, 490 + 529 p.; 1st, 2nd, and 3rd Russian editions, 1944, 1952, and 1964; also available in other languages (1970 English ed. is based on 3rd Russian ed.).

Burnet, F. M., 1971. "Self-recognition" in colonial marine forms and flowering plants in relation to the evolution of immunity. Nature, *232:* 230–235.

Mackie, G. O., 1963. Siphonophores, bud colonies, and superorganisms, pp. 329–337 *in* E. C. Dougherty, ed., The Lower Metazoa, Comparative Biology and Physiology. Univ. California Press, Berkeley, Calif.

Perrier, Edmund, 1880. New views of animal transformations. Popular Sci. Monthly, *16:* 625–640.

——— . 1881. Les Colonies animales et la formation des organismes. Masson, Paris. 785 p.

——— . 1898. *Ibid.,* 2nd ed.

Thompson, J. A., and Geddes, Patrick, 1931. Life: Outlines of General Biology. Williams and Norgate, London. 714 p.

Wilson, E. O., 1953. The origin and evolution of polymorphism in ants. Quart. Rev. Biology, *28:* 136–156.

——— . 1968. The ergonomics of caste in the social insects: Am. Naturalist, *102:* 41–66.

——— . 1971. The Insect Societies. Belknap Press of Harvard Univ. Press, Cambridge, Mass. 548 p.

R. S. Boardman
A. H. Cheetham
W. A. Oliver, Jr.
A. G. Coates
F. M. Bayer

Washington, D.C.
September 1972

Contents

Introduction to Coelenterates

Coelenterates display a greater range of colony morphology and integration than any other phylum. Many zoantharian corals form (or have formed) colonies in which polyps are completely individualized and united only by their continuous skeleton. At the other extreme the siphonophores form polymorphic colonies in which highly specialized individuals are so interdependent that the colony has often been considered an individual of a higher order.

Three major subdivisions of the phylum are commonly recognized but one of these (Scyphozoa) includes no colonial forms. The other two include both colonial and solitary animals. In the following papers the Hydrozoa are mainly represented by the complex siphonophores. "Simple" hydrozoan colonies are not discussed but are similar to coral colonies in general form and in other aspects of coloniality. The Anthozoa include the corals in two main groups, the Zoantharia and the Octocorallia. Zoantharian corals have a good paleontologic record and studies of fossil and living colonies complement each other in several papers. Octocorals are discussed in one paper. In the spectrum of coloniality within the coelenterates, the octocorals are between the zoantharian corals and the siphonophores in astogenetic complexity and degree of integration.

Kinds and levels of integration in fossil and living zoantharian corals are reviewed by **Coates and Oliver,** who note distinct parallels between the evolution of coloniality in corals and of increasing size and complexity of coral reefs. **Wells** briefly defines "colony" and notes the significance of polystomatous forms. Some functional aspects of integration within zoantharian colonies are discussed by **Hubbard** (sediment shedding) and by **Kissling** (circumrotatory movement). The latter describes a peculiar coral way of life that has persisted through much of Phanerozoic time. The first detailed study of astogeny in a rugose coral is provided by **Jull,** who notes unexpected complexities.

Although zoantharian colonies achieve high levels of morphological and functional integration and cooperation, few include specialized members. Among the anthozoans, this line of evolutionary experimentation was left to the octocorals; **Bayer** describes various kinds and degrees of coloniality in this group, which exhibits polymorphism and the monarchial individual.

The hydrozoan siphonophores demonstrate how closely a colony can approach the metazoan individual by the functional integration of specialized zooids. **Mackie** discusses coordinated activities in these and other hydrozoans, and **Phillips** discusses evolutionary strategies in siphonophores and other holopelagic coelenterates.

Coloniality in Zoantharian Corals

A. G. Coates

George Washington University

William A. Oliver, Jr.

U.S. Geological Survey

ABSTRACT

Genera of colonial zoantharian corals have been more numerous than solitary genera through most of Phanerozoic time even though colonial genera were a minority of the Rugosa. Various forms of coralla represent widely different levels of development in terms of both integration within colonies and adaptive success, but the two are not necessarily correlated. Phaceloid rugosans represent a low level of integration with no connection between individuals other than skeletal; in contrast, coenosteoid scleractinians represent a high level of integration with confluent gastrovascular cavities and coordinated skeleton building. But both forms were very successful in terms of generic diversity and longevity, whereas Paleozoic coenosteoid heliolitoidids were relatively unsuccessful.

The evolution of coloniality in zoantharian corals was paralleled by the evolution of the coral-reef building habit. Paleozoic corals built only small reefs, apparently because of slow growth, relative instability, and failure to solve the problem of the disposal of metabolic waste. Mesozoic to Recent corals developed porous skeletons that are strong, stable, and rapidly built. In addition, they have developed an algal symbiosis that has solved the waste problem and increased the rate of $CaCO_3$ deposition. These factors together have made possible the surf-resistant oceanic reefs of late Mesozoic to Recent time.

Authors' addresses: A.G.C., Department of Geology, The George Washington University, Washington, D.C. 20006; W.A.O., Paleontology & Stratigraphy Branch, United States Geological Survey, Washington, D.C., 20244.

INTRODUCTION

In this paper we attempt to interpret the nature and trace the development of coloniality within the zoantharian corals (Tabulata, Heliolitoidea, Rugosa, and Scleractinia) throughout the Phanerozoic. Some discussion of the adaptive strategies that may be represented by coloniality is included and we offer an interpretation of the various colonial forms on the basis of their supposed function and paleontological record.

A prerequisite for this analysis is an understanding of the term "colony." We define it as a group of individuals, structurally bound together in varying degrees of skeletal and physiological integration, all genetically linked by descent from a single founding individual. "Colony," then, includes a range of structures grading from those in which all polyps are completely individualized with independent functions and no soft-part connections, to those in which soft parts, skeleton, and functions are communal. "Degree of integration" expresses the level within this sequence of any given morphological form or taxon.

The basic unit of this discussion is colony growth form. Most morphologic types (Table 1A) are known in several families in two or more of the four groups of corals being discussed as well as in other skeleton-secreting groups of colonial animals. No growth form is limited to one taxon, although a few family and higher-level taxa have a characteristic form. All our analogies and discussions of succession are meant in a functional and adaptational sense; we do not intend to imply evolutionary relationships among the varied and unrelated stocks that at one time or another developed similar growth forms.

The graphs that we use to illustrate the geologic record of coloniality and colonial morphologies are based on an "objective" analysis of the Moore Treatise (Hill, 1956; Wells, 1956) for the Rugosa and Scleractinia and the Orlov Osnovi (Sokolov, 1962) for the Tabulata and Heliolitoidea. The more detailed analysis of the early Paleozoic corals (Fig. 6) is based on Ivanovsky (1965) for the Rugosa, so this graph is not directly comparable with the others. We are aware of imperfections in all these works, but they represent the best available compilations on which this type of analysis can be based.

In this paper we emphasize levels of integration within colonies as a means of understanding the apparent adaptive success of various forms of colonial corals through time. Some of the evidence or results of high-level integration are detailed in other papers in this symposium. Examples of cooperation within colonies of living scleractinia are discussed by both Hubbard (this volume) and Kissling (this volume), while both Hubbard (this volume) and Wells (this volume and 1966) discuss the significance of polystomatous corals, and the question "what is a colony?"

Astogeny and comparative morphology of protocorallite and offset during ontogeny have been studied in recent corals, but Jull (this volume) gives the first detailed description and interpretation of these in a cerioid rugose coral. Data of this kind contribute significantly to our knowledge of coloniality, but much more work is needed before a general understanding can be achieved. Development in colonial zoantharian corals was reviewed by Oliver (1968). He considered various aspects of form, increase, and pattern in colonies and reviewed previous work on the groups under discussion.

Inter- and intracolony variation in corals and its genetics or genetic implications is only indirectly touched upon in this volume, although the generalities of these problems in other animal groups apply to corals as well. The subject also was reviewed by Oliver (1968).

TERMINOLOGY

The terminology used is generally that of the Moore (1956) Treatise on Invertebrate Paleontology, volume F, Coelenterata (especially the Moore, Hill, and Wells glossary, 1956, p. 245–251). Terms describing colony form are illustrated here for convenience (Figs. 5, 7, and 8). We use *septa* to apply only to the skeletal units (sclerosepta of some) and use *mesenteries* for the comparable soft parts. In some discussions it is important to distinguish between the living polyp and bud and the skeletal corallite (or corallum) and offset because in all fossil and most living corals our interpretations are inferred from studies of the skeleton rather than observed in the living animal.

SOLITARY VERSUS COLONIAL

Figure 1 shows the number of solitary and colonial genera per geologic period through the Phanerozoic, based on the Moore Treatise (Rugosa and Scleractina) and the Orlov

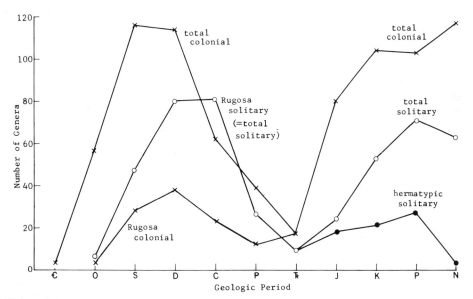

Figure 1

Number of genera of solitary and colonial zoantharian corals per geologic period through Phanerozoic time. Symbols for geologic periods are conventional but Tertiary is divided into Paleogene (P) and Neogene (N). In the Paleozoic (Є-P), only the Rugosa includes solitary corals; Mesozoic (Ŧ-K) and Cenozoic (P, N) corals recorded all belong to the Scleractinia. See Table 2 for raw data.

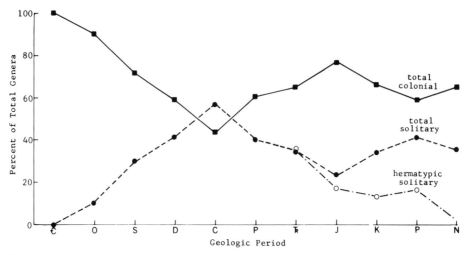

Figure 2

Number of genera of solitary and colonial zoantharian corals per geologic period as percentages of total genera. Hermatypic solitary scleractinian corals are plotted separately and included in total solitary. Period symbols and data base as in Fig. 1.

Osnovi (Tabulata and Heliolitoidea). Apart from a temporary reversal in the Carboniferous, colonial genera appear to have been more numerous than solitary genera throughout their history (also shown in Fig. 2, expressed as percentages of total genera). Solitary and colonial corals have tended to increase and decrease in diversity at the same time and perhaps, on the large scale, in response to the same broad ecologic factors. The curves show that the maxima in the Paleozoic and in the Mesozoic–Cenozoic were not very different, although there are obvious limitations to this kind of data.

The late Paleozoic decrease in numbers of colonial genera reflects the Devonian extinction of heliolitoidids and the post-Devonian scarcity of tabulates; both of these groups of corals are exclusively colonial. In the Rugosa alone, solitary genera outnumbered colonial genera throughout the Paleozoic (Fig. 1). However, solitary percentage of the total increased during the Carboniferous, which helps emphasize the total solitary–colonial reversal shown in Figs. 1 and 2. In the Paleozoic an abundance of "reefal" structures in the Silurian and Devonian and their virtual absence in the Carboniferous may reflect ecologic conditions that resulted in the reduction in numbers and diversity of colonial corals and the temporary dominance of solitary corals. However, the lack of "reefs" could as easily have been the effect as the cause of the decrease in colonial forms.

The decrease in coral diversity at the end of the Paleozoic (Permo-Triassic) is comparable to the decrease at this time of marine invertebrates generally and must be related to a large-scale environmental crisis. Whatever the cause, it seems to have affected solitary and colonial corals similarly (Fig. 1) and can be neglected in a discussion of coloniality.

The relative abundance of solitary and colonial corals in the Mesozoic–Cenozoic is similar to that in the Paleozoic, but probably for different reasons. If, as is generally assumed, the curves for the Paleozoic deal with organisms primarily confined to shallow, marine, shelf seas, then it may be more meaningful to compare these curves to those for hermatypic coral distribution in the Mesozoic–Cenozoic (Fig. 1)* in which case the relative diversity of solitary to colonial goes down markedly. The peak, then, for solitary corals in the post-Paleozoic is largely maintained because of ahermatypic diversification in the deeper sea, nonphotic environments in which most colonial forms have not been successful. We will attempt to show below that coloniality in a broad, generalized way can be related to reef building and that colonial corals became progressively better designed for this environment through time. In this sense the Mesozoic–Cenozoic oceanic reefs are lineal descendents of the much smaller, shelf reefs of the Paleozoic.

It may be possible to suggest further that high parts of the colonial curve (Fig. 2) mark periods of very active reef growth to which colonial corals generally seem highly adaptive. However, this cannot apply to the earliest parts of the curve (Cambrian and Ordovician), where the high ratio of colonial forms simply reflects their earlier origin.

LEVELS OF INTEGRATION

Beklemishev (1964) has discussed colony formation in terms of (1) weakening of individuality of zooids, (2) intensification of the individuality of the colony, and (3) development of cormidia. Cormidia are functional units within colonies comprising polymorphic, highly specialized individuals. Polymorphism is rare in zoantharian corals and cormidia are unknown, although well developed in some other coelenterates.

Beklemishev's first two trends suggest a "cycle" of coloniality with an individual as a starting point, moving to simple structural aggregation, then passing through increasing degrees of integration of both skeletal and soft parts, and culminating in such colony-wide structural unity and communal function that the organism may be considered once more an individual.

Table 1 and Fig. 3 were constructed in an attempt to calibrate this cycle of integration. Table 1A lists coral skeletal morphologies in an order that suggests successive levels of integration. Interpretations of soft-part morphology and integration are given, keyed to separate series of hypothetical and real states (Table 1, B-D). Degrees of skeletal and soft-tissue fusion, interpolyp communication, and amount of nonpolyp tissue (coenosarc) and skeleton (coenosteum) are different aspects of this integration. The skeletal list itself is a complex of series with some morphologies appearing in all or most series and some apparently recurring within a given series. A largely theoretical network of skeletal series is shown in Fig. 3 based on various interpretations of the paleontologic record.

Mode of budding may also be related to "levels" in the integration charts. Wells (in this symposium) states that "extra-tentacular budding results in complete homomorphic individuals, organically or structurally united as corms or true colonies." This type of budding dominates "low" integration levels. Wells further states "intratentacular increase produces incomplete individuals, marked by lack of direct mesenterial couples and often not morphologically individualised." This type would generally characterize

---

*We follow Wells (1956) in assuming fossil scleractinians to be hermatypic or ahermatypic on the basis of their apparent ecologic preference and the known condition of their living descendents. Colonial scleractinians are predominantly hermatypic.

Table 1

Scales of Integration

	A. Skeletal		
	Interpretation of common position on other scales		
	B	C	D
Corallites bounded by imperforate wall with or without epitheca			
1. Phaceloid	1	1	1
2. Cateniform	2	1	1
a. Dimorphic	2		3
3. Cerioid	2	1	1
4. Meandroid	2–4	1	1
Wall perforate			
5. Phaceloid	1–2,3?	1	1
6. Cerioid	2–3	1	1
a. Dimorphic	4		3–4
Separating wall incomplete or lacking			
7. Cerioid	3–4	1	1
8. Cateniform	2–3,4?	1	1(3?)
9. Meandroid	4	1	1
10. Astreoid	2–3	1	1
11. Hydnophoroid	3–4	1	1
12. Thamnasterioid	4	1	1
a. With circumoral offsets	4	1	1
13. Aphroid	4?	1	1
Coenosteoid (see scale C)			
14–17.	4	2–5	1

B. Tissue Level

1. Polyps separated in space
2. Polyp walls in apposition
3. Interpolyp connections by body wall cells or tissue
4. Confluent body cavity

C. Extrapolypoidal Skeleton and Tissue

1. Absent (skeletal types 1–13)
2 (14). Corallite diameter > intercorallite distance; plocoid
3 (15). Corallite diameter < intercorallite distance
4 (16). Corallites vague or hardly distinguishable
5 (17). Meandroid coenosteoid

D. Polymorphism

1. Monomorphic
2. Protocorallite or protopolyp different, all others uniform
3. Dimorphic
4. Dimorphic with functional difference

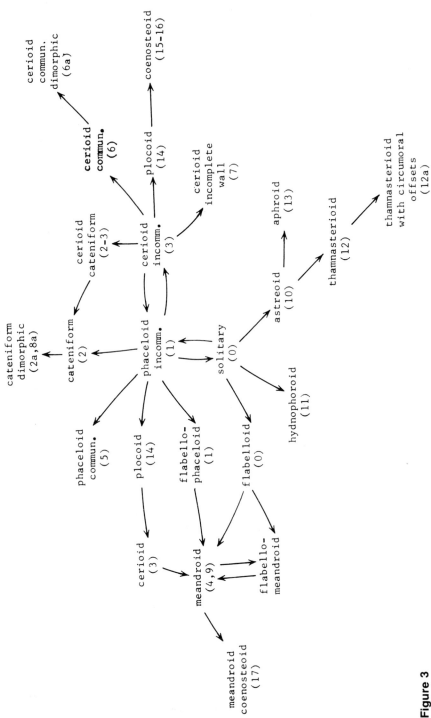

Figure 3

Possible interrelationships of skeletal types. Arrows indicate directions of evolution that have been suggested in literature or that seem logical to us. Most are hypothetical. Numbers in parentheses are skeletal types from Table 1.

the "higher" levels of integration, where individuals are "weakening" with respect to the colony.

The soft-part anatomy of many living scleractinian corals is known and in this group levels of integration can be observed rather than inferred. Matthai (1926, p. 349–351) discussed individuality in living corals and became involved in the rather sterile question of when, in the development or evolution of polystomes, the "normal" individual ends and the "higher-level" individual begins. Wells (this volume; 1966) agrees with Matthai's conclusion that (at least some) polystomes are individuals. Both suggest a twofold division on the basis of budding type, although Wells considers extratentacular budding to result in true colonies composed of individualized polyps, whereas Matthai concluded that all colonies are individuals.

Beklemishev's outline discussion of levels of integration in living scleractinians is sufficiently pertinent to warrant quoting in full (1970, p. 99):

> The development of coloniality in Madreporaria [i.e., Scleractinia] is therefore characterised by the following interrelationships between colonies and zooids: (i) among Madreporaria there are solitary as well as colonial forms; (ii) the zooids of colonial forms differ from solitary corals in progressive reduction of growth [i.e., individual size] within the group and simplification of organisation; (iii) in colonial forms we observe gradual development and strengthening of the role of the coenosarc; (iv) the growth of a colony is achieved both by asexual reproduction and budding of polyps and by budding on the coenosarc; (v) in many forms (meandrina-type colonies) fusion of the calyces of the skeleton and of polyps takes place, with formation of "polyvalent" zooids with several mouths and gullets. In some madrepore corals (e.g. *Hydnophora microcona*) the entire colony constitutes an enormous polypharyngeal "polyp" (G. Matthai, 1926). This is the highest degree of loss of individuality by zooids found in Madreporaria; (vi) there is no polymorphism of zooids.

Most of these "interrelationships" are discussed in the following sections.

INTEGRATION IN LIVING CORALS

One should be able to recognize four general integration states to which the polyps in various kinds of colonies could be assigned. These are as follows (from Table 1B):
1. Polyps separated in space.
2. Polyp walls in apposition.
3. Interpolyp connections by body wall cells or tissue.
4. Confluent body cavities.

In the most extensive modern review of scleractinian anatomy, Vaughan and Wells (1943, p. 16–17) state:

> The coenosarc is that part of a polyp in a colony that lies outside the theca or skeletal wall. The edge-zone is that part that lies over the free portions of the corallite. In simple corals an edge-zone only is present; in colonies the edge-zone is part of the coenosarc, the latter being in effect, a continuation of the edge-zone. The edge-zone (including the coenosarc) is then simply a horizontal fold of the column wall over

the common wall itself. Within the fold is a continuation of the gastrovascular cavity. . . . The edge-zone, and frequently the coenosarc, contains extensions of the mesenteries, but in many corals the mesenteries thus extended are modified into canaliculae. In groups with perforated skeletal walls . . . the interior of the edge-zone and coenosarc and the main body-cavity of the polyp are directly connected through the wall as well as over it.

This would seem to be comprehensive and to assign all living scleractinian corals to state 4 (Table 1B and above), but it is based on relatively few recorded observations or descriptions and some exceptions seem probable.

F. M. Bayer has noted (personal communication, 1972) that on some living phaceloid colonies with imperforate skeletons, the edge zone extends only part way down the branch and that the polyps are, in effect, isolated from each other and must be completely individualized (state 1). It is not known whether polyps were integrated through their edge zones in the early growth stages of the colony or whether isolation took place soon after budding. Many of the Paleozoic rugose corals are of this type, as will be shown later.

Skeletal criteria that may be used to intelrpret polyp integration level in the absence of direct observation are as follows:

1. Presence or absence of epitheca. This skeletal structure is deposited by the outer margin of the edge zone (Wells, 1956, p. 344), so corallites that are separated from each other by epitheca were not occupied by polyps connected to each other by edge zone or coenosarc.

2. Perforate and imperforate skeletons. Perforate (porous) skeletons permit polyp interconnections beneath the surface of the skeleton (Vaughan and Wells, 1943, quoted above). In imperforate skeletons such interconnections can be ruled out.

3. Costae or outer edges of septa. Septa bear a clear relationship to the mesenteries that are within the gastrovascular cavity. If the upper or outer edges of the septa can be traced on the surface of a corallum from one corallite into another, this provides direct evidence for continuity of mesenteries and therefore of gastrovascular cavities.

These criteria can be applied to fossil as well as to recent corals and provide a basis for many of the interpretations that follow.

INTERPRETATION OF COLONY FORM

Seventeen colonial skeletal types are listed in table 1A. Definitions are from the Treatise (Moore, Hill, and Wells, 1956) with only minor modification and simplification. The geologic record of the more common of these is shown on Fig. 4.

Many phaceloid (including dendroid) colonies (Fig. 5A) represent the simplest colonial state, in which polyps are completely separated and are in every way similar to solitary corals except that they occupy the numerous extremities of one skeleton (corallum). If there is any "cooperation" within such colonies it is no different than that which would be found in groups of closely spaced solitary corals and would result from extrinsic factors acting upon the cluster of polyps and not from biologic integration. All dendroid

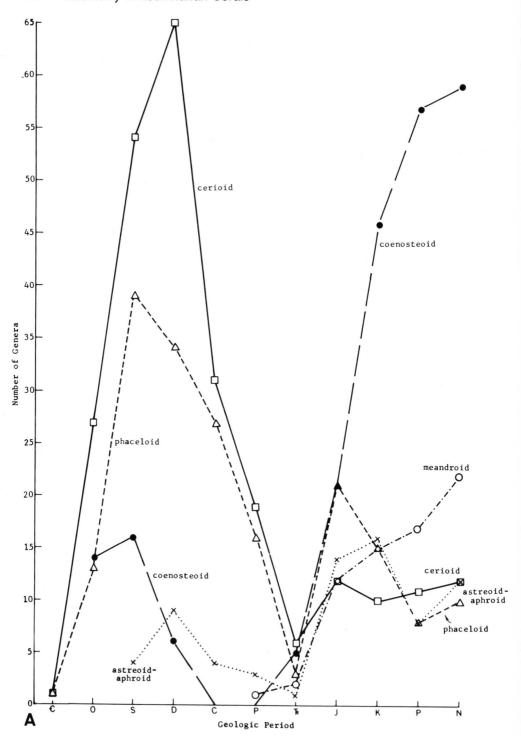

A

and phaceloid rugose corals, many tabulates, and at least some scleractinians are examples of this low level of colonial organization (Table 1B, state 1).

Some phaceloid (or dendroid) scleractinians have a narrow edge zone and imperforate skeletons so that the polyps are completely individualized. Others, however have a continuous conenosarc over the surface of the skeleton and/or a perforate (porous) skeleton. Most of these presumably have polyps with interconnected gastrovascular cavities (state 4), but some may represent lesser degrees of integration (states 2? or 3).

Some phaceloid rugose and tabulate corals have lateral connections or supporting structures between corallites. In rugosans the outgrowths from each corallite are terminated by a normal outer wall (with epitheca) which may abut an outgrowth or the cylindrical wall of an adjacent corallite (Fig. 5B). No common polypal tissue was developed and no colony-wide coordination was necessarily involved. A higher degree of integration is indicated, however, by some phaceloid tabulate corals (e.g., Syringoprorida sensu Sokolov, 1962), in which connecting tubes are occupied by dissepiments (curved skeletal plates) and have no dividing walls (Fig. 5C). The construction of these structures must have involved the cooperative action of two polyps and may indicate temporary soft connections between polyps and even integrated gastrovascular cavities (Table 1B, state 2?, 3, or 4?). Similarly, other tabulates (e.g. Auloporida sensu Sokolov, 1962) have openings between corallites at offsetting points (Fig. 5D). In some of these corals, soft connections between polyp and bud may have persisted through the life of the connected individuals with resulting interdependence and cooperation (state 3? or 4). Laub (1972) described and illustrated auloporoids that could represent state 4 and considers this to have been the common condition of this group (personal communication, 1972).

Number of genera of phaceloid (including dendroid) corals per period are shown in Fig. 4A. This was a common growth form in the Paleozoic for both Rugosa and

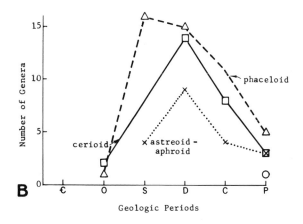

Figure 4

Number of genera per period of selected morphologic forms. The states in Table 1A are combined as follows: cerioid, states 3, 6, 7; phaceloid, states 1, 5; coenosteoid, states 14–17; astreoid–aphroid, states 10, 12, 13; meandroid, states 4, 9. States 2, 8, and 11 are not included. See Table 2 for raw data. A. Total for all zoantharian corals. B. Rugosa only (at same scale).

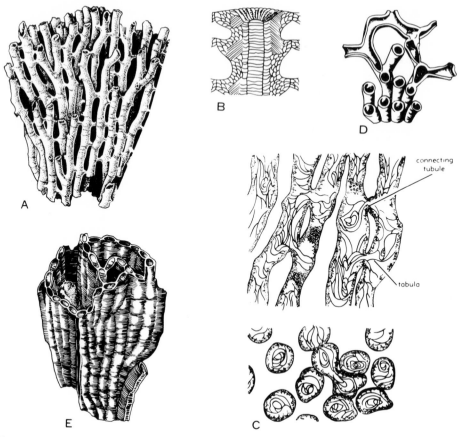

Figure 5

Phaceloid and cateniform colonies. A. Phaceloid corallum with lateral connections (supporting structures). B. Rugosan lateral connection with separating wall (including epitheca). C. Two views of syringoporoid lateral connection with open tubes. D. Auloporoid tabulate coral with completely open skeletal interior. E. Cateniform corallum. From Treatise on Invertebrate Paleontology, courtesy of The Geological Society of America and The University of Kansas Press.)

Tabulata and was the most characteristic form of colonial rugosan genera through most of their history (Fig. 4B).

Figure 6 separates communicate and incommunicate phaceloid (including dendroid) forms for the early Paleozoic. Tabulates are interpreted according to diagnoses in the Osnovi (Sokolov, 1962). Intercommunication between phaceloid corallites was apparently not a significant adaptive advantage. Communicate phaceloid tabulates appeared in the Middle Ordovician and persisted through the Paleozoic but were greatly outnumbered by genera of incommunicate phaceloid tabulates and rugosans.

Cateniform colonies ("chain corals"; Fig. 5E) represent a structural advance on scale 1A in that the corallum is stronger and presumably more resistant to the rigors of the physical environment. It would seem that this form is structurally intermediate between

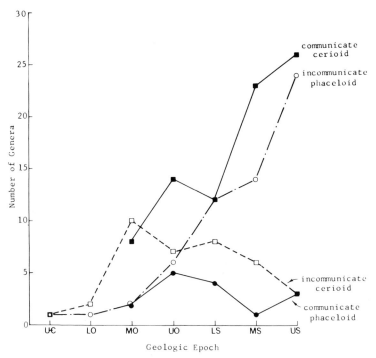

Figure 6

Number of genera of communicate and incommunicate phaceloid and cerioid genera per stage in the lower Paleozoic. [Rugosa data from Ivanovsky (1965); Tabulata data from Sokolov (1962).]

phaceloid and cerioid forms since in it the tubular corallites are in continuous contact with adjacent corallites along two (or three) sides. The paleontologic record, however, shows that cateniform corals appeared in numerous different stocks apparently derived from phaceloid corals in some cases but more frequently from cerioid corals. Cateniform genera or species have been described from all major groups under consideration with the exception of the heliolitoidids.

In most cateniform corals adjacent corallites are separated by a distinct wall. In rugosans the wall includes epitheca and is imperforate (Table 1A, state 2). We interpret this to indicate no significant integration of polyps (Table 1B, state 2). In tabulates, the wall may be with or without epitheca and is commonly if not always imperforate (Table 1A, states 2 or 8). There is no firm basis for interpreting degree of polyp integration, but an intermediate level seems most likely (Table 1B, states 2?, 3, or 4?). These interpretations of the significance of separating walls are more fully explored in the following discussion of cerioid corals.

In one cateniform tabulate genus (*Cystihalysites* Chernychev), adjacent corallites are separated only by a "wall" of dissepiments (Norford, 1962, p. 35) (state 8). This is analogous to the much more common astreoid, thamnasterioid, and aphroid forms discussed below and is more likely to represent a higher level of integration (Table 1B, states 3? or 4).

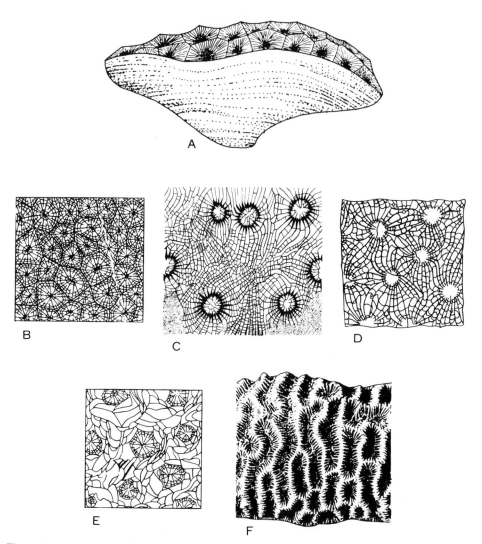

Figure 7

Massive colonies. A. Massive cerioid corallum. B–E. Diagrammatic transverse sections of coralla: B, cerioid; C, astreoid; D, thamnasterioid; E, aphroid. F. Surface of meandroid corallum with several valley systems. (Part A drawn by Elinor Stromberg, modified from D. Hill; B–F from Treatise on Invertebrate Paleontology, courtesy of The Geological Society of America and The University of Kansas Press.)

Cerioid coralla (Fig. 7A, B) are those in which the corallites are closely appressed so as to be in contact on all sides but with a wall separating adjacent individuals. In many Paleozoic corals there is an imperforate double wall (the two parts separated by epitheca) between corallites that may form a high ridge between adjacent calices. These are little more than compact phaceloid coralla. Most likely the polyps were complete individuals, although in physical contact with other polyps on all sides (Table 1B, state 2; most cerioid rugosans). Cerioid coralla with lower physical barriers between polyps or with simpler (single) walls may represent more integrated systems with coenosarc extending over the wall, providing direct interpolyp communication (Table 1B, states 3 and 4; many tabulates and scleractinians).

Communicate cerioid corals are those in which (1) the walls are pierced by relatively large mural pores (as in the common tabulate, *Favosites*), (2) the walls units (commonly trabeculae) are incomplete leaving irregular openings (several tabulates), or (3) the wall is perforated by numerous fine openings (many Scleractinians). In many (most?, or all?) living scleractinians with perforated walls "the interior of the edge-zone and coenosarc and the main body-cavity of the polyps are directly connected through the wall as well as over it" (Vaughan and Wells, 1943, p. 17) (Table 1B, state 4). It is tempting to class all communicate cerioid tabulates as highly integrated by analogy with the scleractinians, but the analogy is weak because the wall openings are few and large rather than small and pervasive. Both mural pores and incomplete walls must have resulted from polyp interconnections, but this may have been temporary or intermittant and at a lower level than in the living scleractinians (Table 1B, states 2, 3, or 4).

Numbers of cerioid genera per period are shown on Fig. 4A. This was the most common colonial form in the Tabulata and very important in the Rugosa (Fig. 4B) and persists to the present day in the Scleractinia (Fig. 4A). Figure 6 separates early Paleozoic communicate and incommunicate cerioid genera and shows that the communicate genera appeared in the Middle Ordovician, outnumbered incommunicate genera by the Late Ordovician, and that diversity continued to increase until Late Silurian time (and on into the Carboniferous; Fig. 4A). Incommunicate cerioid tabulates became extinct at the end of the Silurian. Clearly there was an adaptive advantage to the communicate condition in the Tabulata, but this state was not attained by cerioid Rugosa, all of which are incommunicate.

Meandroid colonies represent a distinct advance on scales A and B (Table 1). The corallum is massive and consists of one to many valley systems (Fig. 7F). If more than one valley system is present, the systems may be integrated to varying degrees (Table 1A, states 4, 9, and 17). There may be a well-developed wall between adjacent valleys (analogous to the cerioid state in nonmeandroid forms); if the wall is incomplete or lacking, the septa may be offset or confluent with those of adjacent series (analogous to astreoid and thamnasterioid states); and there is a coenosteoid state in which an ambulacrum, deposited by coenosarc, separates the valleys. In general, the valleys are physically integrated and interlocking and form a strong corallum, presumably resistant to the rigors of a shallow water situation. The reduced or lacking walls means that less $CaCO_3$ needs to be deposited to attain a given corallum size. This economy may have been important in the success of this group.

Meandroid polyps are polystomes in the sense of Wells (this volume; and 1966). Each valley is occupied in life by a polyp mass with only one gastrointestinal cavity but several (to hundreds) of "mouths," so that the number of polystomes in a colony corresponds to the number of valley systems in the corallum.

Ontogenetically, each polystome is an individual (see Wells in this volume) and the meandroid corallum with only one valley might be considered a solitary coral, whereas the corallum with more polystomes and valleys is colonial. Phylogenetically, it has been suggested that meandroid corals have evolved in two general ways (Wells, 1956, p. 352, fig. 253), from colonial cerioid corals and from solitary flabellate corals. We suggest that the meandroid coral with only one valley and polystome should be considered a solitary coral only if derived (evolved) from a normal solitary coral. Possibly these are incipient colonies in which asexual division has only begun. Some of these ''solitary'' meandroid corals evolved into colonial corals by asexual reproduction, resulting in two or more polystomes. If, as has been suggested, other meandroid corals evolved from cerioid corals through the partial loss of walls, producing chains of individuals within colonies, then these should be considered colonial even if there is only one valley system in the whole colony. In this case polystomes would have developed (or evolved) through integration of polyps and the polystome would represent the highest level of integration known in the zoantharian corals.

Meandroid colonies are very common among the living Scleractinia. The number of genera characterized by this form of growth has steadily increased throughout Mesozoic and Cenozoic time (Fig. 4A), and it seems evident that the form represents a successful adaptation.

Only one meandroid rugosan genus is diagnosed in the Treatise (Hill, 1956), indicating how rare this form is in this subclass. No meandroid coralla are known in either of the other Paleozoic groups. It is worth noting, however, that meandroid ''corallites'' characterize the group Chaetetida, which is common in both the Paleozoic and Mesozoic. These are considered to be corals, hydrozoans, or sponges by different contemporary workers and are not included in our present tabulations.

Astreoid (Fig. 7C), thamnasterioid (Fig. 7D), and aphroid (Fig. 7E) corals are structurally advanced over cerioid and some meandroid corals. They lack walls between corallites so that less $CaCO_3$ is required for a given corallum size, and corallites are more integrated structurally. This may entail some loss in resistance to crushing but presumably in some environmental situations the advantages outweigh the disadvantages. Degree of polyp integration in these forms is unclear, but they are at least in apposition (Table 1B, state 2). In astreoid coralla, septa do not continue from one corallite into another. This suggests that mesenteries did not either and that integration might not be higher than scale B, state 3 (skeletal form known only in Rugosa). Thamnasterioid coralla have septa extending from each corallite into adjacent ones. This clearly implies continuous mesenteries, confluent body cavities, and a high level of integration (scale B, state 4; Rugosa and Scleractinia). Aphroid coralla lack septa in peripheral areas of corallites, so direct evidence of polyp integration is lacking. Clearly, high-level integration of polyps (scale B, state 4) is possible, but perhaps these should be considered as the first step in another series, with the development of extrazooidal sclerenchyme and coenosarc (form known in Rugosa and Scleractinia).

More highly integrated variants of thamnasterioid and, in one sense, parallel developments to meandroid forms are the morphologic forms hydnophoroid and thamnasterioid with circumoral offsets (Table 1A, states 11 and 12a; Scleractinia only). These are budding types with coordinated colony-wide patterns. In colonies with circumoral offsets, most of which are thamnasterioid, a series of successive concentric rings of buds are given off from the central parent stomadeum. In terms of Beklemishev's colonial theory

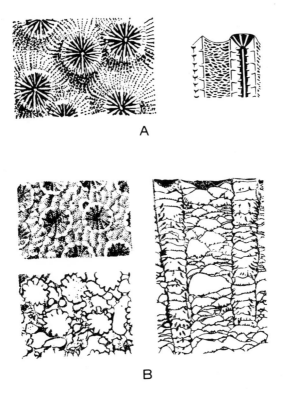

A

B

Figure 8

Coenosteoid colonies. A. Scleractinian. B. Heliolitoidid. From Treatise on Invertebrate Paleontology, courtesy of The Geological Society of America and The University of Kansas Press.)

("strengthening the individuality of the colony"), this degree of coordination represents a higher level of integration than simple thamnasterioid or astreoid types. In circummural budding, a concentric ring of buds forms around a protuberance (monticule), giving rise to the hydnophoroid form. Many rings of buds, each around a monticule, forms a corallum and each ring system probably represents the closest growth form in the Scleractinia to a cormidium.

Thamnasterioid and aphroid coralla represent the highest integration levels achieved by the Rugosa, but in any geologic period these levels were reached by no more than 25 percent of the Treatise rugosan genera, and, in most of these genera, other, "less-advanced" forms are more common. None of the Tabulata reached this level. These forms are known in many scleractinian genera (Fig. 4A) but are overshadowed by the more abundant meandroid and coenosteoid forms.

Coenosteoid corals (Fig. 8A, B) can best be treated as a separate integration series from the other skeletal types discussed (Table 1C), although we have indicated their high position by also placing them at the top of series A (states 14–17). The group includes all corals that developed interpolypoidal tissue (coenosarc) and skeleton (coenosteum). We arbitrarily recognize four states to emphasize diversity and progression in this large group. "Plocoid" describes coenosteoid forms in which corallite diameter is greater than intercorallite distance (Table 1C, state 2). These are more numerous than coenosteoid forms in which intercorallite distance is greater than corallite diameter (Table 1C, state 3). An apparent end member to this series is represented by the family Actinacidae, in which the coenosteum is not only dominent but the corallites, particularly the septa, become difficult to distinguish within the coenosteum (Table 1C, state 4). The meandroid coenosteoid state (1C, state 5) was most likely developed from meandroid rather than from other coenosteoid corals (Fig. 3) but is included here as representing a comparable high level of development and integration.

The coenosteoid form represents the highest level of coloniality achieved by zoantharian corals. The gastrovascular cavity is continuous throughout a colony (as in some of the other forms), but, in addition, skeleton building is a communal project rather than an individual one. This has resulted in greater efficiency and, in general, coenosteoid corals build more skeleton with less polyp than other forms.

These and other factors discussed below have permitted the great success of these forms in the Scleractinia. Figure 4A outlines the history of the group and shows that approximately half of all Tertiary genera are coenosteoid. This would seem to have been the most important form innovation of post-Paleozoic corals.

The Paleozoic Heliolitoidea were entirely coenosteoid (Fig. 4A, Ordovician–Devonian record). They experienced early increase in diversity but then decreased and became extinct at the end of the Middle Devonian. They are commonly found associated with cerioid tabulates but were not nearly as successful as the tabulates in the Paleozoic environments. Possibly the heliolitoidids did not actually achieve the integration level suggested for other coenosteoid corals. More likely, however, other factors made the difference. Many coenosteoid scleractinians build a porous skeleton and, in common with other forms, many have developed a very beneficial symbiotic relationship with certain algae. These adaptations have made possible a much greater efficiency increase than would result from the coenosteoid form alone. The heliolitoidids lacked these extras and simply did not reach the "take-off point."

Coenosteoid coralla are not certainly known in either the Rugosa or Tabulata, although aphroid rugosans and some semiphaceloid tabulates (e.g., Sarcinulida sensu Sokolov, 1962) have relatively small amounts of skeletal material that may be coenosteal.

POLYMORPHISM

Polymorphism, except for minor morphologic differences between initial and later polyps and corallites, is virtually unknown in the four zoantharian coral groups under discussion, so this scale of coloniality (Table 1, scale D) is of limited use.

To some extent, most genera attained state 2 on scale D. Except in the simplest of skeletons, the protocorallite (corallite of initial or founding individual) differs from the offsets (corallites of subsequently budded polyps) in its ontogeny. The earliest morphologic stages are "missing" in offsets (Jull, this volume; Oliver, 1968, p. 21, for review), and this presumably reflects the fact that the protocorallite was built by a sexually produced polyp, whereas offsets were built by asexually produced polyps. Buds originate from a cluster of cells and "skip" the earliest stages of ontogeny. These differences disappear with growth and the mature protocorallite can be separated from its fellows only by tracing it from its point of origin.

Dimorphism in cateniform tabulates (Halysitidae) is well known, although its significance is open to question. In a few genera, tubes ("corallites") of two sizes regularly alternate, suggesting true dimorphism. It seems probable that the two types of tubes were occupied by different types of polyps (Table 1D, state 3) but this is not certain, and the nature of morphologic differences in the polyps or of specialization (if any) is unknown. The genus *Cystihalysites* (see discussion under Interpretation of Colony Form) is commonly considered to be dimorphic, but the "mesocorallites" may simply be a dissepimental wall separating adjacent "autocorallites."

We know of only one example of dimorphism with clear functional separation (Table 1D, state 4). This is in a Silurian tabulate coral in which corallites that produced two offsets alternate at the surface of the colony with corallites that did not offset at all; each corallite of the first type produced one offset of each type (Oliver, 1966). Here, the first corallite type was clearly occupied by a polyp specialized for asexual reproduction, although it may have carried on other functions as well.

Several cerioid tabulates (favositids) have been described as dimorphic but few show convincing differences and most cited "examples" probably represent continuous individual variation within colonies.

Dimorphism is unknown within colonies of Rugosa and Heliolitoidea.

We know of only one possible example of dimorphism in the Scleractinia. Fowler (1887, p. 5–8) described *Madrepora durvillei* as having two types of polyps, both irregularly distributed within colonies. One type of polyp is "normal," the other has certain much-thickened mesenteries. Fowler suggested that the second polyp type was "more digestive" and less reproductive. Brook (1893, p. 15) and Matthai (1926, p. 349) cited Fowler's description as of an example of dimorphism in this group but the significance of Fowler's observations is not clear, and di- or polymorphism in Scleractinia is at least extremely rare.

PATTERNS IN TIME

The distribution in time of the more common morphological states is shown in Fig. 4A. Dominant morphologies are obvious and our interpretation of these is in a previous section. Cerioid and phaceloid genera were most numerous in the Paleozoic, with coenosteoid and thamnasterioid-aphroid genera as important minorities. Meandroid rugosans are known only from the Permian, and this form was of little importance at that time.

In terms of our scales of integration, the coenosteoid and themnasterioid-aphroid states are the highest levels attained by Paleozoic corals. The coenosteoid Heliolitoidea expanded rapidly in the Ordovician but fell off in the Silurian and became extinct in the Middle Devonian. Heliolitoidid corals were similar in form to many of the living scleractinian and octocorals yet they were not successful in competing with the tabulates. Thamnasterioid and aphroid rugose corals appeared in the Early Silurian and rose and fell with the Paleozoic corals as a whole (Fig. 4A) without becoming dominant.

On the scales (Table 1) a few Rugosa would seem to have reached a high level

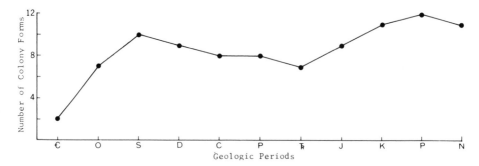

Figure 9

Number of colony forms (as listed in Table 1A) known from each Phanerozoic period. See Table 2 for raw data.

of integration, but the vast majority of rugosan taxa and colonies were incommunicate phaceloid forms with completely individualized polyps or incommunicate cerioid in which the polyps were probably just as individualized.

Most tabulates were communicate and thus probably more highly integrated than most rugosans. It is uncertain, however, what level of polyp integration is here represented. The Tabulata include dimorphic forms and in at least one genus a functional difference between dimorphs has been demonstrated. Such specialization itself indicates a high level of integration and cooperation in the dimorphic species, and suggests that related forms may have been similarly highly developed.

Heliolitoids were entirely coenosteoid. This form suggests a high level of polyp integration and in this sense the group may have been the most advanced of the Paleozoic corals.

Figure 9 shows the diversity of colony forms by geologic period. It is notably symmetrical, as are most of the other graphs. Comparison of Fig. 9 with Fig. 4A illustrates two points: (1) while a few forms dominate, many forms were present during most periods, and (2) form diversity in the Mesozoic–Cenozoic was not significantly greater than in the Paleozoic, although dominant forms were different.

By our integration scales, the Mesozoic–Cenozoic Scleractinia were not necessarily more advanced than the Paleozoic groups, but a much higher percentage of genera show evidence of high-level integration than either the Rugosa or Tabulata. In a general way the cerioid and phaceloid Paleozoic groups were replaced as dominants in the Mesozoic–Cenozoic by coenosteoid and meandroid corals, both representing a high integration level. But cerioid and phaceloid forms persist to the present day, and both coenosteoid and meandroid forms are known in the Paleozoic.

ADAPTIVE SIGNIFICANCE OF COLONIALITY

There are certain advantages of coloniality that would seem to apply to all groups of organisms:

1. Asexual reproduction is an efficient way to rapidly increase numbers of individuals and, in colonies, to increase unit biomass.

2. Skeletal integration provides mechanical strength and stability and larger unit size.

3. Varying degrees of cooperation and integration among individuals in a "population" become feasible.

These are certainly factors in the evolution of colonial corals and in the development of the reef-building habit. In addition, we have noted several distinct steps toward increasing the efficiency of skeleton building. All of these taken together can account for the progressive increase in the role played by corals in reef building through time and explain the apparent "success" or lack of "success" of different forms at different times.

The individual coral colony is, in effect, a miniature reef in which the polyps, to some extent, create their own environment by building a resistant structure up into the water column and away from the substrate. The three general advantages of colonialism cited above were already present in the earliest colonies, and the different colony forms that exist represent different ways to exploit these advantages. The development of coral reefs through time paralleled closely the further evolution of coloniality in corals.

Paleozoic corals generally formed solid skeletons of $CaCO_3$. A few stocks experimented with reducing or eliminating internal walls and this certainly permitted the building of a larger skeleton in less time or for the same expenditure of energy on the part of a given polyp mass. But these forms were not notably successful and do not seem to have had any advantage over "less-advanced" corals. Diverse forms of coralla fall into the same general size range and the small coral reefs of the Paleozoic were more commonly built by phaceloid and cerioid corals than by the skeletally "more-advanced" types. In all probability, both corallum and reef size were limited by morphologic and ecologic factors that the Paleozoic corals could not overcome. Wells (1957, p. 773) emphasized the instability of paleozoic corals; because they lacked an edge zone they were unable to maintain a firm attachment to the substrate as they grew. In addition, they almost certainly had not developed any special ability to efficiently discard the waste products of metabolism. The larger the colony or reef, the more difficult it was for down-current polyps to feed and carry on normal activities.

All Paleozoic coral reefs are small, mound- or lens-shaped bodies (bioherms) generally limited to shelf seaways or shelf edges, most commonly in areas of general carbonate deposition. At various times and places, representatives of each of the three Paleozoic groups were essential elements in such reefs. They provided the framework and much of the bulk that created local centers of activity for relatively rich and varied biotas. They controlled their own environment in the sense that activity was great enough to build well above the general level of sedimentation, but all such reefs were small and there is little evidence of growth into the shallow, turbulent zone.

The larger reefs of the Paleozoic were built by animals other than corals and by $CaCO_3$-secreting algae, and on these reefs corals tended to play only a minor role, in which coloniality was not an apparent factor.

Mesozoic–Cenozoic corals formed $CaCO_3$ skeletons which, in many groups, are very porous. This is another step in increasing the mechanical efficiency of skeleton building. More important, a symbiotic relationship with photosynthetic algae (zooxanthellae) in hermatypic (including most reef) corals, has allowed the rate of precipitation of $CaCO_3$ to be greatly increased and has, to a large extent, solved the problem of waste disposal (see Yonge, 1968, p. 333–337, for general review and discussion). The result has been the development of the modern oceanic reef in which corals with their zooxanthellae are able to construct a limestone framework with such speed and efficiency as to counter the destructive effect of wind-driven sea and surf, and to do this over an immense area and for a long period of time.

Porous skeletons and zooxanthellae have undoubtedly had an effect on the evolution of scleractinian colony form as well as on the speed and efficiency of corallum building. To some extent, form is a compromise between the need for strength and bulk and the energy available for $CaCO_3$ production. Meandroid coralla are mechanically strong and can be very large, but the amount of $CaCO_3$ deposited per polyp is less than in most other massive types. Coenosteoid skeletons allow even more skeleton per polyp.

There is a ceiling to this trend, however. Plocoid corals are more common than coenosteoid forms in which the proportion of common tissue (coenosarc and coenosteum) is greater. The extreme coenosteoid type, represented by the Actinacidae, with few and ill-defined polyps or corallites, generally have small coralla. Presumably, with too much common tissue and too few polyps, the energy base of the coral system is too low to produce large quantities of $CaCO_3$.

All coralla tend to lift polyps away from the substrate and up into the water column. Perhaps this is all that was needed for the relative success of cerioid and phaceloid forms in the Paleozoic. Slow growth rates, relative instability, and failure to solve the waste problem prevented the growth of very large colonies or reefs and prevented colonization of the photic, food-rich, but very high stress wave zone. Paleozoic coenosteoid and meandroid froms were apparently not adaptive under the limitations indicated.

With the advent of the symbiotic zooxanthellae and the development of "loose" porous skeletons, the mechanically strong, rapidly constructed, highly stable, large skeletons of the Mesozoic–Cenozoic–Recent became feasible within the energy budget of the organisms. These provided strong, wave-resistant frameworks that created additional environments for back-reef, lagoonal, and patch-reef growth. In this situation high levels of integration and large biomass per polyp became critical factors and coenosteoid and meandroid growth were apparently highly adaptive.

It is logical to assume that the colony as a whole is responding to environmental pressure. Increased integration, with its provision for colony-wide metabolic activities, will therefore be generally adaptive. This may explain the development of a number of morphotypes, such as thamnasterioid and hydnophoroid, which have provided this advantage but have not had the precise mechanical properties, growth rates, etc., which the most successful forms evidently possessed.

In summary, the highly stressed shallow wave zones have been the environment in which colonial growth has been preeminently successful, and in these zones, rate of growth, production of $CaCO_3$, and mechanical efficiency are the most important factors. In whatever particular structural unit these factors are accommodated, a high degree of colony-wide cooperation, i.e., a high level of integration, is evidently needed.

REFERENCES

Beklemishev, V. N. 1964. Principles of comparative anatomy of invertebrates, v. 1, Promorphology, 3rd ed., Moscow. (Engl. transl. 1970, Oliver & Boyd, Edinburgh, and Univ. Chicago Press. 490 p.)

Brook, George. 1893. The genus *Madrepora, in* Catalogue of the madreporarian corals in the British Museum (Natural History), v. 1, 212 p.

Fowler, G. H. 1887. The anatomy of Madreporaria, II, *Madrepora.* Quart. Jour. Micros. Sci., *17*: 1–16.

Hill, Dorothy. 1956. Rugosa, p. 233–324, *in* R. C. Moore, ed. q.v.

Ivanovsky, A. B. 1965. The ancient rugosans. Acad. Sci. USSR, Siberian Branch, Inst. Geol. and Geophys., 152 p.

Laub, R. S. 1972. The auloporid genus *Cladochonus* M'Coy, 1847: New data from the New York Devonian. Jour. Paleont., *46*: 364–370.

Matthai, George. 1926. Colony formation in astreid corals. Roy. Soc. London, Philos. Trans., ser. B, *214*: 313–356.

Moore, R. C., ed. 1956. Treatise on Invertebrate Paleontology, pt. F, Coelenterata. Geol. Soc. America, and Univ. Kansas Press. 498 p.

———, Dorothy Hill, and J. W. Wells. 1956. Glossary of morphological terms applied to corals, p. 245–251, *in* R. C. Moore, ed. q.v.

Norford, B. S. 1962. The Silurian fauna of the Sandpile Group of northern British Columbia. Geol. Survey Canada, Bull., *78*: 51 p.

Oliver, W. A., Jr. 1966. Description of dimorphism in *Striatopora flexuosa* Hall. Paleontology, *9*: 448–454.

———. 1968. Some aspects of colony development in corals. Paleont. Soc. Mem., *2*: 16–34 [Jour. Paleont., *42* (5 II)].

Sokolov, B. S. 1962. Subclass Tabulata *and* Subclass Heliolitoidea, *in* Yu. A. Orlov, ed., Fundamentals of Paleontology, v. II, Acad. Sci. USSR, p. 192–285. (Engl. trans., 1971, Israel Program for Scientific Translation. p. 293–438.)

Vaughan, T. W., and J. W. Wells. 1943. Revision of the suborders, families, and genera of the Scleractinia. Geol. Soc. America, Spec. Paper, *44*: 363 p.

Wells, J. W. 1956. Scleractinia, p. 328–444, *in* R. C. Moore, ed. q.v.

———. 1957. Corals, *in* H. S. Ladd, ed., Treatise on Marine Ecology and Paleoecology, v. 2, Paleoecology. Geol. Soc. America, Mem, *67*: 773–782.

———. 1966. Evolutionary Development in the Scleractinian Family Fungiidae p. 223–246, *in* W. J. Rees, ed., The Cnidaria and their Evolution, Zoological Society London Symposium no. 16.

Yonge, C. M. 1968. Living corals. Roy. Soc. London, Proc., ser. B, *169*: 329–344.

Table 2

Numbers of Genera per Period of Each Form Listed on Table 1, for Each of the Four Orders (Treatise) or Subclasses (Osnovi) Under Consideration

	A. Heliolitoidea (H) and Tabulata (T) (Based on classification and diagnoses in Osnovi; Sokolov, 1962)								
	Є	O	S	D	C	P	Ř	Total	
Solitary								none	
Total colonial	2	53	88	76	39	27	1?	285	
1	1	5	13	15	9	8	—	51	
2	—	3	1	—	—	—	—	4	
2a	—	—	1	—	—	—	—	1	
3	1	9	1	—	—	—	—	11	
4								—	T
5	—	7	10	4	7	3	—	31	
6	—	15	45	49	17	16	1?	143	
6a	—	—	1	2	—	—	—	3	
7	—	1	—	2	6	—	—	9	
8	—	—	1	—	—	—	—	1	
9									

A. Heliolitoidea (H) and Tabulata (T)
(Based on classification and diagnoses in Osnovi; Sokolov, 1962)

	€	O	S	D	C	P	Ɍ	Total	
Solitary								none	
Total colonial	2	53	88	76	39	27	1?	285	
10									
11									
12									
12a									
13									
14	—	13	14	4	—	—	—	31	⎫
15	—	3	4	2	—	—	—	9	⎬ H
16	—	1	2	—	—	—	—	3	⎭
17								—	

B. Rugosa
(Based on classification and diagnoses in Treatise; Hill, 1956)

	€	O	S	D	C	P	Total
Solitary	—	6	52	86	82	26	252
Total colonial	—	3	28	38	23	12	104
1	—	1	16	19	16	6	58
2	—	1	—	—	—	—	1
2a							—
3	—	2	8	16	8	4	40
4	—	—	—	—	—	1	1
5							—
6							—
6a							—
7							—
8							—
9							—
10	—	—	2	7	3	1	13
11							—
12	—	—	1	6	2	2	11
12a							—
13	—	—	2	3	3	3	11
14							—
15							—
16							—
17							—

C. Scleractinia
(Based on classification and diagnoses in Treatise; Wells, 1956)

	R	J	K	P	N	Total
Ahermatypic solitary	—	6	32	44	60	142
Hermatypic solitary	9	18	21	27	3	78
Total colonial	17	80	104	103	117	421
1	3	21	15	8	10	57
2	—	1	—	—	—	1
2a						
3	6	12	9	10	10	47
4	2	9	13	16	16	56
5						
6						
6a						
7	—	—	1	1	2	4
8						
9	—	3	2	1	6	12
10	—	—	2	2	2	6
11	—	—	2	2	2	6
12	1	5	9	1	1	17
12a	—	9	5	5	9	28
13						
14	4	11	23	31	30	99
15	1	3	15	16	19	54
16	—	—	—	1	—	1
17	—	7	8	9	10	34

Counts are based on the genera recognized in the authority cited and our interpretation of the diagnoses in the same work. Genera that include more than one form or that range through more than one period are entered for each form and period of occurrence. Thus the total number of genera known from a period may be less than the sum of the numbers lower in the tables, and the total number of genera listed in the Treatise or Osnovi as having a given form will be less than the sum of the corresponding horizontal row in the table.

We have not attempted to update the tables by adding genera not included in the Treatise or Osnovi except to add examples of some of the uncommon forms so as to indicate the greater diversity that does exist. Most such additions have been in the cateniform and dimorphic groups.

What Is a Colony in Anthozoan Corals?*

J. W. Wells   Cornell University

ABSTRACT

Animal colonies are groups of individuals either structurally bound together or structurally separated but bound by behavior. The first condition only is applicable in the
case of anthozoan corals, and thus a coral colony appears to be a definable entity.
It is a clone in the sense that it comprises individuals all descended apomictically
from a single founding ancestor, although all coral clones are not colonies. The rugose
corals increased asexually exclusively (except for rare abnormalities) by extratentacular
budding followed by separation of the corallites as individuals, but in the scleractinian
corals intratentacular budding is the rule in more than half the known living and
extinct families. Extratentacular budding results in complete homomorphic individuals,
organically or structurally united as corms or true colonies. But intratentacular increase
produces incomplete individuals marked by lack of directive mesenterial couples and
often not morphologically individualized. The resulting growth has not the morphological value of a colony in the sense of being constituted of distinct or complete individuals
and must be regarded itself as an individual to which the term "cerberoid" or
"polystome" is most applicable.

Author's address: Department of Geological Sciences, Cornell University, Ithaca, New York 14850.

*Reprinted from Geological Society of America, Abstracts with Programs (1971 Annual Meeting, Washing-
ton), v. 3, no. 7, p. 748.

Sediment-Shifting Experiments: A Guide to Functional Behavior in Colonial Corals

J. A. E. B. Hubbard   King's College, London University

ABSTRACT

Four species of *Porites* were selected from the Florida reef tract for experimental studies designed to determine the functional relationship of the polyp to the calcareous skeleton. Each species was subjected to systematic sediment-shedding and cursory feeding and calibrated current-flow experiments, crude lethal- limit studies of their temperature and salinity tolerances, and time-marked growth studies under laboratory conditions. At the end of these experiments branches of *Porites furcata* Lamarck and *P. divaricata* Lesueur were prepared for standard histological and petrographical observations. An analysis of these results shows that *Porites* is neither truly colonial nor truly polystomous; it possesses diagnostic characteristics of both these types of organization and is therefore designated quasi-polystomous. It is a colonial coral in which the polyps comprise incomplete individuals with a common enteron; but they also have distinct stomodea and tentacular rings which are capable of independent action. *Porites* also possesses unique regenerative properties which probably help it to survive adverse conditions.

Examples are given from other Floridian reef corals from known environments of the gradations in coloniality indicated by Wells (this volume). Thus structurally linked colonial corals show behavioral responses which range from solitary reactions in *Mussa angulosa* (Pallas) through partial coordination in *Porites* to complete coordination in *Diploria* and "brain" corals.

Author's address: Department of Geology, King's College, Strand, London, England.

Complete confidence as to the behavioral nature of coloniality can only be ascertained from combining histological and behavioral techniques with observations on skeletal anatomy.

INTRODUCTION

During the course of an investigation into the potential use of fossil corals as environmental indicators (Hubbard and Pocock, 1972) it became apparent that the functional relationship between living tissues and the morphologies of their surfaces of attachment was not understood. The scleractinian corals are usually regarded as passive sedentary animals typified by a radial calical plan (Hubbard, 1972a), but there is not a single species or genus which has been fully studied and there is therefore no standard with which other corals may be compared. Also the recent literature pertaining to the group is widely scattered and often published in obscure journals; consequently crucial information is easily overlooked and very little of the information available is found to relate to one family. This, in turn, is seldom related to one location, genus, species, or specimen. It therefore seemed desirable that several species of an easily accessible, biologically well known genus should be studied from as many aspects as possible in order that the behavioral needs and responses of the soft tissues could be related to its skeleton. These findings could then be used as a basis for functional morphological studies of other corals from known environments which exhibit varying degrees of coloniality.

The genus *Porites* was selected for this project for several reasons: (1) It is tolerant of laboratory conditions; (2) it is a comparatively common coral with a wide distribution on local, regional, and global scales; (3) its species cover a wide range of growth forms, ranging from hemispherical, through crustose, to branching types; (4) it has a distinctively fenestrate skeletal architecture which was presumed to have functional significance; (5) it commonly occurs in the same areas as *Manicina areolata* (Linnaeus) and therefore must have many of the same environmental requirements and tolerances (see D. F. Squires, 1958, unpublished biological report, Smithsonian Institution, Washington, D.C.); and (6) it has also been studied by biologists (Roos, 1967; Franzisket, 1969).

The specimens were studied in three main ways: (1) cine-filmed sediment-shedding behavioral studies (Hubbard and Pocock, 1972); (2) time-marked growth studies to detect the relationship of mineralogical changes within the skeleton over a period of one calendar year (Hubbard, 1972b); and (3) anatomical studies of the histology and hard parts in order to relate their functional responses to their skeletons and probable states of preservation. They were also subjected to subsidiary observations concerned with (1) feeding responses, (2) current responses, (3) destruction by bioerosion, (4) lethal limits (rapid experimental extremes of temperature and salinity), and (5) regeneration experiments. All these studies, with the exception of the regenerative experiments, were synchronously applied to specimens of 18 genera (Table 1) which had been selected from ecologically distinct communities in the Florida reef tract. As the histological results are not yet available for 17 of these genera, the discussion of function and coloniality has been confined to the behavioral and structural findings concerning *Porites,* with which selected corals could be compared and contrasted.

Table 1

List of Species Studied from the Florida Reef Tract

Suborder Astrocoeniida
    Family Acroporidae
        Acropora cervicornis (Lamarck)
        Acropora palmata (Lamarck)
    Family Agaricidae
        Agaricia agaricites (Linnaeus)
        Agaricia fragilis Dana
    Family Siderastreidae
        Siderastrea siderea (Ellis and Solander)
    Family Poritidae
        Porites astreoides Lamarck
        Porites divaricata Lesueur
        Porites furcata Lamarck
        Porites porites (Pallas)
Suborder Faviida
    Family Faviidae
        Diploria clivosa (Ellis and Solander)
        Diploria labyrinthiformis (Linnaeus)
        Diploria strigosa (Dana)
        Colpophyllia natans (Houttuyn)
        Manicena areolata (Linnaeus)
        Cladocora arbuscula Lesueur
        Solenastrea hyades (Dana)
        Montastrea annularis (Ellis and Solander)
        Montastrea cavernosa (Linnaeus)
    Family Oculinidae
        Oculina diffusa Lamarck
    Family Trochosmiliidae
        Meandrina meandrites (Linnaeus)
        Dichocoenia stokessi Edwards and Haim
    Family Mussidae
        Mussa angulosa (Pallas)
        Isophyllastrea rigida (Dana)
        Mycetophyllia lamarckiana (Dana)
        Isophyllia sinuosa (Ellis and Solander)
Suborder Caryophyllida
    Family Caryophyllidae
        Eusmilia fastigiata (Pallas)

The corals' ecological distribution and functional responses to sediment rejection are summarized in Hubbard and Pocock (1972).

OBSERVATIONS ON *Porites* WITH PARTICULAR REFERENCE TO
Porites furcata Lamarck AND *Porites divaricata* Lesueur

Four species of *Porites* Link 1807 were kept in a circulating open seawater aquarium in Miami for 12 months. These species were selected on the basis of their availability and variety of growth forms. Crustose and subhemispherical growth forms were represented by *P. astraeoides* Lamarck; heavily branched forms by *P. divaricata* Lesueur. In nature these species tend to occupy slightly different niches, although their communities are never mutually exclusive, and frequently overlap. Preliminary calibrated current-flow experiments yield results which allow the correlation of these distributions with the species' sensitivities to different flow regimes. *P. astraeoides* appears to be the most catholic species and is found in both clear and turbid waters, where there is a hard substrate. *P. divaricata* frequently occurs in *Thalassia* beds, and in Curaçao around Awa di Oostpunt it is conspicuous in areas of high salinity which are subject to sporadic exposure associated with extreme spring tides. It also dominates the coral community at Soldier Key in Florida, which is another marginal environment subject to harsh extremes of temperature. *P. porites* and *P. furcata* are probably better known for their occurrences in more open waters, e.g., around Piscadera Baai and Santa Martha Baai (Curaçao), and around such areas as Bache Shoal in the Florida Keys and Cayo Grande off Mayaguez, southwestern Puerto Rico. They are also found in more restricted areas, such as Key Biscayne, Florida, and Castle Harbour, Bermuda. Thus *P. porites* and *P. furcata* would seem to occupy both an intermediate growth form between *P. astraeoides* and *P. divaricata,* and an intermediate environmental regime. But preliminary quick-killing experiments, using cessation of ciliary action as indication of death, do not show marked differences in salinity or temperature tolerances. This may indicate that the cause of their distribution either relates to variations in the sturdiness of the young planula or variations in food supply, neither of which have been tested.

In the laboratory all four species show behavioral reactions indistinguishable from each other when teased with a knitting needle, fed zooplankton, or inundated with sediment. But they are readily differentiated by the varying velocities at which they respond to current-flow experiments. In all cases the tentacles of the polyp are extremely active when judged by scleractinian standards. When plankton is introduced into the water, the polyps distend to their maximal elongation and sway in a coordinated manner while their tentacles reach out for the plankton. On capturing their prey they respond instantaneously and often wrestle with it. This response is sufficiently strong that it can be sensed when a knitting needle is placed within reach of a tentacle. Similarly, when a 3-cc sample, comprising equal volumetric proportions of all Wentworth's grades of sand and granules, is deposited evenly on a 20-cm^2 area of *Porites* polyps, they show the same basic response irrespective of species. At first all tentacles are withdrawn and the polyps retract by exuding water from the stomodea; then, after a matter of seconds, or at most minutes, an irregular distension takes place which causes the inundated area to swell and displace the particles by changing their angles of rest. This results from the stomodeal uptake of water (Hubbard, 1972a; Hubbard and Pocock, 1972). They then free their tentacles, which become instrumental in passing large particles from one polyp to deposit them on the next polyp, thus systematically moving the particles away from the center of irritation. Thus in terms of external coordination there is an initially uniform distension of the inundated polypal area, and this is then followed by periodic regional swellings. However, superposed on this coordinated action *Porites* also shows a considerable degree of polypal independence. Cine-filmed observations show that adjacent polyps frequently distend and contract at random. This independence of action appears to be a somewhat anomalous behavioral reaction by comparison with Horridge's (1957) observations, which showed that there were definite paths along which responses traveled.

In order to elucidate these behavioral responses the internal anatomy of *Porites* was studied by means of (1) etching by decalcification in *Calex;* (2) light microscopic studies of serial histological sections of decalcified skeletons; (3) light microscopic studies of thin sections of skeletons both with and without the associated tissues; (4) binocular biological microscope studies of skeletons and their associated tissues before, during, and after decalcification; and (5) selected scanning electron microscope studies of partially tissue clad skeletons. These observations show that each polyp appears at the surface of the corallum as a discrete individual which is capable of independent action and is endowed with a complete ring of tentacles. But the internal calcareous skeleton and histological anatomy is both fenestrate and totally confluent (Fig. 1). Thus not only is the skeleton fenstrate but all three layers of tissue pass through the fenestrae, which

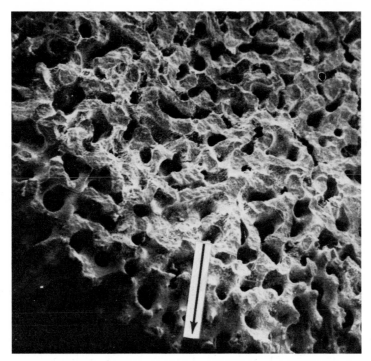

Figure 1

Detail of fenestrate skeletal fabric in a transverse section through a branch of *Porites.* The calical surfaces project downward in the direction of the arrow, which is 10 mm long.

suggests that there is only one enteron which is common to the whole organism. This means that the multitude of polyps with independent stomodea all lead into one large confluent enteron which ramifies the whole skeleton. Thus there appear to be both individual and colonial anatomical and behavioral elements combined within the same animal. But it has not proved possible to analyze what controls these responses, as no trace of any type of "nervous" system has been found. Therefore, for lack of evidence to the contrary, I am forced to conclude that some form of neuroid conduction (Mackie, 1970) must be responsible for the coordinated behavioral responses. But it may well be that these anatomical peculiarities account for much of *Porites'* success in withstanding adverse conditions. Thus *Porites* is not only able to survive periods of exposure, extremes of temperature and salinity, and remove large quantities of mixed-sized sediments, but it also has an unique capacity for regeneration after skeletal damage. For instance, when a branch was broken, polyps with complete rings of tentacles developed from each of the tissue-clad fenestrae exposed at the broken surface; this was at right angles to the original growth surface. Thus, in adversity, *Porites* is capable of generating feeding areas in random planes tangential to its original feeding surface.

THE "COLONIAL" STATUS OF *Porites*

The above-mentioned combination of behavioral and anatomical studies of the same specimens of *Porites* highlights several important issues pertaining to the degree of coloniality which is implied. This is best discussed in the light of Wells' abstract (this volume), which concisely defines the meaning of the term "colony."

First, experimental growth and regenerative studies confirm Vaughan and Wells' (1943, p. 151) statement that *Porites* reproduces by extratentacular budding. This, by definition (Wells, this volume), places *Porites* among the reproductive minority in scleractinian corals. Second, if my interpretation that *Porites* has a communal enteron shared by many polyps with individual stomodea is correct, *Porites* would appear to be anomalous, as its polyps are not "complete homomorphic individuals, organically or structurally united as corms or true colonies." The continuity of the enteron between polyps would suggest an organization which might be defined as quasi-polystomal. That is, the polyps are "incomplete individuals" although they do not completely conform to the definition of a polystome (Wells, this volume). Thus *Porites* defies this classification in that it is unacceptable as either a true colony or a polystome; it possesses diagnostic characters of both.

THE RELATIONSHIP BETWEEN CORAL TISSUES AND SKELETONS

Apart from the detailed skeletogenetic studies by Barnes (1972), Goreau and Goreau (1959), and Bryan and Hill (1940) there is little information on the precise mechanism which governs the deposition of a coral skeleton. But the relationship of the tissues to the skeleton is even less well known, although both skeletons and living tissues have been fairly well described individually. The skeletal anatomy is treated comprehensively by Vaughan and Wells (1943), and the histological studies are largely those of Matthai (1928, pls. I–IV; 1923). Speculations have been made by Wise (1970) and

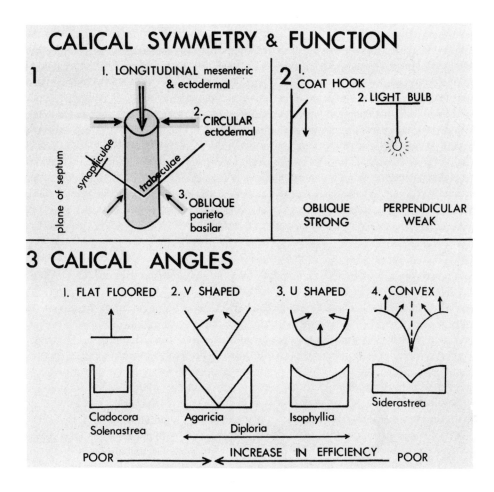

CALICAL SYMMETRY & FUNCTION

1

1. LONGITUDINAL mesenteric & ectodermal

2. CIRCULAR ectodermal

plane of septum

synapticulae

trabeculae

3. OBLIQUE parieto basilar

2

1. COAT HOOK

2. LIGHT BULB

OBLIQUE STRONG   PERPENDICULAR WEAK

3 CALICAL ANGLES

1. FLAT FLOORED   2. V SHAPED   3. U SHAPED   4. CONVEX

Cladocora
Solenastrea

Agaricia
Diploria

Isophyllia

Siderastrea

POOR ⟶ ← INCREASE IN EFFICIENCY   POOR

Figure 2

Diagrammatic synopsis of the functional relationships of polypal tissues to skeletons. (Modified from Hubbard and Pocock 1972.) 1. Shows the angles of intersection between the muscle tissue sheets and the skeletal planes. The comparative strength of the muscular action is indicated by the width of the arrows; the polyp is represented as a hollow cylinder and the skeletal elements by straight lines. 2. Domestic analogies of the importance of the angle of attachment to the purpose required. 3. Schematic longitudinal sections through the chief coral calical geometrics which appear to have functional significance.

Hubbard and Pocock (1972), who interpret the skeletons as a functional support. The main thesis developed by Hubbard and Pocock is summarized in Fig. 2. This demonstrates that the structural strength depends on the angle of intersection between the calical elements and the directions of tension which may be exerted by the three sheets of polypal muscle tissue. The skeletal complexity is thus viewed as a means of stabilizing active polyps, not as a direct muscle attachment scar for which there is no histological evidence. This argument is extended to show how variations in calical symmetry can have effects on the behavioral efficiency of a range of scleractinian genera. Hence, despite the fact that all species tested have an approximately standard distensional potential of about 400 percent, some species are more versatile than others, and these tend to have the greater ecological distribution. Thus of the species tested, those with fenestrate or ornate skeletons tend to have more active polyps than those with smoother septa. *Porites* epitomizes the end products of two complementary scales. It has both a completely fenestrated skeleton and highly active tentacles. Yet, from the point of view of fossil studies it is interesting to note that unless the tissues are present, it is extraordinarily difficult to trace the calical floor of the last living polyp. Moreover, experimental abrasion may render specimens with known orientations unreorientable unless the trabecular structures are preserved. Thus fossil forms of *Porites* and analogous corals will probably present the geologist with many functional, diagenetic, and reconstructional problems (Hubbard, 1972b) in addition to the well-known taxonomic ones.

COMPARATIVE RESPONSES OF DISTINCT COLONIAL SKELETAL TYPES

Wells' (this volume) abstract provides a neat conceptual framework to which the various colonial corals' behavioral patterns may be ascribed. This can be shown to clearly demonstrate why some corals, such as *Mussa angulosa* (Pallas) (Fig. 3), behave essentially as clumps of individuals, while others (Fig. 4) show progressive degrees of coloniality which ultimately pass through quasi-polystomous forms like *Porites* to culminate in truly polystomous forms like *Diploria*. In the first instance sediment is more readily shed when the calice is already at an angle (Fig. 3) than when it is horizontal. This results from the fact that the angle of rest of the individual particles is more rapidly increased to a level of instability by distension of an already inclined surface than a horizontal one. Similarly, the calices near the substrate are in constant danger of being inundated or ''suffocated'' by migrating sediments (Marshall and Orr, 1931, pp. 130–131). Conversely, the polyps at an angle to the substrate are in the optimum position for filtering nutrient from the circulating waters. Apart from the fundamental control of the budding pattern, the distribution of polyps has a considerable bearing on their environmental strategy which is most readily tested by means of simple, cine-filmed, sediment-shedding experiments.

CONCLUSIONS

This study of *Porites* has shown that sediment-shedding experiments are vital to the recognition of behavioral responses in living scleractinian corals. But they are not adequate proof, in themselves, of the behavioral nature of coloniality. This can only be ascertained from interpreting the combined results of behavioral experiments with detailed serial histological and skeletal studies.

CALICAL ORIENTATION & FUNCTION

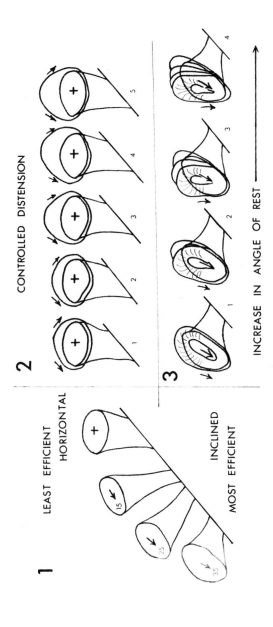

Figure 3

Diagrammatic sketches drawn from cine-filmed observations of sediment rejection by *Mussa angulosa* (Pallas). This species, although a structurally colonial coral, exhibits solitary behavior. 1. Tangential section through a fasciculate colony to show the effects of varying calical orientations. 2 and 3. Show a series of stages in sediment shedding by controlled distension. In 2 the calices are horizontal and in 3 they are inclined.

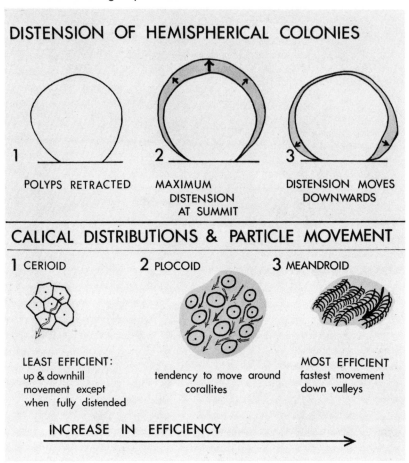

Figure 4

Diagrammatic sections through hemispherical colonies to show the downward migrating paths of controlled distension shown by most hemispherical forms. The lower diagram shows tangential view of hemispherical coral surfaces to demonstrate the relationship between the main calical distributions found in such colonies and their relationship to the movement of sedimentary particles. Cerioid forms present an inconvenient undulating topography for movement of particles; plocoid calices periodically block the path of movement but are generally more efficient; meandroid forms show a totally coordinated polystomous condition.

ACKNOWLEDGMENTS

I thank F. M. Bayer and A. G. Coates for constructive criticisms of the manuscript; J. W. Wells and W. A. Oliver, Jr., for their infectious curiosity; C. E. Lane and the anatomy technicians of Kings College for histological preparations; C. E. Emiliani and

CALICAL ORIENTATION & FUNCTION
IN FOLIACEOUS MONTASTREA

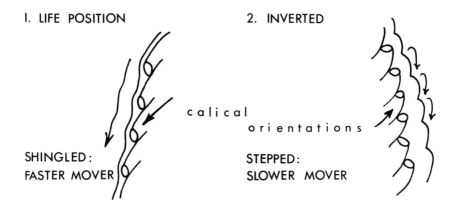

1. LIFE POSITION

2. INVERTED

c a l i c a l

o r i e n t a t i o n s

SHINGLED :
FASTER MOVER

STEPPED:
SLOWER MOVER

Figure 5

Sketches of *Montastrea* shedding sediment in life position and where inverted which emphasize the importance of calical orientation with respect to effective functional activity.

H. B. Moore of R.S.M.A.S., Miami, and many Caribbean and Atlantic marine laboratories for facilities; and the Royal Society, Esmée Faribairn Charitable Trust, Bermuda Biological Station, and University of London Central Research Fund for financial assistance.

REFERENCES

Barnes, D. J. 1972. The structure and formation of growth-ridges in scleractinian coral skeletons. Roy. Soc. London, Proc., ser. B, *182:* 331–350, pls. 18–21, 7 figs.

Bryan, W. H., and D. Hill. 1940. Spherulitic crystallization as a mechanism of skeletal growth in the hexacorals. Roy. Soc. Queensland, *LII* (9): 78–91.

Franzisket, L. 1969. Riffkorallen können autotroph leben, *in* Die Naturwissenschaften. Springer-Verlag, Berlin.

Goreau, T. F., and N. I. Goreau. 1959. The physiology of skeleton formation in corals, II, Calcium deposition by hermatypic corals under various conditions in the reef. Biol. Bull., *117:* 239–250.

Horridge, G. A. 1957. The co-ordination of the protective retraction of coral polyps. Roy. Soc. London, Phil. Trans., ser. B, *240:* 495–529.

Hubbard, J. A. E. B. 1972a. *Diaseris distorta:* an 'acrobatic' coral. Nature, *236:* 457–459.

———, 1972b. Cavity formation in living scleractinian reef corals and their recognition in fossil analogues. Geol. Rundsch., *61:* 551–564.

———, and Y. P. Pocock. 1972. Sediment rejection by recent scleractinian corals; a key to palaeoenvironmental reconstruction. Geol. Rundsch., *61:* 598–626.

Mackie, G. O. 1970. Neuroid conduction and the evolution of conducting tissues. Quart. Rev. Biol., *45:* 319–332.

Marshall, S. M., and A. P. Orr. 1931. Sedimentation on Low Isles Reef and its relation to coral growth. Great Barrier Reef Exped. 1928–29. Brit. Mus. (Nat. Hist.) Sci. Rept., 1(5): 94–133, 7 figs.

Matthai, G. 1923. Histology of the soft parts of astreid corals. Quart. Jour. Micros. Sci., *67:* 101.

———. 1928. Catalogue of the Madreporarian corals in the British Museum (Natural History), VII, A monograph of the Recent Meandroid Astreidae. Brit. Mus. (Nat. Hist.), London, 288 p., 72 pls.

Roos, P. J. R. 1967. Growth and Occurrence of the Reef Coral *Porites astreoides* Lamarck in Relation to Submarine Radiance Distributions. Academic proefschrift, Utrecht.

Squires, D. F. 1958. Biomechanics of *Manicina areolata.* 92 p., 7 pls., 13 figs., unpublished manuscript deposited with Smithsonian Institution, Washington, D.C.

Vaughan, T. W., and J. W. Wells. 1943. Revision of the suborders, families and genera of the Scleractinia. Geol. Soc. America, Spec. Paper, Baltimore, *44:* 363 p., 51 pls., 39 figs.

Wise, S. M. 1970. Scleractinian coral exoskeletons: surface micro-architecture and attachment scar patterns. Science, *169:* 978–980.

Circumrotatory Growth Form in Recent and Silurian Corals

Don L. Kissling   State University of New York, Binghamton

ABSTRACT

Unattached spheroidal coral colonies which have radially disposed corallites resulting from growth outward in all directions are designated as *circumrotatory*. Frequent rotation during growth permitted complete envelopment by living polyps. Abundant circumrotatory colonies of the modern scleractinian *Siderastrea radians* from the Florida Keys and the Silurian tabulate *Favosites favosus* in the Brassfield Formation of southwestern Ohio are found only where exceptional environmental conditions prevail or had prevailed. Typical *S. radians* exhibit hemispheroidal colonies cemented rigidly to firm substrata and are absent where sediment cover is appreciable. However, on the windward calcirudite margins of elevated carbonate mudbanks, where oncoming waves create a constant surf, *S. radians* colonies are invariably circumrotatory and are rolled about frequently. Circumrotatory growth forms are represented by 30 percent of numerous *F. favosus* colonies found on the sloping flanks of small bioherms or ancient mudbanks whose crests, during time of deposition, were elevated above the surrounding seafloor and inhabited by rooted crinoids and foliate bryozoans. Detrital flanks were subjected to wave action, especially on south-facing, windward slopes, where *F. favosus* colonies were most abundant. Only 6 percent of *F. favosus* colonies from contemporaneous interbiohermal strata are circumrotatory. Nearly all specimens collected from an unconformable surface believed to represent a wave-swept lithified substratum during deposition exhibited typical hemispheroidal colonies having turbinate or encrusting bases.

Author's address: Department of Geological Sciences, State University of New York, Binghamton, New York 13901.

Founder polyps of circumrotatory *Siderastrea radians* encrust inert fragments of *Porites porites*. Rapid budding of the daughter polyps around these elongate nuclei results in radial growth of corallites and a tendency for prolate forms among circumrotatory colonies. Growth cessation is marked by concentric arcs of skeletal discontinuity and record prolonged burial of colony surfaces. While free-living colonies are capable of some degree of self-rotation, only frequent rotation by waves ensures development of circumrotatory forms. Ultimately waves also lose their efficacy as colonies increase in mass. Cementation was not essential for the welfare of fully grown *Favosites favosus* colonies regardless of form, nor perhaps for the metamorphosis of their larvae. In most circumrotatory *F. favosus* the growth nucleus consists of a protocorallite from which daughter corallites extend in tangential and radial directions. The predominance of oblate colony forms is an artifact of compaction and the crushing of thin skeletal tissues. In life *F. favosus* colonies must have weighed far less than *S. radians* colonies of comparable sizes; and as almost no evidence for growth cessation exists in *F. favosus,* probably less turbulence was required to roll them about. Recent and Silurian circumrotatory corals evidently occupied corresponding ecological stations. Perhaps circumrotatory colonies of any species, in which some other growth form is typical, may serve as environmental indicators for wave-swept margins of mudbanks and sandy shoals.

INTRODUCTION

Free-living, spheroidal, or ellipsoidal coral colonies which have radially disposed corallites resulting from growth outward in all directions from a center are designated herein as "circumrotatory" colonies. Unfortunately, this term begs the criticisms so often directed to words newly applied to specific natural phenomena, resembling as it does the word "circumlocutory" (evasive speech), made the more infamous by Charles Dickens's indictment of nineteenth-century English bureaucracy. Nevertheless, it seems a useful label for preservable colonies which undergo or have undergone frequent rotation during growth, thereby permitting complete surficial envelopment by living polyps. As the term "circumrotatory" alludes to the morphogenesis of colonies so designated, the term departs from the descriptive nomenclature usually applied to colony form in corals, recently compiled by Oliver (1968, p. 18). However, the resultant spheroidal or ellipsoidal (massive) colony form is explicit in the definition.

This growth phenomenon in corals was discovered nearly a century ago by John Murray during a visit of H.M.S. *Challenger* to the New Hebrides reefs. Murray's terse observations (1885, p. 517) express the essential effects and cause of circumrotatory form. "Some specimens of *Porites* were unattached," he stated, "though living, being in the form of rounded masses entirely covered by living polyps, and probably from time to time rolled over by the waves." Growth habits necessary to produce circumrotatory colonies of *Porites* and *Astreopora* on the Cocos-Keeling Atoll were postulated by Wood-Jones (1907, p. 520, 539). He stated that small colonies may attain spherical forms by initially adhering to and then engulfing as nuclei isolated skeletal fragments, but that further growth produces hemispherical colonies, as increased weight inevitably causes polyps on the undersurface to die while the remainder of the colony continues to grow.

The present paper compares circumrotatory growth form in the modern scleractinian

Siderastrea radians (Pallas) from the Florida Keys and the Silurian tabulate *Favosites favosus* (Goldfuss) from southwestern Ohio. An attempt is made to document the nature of attachment and subsequent growth and the life habitats of living and fossil circumrotatory corals. Finally, I hope to demonstrate the nonrandom development of this growth form; that the influence of external factors is such that abundant circumrotatory colonies are found only where exceptional environmental conditions prevail or had prevailed.

HABITAT OF MODERN CIRCUMROTATORY CORALS

Siderastrea radians is ubiquitous throughout the Florida Keys and Dry Tortugas, and indeed in most shallow waters of the tropical western Atlantic. Its presence in a great variety of hard-substrate habitats gives testimony to its exceptional hardiness; it is more tolerant to exposure, sediment abrasion, and extremes of temperature and salinity than other West Indian hermatypic coral species. Almost invariably its hemispheroidal to flattened colonies are cemented rigidly to firm substrata, most commonly eroded limestone surfaces or coral boulders, but occasionally on erect branches of alcyonarians and *Porites porites* (Pallas) and old bottles (Fig. 3C). One memorable specimen was found attached to the dorsal carapace of a living *Limulus polyphemus* Linnaeus. But these are the typical encrusting forms of the species and are not circumrotatory. During several years of diving I have found fewer than 10 circumrotatory colonies in hard-substrate habitats, while several thousand attached specimens were observed. A single circumrotatory specimen was reported by Squires (1958, p. 249) on hard substrata at Turtle Rocks, Bimini, Bahamas. Several such specimens living unattached on sand and gravel substrata populated by marine grasses were reported by Verrill (1901, p. 153) at Bermuda, by Duerden (1904, p. 3) in Kingston Harbor, Jamaica, and by Yonge (1935, p. 202) on Bird Key Shoal, Dry Tortugas. In general, however, this coral is absent wherever sediment cover exceeds a few centimeters; its distribution is predictable in most instances on the basis of sediment thickness (Kissling, 1965).

The principal exceptions to this are the windward, wave-swept margins of elevated carbonate banks, such as Rodriguez Bank (Fig. 1) lying 2 kilometers south of Rock Harbor, Key Largo. Here, a 4-meter-thick accumulation of carbonate mud forms a broad, flat-surfaced shoal supporting a low mangrove island. Most of the bank surface is covered by the marine grass *Thalassia testudinum* König, whose dense rhizomes help to trap and retain the sediment. Well-defined bank margins of skeletal sand and gravel slope out to surrounding depths of 2 to 3 m. The northern and western margins of the bank, facing toward Key Largo, lie relatively protected from waves. However, wind-driven waves originating on the adjacent open shelf break across the southeastern margin of Rodriguez Bank during all but the calmest days. In their discussion of Rodriguez Bank, Turmel and Swanson (1964) demonstrated that easterly and southeasterly winds predominate regionally.

A distinct biotic zonation, evident because of its conformity with changes in topography, exists across the southeastern margin of the bank. While many prominent species extend beyond the zones they characterize, relative abundances of most species change markedly across the bank margin and manifest the influence of wave turbulence and water depth

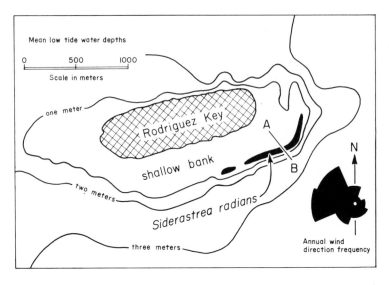

Figure 1

Distribution of *Siderastrea radians* (in black) and location of Traverse A-B on Rodriguez Bank, Florida Keys. Bathymetry and wind-direction frequency data for the years 1918–1947 modified from Turmel and Swanson (1964).

upon their distribution (Fig. 2). The windward slope between 0.8 and 2.3 m depth is characterized by abundant marine grasses, *Thalassia testudinum* and *Diplanthera wrightii* Aschers; by numerous alcyonarians, including *Plexaurella dichotoma* (Esper), *Pseudopterogorgia acerosa* (Pallas), *Eunicea* spp., and *Pterogorgia anceps* (Pallas); by numerous echinoids, of which *Clypeaster rosaceus* (Linnaeus) predominates; and by several scleractinian species, particularly *Cladocora arbuscula* Lesueur, rare or absent in the other zones. The substratum consists of fine calcarenite. The *Goniolithon* zone, occupying the crest of the bank margin and partly exposed during low tides, is covered by dense clumps of the coralline alga *Goniolithon strictum* Foslie, which provide shelter for abundant holothurians, ophiuroids, and decapod crustaceans. A substratum of muddy coarse calcarenite lies beneath the *Goniolithon*. The shallow leeward lagoon, floored by sandy calcilutite, is populated chiefly by marine grasses, sponges, mollusks, and holothurians.

The ramose scleractinian *Porites porites* is the most conspicuous species inhabiting the windward bank edge. Marked differences in growth habit exist between these abundant colonies and the scarce specimens occupying the lagoon and windward slope. The latter possess slender branches which bifurcate upward from an inert portion of the colony buried in the substratum, giving support to the erect living portion. In contrast, most *P. porites* in the *Porites* zone exhibit peculiarly flattened colonies having thicker branches intertwined and arrayed horizontally and lie unburied upon the substratum. Upper and lower surfaces of many colonies are so equally developed and so lacking in moribund portions that it seems likely that these, too, especially the smaller colonies, are subjected occasionally to mechanical overturning. These branching colonies, together with short-

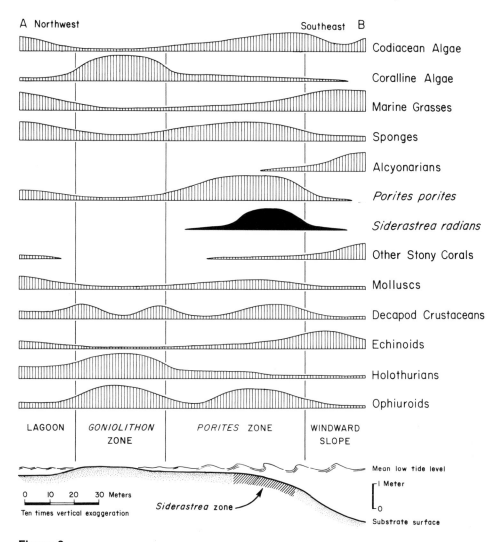

Figure 2

Zonation and relative abundances of organisms inhabiting the southeastern margin of Rodriguez Bank along Traverse A-B.

leafed *Thalassia testudinum* and clumps of *Halimeda opuntia* (Linnaeus) and other codiacean algae, provide homes and substrata for numerous sponges, ophiuroids, decapods and bivalve mollusks. Abundant broken fragments of *P. porites* produce a substratum of coarse sandy calcirudite and serve as sites for larval attachment and growth by *Siderastrea radians*.

    The abundant colonies of *Siderastrea radians* on Rodriguez Bank are circumrotatory almost without exception. Their shapes range from nearly perfect spheres to prolate ellipsoids, and the diameters range from 2 to 12 cm (Fig. 3). They are found only

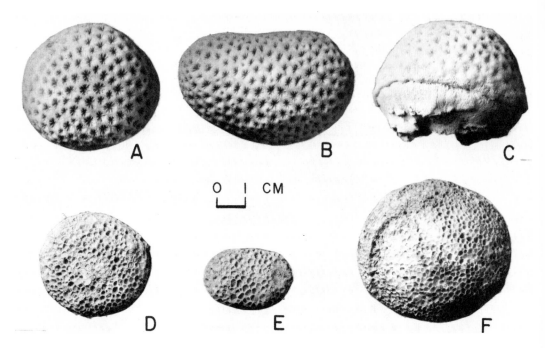

Figure 3

Siderastrea radians and *Favosites favosus*. A, B, circumrotatory *S. radians* colonies; C, hemispheroidal, encrusting colony of *S. radians;* D–F, circumrotatory *F. favosus* colonies.

along the southeastern margin of the bank, where surf action is most vigorous (Figs. 1 and 2). *S. radians* colonies diminish sharply in number and become progressively more irregular in shape toward the northern and western extremities of their distribution and toward the base of the windward slope. The most perfectly spheroidal colonies are found at the center of their range, where they are most numerous. The majority of colonies lying free upon the coarse substratum is wholly enveloped by living polyps pigmented by included zooxanthellae. Polyps on the upper surfaces of colonies remain retracted during daylight hours; however, on nearly all specimens examined polyps in a narrow ring around the base—those polyps in contact with the substratum—were expanded 4 mm beyond the skeletal surface. On many larger specimens polyps on the lower surfaces were withdrawn deeply and appeared somewhat bleached, owing to loss of the symbiotic zooxanthellae. Relatively calm weather had prevailed for three weeks prior to these observations, and perhaps these latter colonies had not been overturned during that time. Colonies larger than 8 cm in diameter were rare and most of these had dead and bored bottom surfaces. A few large circumrotatory colonies of *Solenastrea bournoni* Milne-Edwards and Haime and *Dichocoenia stokesii* (Milne-Edwards and Haime) were present. Except for a single *Solenastrea,* these also possessed lifeless bottom surfaces.

Precisely the same physiographic and hydrographic conditions and the same biotic zonation are embodied in Tavernier Bank, located 7 km southwest of Rodriguez Bank, adjacent to Tavernier, Key Largo. The mangrove island here is somewhat smaller, the bank lagoon somewhat less developed; nevertheless, abundant circumrotatory *Siderastrea radians* colonies inhabit the outer half of the *Porites* zone along the southeastern margin of Tavernier Bank. One additional but dissimilar occurrence is known. Circumrotatory and typical encrusting colonies of *S. radians* coinhabit shallow seagrass meadows of *Thalassia testudinum* and *Syringodium filiforme* Kutz on the lee of Pelican Shoal Reef, situated upon the shelf edge 8 km south of Boca Chica Key. Circumrotatory colonies repose unattached on a coarse calcarenite substratum, while typical hemispheroidal colonies encrust scattered inert coral rubble cast there by storms. Although protected on the south and west by reef buttresses and elevated boulder ramparts, this leeward shoal lies open to the east, so that heavy swell and wave trains generated by easterly winds pass over it before breaking against the boulder ramparts. Ecological surveys were made across four other *Porites*-fringed, elevated carbonate banks in the lower Florida Keys. These are located at the entrance of Coupon Bight, east of Pye Key, south of Pye Key, and south of Summerland Key. Two features in common to all four banks and in contrast to Rodriguez and Tavernier Banks are (1) the absence of appreciable surf due to relative protection, and (2) the complete absence of *S. radians,* whether circumrotatory or otherwise, despite the conspicuous presence of *Porites porites, Cladocora arbuscula,* and *Manicina areolata* (Linnaeus), which seldom encrust large objects.

HABITAT OF SILURIAN CIRCUMROTATORY CORALS

Circumrotatory colonies of the tabulate coral *Favosites favosus* were found associated with several small bioherms exposed in quarry walls at Fairborn, Ohio (Fig. 4). These bioherms, situated in the upper 4 m of the Brassfield Formation of late Llandovery age (Early Silurian), range from 2 to 8 m in diameter and 1 to 3 m in height. These structures are apparently circular or elliptical in plan and moundlike in vertical section, and consist of a central massive core from which detrital flanking strata slope steeply in all directions. Bioherm cores are poorly bedded, argillaceous, mud-supported biomicrites containing an intertwined profusion of articulated crinoid holdfasts and intact zoaria of foliate and fenestrate bryozoans. Bioherm flank strata are thin-bedded, grain-supported biomicrites comprised primarily of disarticulated crinoid and bryozoan skeletons. Within short distances of bioherm cores the flank strata become thicker bedded, less muddy, and nearly horizontal as they grade into interbiohermal strata. The latter, representing the bulk of limestones exposed in the quarries, are cleanly washed biosparites consisting of the broken and abraded debris of crinoids and bryozoans. Interbiohermal strata display current-ripple bedding throughout.

I view these bioherms as having been stable carbonate mudbanks, elevated 50 to 150 cm above the surrounding level seafloor, and formed by the sediment-baffling and binding effects of dense stands of rooted crinoids and foliate bryozoans. The cores were rimmed by detrital calcarenite flanks subjected all around to wave action. Windward and lee slopes of bioherms may be distinguished on structural and paleontological criteria. Measurements of four bioherms that are exposed adequately revealed uniformly steeper flank slopes on the north sides than on the south sides, with maximum dips averaging

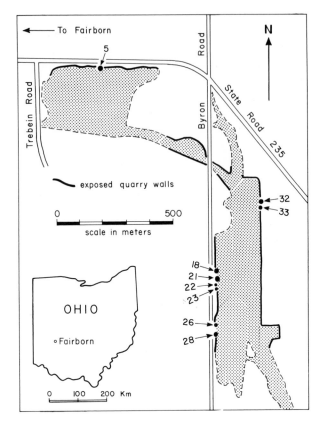

Figure 4

Bioherms and interbiohermal strata exposed in the Southwestern Portland Cement Co. quarries at Fairborn, Ohio. The numbered dots indicate the locations and approximate sizes of bioherms.

33° and 23°, respectively. Because this consistency in asymmetry is matched in these bioherms by notably larger coelenterate populations on the south flanks, I assume that wind-driven waves from the south had prevailed or at least were dominant, and that the south-facing flanks represented windward slopes. This interpretation is substantiated by the known Silurian paleoslope direction toward the southeast. Despite disparity in size, probable water depth, and composition of the biota, the bioherms, if diagnosed correctly, must have resembled Rodriquez Bank in many important attributes. Details of the structure and ecology of these Silurian bioherms will be reported elsewhere (Kissling, manuscript in preparation).

Just as at Rodriguez Bank, pronounced zonation of the fossil biota is exhibited across these densely populated ancient mudbanks (Fig. 5). Crinoids possessing rooted holdfasts, particularly *Botryocrinus* sp. and *Clidochirus* sp., evidently flourished upon the bioherm crests, as did ostracodes, sponges, wormlike burrowers, gastropods, and holothurians(?). The latter two were apparently more numerous toward the lee sides of crests. Foliate and fenestrate bryozoans, represented by many species, were abundant in all facies

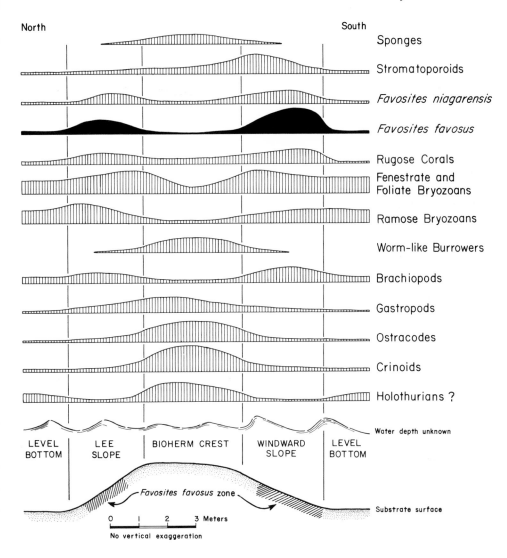

Figure 5

Zonation and relative abundances of fossil taxa across a Silurian bioherm reconstructed to appear as it might have during deposition. Modeled after Bioherm 33 with supplementary data from other bioherms.

but reached their acme on the peripheries of bioherm crests and on the upper flank slopes. The ramose bryozoans *Hallopora magnopora* (Foerste) and *Hennigopora* sp. were abundant on the lower flank slopes, especially the lower portions of lee slopes, but extremely rare on bioherm crests. The few massive zoaria of *Fistulipora* sp. were confined to the bioherm crests. *Platystrophia reversata* (Foerste), *Dolerorthis flabellites* (Foerste), and other brachiopod species were twice as numerous on the windward slopes of bioherms as on the lee slopes. Although containing far fewer obviously indigenous specimens, interbiohermal level bottoms apparently were populated sparingly at least by brachiopods, holothurians (?), and all bryozoans, except those having massive zoaria.

Generally, similar distribution patterns are exhibited by all coelenterate groups: that is, relatively sparse on bioherm crests, abundant on bioherm flanks—especially the windward slopes, and rare on interbiohermal level bottoms. The stromatoporoid *Clathrodictyon vesiculosum* Nicholson and Murie displays the greatest asymmetry in distribution; most specimens evidently inhabited the outer crests and upper slopes of the windward sides of bioherms. Habitat preference was somewhat less pronounced in *Favosites niagarensis* Hall and rugose corals, although these also tended to favor the windward slopes and rarely colonized the broad level bottoms. *F. favosus*, on the other hand, displays particularly marked zonation, being concentrated on the medial and lower reaches of bioherm flanks and rarely found in bioherm cores. They were nearly twice as abundant on windward slopes as on lee slopes, and 95 times more abundant per unit area of exposure on bioherm flanks than in interbiohermal strata.

Most *Favosites favosus* colonies are hemispheroidal or nearly discoidal. They possess broad turbinate bases. Most colonies appear to have acquired stability by tolerating burial of their proximal portions while maintaining their centers of gravity at or just below the substrate surface, rather than by cementing firmly to the substratum. Similar hemispheroidal or discoidal forms are typical for most species of the genus, including *F. niagarensis*. Although many specimens of the Silurian *F. forbesi* Milne-Edwards and Haime and the Devonian *F. alpenensis* Winchell (Swann, 1947) display nearly spherical colonies, invariably these either possess points of firm basal attachment or had encrusted and encircled erect crinoid stems, and are not circumrotatory. Indeed, a number of *F. favosus* colonies found at Bioherms 18 and 21 (as with certain *Siderastrea radians* colonies mentioned earlier) had encrusted and encircled some erect, stemlike structure, but the host organisms had left no trace. Circumrotatory colonies of *F. favosus* are concentrated on the bioherm flanks, where 30.4 percent of 594 specimens collected exhibit this growth form. Only 6.3 percent of 96 specimens collected from contemporaneous interbiohermal strata show evidence of having rotated during growth. Circumrotatory colony shapes range from nearly perfect spheres to prolate, especially oblate ellipsoids (Fig. 3). Diameters range from 2 to 12 cm. On several oblate colonies opposing surfaces are deeply depressed.

The bioherms examined occur in discernable groups or clusters separated by distances of 150 to 350 m (Fig. 4). From one bioherm cluster to the next, relative abundances of different coelenterate taxa vary considerably; for example, rugose corals are proportionately high at Bioherm 5, *Clathrodictyon vesiculosum* are proportionately high at Bioherms 32 and 33. Partly this is an artifact of fortuitous exposure of core or flank facies, but as population variance is smaller between bioherms belonging to the same cluster than between clusters, some part must express real differences in coelenterate distribution. The most systematic differences between bioherm clusters belong to *Favosites favosus*. From north to south that species comprises 10 percent of the coelenterate fauna at Bioherm 5, 56 percent at Bioherms 32 and 33, 67 percent at Bioherms 18–23, and 72 percent at Bioherms 26–28. Moreover, the incidence of circumrotatory colonies among *F. favosus* increases progressively southward: none at Bioherm 5, 15 percent of *F. favosus* specimens

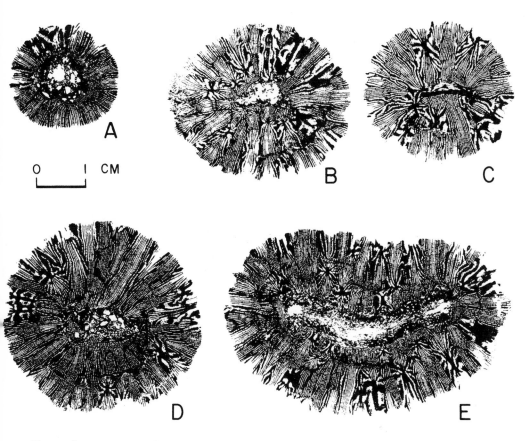

0 ⊢——⊣ I   CM

Figure 6

Sections through the centers of circumrotatory colonies of *Siderastrea radians*. See discussion in text.

at Bioherms 32 and 33, 24 percent at Bioherms 18–23, and 44 percent at Bioherms 26–28. Seemingly, some environmental gradient promoted or inhibited the development of circumrotatory colonies in *F. favosus,* but what this north–south vectorial effect was is beyond reasonable conjecture.

A narrow shaly zone bracketed between two erosional unconformities and containing abundant coelenterates lies at the top of the Brassfield Formation, beneath the Dayton Limestone. This zone extends along the eastern wall of the southern quarry (Fig. 4) but is absent at places where the unconformities converge. The juxtaposition of erosional surfaces suggests that corals in this zone had inhabited a lithified limestone substratum in relatively shallow, wave-swept waters. This interpretation is strengthened greatly by the fact that *only in this zone* of the Brassfield are lithoclasts found, are crinoids absent, are the skeletons of solitary rugose corals abraded, are *Favosites niagarensis* colonies almost exclusively discoidal, and are some *F. favosus* colonies apparently cemented to the substratum. Significantly, in this zone, patently analogous to the hard-substrate

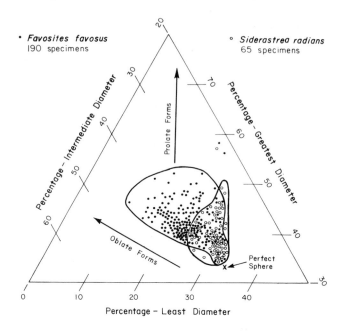

Figure 7

Frequency distribution of colony shapes for 255 *Favosites favosus* and *Siderastrea radians* specimens. Percentage diameter refers to the percentage that the length of each of three principal axes of a corallum comprises of the sum of lengths of all three axes, calculated for each circumrotatory colony.

habitats of typical encrusting *Siderastrea radians* colonies, only 2.2 percent of 137 *F. favosus* colonies examined possessed circumrotatory growth forms.

DEVELOPMENT OF CIRCUMROTATORY GROWTH FORM

Upon settling, the larvae of *Siderastrea radians* must encrust some foreign solid object if they are to undergo metamorphosis successfully. Of 25 circumrotatory colonies examined in thin section, 23 had encrusted inert, in some cases algal-coated fragments, of the branching coralla of *Porites porites* (Fig. 6). The nuclei of two colonies consist of molluscan shell fragments (Fig. 6C). Asexual budding in *S. radians* is extratentacular and new corallites are inserted more or less regularly by lateral increase during growth (Duerden, 1904). Additional individuals develop rapidly from the founder polyp so that the entire nucleus soon becomes enveloped by fledgling corallites. Unimpeded growth over the entire surface when accompanied by frequent rotation ensures radial symmetry of corallites and ideally results in spheroidal colonies. The strong tendency for prolate forms in circumrotatory *S. radians* (Fig. 7) reflects the elongate shapes of *Porites* skeletal fragments which they encrust (Fig. 6E). Colony growth, however, is not unimpeded despite the remarkable sphericity of many specimens. Approximately half of the colonies sectioned reveal one or more intervals of growth cessation, marked by concentric arcs of discontinuity (Fig. 6B, D, E). These discontinuities record the death of polyps buried too long to endure, followed by the reestablishment of living polyps over the dead surface, presumably after the colony was rolled over. The high density of *S. radians* coralla would tend to promote stability and lead to intermittent growth. In contrast, a few unusual specimens are deceptively lightweight. These possess hollow cores, caused perhaps by biogenic destruction of the nucleus and inner parts of the corallum. Those penetrated by one or more small entrances are inhabited frequently by mantis shrimp or clinid fishes.

The presence of bands of distended polyps around the bases of circumrotatory colonies on Rodriguez Bank suggests that these corals possess an innate mechanism for rotating. Perhaps countless feeble muscular contractions of polyps and tentacles enable a colony to rotate spontaneously in a random manner. Fabricius (1964) has shown that *Manicina areolata* colonies are capable of rotating 180° if they are overturned inadvertently; but polyps of that species are far larger. To test this capacity in *Siderastrea radians* eight circumrotatory colonies were marked and placed in an aquarium. During the course of several days two of the smaller, more spherical colonies rotated through 30° of arc without changing location. The others remained perfectly stationary. Thus, while some measure of self-rotation exists, it nevertheless seems inadequate and too inconsistent to maintain the vitality of polyps temporarily buried without the intervention of some external agency.

Probably the conditions and substrata that induced larval settlement in *Favosites favosus* were different from those in *Siderastrea*. Unlike *S. radians,* most normal colonies of *F. favosus* lived free on sandy substrata, at places in association with circumrotatory colonies. Evidently cementation was not essential for the welfare of fully grown colonies, or perhaps for the metamorphosis of their larvae. The growth nucleus in most circumrotatory *F. favosus* colonies (70 percent of those not obscured by crushing) consists of a polygonal protocorallite lacking tabulae. Protocorallite diameters range from 2.8 to 3.5 mm; the largest daughter corallites range from 2.2 to 3.0 mm in diameter. Unidentified, thin-walled, hollow spheres, 4 to 7 mm in diameter, represent foreign nuclei in the remaining circumrotatory colonies (Fig. 8B). Other exotic skeletal material has not been observed. While virtually all protocorallites are located near the exact centers of circumrotatory colonies, the hollow spheres are most commonly acentric. Daughter corallites extend from the protocorallite surface in tangential and radial directions . Further colony growth was strictly radial, accompanied by lateral increase in which new corallites were inserted between existing corallites. Corallite thecae are thickened abnormally at the peripheries of most colonies, and in many the last-formed, distal tabulae are crowded together closely. These eleventh-hour skeletal innovations probably reflect metabolic changes in response to burial or deterioration preceding death.

Relatively few circumrotatory colonies of *Favosites favosus* approach spherical forms (Fig. 7). The predominance of oblate forms is largely an artifact of compaction; the colonies, consisting of thin thecae and tabulae, were compressed and crushed in the plane parallel to bedding (Fig. 8C, D, E). In life these colonies must have weighed far less than *Siderastrea radians* colonies of comparable sizes. Not surprisingly, I have found almost no evidence for temporary growth cessation in *F. favosus* colonies, although

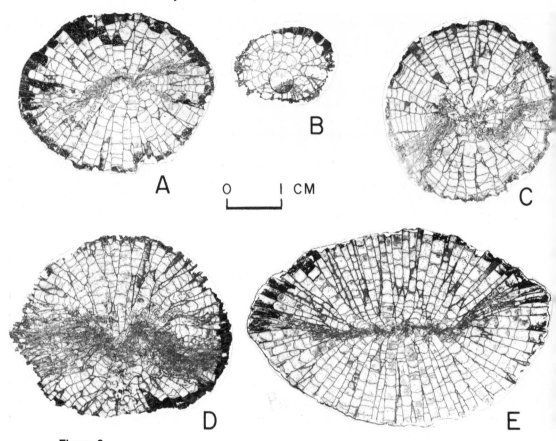

Figure 8
Sections through the centers of circumrotatory colonies of *Favosites favosus*. See discussion in text.

this phenomenon is seen commonly in sympatric *F. niagarensis* colonies in which dead, mud-covered portions had become repopulated by lateral growth of corallites from viable portions. A concentric zone of abnormally thickened thecae was found within a single circumrotatory colony, but its burial was probably temporary, as all the same corallites continued without reorientation in subsequent growth. Presumably, less turbulence was required to roll *F. favosus* colonies about, and turbulence must have been constant enough to prevent marked growth cessation.

CONCLUSIONS

Abundant circumrotatory colonies of *Siderastrea radians* are found only on the windward margins of elevated carbonate banks where hard substrata are buried deeply and where surf action is appreciable. In hard-substrate habitats or on carbonate banks not subjected to surf action, these corals either assume the typical hemispheroidal-encrusting forms or they are absent altogether. Although free-living colonies are capable of some degree of self-rotation, frequent rotation by waves is necessary for the development of circumrotatory colonies. Ultimately, the motive force of waves loses its efficacy as colonies become too massive or too steadfastly buried, whereupon the colonies expire or, less commonly, undergo a modification of form. The development of circumrotatory colonies on Rodriguez Bank is a function of three factors: (1) the presence of abundant dead *Porites* fragments to serve as nuclei for attachment, (2) the ability of *S. radians* to encompass these fragments during early growth stages, and (3) the presence of surf action to reorient the colonies at times during growth. The development of circumrotatory *Favosites favosus* colonies was dependent upon two factors: (1) the ability of asexually produced daughter corallites to envelop the protocorallite when colony reorientation accompanied colony growth, and (2) the preferred habitation of *F. favosus* on the lower slopes of elevated Silurian mudbanks, where wave-generated turbulence rolled these colonies about without disturbing others. Because cementation was not essential and because they were lightweight, circumrotatory colonies of *F. favosus* developed on the lee slopes of banks in proportions equivalent to those on windward slopes, but not in equal numbers.

The Recent and Silurian corals discussed are believed to have occupied corresponding ecological stations, but only where these species exhibit circumrotatory growth forms. Perhaps circumrotatory colonies of *any* species, in which some other growth form is manifestly typical, may serve as environmental indicators for wave-swept margins of mudbanks and sandy shoals. Certainly that relationship and its evaluation need not be limited to corals. Species of the bryozoan *Mastigophorella* normally display encrusting forms. However, I have examined numerous spheroidal to biscuit-shaped zoaria of *Mastigophorella* sp., 2 to 6 cm in diameter and wholly enveloped by zooecia, recovered from coarse shelly marl of the Waccamaw Formation at Old Dock, North Carolina. These particular strata are believed to represent the seaward slope of a Pliocene offshore bar. Many other examples undoubtedly exist.

ACKNOWLEDGMENTS

Susan F. Loucks, Laurie S. Thomas, and Philip Childs aided in various parts of the field studies and they merit grateful acknowledgment. Richard D. Anderegg, Superintendent of the Southwestern Portland Cement Company at Fairborn, Ohio, permitted me uncustomary access to company property and provided a base map of the quarries. I am indebted to him for his help and kind interest. The field studies were supported by a Grant-In-Aid awarded by the Research Foundation, State University of New York. This paper was presented orally at the Geological Society of America, Southeastern Section Meeting, 1970.

REFERENCES

Duerden, J. E. 1904. The coral *Siderastrea radians* and its postlarval development. Papers Dept. Mar. Biol., Carnegie Inst. Washington Pub., *20:* 130 p.

Fabricius, Frank. 1964. Aktive Lage- und Ortsveränderung bei der Koloniekoralle *Manicina areolata* und irhe paläoökologische Bedeutung. Senck. Lethaea, *45:* 299–323.

Kissling, D. L. 1965. Coral distribution on a shoal in Spanish Harbor, Florida Keys. Bull. Mar. Sci., *15:* 599–611.

Murray, John. 1885. Report of the scientific results of the voyage of H.M.S. *Challenger* during the years 1873–1876. Narrative, *1,* pt. 2: 511–1110.

Oliver, W. A., Jr. 1968. Some aspects of colony development in corals, *in* Paleobiological aspects of growth and development, a symposium. Paleont. Soc., Mem., *2:* 16–34 (Jour. Paleont., v. 42, supp.).

Squires, D. F. 1958. Stony corals from the vicinity of Bimini, Bahamas, British West Indies. Am. Mus. Nat. Hist., Bull., *115*(4): 217–262.

Swann, D. H. 1947. The *Favosites alpenensis* lineage in the Middle Devonian Traverse Group of Michigan. Univ. Mich. Mus. Paleont., Contr., *6:* 235–318.

Turmel, R., and R. Swanson. 1964. Rodriguez Bank, *in* South Florida carbonate sediments. Geol. Soc. America, Field Trip Guidebook no. 1, Ann. Conv. Miami Beach, p. 26–33.

Verrill, A. E. 1901. Variations and nomenclature of Bermudian, West Indian and Brazilian reef corals, with notes on various Indo-Pacific corals. Connecticut Acad. Arts and Sci., Trans., *11:* 63–168.

Wood-Jones, F. 1907. On the growth forms and supposed species in corals. Zool. Soc. London, Proc., *77:* 518–556.

Yonge, C. M. 1935. Studies on the biology of Tortugas corals, II, Variation in the genus *Siderastrea*. Papers Tortugas Lab., Carnegie Inst. Washington Pub. 452, *29*(9): 199–208.

Ontogeny and Hystero-Ontogeny in the Middle Devonian Rugose Coral *Hexagonaria anna* (Whitfield)

R. K. Jull   University of Windsor

ABSTRACT

Protocorallite and hysterocorallite development in colonies of *Hexagonaria anna* are described in detail. The specimens originate from the Givetian Bell Shale and Ferron Point Formation in Michigan. Comparison of the characters of growth between the two types of corallites shows that the very early stage of protocorallite development when it is initially infilled with sclerenchyme, followed by an aseptate phase, is lacking in the hysterocorallite. Otherwise details of development in both types of corallites compare closely, especially with regard to the inobvious nature of protosepta throughout development and the lack of typical patterns of rugosan septal insertion.

INTRODUCTION

Study of corallite development is well established as a valuable aid in determining the taxonomy of rugose corals, with the ontogeny of solitary corals receiving the greatest degree of attention to date. Many studies have been made of corallite development in colonial species, but since the protocorallite (the first corallite of the colony) is only rarely preserved intact, all these studies, with the exception of one by Stumm (1967), have been concerned only with development in hysterocorallites (or offsets, which are corallites formed subsequent to the protocorallite). The phylogenetic value of these studies

Author's address: Department of Geology, University of Windsor, Windsor, Ontario, Canada.

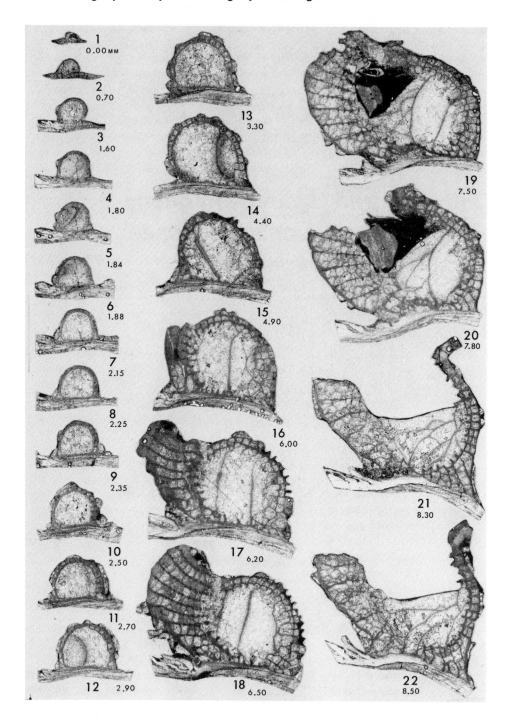

is tempered by the observation of Smith and Ryder (1926, p. 155) that the brephic stage of ontogeny in the protocorallite is omitted in the hysterocorallites of some genera. For this reason, these authors proposed the term "hystero-ontogeny" to describe development in the hysterocorallite. Oliver (1968, p. 21), in a review of coral colony development, considered that this acceleration of early development of the offset occurs in all modes of increase and that "offsets commonly skip the brephic and early neanic stages."

One of the main purposes of the present study is to compare the details of development of proto- and hysterocorallites in *Hexagonaria anna*. The observed differences should indicate the correctness of the above statements with regard to the single species here in question. Earlier studies, also reviewed by Oliver (1968, p. 22), have been made of protocorallite–hysterocorallite relationships in Devonian tabulate species of *Pleurodictyum* and *Favosites*. However, they are devoted more to patterns of colony formation than to details of corallite development and contribute little to an understanding of the sequences of changes between the founding of a corallite and the acquisition of its mature characters.

Colonies for this study originate from the Bell Shale near Rogers City, Michigan, and the Ferron Point Formation at Rockport, near Alpena, Michigan. Both horizons are Givetian in age and are included in the Traverse Group. Specimens were prepared by means of parallel grinding, with acetate peel impressions being taken at varying intervals, ranging from 0.01 to 0.2 mm. Colonies with intact protocorallites were embedded in blocks of epoxy resin for both structural support and protection of the protocorallite walls during repeated immersion in a dilute hydrochloric acid etching bath. The terminology used follows that of Hill (1935, 1956) and Oliver (1968).

Abbreviation of specimen catalogue numbers is as follows: USNM, U.S. National Museum of Natural History, Washington; UW, Department of Geology, University of Windsor, Ontario.

ONTOGENY IN *Hexagonaria anna*

Six relatively complete sequences of protocorallite ontogeny were studied. These are USNM 170295 and USNM 170300 from the Bell Shale and USNM 170297-98-99 and USNM 113766a from the Ferron Point Formation. Also, two incomplete series were examined from the following specimens: USNM 113494 and USNM 170296 from the Bell Shale. Details of development are illustrated in Fig. 1.

Figure 1

Ontogeny in *Hexagonaria anna* (Whitfield), USNM 170299/2, 3, 6, 9, 11, 13, 17, 19, 22, 26, 30, 33, 40, 48, 52, 63, 65, 67, 73, 76, 82, 84, Ferron Point Formation at Rockport, Alpena County, Michigan; ×5. The cumulative distance of distal growth in millimeters is listed below each figure number. The protocorallite is attached to the dorsal valve of *Atrypa*. 1–4, Brephic stage, showing in 1, a small area of tabularium clear of sclerenchyme, and in 4, the first tabula; 5–15, early neanic stage, showing in 5–9, septa faintly visible in sclerenchyme around part of the corallite, and in 10–15, septa and a single row of dissepiments clearly visible around part of the corallite; 16–18, late neanic stage, showing in 16, the appearance of the first carinae in a zone of thickening; 19–22, ephebic stage, showing in 22 the start of increase of the first offset in the lower right-hand corner of the parent.

Of the six relatively complete series studied, two are free of basal attachment, one is attached to the wall of the solitary coral *Cystiphylloides,* and three are attached to a valve of the brachiopod *Atrypa.* Stumm (1967, pl. 1, figs. 16–21) illustrated colonies similar to those here studied. All the present colonies are small, comprising only the protocorallite and a few hysterocorallites, although one (USNM 170300) has 18 hysterocorallites.

The proximal parts of attached protocorallites have a flattened side against the object to which they are attached, whereas unattached protocorallites have a somewhat irregular shape.

Brephic Stage

The tip of the protocorallite is formed by a solid mass of sclerenchyme enclosed by a wall. The sclerenchyme commences to clear when the corallite is about 1 mm in diameter, exposing an aseptate interior surrounded by a wall thickened by sclerenchyme. This thickening is irregular in some corallites, with axially directed projections which resemble septa temporarily present.

One or two tabulae are present in the corallite before the appearance of septa. Stumm (1967, p. 106) mistakenly identified the early tabulae as united cardinal–counter septa.

Early Neanic Stage

Very short septa are first faintly evident within the zone of sclerenchyme lining the wall when the corallite is some 2 to 2.5 mm in diameter. Shortly after their appearance, the sclerenchyme is diminished in extent, revealing a series of short noncarinate septa around part of the corallite. These septa are of approximately the same length in any one part of the corallite and are confined to the very narrow dissepimentarium of a single row of dissepiments. Septa in four corallites studied are initially about 0.2 mm in length, but in two others they are 0.28, and 0.32 to 0.40 mm, in length, respectively. Detailed tracing of septal insertion in the closely spaced serial peel sections showed that septa are not inserted in an orderly fashion in the cardinal and alar fossulae according to typical patterns of rugosan insertion. The position of the protosepta could not be established with certainty in any of the corallites studied. Both major and minor septa are inserted in groups where space is available around the corallite and both are of the same length during early development. In attached protocorallites, the last area of the corallite to acquire septa is adjacent to the area of attachment.

During the latter part of this stage, septa are lengthened axially and major septa become longer than the minors in parts of the corallite. Associated with the septal lengthening is the development of additional rows of dissepiments.

Late Neanic Stage

This stage is taken to commence with the first appearance of carinae on some septa. Corallites are about 3.5 to 5 mm in diameter and have nearly the adult complement of 30 to 36 septa at the beginning of this stage. Less than half of the septa are usually modified to bear carinae, and commonly these are the longest septa in the corallite.

Carinae are formed in a region in which the sclerenchyme is thickened to either line the septa or totally infill the interseptal loculi. After carinae are formed, the deposition of sclerenchyme is diminished or terminated in that particular part of the corallite.

Subsequent development is slower than that of earlier stages and involves mainly an increase in corallite diameter and perhaps the insertion of a few septa. Additional septa may become carinate, but it is not characteristic of the species for all septa to be carinate.

HYSTERO-ONTOGENY IN *Hexagonaria anna*

The following description is based on detailed examination of 30 examples of increase arising from hysterocorallites in five colonies. These colonies are USNM 170301 and UW F1455-56-57 from the Ferron Point Formation, from which 10, 7, 2, and 2 corallites respectively were studied, and USNM 170302 from the Bell Shale, from which 7 corallites were studied. Examination was also made of offsets arising from protocorallites in the six relatively complete sequences listed above under protocorallite ontogeny. All observed examples of increase are lateral in nature, and a typical sequence of development is illustrated in Fig. 2.

Hysterobrephic Stage

Increase commences with the appearance of sclerenchyme in the zone of increase of the parent corallite. New septa are inserted in the inner part of the dissepimentarium in this zone, one in each preexisting interseptal space. Three to seven new septa are formed, with four or five being most common; often they are not inserted at the same time. They are shorter and thicker than the normal septa in the corallite and they terminate peripherally in the middle to outer part of the dissepimentarium. Simultaneous with, or shortly after the insertion of these septa, a similar number of preexisting septa are modified in either one of two ways (Fig. 3). In pattern 1, which is most common, they are progressively shortened from their axial ends until they are only short projections on the corallite wall. In pattern 2, each preexisting septum becomes discontinuous in two parts; a short peripheral part is left at the wall, and the axial part is shortened and normally is eventually suppressed. One variation of pattern 2 was noted in which the remaining parts of the preexisting septa, rather than being suppressed, are united to the peripheral ends of the newly inserted septa. The result of these two patterns of septal modification in the parent corallite is that preexisting septa in the zone of increase are altered to become the first septa of the offset, and these septa which are lost to the parent are replaced by newly inserted septa. Of the 30 examples of increase from hysterocorallites examined, only 7 or possibly 8 are of pattern 2, and all of them occur in coralla having also the more common pattern 1. Of 10 offsets arising from protocorallites examined, 2 showed pattern 1, 2 were somewhat like pattern 2, and the remaining 6 were problematical mainly because of insufficient data.

Subsequent development is similar in all observed hysterocorallites. A wall dividing offset from parent corallite is progressively developed so as to leave the peripheral

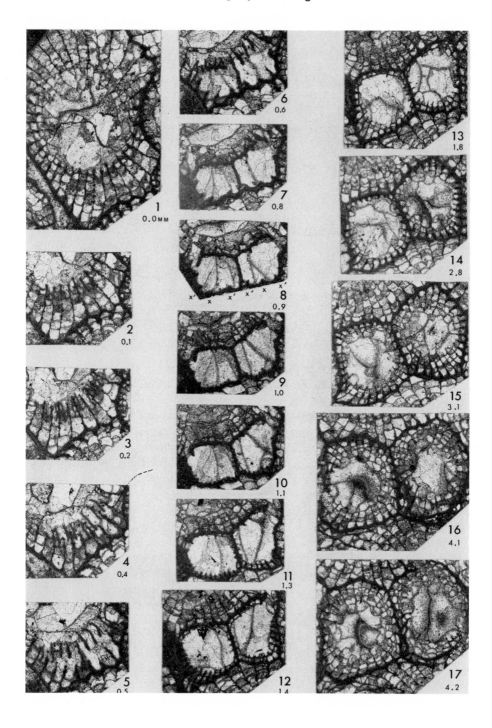

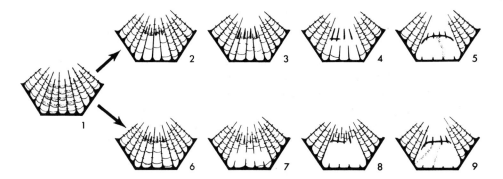

Figure 3

Diagrammatic illustrations of the two basic patterns of septal modification during the initial stages of hystero-ontogeny in *Hexagonaria anna* (Whitfield). 1, Parent corallite prior to the start of increase; 2–5, pattern 1, in which insertion of septa is followed by shortening of preexisting septa; 6–9, pattern 2, in which insertion of septa is succeeded by the preexisting septa becoming discontinuous into two parts with the inner part normally being suppressed. In both patterns, preexisting septa are inherited by the offset and newly inserted septa replace the earlier septa in the parent corallite.

ends of septa in this region within the offset (Fig. 2: 6, 7). A pause occurs in the development of this wall, leaving an opening between the parent and a corner of the offset (Fig. 2: 8–12). Septa in the offset are shortened and a number are suppressed, especially those on the dividing wall.

Throughout this stage of development, dissepiments are reduced in the zone of increase until all are lost. Tabulae are first evident in the offset when the dividing wall commences to develop.

Early Neanic Stage

At the beginning of this stage, the offset is nearly closed off from the parent and lacks dissepiments. It has tabulae and usually four to six septa. The position of the protosepta cannot be established with certainty since they are not of a distinctive nature

Figure 2

Hystero-ontogeny in *Hexagonaria anna* (Whitfield), USNM 170301/43-32, 30, 25, 24, 20, 19, Ferron Point Formation at Rockport, Alpena County, Michigan. ×5. The cumulative distance of distal growth in millimeters is listed below each figure number. 1, Parent corallite prior to the start of increase; 2–8. hysterobrephic stage, showing in 8, septa marked by x and x', which might be cardinal and alar septa, respectively; 9–17, neanic stage, showing in 9, development of a number of short new septa in both corallites, and in 15–17, irregular development of carinae on a few septa and temporary development of lonsdaleoid dissepiments in part of the right-hand corallite.

during development and septal insertion in the species is quite irregular. There is some indication that the septa marked x and x' in Fig. 2: 8 are, respectively, the cardinal and two alar septa in each corallite. These are among the first septa in the offset, and the earliest insertion of metasepta is irregularly in the positions which might be cardinal and alar fossulae. A similar position of the cardinal or counter septum in *H. namnetensis* (Barrois) was indicated by Sorauf (1969, p. 183).

Metasepta appear very rapidly around all or most of the corallite, although a few corallites were observed which remained nearly aseptate for a lengthy period of development. Normally some 30 septa are present in the corallite when it measures 3 to 3.5 mm in diameter. All septa during the early part of this stage are short, noncarinate, and of nearly the same length. A single row of dissepiments is developed around the corallite as septa are inserted. With further corallite development, septa are lengthened to reach halfway or more to the axis, and additional rows of dissepiments are progressively formed. Although the corallite diameter increases rapidly, only a very few septa are inserted, and in some corallites, a few septa are temporarily suppressed (Fig. 2: 15–17). Around parts of some corallites, minor septa become slightly shorter than the major septa.

Late Neanic Stage

This stage is taken to commence with the appearance of carinae on the septa. With this development, apart from the smaller diameter and number of septa, the corallite resembles the adult morphology. Normally in this species, carinae appear on septa when the corallite has from 32 to 37 septa and measures from 3.5 to 6 mm in diameter. Carinae are developed on only a few septa in the corallite, and one corallum was noted in which they are absent throughout development. Sclerenchyme is present in the region in which carinae are formed. Many minor septa are of the same length as major septa or only very slightly shorter.

Development through the late neanic to the ephebic stage is much slower than earlier development and involves mainly an increase in corallite diameter with the addition of a few septa.

Remarks

Lateral increase has been studied in two other species of *Hexagonaria,* these being the Givetian *H. laxa* Gürich described by Różkowska (1960) and the Siegenian to Emsian *H. namnetensis* (Barrois) described by Sorauf (1969). It is interesting that new septa are not inserted in either species during the initial stages of increase as they are in *H. anna,* although the preexisting septa become discontinuous in a way similar to that described above in pattern 2. Also worth noting is the presence of an obviously long cardinal (or counter) septum in *H. namnetensis,* a character quite lacking in *H. anna.*

PROTOCORALLITE ONTOGENY COMPARED TO HYSTERO-ONTOGENY

The foregoing descriptions show that much of the brephic stage of ontogeny in protocorallites of *H. anna* is lacking in the hysterobrephic stage. Missing is the earliest part of protocorallite development when the corallite is filled with sclerenchyme, followed by a short stage in which it lacks septa. Since the offset in this species inherits its first septa from the parent, it does not lack septa at any stage of its development, although commonly some of these septa are suppressed. Allowing for the mechanics of separation of the offset from the parent, subsequent development in both types of corallites are so closely comparable within the limits of variation as to be nearly identical. Hystero-ontogeny in *H. anna* accurately reflects early neanic and subsequent development and approximates the latter part of the brephic stage.

An interesting aspect of this species is the character of septal insertion. Protoseptal development characteristic of that known in solitary Rugosa, and presumed by earlier authors to occur in protocorallites of colonial forms, is absent or possibly very obscure in both types of corallites of *H. anna*. Furthermore, an ordered rugosan pattern of metaseptal insertion is also lacking in this species.

CONCLUSIONS

The septal development of rugose corals, which is characterized by formation of the six protosepta, followed by insertion of metasepta in four positions in the corallite, is apparently lacking in *H. anna*. It is interesting that this species, the first colonial rugose coral with a known protocorallite ontogeny, should seem to lack the rugose pattern of septal insertion. No doubt this is not applicable to all colonial rugose corals. More or less regular patterns of rugosan insertion are known, for instance, in hysterocorallites of some Lower Carboniferous species of *Lithostrotion* Fleming and *Lonsdaleia* McCoy (Jull, 1965, 1967) and these would probably reflect similar insertion patterns in the protocorallites. However, septal insertion can no longer be regarded as basically being of an obvious single ordered pattern in all presumed sexually produced corallites of the Rugosa. Observations have been made of irregular metaseptal development in other species, such as in hysterocorallites of *Thysanophyllum orientale* Thomson (Jull, 1967), and these may indicate similar irregular patterns in the protocorallites.

The opinion of earlier authors that the early stage of development in the protocorallite is missed in the hysterocorallite is confirmed for the first time. However, since the characters of development in the two types of corallites in *H. anna* are comparable from the early neanic stage onward, less is missing than was suggested by Oliver (1968). This is also indicated by other recent studies of increase in rugose corals (Fedorowski, 1965; Jull, 1967). Early studies of increase, such as the description of offset development in *Lonsdaleia* made by Smith (1916), lacked the more contemporary serial sectioning techniques and are very likely to have missed some significant details of the initial stages of corallite development. This might have lead to the opinion that hystero-ontogeny commonly skips more of the details of protocorallite development than is here suggested.

Lateral increase in colonial species involves offsets which either inherit their first septa from the parent, as they do in *H. anna,* or have all septa independently formed, as has been described by Fedorowski (1965) in Lower Permian forms of *Tschussovskenia* Dobrolyubova, and by Jull (1967) in *Lonsdaleia*. Offsets of the latter type exhibit the greater degree of independent development, and it appears that species with such offsets possess the "lower level of colonialism" of the two. Development in such offsets probably lacks very little indeed of the protocorallite development and should prove to be of the greatest value in phylogenetic considerations.

ACKNOWLEDGMENTS

I am very grateful to W. A. Oliver, Jr., U. S. Geological Survey, Washington, for his criticism of the manuscript of this paper and his considerable help during the initial stages of this study. S. W. Mitchell, George Washington University, Washington, provided much help with field collecting and donation of material, and I am thankful for the use of photographic facilities at Wayne State University, Detroit. Material for this study was loaned from the U.S. National Museum, and financial support for the project was provided by National Research Council of Canada Grant A7048.

REFERENCES

Fedorowski, J. 1965. Lower Permian Tetracoralla of Hornsund, Vestspitsbergen. Studia Geol. Polon., *17*: 7–173.

Hill, D. 1935. British Terminology for rugose corals. Geol. Mag., *72*: 481–519.

———. 1956. Rugosa, *in* R. C. Moore, ed., Treatise on Invertebrate Paleontology, pt. F, Coelenterata. Geol. Soc. America, and Kansas Univ. Press, p. 233–324

Jull, R. K. 1965. Corallum increase in *Lithostrotion*. Palaeontology, *8*(2): 204–25.

———. 1967. The hystero-ontogeny of *Lonsdaleia* McCoy and *Thysanophyllum orientale* Thomson. Palaeontology, *10*(4): 617–628.

Oliver, W. A. Jr. 1968. Some aspects of colony development in corals. Jour. Paleont., *42*(5), pt. 2: 16–34.

Różkowska, M. 1960. Blastogeny and individual variations in tetracoral colonies from the Devonian of Poland. Acta Paleont. Polon., *5*(1): 3–64.

Smith, S. 1916. The genus *Lonsdaleia* and *Dibunophyllum rugosum* (McCoy). Geol. Soc. London, Quart. Jour., *71*: 218–272.

———, and T. A. Ryder, 1926. The genus *Corwenia*, gen. nov. Ann. Mag. Nat. History, ser. 9, *17*: 149–159.

Sorauf, J. E. 1969. Lower Devonian *Hexagonaria* (Rugosa) from the Amorican Massif of western France. Palaeontology, *12*(2): 178–188.

Stumm, E. C. 1967. Growth stages in the Middle Devonian rugose coral species *Hexagonaria anna* (Whitfield) from the Traverse Group of Michigan. Univ. Michigan, Mus. Paleont. Contr., *21*(5): 105–108.

Colonial Organization in Octocorals

Frederick M. Bayer   University of Miami

ABSTRACT

This summary reviews the range of complexity of form resulting from vegetative reproduction of zooids in the anthozoan subclass Octocorallia (= Alcyonaria). Three sharply delimited octocoral groups are recognized: the order Coenothecalia (with only one surviving species), which have no spicules but produce a massive, madrepore-like skeleton; the order Pennatulacea, in which polymorphic colonies with hierarchical dominance are the rule; and all others (orders Stolonifera, Telestacea, Alcyonacea, and Gorgonacea), in which all degrees of colonial form and of zooidal integration are found, from simple, loosely united groups of monomorphic zooids arising from encrusting stolons (*Clavularia*), to highly integrated, dimorphic colonies whose component zooids share functions in such a way as to preclude independent existence.

It is considered that colonial integration is expressed both in the division of labor (e.g., feeding and digestion, water transport and circulation, sexual reproduction) between dimorphic types of zooids, and in the coordinated colonial functions such as anchoring and support, regularity of branching, and response to epizoites and commensals, which are shared among many or all of the zooids in a colony.

Although paleontological evidence is scanty, present interpretation of known fossils suggests that complex colonial forms similar to modern pennatulaceans with a high degree of colonial integration were already flourishing in Precambrian times. Several of the Recent groups are clearly recognizable in Tertiary deposits in various parts

Author's address: Rosenstiel School of Marine and Atmospheric Sciences, University of Miami, Miami, Florida 33149.

of the world. It seems likely that Octocorallia with diverse degrees of complexity and integration have been in existence over a long span of geological time.

INTRODUCTION

The octocorals are typically anthozoan in the structure of their zooids, which have a tubular pharynx (sometimes also called "esophagus" or "stomodaeum") extending from the mouth into the gastrovascular cavity, radial septa (called "mesenteries" by some authors) extending from body wall to pharynx and partitioning the gastrovascular cavity, and tentacles at the upper end of the column, positioned between the septa and surrounding the mouth. Among the anthozoans they are unique in having no more than eight septa and eight pinnately branched tentacles, one for each of the chambers formed by the septa. Most of them produce a calcareous skeleton, totally lacking in a few cases, mostly in the form of calcareous spicules (occasionally inseparably interlocked or solidly fused), rarely massive (Coenothecalia only). A thin proteinous cuticle is produced in many, if not all, stoloniferans and telestaceans, and a proteinous axial skeleton is found in all holaxonian Gorgonacea and in most Pennatulacea. Octocorals have a simple life history without any complex larval stages or alternation of generations, and all produce "colonies" vegetatively by budding from the sexually produced "founder" zooid (Fig. 1).

Although divided into six orders, the subclass Octocorallia shows only two clear lines of subdivision. One of these separates off the Coenothecalia, which produce a

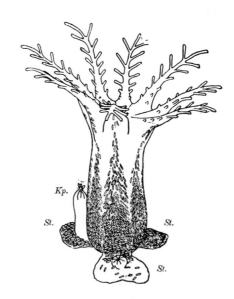

Figure 1

Young colony of *Anthoplexaura dimorpha* Kükenthal, with founder zooid and two vegetatively produced daughter zooids. (From Kinoshita, 1910.)

massive skeleton similar to that of scleractinians, not composed of fused spicules. The other line separates off the Pennatulacea, whose colonies are always dimorphic, asexually produced in a different way from that occurring in other groups, and adapted for life on a soft (sandy or muddy) substrate. The orders Stolonifera, Telestacea, and Gorgonacea (the latter with two suborders, Scleraxonia and Holaxonia) have highly characteristic colonial architecture in the "typical" species, but are linked by intergrading forms that make hard-and-fast ordinal and subordinal subdivision very difficult indeed.

In modern seas, the octocorals are a group of moderate diversity (on the order of 2000 species) and wide distribution. Lacking any protection against desiccation, they extend from roughly the level of mean low water down to abyssal depths (deepest recorded occurrence is an unidentified *Umbellula* at 6620 m; Madsen, 1956). Few species have any tolerance for reduced salinity, so the group is poorly represented in estuarine habitats, generally absent. Geographically, they occur from Arctic to Antarctic and around the world. They may form a conspicuous part of the marine environment as they do on reefs in the tropical Atlantic (order Gorgonacea) and Indo-west Pacific (order Alcyonacea), where in some localities they dominate the sessile community. The fossil record for octocorals is poor because few of them have skeletal structures suitable for preservation, but a number of finds (D'Achiardi, 1868; Bayer, 1955; Duncan, 1880; Hickson, 1938; Kugler, 1971) suggest that by the Tertiary the major groups as we know them were present and probably as well represented as they are now. Moreover, some problematical fossils from the Precambrian in Australia are strongly suggestive of pennatulaceans, and if they actually represent that group, then the most complex of all octocorals already existed over 500 million years ago (Glaessner, 1959, 1961).

Atlantic coral reefs are built for the most part by a few species of hermatypic scleractinians, but may support a community of perhaps 60–70 gorgonian species, some of which occur in vast numbers. By contrast, the soft corals are the important octocorals of Indo-Pacific reefs, where they compete with a scleractinian fauna of 200–300 species and do not form a conspicuous part of the reef community. With the possible exception of *Heliopora* (the blue coral), a few species of Alcyonacea (soft corals), and *Tubipora* (organ-pipe coral), they are not significant reef formers, although most of them contribute to the sediment through their calcareous spicules.

As in the scleractinian corals, the octocorals have developed an association with the unicellular algae known as zooxanthellae. This symbiotic association varies in its intimacy from cases in which the coelenterate apparently cannot exist without zooxanthellae (Gohar, 1940, 1948), having lost the ability to feed, to those in which feeding and digestion are still possible and the host can live without its algae (but probably do not do so in nature). Association with zooxanthellae is a preponderantly tropical phenomenon and obviously is confined to shallow water (although not exclusively to reefs). It has elsewhere been speculated that it was the development of the symbiotic relationship with zooxanthellae which enabled zoantharian corals to reach a sufficient size to produce reefs. Among the octocorals, however, some of the largest forms (e.g., *Paragorgia arborea*, *Primnoa reseda*) are not associated with zooxanthellae or with the reef habitat. It may be significant that one of these (*Paragorgia*) has dimorphic zooids that form a water-transport system, and that most of the tropical soft corals that reach large size have both zooxanthellae and a special water-transport system formed by siphonozooids. It should be noted, how-

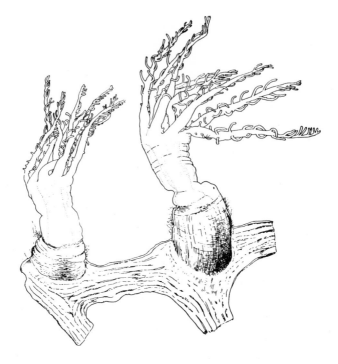

Figure 2

Part of a colony of *Cornularia sagamiensis* Utinomi with two fully expanded polyps. (From Utinomi, 1955.)

ever, that large size is not always associated with dimorphism. In any case, symbiosis with zooxanthellae seems to have made reef formation possible by enhancing the ability of zoantharian corals to deposit $CaCO_3$, but the octocorals mostly have no potential in this direction, owing to the spicular nature of their skeletons. There is no evidence that zooxanthellae play any role in the calcification of spicules.

LEVELS OF COLONIAL ORGANIZATION AND INTEGRATION

Although colonial ontogeny is for the most part unknown and the stages in colonial complexity are inferred from the morphological evidence provided by the fully developed colonies, the increasing degrees of complexity and regularity show more and more interdependence of individual polyps, functional specialization, and more highly developed mechanisms controlling asexual reproduction, hence increasing degrees of colonial integration.

Levels of colonial organization range from quite simple to extremely complex. In the simplest forms, zooids are of only one kind and are not united laterally by thick layers of mesogloea. They reproduce asexually from stolonic outgrowths from the base by which they are attached to the substrate. Typical of this level are the members of

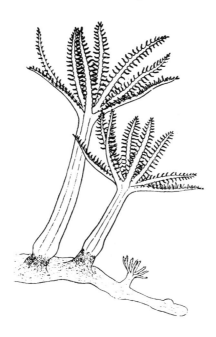

Figure 3

Clavularia hamra Gohar. A terminal part of the stolon, ribbon-like and showing three zooids of different ages. A bud is seen near the end. (From Gohar, 1948.)

the family Cornulariidae (Fig. 2) and some species of the family Clavulariidae (Fig. 3) in the order Stolonifera. Also included are some species now assigned to the order Telestacea, for example, a few species in the genus *Telestula* (Fig. 4). At this level of organization, the zooids are mostly tall and slender, cylindrical, or somewhat clavate. The fully developed individuals in a colony are of nearly uniform height, suggesting that the height of zooids is limited by a physiological or structural feature and may be genetically controlled.

Within the Stolonifera there also are species whose colonies are formed as above but with stolonic outgrowths proceeding laterally from the body wall of the zooids in addition to basally, so that new zooids arise at various levels above the substrate. When the zooids have reached a given height limit they produce lateral stolons, from which new zooids arise. The daughter individuals are physiologically interconnected via the solenia (entodermal canals) of both lateral and basal stolons, but the gastric cavities of adjacent zooids are not directly joined. This type of colonial structure is seen in some species of *Clavularia*, such as the type species of the genus, *C. viridis* Quoy and Gaimard (Fig. 5). Organizationally not very different from this condition is the colonial structure found in *Tubipora*, also allocated to the Stolonifera. Here the stolons are flat, expanded platforms extending between the zooids, containing a solenial network from which new polyps arise as the spacing between older zooids increases with upward growth. The platforms occur at rather regular intervals throughout the colony, and the

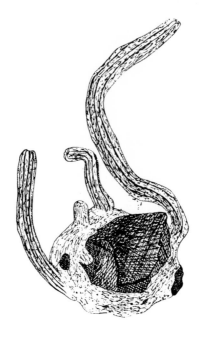

Figure 4

Telestula septentrionalis Madsen. Small colony without lateral daughter zooids. (From Madsen, 1944.)

lower ends of the elongated gastric cavities are sealed off by a succession of funnel-shaped tabulae as growth proceeds upward (Fig. 6).

The various species in the nominal order Telestacea have a range of organization comparable with that found in the Stolonifera just discussed. They have tall zooids, for the most part much taller than those of the Stolonifera, and they produce narrow, often anastomosing stolons adherent to the substrate, from which new zooids arise. In some, but not all, the lower part of the gastric cavity becomes partially filled in with mesogloea containing spicules and penetrated by entodermal canals communicating directly with the upper, functional part of the coelenteron. Tabulae are not formed as in *Tubipora*. The simplest telestaceans differ from the simple stoloniferans only in the infilling of the gastric cavity by spicule-containing mesogloea and in the form of the spicules. The more complex forms differ in the production of daughter zooids directly from the wall of the parent. No lateral stolons are produced as in *Clavularia* and *Tubipora*, but a conspicuous network of solenia is developed in the body walls of the zooids, and it is from this system that new zooids arise vegetatively. Two patterns of ramification (i.e., of budding) occur in this order: monopodial, in which tall axial zooids produce lateral daughters in a more or less regular pinnate arrangement (Fig. 7), sometimes to the third or fourth order of branching; and sympodial, in which each zooid produces one or two daughters of similar size from the distal part of the zooid wall just below the anthocodia, thus forming zigzag stems that dichotomize here and there to produce small, bushy colonies that have no dominant axial zooid (Fig. 8).

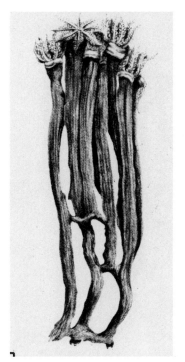

Figure 5

Clavularia viridis Q. and G. (From Thomson and Dean, 1931.)

The next level of colonial integration is the development of extensive coenenchyme, which unites adjacent zooids. In octocorals, coenenchyme is a common colonial mesogloea containing calcareous spicules, in which the gastrovascular cavities of the zooids are embedded. No boundary separates the coenenchyme of one zooid from that of its neighbors, and the gastrovascular cavities of adjacent zooids may lie very close together. Only the short distal part of each zooid, bearing the tentacles, remains independent of its neighbors. The coenenchyme is permeated by an anastomosing solenial system which serves to interconnect all the zooids of the colony. New zooids arise between the old individuals, developing from the superficial solenia, and become deeply embedded as the colony enlarges. Production of new zooids continues throughout the life of the colony, which thus contains zooids of various lengths. All members of the order Alcyonacea conform to this general plan, although the colonial form varies from massive, with more or less distinct lobes and plications, to arborescent, but even in the latter case (which may simulate gorgonaceans) the trunk and branches consist of bundles of very elongated zooids whose anthocodiae emerge on the branchlets at various levels in the colony.

Dimorphism of zooids appears for the first time in the order Alcyonacea. In all the groups mentioned heretofore, all zooids are exactly alike. Even the primary zooid arising from a fertilized egg is identical with the vegetative daughters. In the Alcyonacea, however, some of the zooids are anatomically quite different from the primary type, which is termed the "autozooid." These different individuals have an enormously developed

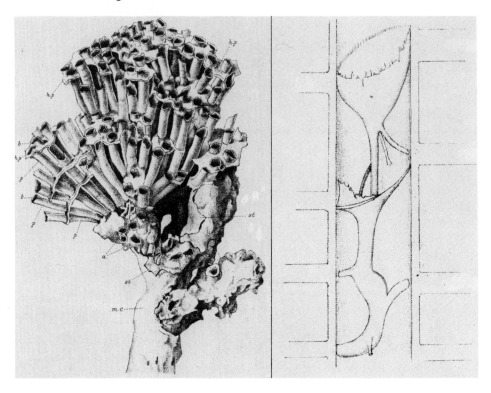

Figure 6

Tubipora musica Linn. a, Part of colony; b, diagram of tabulae. (From Hickson, 1883.)

siphonoglyph (i.e., ciliated groove along one edge of the pharynx), drastically reduced tentacles, and they bear gonads. These zooids, termed "siphonozooids," are specialized for moving water (by means of the siphonoglyph) and for reproduction. They have lost the ability to feed and to digest food, which are the function of the autozooids. The fact that all the branched, lobed, and arborescent alcyonaceans are monomorphic, whereas most of the large and massive forms are dimorphic, suggests that dimorphism arose out of a need to transport water more efficiently into the large colonial structure. Relegation of reproductive functions to the siphonozooids may be related to the abundance of water passing through them.

One group at present assigned to the Alcyonacea, the family Xeniidae, includes both monomorphic and dimorphic species, some of which seem to differ from one another in no other way. The colonies commonly are mushroom-shaped, with a sterile trunk and a rounded capitulum bearing the zooids, but in some species the trunk is subdivided into several lobes covered with zooids, and others are merely mat-like encrustations. All of the xeniids have peculiar, corpuscle-like spicules of very small size, entirely different from other alcyonacean spicules, and some species have the remarkable ability to pulsate the autozooids in unison, rhythmically opening and closing the tentacles. On the whole, this family is rather unlike the other alcyonaceans (Fig. 9).

Figure 7

Telesto colony. Monopodial budding in pinnate arrangement. (Diagrammatic; original.)

Figure 8

Pseudocladochonus colony. Sympodial budding in dichotomous arrangement. (Diagrammatic; original.)

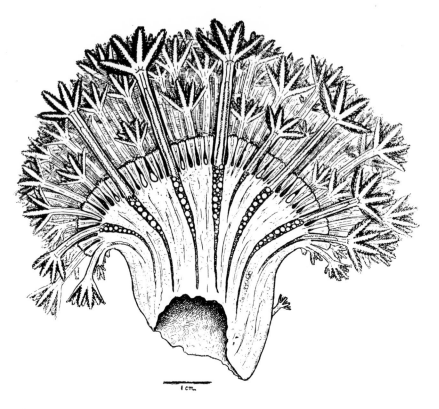

Figure 9

Colony of *Heteroxenia fuscescens* cut in half, to show the relation of the coelenteric cavities of the autozooids with gonads to those of the siphonozooids without. (From Gohar, 1940.)

The arborescent alcyonaceans fall into two groups on the basis of colonial structure. In the family Nephtheidae, colonies reach a large size, but the trunk is permeated by spacious gastrovascular canals and contains little coenenchyme. When the gastrovascular spaces expel their water, the entire colony collapses and shrinks to a much smaller size. Zooids are arranged in clusters at the ends of branchlets, and in some species the proximal ones form frilly or collar-like groups around the trunk. The arrangement of zooids on the branchlets, and the arrangement of the larger branches in the colony are characters that are used to subdivide the family, as they seem to follow consistent patterns (Figs. 10, 11, and 12).

In the family Siphonogorgiidae, the arborescent growth form is even more pronounced, the main trunk is narrower and more rigid, and the branches subdivided to a greater extent. The form of the colony and the arrangement of zooids on the branches is consistent within species.

In the remarkable family Maasellidae (= Fasciculariidae), the branches bearing zooids form a terminal tuft, and in contraction can be withdrawn as a group within the trunk, much as an actinian retracts its tentacles into the column (Fig. 13).

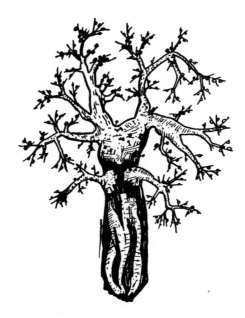

Figure 10
Diagram of divaricate branching in Nephtheidae. (From Thomson and Dean, 1931.)

Figure 11
Diagram of umbellate branching in Nephtheidae. (From Thomson and Dean, 1931.)

Figure 12
Diagram of glomerate branching in Nephtheidae. (From Thomson and Dean, 1931.)

Figure 13
Studeriotes longiramosa Kükenthal. Zooid-bearing branches retractile into trunk. (From Thomson and Dean, 1931.)

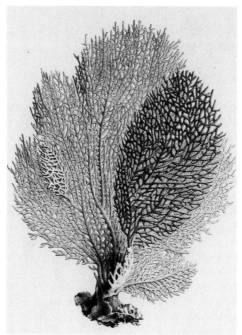

Figure 14
Gorgonia flabellum Linn. (From L. Agassiz, 1880.)

The most intimately integrated colonial organization in the order Alcyonacea is the remarkable species known as *Bathyalcyon robustum* Versluys. The fully developed colonies consist of a single giant autozooid in whose body wall are embedded numerous siphonozooids (Versluys, 1906; Bock, 1938). The gonads are located in the siphonozooids, which as usual have a very well developed siphonoglyph. The autozooid itself is sterile. Clearly, here is a case in which the siphonozooids function as ''organs'' of their vegetative parent; neither could exist without the other.

In the order Gorgonacea, colonies are predominantly arborescent, and colonial integration is demonstrated chiefly by the regularity of vegetative growth. Although the order is unified by the very consistent morphological form of the zooids, it is divided into two broad groups by the nature of the axial skeleton. The boundary between these two groups, the Scleraxonia and the Holaxonia, is still not agreed upon, and some workers remove the family Paragorgiidae from the order altogether. The zooids are laterally united by coenenchyme only basally (roughly half the length of the fully extended gastrovascular cavity, or less—often much less). The common colonial coenenchyme is supported on an axial structure composed of a horny proteinaceous material (gorgonin) that may be more or less extensively permeated by calcium carbonate—sometimes in the form of spicules, sometimes not. This axial skeleton, whatever its structure, is the product not of any individual zooids but of the colony as a whole. It is, in fact, a specialized part of the coenenchyme.

Division of labor, as reflected by dimorphism, is very rare in the Gorgonacea. It occurs only in the Scleraxonian families Coralliidae and Paragorgiidae, and the latter

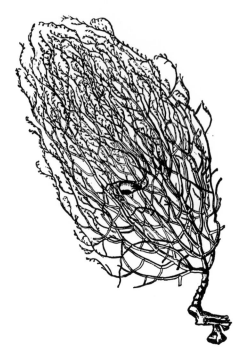

Figure 15

Acanella arbuscula (Johnson). (From A. Agassiz, 1888.)

is considered by some to be alcyonacean rather than gorgonacean in its affinities. Nevertheless, *Corallium* must be considered a gorgonacean, and dimorphism certainly is the rule in this genus. The gonads are produced by the autozooids rather than by the siphonozooids. In *Paragorgia* functional specialization is apparently still in a formative state, as both autozooids and siphonozooids bear gonads, but with more in the latter.

In the holaxonian families of the Gorgonacea, which have an axial skeleton of scleroprotein with nonspicular calcareous deposits, increasing degrees of colonial integration are expressed in the increasing regularity of branching and occurrence of zooids. No functional specialization occurs, as far as is known. Branching is generally consistent in pattern but variable in details, which may be influenced by environmental factors such as water currents. In the more generalized types, the zooids are randomly distributed and branching is bushy—either pinnate or dichotomous. Orientation of the branches in a single plane, and location of the zooids on one side of the resulting flat colony, are colonial responses to unidirectional water movement.* The zooids are placed in a more favorable position to filter plankton from the passing current. Some flat colonies, as in the family Gorgoniidae, have zooids on both surfaces, and the branchlets may fuse into a quite regular network (Fig. 14). In other families, such as Paramuriceidae and Acanthogorgiidae, the zooids are either confined to one surface or are directed toward one surface.

_____

*The same kind of flattening occurs in sponges, hydroids, hydrocorals, scleractinians, bryozoans, and in some species of algae.

Figure 16

Plumarella pourtalesii (Verrill). Pinnate branching, zooids biserial. (From Verrill, 1883.)

In three families having strong calcification in the axial skeleton, the Melithaeidae and Parisididae in the suborder Scleraxonia, and the Isididae in the suborder Holaxonia, the brittle, calcareous axis is interrupted by horny nodes that provide enough flexibility to withstand environmental turbulence (Fig. 15). This is clearly a colonial response to environmental forces. The modification is coenenchymal, not zooidal.

In one family, Primnoidae, with predominantly pinnate branching (Fig. 16), distribution of the zooids over the surface ranges from random through biserial and paired to regularly whorled. In any given species, the number of zooids in a whorl and the spacing of whorls along the branches varies only within circumscribed limits. Vegetative production of new zooids, either terminally or interstitially between older whorls, therefore is controlled with respect to time, place, and number and must be colonially, not zooidally, coordinated. In this family the sequence of branching may be very regular. In some species of *Calyptrophora,* for example, the first two branchings of the primary axis are dichotomous, in planes at right angles to each other, and in quick succession; subsequent branchings are unilaterally pinnate and in one plane. The result is two fan-like groups of parallel branches arising in parallel planes at the top of a supporting stalk.   Members of the family Primnoidae, like gorgonaceans in general, are prone to infestation by a variety of commensals, predators, parasites, and epizooites, and some of the species show a colonial response to the foreign organism. Polychaete worms are common inhabi-

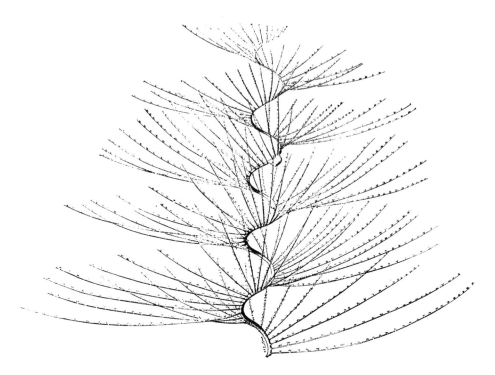

Figure 17

Iridogorgia pourtalesii Verrill. Unilateral spiral branching. (From A. Agassiz, 1888.)

tants of octocorals, and primnoids such as *Narella,* when infested with them, respond by producing enormously enlarged spicules in the affected region. The worm lives along the main axis, traveling along it between the zooids. Each zooid along the worm's path produces one or two broadly arched, flattened spicules that enclose a tunnel limiting the worm to a more or less definite path. This spicular modification is the same in all polyps, and the modified spicules are as distinctive for the given gorgonian species as are the normal ones. The gorgonian is in some degree protected from the polychaete, which itself is protected by the modified spicules. It is not known whether the worm is a commensal, parasite, or predator on the gorgonian, but it probably is not the last, as no damage to the zooids is evident. In other families, normally having spicules much smaller than in the primnoids, infesting polychaetes induce the growth of a web-like expansion of coenenchyme along the affected branch. The individual spicules are too small to be modified into protective structures, so the coenenchyme as a whole carries out this function.

Of all the holaxonian gorgonians, members of the family Chrysogorgiidae show the greatest regularity in branching. Most are inhabitants of deep water. The few exceptions show branching in one plane, similar to that seen in other families. Those living in deep, still water assume a multiplanar form arranged around an essentially spiral axis. Some are unbranched (genus *Radicipes*), but even in these, the long, whiplike colony

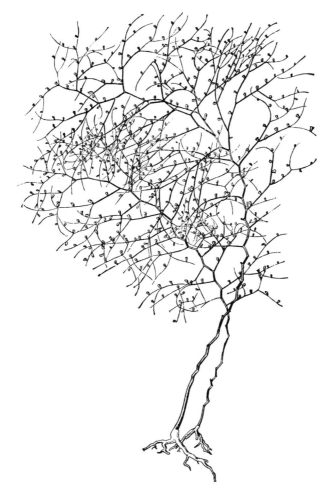

Figure 18

Chrysogorgia sp. Dichotomous branches arising spirally around main trunk. (From A. Agassiz, 1888.)

is spirally twisted—a form found also in other gorgonian families (Isididae and Ellisellidae) and in other coelenterates (e.g., Antipatharia). The branched forms may produce long, simple branchlets from the outer side of an ascending spiral main axis (*Iridogorgia*, Fig. 17). The branchlets are regularly spaced along the spiral, and the zooids are regularly spaced along the branchlets. More commonly, however, the main axis is a tighter spiral, producing a fixed number of lateral branches in each revolution. The direction of the spiral is clockwise in some species, counterclockwise in others; in some, one complete turn around the spiral brings the fourth branch directly above the first, but in other species, the first branch directly aligned with that at the beginning occurs at the completion of the second revolution, and very commonly this is the sixth. The lateral branches

Figure 19

Policella manillensis Kölliker. (From Kölliker, 1870–1872.)

then subdivide dichotomously a constant number of times, with a fixed number of zooids occurring between each dichotomy (Fig. 18).

Most members of the family Chrysogorgiidae inhabit soft bottom, where there are few solid objects for attachment. As the colonies are rather delicate, even small objects such as mollusk shells, bits of echinoid test, and the like are adequate supports. The holdfast established by the founding zooid eventually grows into a thickened, discoidal calcareous button to which is attached the proteinous axis. If the support is very small, the coenenchyme has the ability to produce branching, calcareous, root-like anchoring processes (Fig. 18) that penetrate the mud and hold the colony in position. This is covered by coenenchyme containing spicules, but no zooids are formed.

Among the octocorals, the pennatulaceans as a group are the most advanced in terms of colonial complexity, polymorphism, functional specialization of zooids, and colonial integration. In fact, the pennatulacean colony acts as a superorganism whose various kinds of zooids perform respiratory and circulatory, feeding and digestive, sexual, and locomotory functions.

Within the order Pennatulacea, several further intergrading levels of organization can

Figure 20
Sclerobelemnon schmeltzii Kölliker. (From Kölliker, 1870–1872.)

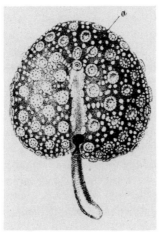

Figure 21
Renilla reniformis Pallas. (From Kölliker, 1870–1872.)

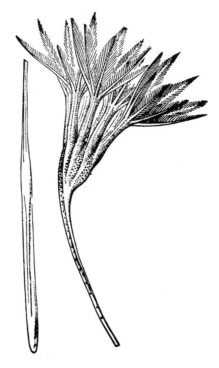

Figure 22

Umbellula lindahlii Kölliker. (From Verrill, 1885.)

be recognized. In the simplest, the daughter zooids (autozooids and siphonozooids) are arranged more or less generally around the primary individual (oözoöid), as in *Cavernularia, Policella, Veretillum,* and a few other genera (Fig. 19). They become progressively more regularly arranged in the direction of bilaterality and in the relative positions of autozooids and siphonozooids. A longitudinal naked tract along the rhachis (body of the oözoöid) divides the other zooids into two lateral fields (Fig. 20) as in *Kophobelemnon, Sclerobelemnon,* and *Mesobelemnon,* and the zooids become limited to one face of a flattened colony (*Renilla*) (Fig. 21) or to the distal end of a tassel-like structure (*Umbellula*) (Fig. 22). In the latter arrangement, the tassels of zooids may be repeated as widely spaced pairs or whorls of zooids along the rhachis (*Chunella*).

In the biserial colonies, the autozooids are placed in more or less distinct transverse rows and more or less completely united by their bases to form flattened, leaf-like structures with a marginal fringe of zooids. The siphonozooids may be located on the sides of the autozooid leaves, between them, or in more or less restricted areas along the naked tract separating the two rows of leaves (Fig. 23). Further, each autozooid may produce sharply projecting spicules, probably defensive in function, always located in the same position.

In many (perhaps all) species of Pennatulacea, the colonies are capable of limited movement by means of the muscular and distensible proximal end of the stalk. Specimens

Figure 23

Pennatula rubra Ellis. (From Kölliker, 1870–1872.)

of *Renilla reniformis* (the sea pansy) can move from place to place by withdrawing the stalk from the substrate and reinserting it a short distance away, subsequently pulling the frond into a new position. Slender, rod-like forms, such as *Stylatula* and *Virgularia,* having a stiff, calcified axis that affords the colony little flexibility, nevertheless move up and down in the sandy or muddy substrate with the ebb and flow of the tide, as was reported more than 200 years ago by Rumphius (1705).

SUMMARY

The octocorals display a wide and virtually continuous range of colonial organization. In this one subclass are found extremely simple colonies founded by an individual produced from a fertilized egg and composed of asexually produced "daughters" exactly like itself and retaining a loose organic connection by way of a system of entodermal canals permeating the narrow, meandering stolons from which the individuals arise. Each of these individual zooids is essentially an independent organism in spite of its organic connections with its vegetative siblings. No functional specialization has developed, and presumably every zooid retains the capacity for vegetative reproduction and could reestablish an entire colony. No two colonies are exactly alike, but distribution of stolons over the substrate always has the same pattern in any given species so the spatial relationships among zooids in different colonies have no more than a general similarity. Although each zooid is equipped to carry on all the vital processes, some advantages must result from the persistence of organic connection—such as the distribution among several (or all) individuals of food ingested by one or a few—to give colonial species a selective advantage.

At the other extreme, there are octocoral species whose colonies are highly organized and composed of functionally specialized individuals distributed in a very orderly sequence to form a structure highly characteristic for each species. In this case, the original founding zooid loses the ability to feed and take in water, never achieves the ability to reproduce sexually, and serves mainly as a support for its vegetative offspring. These are arranged in more or less orderly groups which, in the more complex forms, are serially repeated along much of the length of the rhachis (i.e., the body of the primary zooid); some of them are structurally modified to form a water-transport system through hypertrophy of the ciliated groove and reduction or loss of other structures, while others are adapted primarily for feeding and reproduction. Here, all the specialized zooids function together as a unit—a sort of superindividual—but cannot function alone. Within a species, all colonies are remarkably similar and the spatial relationships of their component individuals are always essentially the same. Between these two extremes—the simple stoloniferans and the complex pennatulaceans—there are innumerable subtle intermediates, but no one order includes the full range.

Two of the six orders stand clearly apart. The Coenothecalia, which has a long fossil history, is now represented by a single surviving species (*Heliopora coerulea*) and clearly is a relict. Although unmistakably octocorallian in anatomy, its skeleton is massive in the fashion of hermatypic scleractinians and its colonial organization is like that of scleractinians, with similar growth form. Its level of colonial integration is low. On the other hand, the Pennatulacea, with a high degree of colonial integration, is a rather diverse group with wide bathymetric and geographic distribution. Whether it was previously more prolific than now is impossible to say because of its fragmentary geological record, but it certainly was extant in the Eocene (*Graphularia*), and possibly in the Silurian (*Spirophyton, Alectorurus*) or even earlier (*Charnia, Rangea*; Glaessner, 1959, 1961). There are no known intermediates linking the Pennatulacea with any other order. A few gorgonaceans simulate certain of the simpler pennatulacean colonies, but they are unquestionable gorgonaceans and the resemblance is superficial. The order showing the most significant similarity to the Pennatulacea is the Telestacea, in some of which the tall axial zooid produces lateral daughters from its own body wall in a more or less regular organization. Even here the resemblance is superficial, as the axial zooids are not necessarily the sexual founder zooids (although one of them in each colony certainly is), the lateral daughters may also produce lateral daughters, all zooids are functionally monomorphic, the colonies have a low grade of integration, and the organization is basically Stoloniferan.

The orders Stolonifera, Telestacea, Alcyonacea, and Gorgonacea (with two suborders, Scleraxonia and Holaxonia) overlap to some extent so that drawing hard and fast boundaries between them is difficult. As these orders are based, in part, on aspects of colonial complexity, it follows that none has a very wide range in this character. The stoloniferans are at the lowest level, the monomorphic polyps reproducing vegetatively only from lateral stolons, never from the body walls. In certain species, these stolons arise from the body wall some distance above the substrate, so their condition approaches that found in the telestaceans, where the daughters arise directly from the parent body wall. In fact, there seems to be little justification for maintaining these two groups as separate orders.

In the arborescent genus *Coelogorgia,* the telestaceans overlap the arborescent alcyonaceans (e.g., *Siphonogorgia*), and more extensive anatomical research will be required to determine whether they are superficially or fundamentally similar in structure. In turn, the arborescent Alcyonacea overlap the arborescent scleraxonian Gorgonacea such as *Paragorgia,* which has been considered an alcyonacean on anatomical grounds (Verseveldt, 1940). Hence it seems unrealistic to consider the levels of colonial organization in these four orders from the standpoint of traditional classification. Two of the groups have no trace of functional specialization and are strictly monomorphic (except that the axial zooids in *Telesto* are longer than the laterals and thus might be considered a morphologically distinct type), one shows it to a more or less limited extent according to where its taxonomic limits are drawn, and one has a substantial number of genera with some or all species dimorphic and functionally specialized. Where dimorphism occurs, it commonly is in the form of zooids specialized for moving water through the colony, and thus is often found in groups having species whose colonies attain massive dimensions. Further, these same pumping zooids (''siphonozooids'') may also assume the reproductive functions of the colony, leaving the autozooids with feeding as their principal activity. Such functional specialization does not always depend upon size, however. Some species whose colonies never reach large size (at most roughly 15 cm high and 15 cm in diameter) are dimorphic, whereas some that grow quite large (1 m or more) are not. The greatest degree of colonial integration is found in species whose colonies contain only a single autozooid in whose walls are embedded the numerous siphonozooids which control the hydrostatics of the parent autozooid and perform the reproductive functions. The aggregation of autozooid and siphonozooids behaves as a single biological entity.

A high level of colonial integration, or at least of coordinated vegetative growth, is achieved without the development of dimorphism. In most gorgonaceans, few of which show dimorphism, vegetative reproduction of the zooids proceeds in a manner that results in a colonial structure characteristic of the species. This could not occur without a control mechanism. In two families (Chrysogorgiidae, Primnoidae), colonial growth is so regular that zooids are arranged in a relationship to one another that varies only within rather narrow limits, the position of each branch in the colony is more or less rigidly predetermined, and extrazooidal tissues produce supporting structures that affix the entire colony to (or in) the substrate.

Although it is easy enough to postulate that in each species the zooids develop to a predetermined (genetically? physiologically? structurally?) height, extrazooidal growth proceeds to a predetermined limit before zooidal replication occurs, and axial growth continues only for a certain distance before bifurcation or lateral ramification of the axis is triggered, it is less easy to imagine what forces cause ramification to follow a clockwise or counterclockwise spiral, why branches should be regularly aligned within these spirals, why adjacent branchlets should in some cases fuse when their courses intersect but in other instances remain free. The end result is that the zooids of a colony are deployed in the environment in the most effective arrangement for feeding, and situated with respect to one another in a manner that permits the most efficient distribution of nutriment. The control mechanism that brings it about must be a property of the colony as a whole, not of the individual zooids.

Although the ancestral octocoral probably was solitary, the advantages of coloniality must have been so great that the solitary state vanished completely. No undisputed trace of it exists in the Recent fauna, as the few putative examples of solitary forms (e.g., *Haimea*) eventually have been shown to be the primary founder zooids of colonial forms. The advantages of coloniality were probably related to reproduction, nutrition, and, perhaps, respiration. The largest octocoral zooids are rather small in comparison with a large actinian or even scleractinian zooid, and the average octocoral zooid is very small in this comparison. The female zooid is capable of producing several eggs, but their number is limited by the size of the adult, and of the eggs produced in one breeding season not all will mature and be fertilized (Grigg, 1970, p. 141). Thus a vegetative repetition of the adult increases its reproductive potential. This same repetition of adult zooids also increases the feeding capacity of the colony, especially if the colonial form places some of the individuals in positions more advantageous for capturing food than others. Moreover, the maintenance of gastrovascular connection between individuals permits many members of a colony to benefit when a few localized individuals feed.

Increasing colonial size, especially if massive coenenchyme is produced, raises problems of respiration because deeper-lying tissues are far removed from the surrounding seawater. The same interzooidal gastrovascular connections permit the distribution of water throughout the colony, and the development of zooids specialized for pumping water (siphonozooids with powerful ciliated groove) enhances this capacity.

REFERENCES

D'Achiardi, Antonio. 1868. Studio compartivo fra i coralli dei terreni terziari del Piemonte e dell' Alpi Venete. Ann. Univ. Toscane, *10*: 73–144, pls. 1, 2.

Agassiz, Alexander. 1888. Three cruises of the United States Coast and Geodetic Survey Steamer "Blake" . . . v. 2. Bull. Mus. Comp. Zool. Harvard, *15*: 1–220, figs. 195–545.

Agassiz, Louis. 1880. Report on the Florida Reefs. Mem. Mus. Comp. Zool. Harvard, *7*(1): (i–iv) + 1–61, pls. 1–22.

Bayer, F. M. 1955. Remarkably preserved fossil sea-pens and their Recent counterparts. Jour. Wash. Acad. Sci., *45*(9): 294–300, figs. 1, 2.

Bock, Sixten. 1938. The alcyonarian genus *Bathyalcyon*. Kungl. Svenska Vetenskapsakad. Handl. (Tredje Serien), *16*(5): 1–54, pls. 1, 2.

Duncan, P. M. 1880. A monograph of the fossil corals and Alcyonaria of Sind. Mem. Geol. Surv. India, Palaeont. Indica, *14*(1): 1–110, pls. 1–28.

Glaessner, M. F. 1959. Precambrian Coelenterata from Australia, Africa and England. Nature (London), *183*(4673): 1472–1473, fig. 1.

———. 1961. Pre-Cambrian animals. Sci. Am., *204*(3): 72–78, illustr.

Gohar, H. A. F. 1940. Studies on the Xeniidae of the Red Sea. Publ. Mar. Biol. Station Ghardaqa, *2*: 23–118, pls. 1–7, figs. 1–10.

———. 1948. A description and some biological studies of a new alcyonarian species *Clavularia hamra* Gohar. Publ. Mar. Biol. Station, Ghardaqa, *6*: 3–33, pls. 1–3.

Grigg, R. W. 1970. Ecology and population dynamics of the gorgonians, *Muricea californica* and *Muricea fruticosa*—Coelenterata: Anthozoa. Unpublished Ph.D. dissertation, Univ. Calif., San Diego. xvii + 261 p.

Hickson, S. J. 1883. The structure and relations of Tubipora. Quart. Jour. Micros. Sci., ns., *23*: 556–578, pls. 39–40.

———. 1938. An alcyonarian from the Eocene of Mississippi. Jour. Wash. Acad. Sci., *28*(2): 49–51, figs. 1–4.

Kinoshita, Kumao. 1910. Ueber die postembryonale Entwicklung von *Anthoplexaura dimorpha* Kükenthal. Jour. Coll. Sci. Imp. Univ. Tokyo, *27*(14): 1–[16], figs. 1–3.

Kölliker, Albert. 1870–1872. Anatomisch-systematische Beschreibung der Alcyonaria. Erste Abtheilung: die Pennatuliden. Abhandl. Senckenb. Naturf. Ges., *7*: 111–255, 487–602; *8*: 85–275, pls. 1–24.

Kugler, H. G. 1971. An enigmatic gorgonian remnant. Eclogae Geologicae Helvetiae, *64*(3): 635–636, pls. 1–3.

Madsen, F. J., 1944. Octocorallia. The Danish Ingolf-Expedition, *5*(13): 1–65, 1 pl.

———. 1956. *Primnoella krampi* n. sp. a new deep-sea octocoral. Galathea Rept., *2*: 21–22, fig. 1.

Rumphius, G. E. 1705. D'Amboinsche Rariteitkamer . . . Fr. Halma, Amsterdam, [28] + 340 + [43] p., frontisp., portr., 60 pls.

Thomson, J. A., and L. M. I. Dean. 1931. The Alcyonacea of the Siboga Expedition, with an addendum to the Gorgonacea. Siboga-Exped. Monogr., *13d*: 1–227, pls. 1–28.

Utinomi, Huzio. 1955. On five new stoloniferans from Sagami Bay, collected by His Majesty the Emperor of Japan. Jap. Jour. Zool., *11*(3): 121–135, figs. 1–11.

Verrill, A. E. 1883. Report on the Anthozoa, and on some additional species dredged by the "Blake" in 1877–1879, and by the U. S. Fish Commission Steamer "Fish Hawk" in 1880–82. Bull. Mus. Comp. Zool., Harvard, *11*(1): 1–72, pls. 1–8.

———. 1885. Results of the explorations made by the steamer "Albatross," off the northern coast of the United States, in 1883. U. S. Fish Comm. Rept., *1883*: 503–699, pls. 1–44.

Verseveldt, J. 1940. Studies on Octocorallia of the families Briareidae, Paragorgiidae and Anthothelidae. Dissertation. E. J. Brill, Leiden. [i–x] + 1–142 + I–V p.

Versluys, J. 1906. *Bathyalcyon robustum* nov. gen. nov. spec. Zool. Anz., *30*(17/18): 549–553, figs. 1–4.

Coordinated Behavior in Hydrozoan Colonies

G. O. Mackie   University of Victoria

ABSTRACT

Coordinated activities in thecate and athecate hydroid colonies and in siphonophores are reviewed. Coordination may be achieved by means of electrical impulses propagated between connected zooids in nerves or in nonnervous epithelia. In some cases behavioral coordination may result from pressure changes transmitted in the canal system or from some other purely mechanical process. Coordination by chemical means, theoretically possible, is not known to occur.

Feeding activities are generally carried on independently in hydrozoan colonies. Digestive movements may be synchronized. Protective and escape movements are almost universally found to be coordinated between the zooids in hydrozoan colonies. Some types of nonescape locomotory activity also involve coordinated activities between different zooids in the colony.

INTRODUCTION

All invertebrate bud-colonies are integrated. Their component zooids are connected by living tissues, materials pass between them, and the whole morphology of the colony is the product of a single morphogenesis. To review the subject of integration of hydrozoan colonies would require coverage of growth, form, metabolism, physiology, and

Author's address: Department of Biology, University of Victoria, Victoria, B.C., Canada.

behavior—in short all aspects of the biology. Instead, this review will deal with one particular aspect of colonial integration: coordination of behavior.

Under the heading "behavior" we will consider only those activities involving muscular contraction. Morphogenetic movements, some of which seem to be controlled by diffusible metabolites (e.g., Rose, 1967) are excluded from the present discussion. There may, however, be a "gray area" here, for some activities such as the rhythmical extension and retraction of stolon tips in hydroids (e.g., Wyttenbach, 1968), might represent either muscular movements or nonmuscular contractile changes in the meristematic cells. Contractility is a property of cells other than muscle, and on the biophysical level the differences between muscular and other types of contractility need not be profound.

As possible mechanisms whereby behavior is coordinated, we will consider coordination by mechanical interaction, by specific chemical signals, and by spreading electrical impulses; an attempt will be made to classify the responses on this basis.

A final point in setting the terms of reference for this paper concerns the distinction between collaborative and coordinated behavior. If the zooids in the colony are active together due to their independent reception of similar stimuli originating from some source outside the colony, their behavior cannot be regarded as coordinated, even if they appear to be collaborating in the execution of a common behavioral task. Although this distinction is clear-cut in semantic terms, it is a hard one to draw in practice, as we often simply do not know enough about the behavior to be sure which phenomenon we are dealing with.

Three groups of colonial hydrozoans will be considered: the thecate and athecate hydroids and the siphonophores.

THECATE HYDROIDS

Until recently it was considered that the polyps in thecate hydroid colonies were behaviorally isolated. Naumov (1969) stated that stimuli are not transmitted between polyps in the colony. However, Josephson (1961a) briefly noted observations to the opposite effect in the case of *Obelia,* and Morin and Cooke (1971a, 1971b, 1971c) found that stimulation evokes spreading hydranth contractions in *O. geniculata* and related campanularians. Flashes of light accompany the spread of the withdrawal response. The bioluminescent flashes are always preceded by electrical pulses ("luminescent potentials") which are propagated singly or in bursts through the colony at a velocity of about 22 cm/sec. The spreading hydranth withdrawals can reasonably be regarded as protective behavior and, although there is no direct evidence regarding the utility of flashing, the fact that it accompanies hydranth withdrawal suggests that it may also somehow serve a defensive role.

Hydranths contract spontaneously, move their tentacles, and open their mouths and ingest food, but these activities are not coordinated among different hydranths.

ATHECATE HYDROIDS

Because of their relatively large size and the accessibility of their hydranths for attachment of recording electrodes, athecate hydroids have received more attention than thecates.

Josephson (1961a) reviewed early evidence on coordinated responses in athecate colonies and added new findings on several species. A considerable body of work is now available, particularly on *Tubularia*. The various athecate forms show sufficiently important differences to require separate descriptions.

Tubularia

Spontaneous activity cycles known as "concerts" occur in all hydranths of *Tubularia*, consisting basically of a pumping sequence by which fluids are shifted from one end of the proboscis cavity to the other, or from the stalk cavity to that of the proboscis and back (Beutler, 1925; Josephson and Mackie, 1965). Concerted tentacular flexions accompany the pumping movements.

Concerts are accompanied by a characteristic pattern of electrical events originating from or spreading within cells or groups of cells of uncertain identity, known as pacemakers (Josephson, 1961b, 1962, 1965; Josephson and Mackie, 1965). Major interacting pacemaker systems are located in the distal stalk region (neck potential system) and in the hydranth body (hydranth potential system).

Polyps which are connected by stolons containing living tissue often show synchronized concert activity, which is attributable to coupling between their neck potential pacemakers. Josephson (1965) called the coupling system the "triggering system." Josephson and Mackie (1965) attribute functional advantages to concert synchronization in terms of improved "tidal" ebb and flow of fluids between hydranths and stems,

In addition to the triggering system, the stem possesses two additional conducting systems (Josephson, 1965). One, the "slow system," controls opening of the distal tentacles, a response of unknown behavioral significance. Stimulation of the stem between connected polyps can, but does not usually, elicit the opening response of the distal tentacles in both polyps (Josephson, 1961a). The system conducts at about 15 cm/sec, compared with 17 cm/sec for the triggering system, and has a higher threshold of excitability. Continued low-frequency stimulation of the stem tends to inhibit the neck potential system in connected polyps and to suppress concerts. The inhibition is probably due to impulses traveling in the distal opener system (Josephson and Uhrich, 1969).

There is some indirect evidence that the distal opener system involves nervous pathways in the stem and polyps, but the histological identity of the three stem systems has not yet been satisfactorily established.

Cordylophora

If a polyp is pinched or a stolon prodded, a wave of polyp contraction spreads across the colony (Josephson, 1961a). The response appears to resemble that described in the related form *Clava* (Föyn, 1927).

Josephson (1961b) recorded electrical signals from the stolons following stimulation. These events propagate at about 2.7 cm/sec at 22°C. Single pulses or bursts may be produced by single shocks. Repeated stimulation leads to an increase in the number of pulses in the bursts; with longer bursts the greater and faster is polyp contraction in the vicinity. Not all pulses spread throughout the colony. Whether or not distant polyps contract and to what degree depends on the number of pulses which get through to them, and this is related to intensity of stimulation.

Fulton (1963) found that *Cordylophora* polyps undergo regular cycles of peristaltic activity, and he showed that these were roughly synchronized between connected polyps. There is no evidence of any conducting system operating to integrate the activity of different polyps other than the one described by Josephson, and this is not involved in the coordination of peristaltic activity (Mackie, 1968). Coordination might be "hydraulic," i.e., mediated by pressure changes in the fluid column itself.

Syncoryne

Stimulation of the stem adjacent to a hydranth evokes some degree of polyp response but only if the stimulus is applied close to a polyp. There is little, if any, evidence for spreading responses between adjacent polyps (Josephson, 1961a). *Syncoryne* is exceptional in lacking interpolyp coordination or exhibiting it only in a rudimentary form.

Pennaria

Josephson (161a) confirms Zoja's (1891) observation that polyp contractions spread from points of strong stimulation. Spread is very slow (about 1 cm/sec). Both the magnitude of the contraction and the distance of spread vary with stimulus strength.

Hydractinia and Podocoryne

Hydractinia lives on the outside of gastropod shells inhabited by hermit crabs. Presence of the crab appears to be essential for maintenance of the colony in a fully differentiated form, but the factors responsible for this dependence are unclear (Cazaux, 1958; Burnett et al., 1967). Four polyp types are present: gastrozooids, gonozooids, tentaculozooids, and dactylozooids. The medusae are not set free in most species but are reduced to sessile sporosacs.

In *H. echinata,* Josephson (1961a) found evidence of a local conducting system spreading polyp contractions over a limited area and of a second, through-conducting system activated by strong stimulation and involving the whole colony on an all-or-none basis. Stokes (1972) has recorded the electrical correlates of these two systems and of a third possible system. The local system ("incrementing system") conducts at 4.8 cm/sec, the general or "through system" at 12 cm/sec. The incrementing system mediates degrees of local activity from the single polyp up to small groups. Stokes gives unequivocal evidence for the presence of nerves within and between the polyps in the colony. The nervous system in whole or part would seem to be a likely candidate for the incrementing system.

Although the gastrozooids respond in the same way to excitation transmitted via the two systems, the dactylozooid response differs. Following stimulation received from the incrementing system the muscle response is unilateral, while after activation from the through system it takes the form of bilateral thrashing movements. In spite of this difference, responses are probably protective in both cases.

The related *Podocoryne* appears to lack a local system, showing the colony-wide response only (Josephson, 1961a).

Proboscidactyla

The Limnohydrina include a number of small, often very specialized, athecate hydroids, of which *Proboscidactyla* shows the highest degree of colonial development. It lives commensally with sabellid polychaetes. Recent evidence (Strickland, 1971; Donaldson, 1971) indicates that contact with the worm's radioles is essential for the hydroid to be maintained in the differentiated state.

Spencer (1971) discovered a protective response in *Proboscidactyla* colonies in which all three polyp types (gastrozooids, dactylozooids, and gonozooids), as well as young attached medusae, contract. In nature the response is seen when the worm retracts its tentacular crown or when small invertebrates clamber over the colony. Excitation at any point can cause contraction of all members by conduction of electrical signals ("colonial pulses") through the stem. These pulses spread at about 9 cm/sec and are through-conducted to all points, like those in the *Hydractinia* through system.

The medusae of *Proboscidactyla* achieve swimming ability when about 0.5 mm in diameter and while still attached to the colony. They swim in bursts like free medusae, an activity attributable to pacemakers, presumably nerves, in the bell margin. "Crumpling," a protective closure response involving excitable epithelial tissues, is developed later. Colonial pulses reaching attached medusae elicit swimming while the medusae are small but later evoke crumpling. This remains the response to colonial excitation until the medusae, now measuring 0.8 mm, are released.

Functionally speaking, then, the medusae are "part of the colony" prior to their release, as their action systems are controlled by impulses arriving from the colony.

SIPHONOPHORES

Siphonophores are the most complicated and highly evolved animal colonies, showing a higher degree of polymorphic differentiation and functional specialization of component zooids than any other colonial group (Mackie, 1963). They inhabit the competitive world of the plankton and have developed sophisticated response capabilities in the areas of locomotion and protective behavior. There is a fair amount of scattered information on siphonophore behavior but this account will stress recent work on a few forms which have received special study from the behavioral point of view.

The chondrophores *Porpita* and *Velella* are excluded, as they are best regarded as large, individual, tubularian hydranths and consequently have no place in discussions of coloniality (Fields and Mackie, 1971, and references therein).

Siphonophore behavior is often too complex or too little understood to be easily analyzable in terms of local versus coordinated behavior. For example, some physonectid siphonophores show a diffuse photosensitivity of the hinder parts and swim in response to sudden illumination. They also show color changes in response to light but these appear to be locally induced, and there is no correlation between the distribution of chromatophores and the distribution of light-sensitive areas for the swimming response.

These siphonophores carry out diurnal vertical migrations, presumably in response to light, but whether they do so by swimming or by density regulation is unclear. Density regulation involves muscular elimination of gas bubbles from the float and resecretion of gas in the gas gland, both seemingly local events. The float is not one of the regions showing photosensitivity as far as swimming is concerned. However, the float tissues involved in elimination and secretion of gas might be independently photosensitive. The whole complex of light-induced activities, buoyancy control, and vertical movements requires much more study.

Physalia

The Portuguese Man-of-War consists of seven different zooid types (Totton, 1960). The float (part of the primary zooid) is capable of rather complex muscular movements concerned with erection of the crest and adoption of a curved sailing posture. The secondary zooids show various local activities. The only clear example of a coordinated response is the generalized, presumably protective contraction which spreads through the colony when any part is abruptly stimulated (Bigelow, 1891; Mackie, 1960). Nerves are present in the ectoderm throughout most regions, but it is not clear whether they are responsible for the through-conducted response or organize local activities. The latter appears most likely from what is known about hydrozoan conducting systems generally (Mackie, 1970).

Physalia can catch large fish by collaborative action of the many tentacles and gastrozooids (Wilson, 1947), but there is nothing to suggest that these activities are coordinated. Likewise, pumping movements are performed by the digestive members, distributing food around the colony, but pumping is probably locally organized.

Hippopodius

In sessile hydroids and in *Physalia* the swimming activity of attached medusoids contributes nothing toward locomotion of the colony. In physonectid and calycophoran siphonophores, asexual medusoids have been evolved which serve specifically for locomotion. These nectophores or swimming bells are dependent on the colony for food, as they lack mouths and tentacles. They may be shed by autotomy but die soon after release.

Hippopodius has two columns of nectophores. In each column, the velar apertures of the upper nectophores open against the backs of the nectophores beneath. Thus, with the exception of the bottom two, their swimming activity has no locomotory effect. The upper ones apparently serve as a stock of replacements for ones lost from below. They also provide buoyancy for the colony as a whole (Jacobs, 1937) and enclose a sheltered central space into which the stem can be withdrawn.

Swimming is not obviously coordinated. It does not start or stop simultaneously in the different nectophores and each nectophore has its own rhythm. However, once activity begins (usually in an upper one) it tends to spread to others, probably because of the mechanical agitation set up. The marginal swimming centers in the nectophores are not connected by nerves with the stem (Mackie, 1964).

Hippopodius shows what is held to be a photokinetic response, aggregating in relation to illumination (Mackie and Boag, 1963). The location of the photoreceptors is not known, but in view of the lack of swimming coordination, it seems likely that each nectophore is independently photosensitive.

When a *Hippopodius* is touched or abruptly stimulated it withdraws the stem and tentacles into the interior of the nectophore column, curls up the margins of the nectophores, becomes opaque due to formation of light-scattering granules adjacent to the covering epithelium of the nectophores, and emits a flash of light from the same epithelium (Mackie, 1965; Mackie and Mackie, 1967). This syndrome is considered to represent a complex of protective responses. Excitation spreads primarily in conducting epithelia and can be recorded in the form of electrical potentials. Epithelial or "neuroid" conduction is now known to be widespread in siphonophores and hydromedusae, functioning particularly in protective types of response (Mackie, 1970).

Hippopodius differs from most siphonophores in lacking an escape swimming response. The epithelial potentials transmitted through the colony do not evoke contractions in the swimming muscles but only in the radial fibers concerned with curling up the margin. Locomotion is so slow in this form that swimming would probably be ineffective in removing the colony from dangerous situations.

Muggiaea and the Family Diphyidae

Typical diphyids have a pointed anterior nectophore and a smaller posterior one which fits into the back of the anterior one but not in such a way as to impede swimming activity. The stem streams out behind but can be withdrawn into the hydroecial folds of the nectophores. Diphyids dart around rapidly using the anterior or both nectophores or cruise slowly using the small posterior one. *Muggiaea* has only the anterior nectophore but is otherwise similar to the diphyids.

Locomotion serves either for escape, as when a nectophore or the stem is touched, or for spreading of the stem and tentacles ("fishing behavior"). Stimulation of a nectophore of *Chelophyes* excites the epithelial system and elicits rapid stem contraction and escape swimming. The epithelial pulses are propagated at velocities of at least 35 cm/sec (Mackie, 1965).

In addition to rapid stem contraction, a slower spreading response can be obtained by weak stimulation (Ebbecke, 1957). Fishing behavior is described in *Muggiaea* and involves a coordinated sequence of events in the stem and nectophore, wherein the colony swims in a circle, releasing the tentacles, which spread out centrifugally (Mackie and Boag, 1963).

A nerve tract connects the stem with the marginal nerves of the nectophore (Mackie, 1964), providing the potentiality for one of two possible pathways of communication between the stem and nectophore, the other being the epithelial route. It is clear that locomotion is under colonial control in diphyids, contrary to the situation in *Hippopodius*.

An interesting feature of this group from the point of view of coloniality is their ability to strobilate subcolonies from the posterior end of the stem. These "eudoxids" are fully equipped for swimming, feeding, and reproduction. Beklemischev (1958) called them "colonies within the colony."

Nanomia

Physonectid siphonophores such as *Nanomia* have an apical gas-filled float below which the nectophores are arranged in two regular columns on either side of the stem. The lower part of the stem bears the gastrozooids, palpons, bracts, and sexual medusoids. The food-catching tentacles are parts of either palpons or gastrozooids and are not separate zooids, although their behavior is largely independent of the zooids they belong to. The parts of the colony showing behavioral activity are the float, stem, and all the secondary zooids named above with the exception of the sexual medusoids and the bracts. Bracts undergo autotomy but are otherwise not active.

Food capture appears to be essentially a locally organized process, as in hydroids. Tentacles making contact with food paralyze it with their nematocysts and haul it up to the stem, where it is ingested by gastrozooids. Following food capture by the tentacles gastrozooids in the vicinity and for some distance along the stem begin to writhe and elongate (Mackie and Boag, 1963). This activity increases the likelihood of contact being made with the food object. Mackie (1964) suggested that the spread of the response was due to activation of nervous pathways in the stem. An alternative explanation in terms of the spread of a chemical activating substance from the site of food capture is possible, as it is known that crustacean tissue fluids and glutathione can evoke writhing behavior. As in *Physalia,* several gastrozooids may collaborate in ingesting the prey, but this behavior does not appear to be integrated. Electrical pulse patterns have been recorded (Mackie, unpublished) from gastrozooids during collaborative ingestion of food, and these patterns show no correlation with one another.

After the food has undergone preliminary digestion in the gastrozooids it is passed into the stem and flushed to and fro by rhythmic peristaltic movements performed by the gastrozooids and palpons. Gastozooids which have *not* fed as well as those which have take part in these pumping movements. They show independent pumping rhythms, as do the palpons. The latter act as accessory digestive organs, carrying on intracellular digestion, releasing the end products into the coelenteron for dispersion through the colony, and eliminating wastes.

Sometimes a gastrozooid may fall into a reciprocating pumping pattern with a group of adjacent palpons, filling them when it empties and emptying them when it fills. It would appear likely that pressure changes transmitted via the coelenteric fluid are responsible for this type of coordination. Such collaborations are temporary. Nothing comparable to the triggering system in *Tubularia* has come to light in *Nanomia*.

Swimming ability is well developed in *Nanomia*. Each nectophore has its own pacemaker system which determines its swimming frequency. The direction in which the water jet is emitted from the nectophore is controlled by a pair of muscle bundles (Claus's fibers). When the fibers are relaxed, the water jet is emitted downward and the colony swims upward. When they contract, they deflect the water jet upward and reverse swimming results (Mackie, 1964).

The nectophores are quiescent except when ''aroused'' by excitation arriving from the colony. With weak tactile stimulation of the hinder regions locomotion begins gradually in the forward direction. One or a few nectophores start to pulsate and then others join in. Groups on the two sides tend to alternate, so that the colony moves in a zigzag pathway. Before locomotion begins, the stem and tentacles usually contract slightly, indicating the buildup of a state of excitation in the stem.

When the hinder regions of the stem are stimulated more strongly, forward swimming ensues suddenly in the form of one or a few synchronous pulsations of all the nectophores in the column, driving the colony straight ahead for a considerable distance; at the same time the stem and appendages contract strongly. The ability to perform this coordinated forward swimming depends on the integrity of the nerve tracts connecting the nectophore margins with the stem. Following section of the nerves, the ability is lost (Mackie, 1964).

Reverse swimming also involves synchronous action of all the nectophores. It is elicited by stimulation of the float or anterior nectophores and, like forward swimming, is accompanied by strong stem contraction. Integrity of the connecting nerves in the nectophores is not essential for the response, and this, and other evidence, including electrical recordings, identifies the covering epithelium of the nectophores with the conduction pathway for the response. It appears that excitation in the epithelial system activates both the swimming muscle and Claus's fibers, whereas nervous excitation activates only the swimming muscle. Claus's fibers are homologous to the radial fibers which produce marginal curling in *Hippopodius* nectophores and crumpling in free medusae. *Nanomia* thus employs the protective closure mechanism inherited from a primitive ancestor but modifies it to achieve reverse locomotion, likewise protective in nature.

The stem contractions which accompany escape swimming in *Nanomia* have been explored from the neuromuscular viewpoint (Mackie, unpublished). Giant axons run the length of the stem, making synaptic connections with elements of the general nerve plexus. The giant axons do not penetrate the secondary zooids but are connected with them via the nerve plexus.

"Spasmic" stem contraction involves flurries of signals in the giant axons and, as these travel rapidly up the stem, they evoke contractions successively along the stem musculature. The spread of the contractile response is rapid, corresponding to the conduction velocity of the giant axons (up to 3 m/sec). Thus in a *Nanomia* 30 cm long, it takes only a few hundred milliseconds for the contraction wave to pass all the way along the stem, for the stem to contract, and for the nectophores to swim synchronously forward.

In addition to rapidly conducted, spasmic contractions, slowly conducted contractions of the stem and appendages can occur. Conduction velocities are here about 0.3 m/second. The general tonus of the stem is related to the level of activity in this slowly conducting "contraction pulse system." Changes in light intensity can affect contraction pulse frequency.

Stem-contraction pulses do not penetrate the secondary zooids. The zooids do, however, contract during spasmic stem contraction, and presumably do so in response to intensive nervous activity arriving via the nerve plexus. Transmission between the stem and appendages is not a simple, one-to-one process but involves conduction "blocks" which allow only relatively strong types of excitation to pass through.

Thus, from the point of view of coordination, gastrozooids and their tentacles and palpons lead a fairly autonomous existence but come under colonial control when excitation levels in the stem rise to a certain threshold. This is most clearly seen in the spasmic contractions accompanying escape behavior. Nectophores, by contrast, are normally inactive but respond to excitation from the colony. They show autonomy only in the sense that once excited they beat at their own rhythms.

CLASSIFICATION OF COORDINATED RESPONSES

A. *Classification by mechanism*
  1. Mechanical
     a. Agitation: Movements of a zooid or water currents resulting from such movements agitate adjacent zooids and arouse activity in them. A possible example is the spread of swimming activity among the nectophores in *Hippopodius*.
     b. Pressure changes or fluid movements within the canal system trigger behavior within groups of connected zooids. Possible examples include the coordinated peristaltic movements of *Cordylophora* hydranths and the coordinated swellings and contractions sometimes seen in groups of palpons and gastrozooids in *Nanomia*.
  2. Chemical
     a. Diffusion through the external medium of triggering substances released by the organism itself: No examples are known. Inclusion of this category is prompted by the suggestion of Burnett et al. (1963) that mouth opening in *Hydra* may be caused by release of a chemical "retrohormone" from the polyp's own nematocysts when these discharge. The hypothesis remains unproved for *Hydra* and untested for colonies.
     b. Diffusion of triggering substances via the canal system: No examples are known but in *Hydra* the response known as "neck formation," which occurs following ingestion of food, involves activation of an internal tyrosine receptor (Blanquet and Lenhoff, 1968). Thus food breakdown components might serve as behavioral triggers in colonies.
     c. Spread of a triggering substance from cell to cell in the tissues connecting zooids: No examples are known, but quite large molecules can pass between cells in some tissues, and passage of such materials or "propagated" chemical changes involving such chemical activators might serve to coordinate activities of connected zooids over short distances.
  3. Electrical
     a. Impulses in the nervous system: Spasmic stem contraction and accompanying contraction of zooids in *Nanomia* involves conduction in nerves. Coordinated forward swimming involves a specific nerve pathway. The distal opener system in *Tubularia* is thought to depend on nerves, and the incrementing system in *Hydractinia* may tentatively be considered in this category.
     b. Nonnervous impulses: The system coordinating polyp contractions in *Cordylophora* is probably epithelial, as nerves appear to be absent from the stem. The protective responses described in *Hippopodius* nectophores, and reverse swimming in *Nanomia* nectophores, are epithelially conducted as far as the nectophores are concerned, but it is not clear whether the spread of these

responses between the active zooids is epithelial or nervous. In forms such as *Hydractinia* and *Physalia,* where two sorts of conduction occur, nerves might be responsible for one sort of conduction (presumably incremental conduction) and epithelia for the other. By the same reasoning, the triggering system of *Tubularia* is likely to be epithelial.

B. *Classification by function*

1. Coordinated feeding behavior: With the possible exception of gastrozooid writhing activity in *Nanomia,* feeding behavior is not coordinated in hydrozoan colonies.
2. Coordinated digestive movements: Synchronized patterns of digestive movements are described in *Cordylophora, Nanomia,* and *Tubularia.*
3. Coordinated protective responses and escape behavior: Spreading zooid contractions or withdrawals are described in all hydroids except *Syncoryne.* In *Obelia* they are accompanied by luminescence. In *Hippopodius* stem withdrawal is accompanied by various changes in the nectophores, notably opacification, luminescence, and protective rolling up of the margin. In the diphyids and *Nanomia* escape swimming (bidirectional in *Nanomia*) accompanies the contraction of stem and attached zooids.
4. Nonescape locomotion: Fishing behavior in *Muggiaea* is a coordinated activity, as apparently is zigzag locomotion in *Nanomia.*

REFERENCES

Beklemishev, W. N. 1958. Die Grundlagen der vergleichenden Anatomie der Wirbellosen, v. I, Dtsch. Verlag. Wissensch. Berlin.

Beutler, R. 1925. Beobachtungen an gefütterten Hydroidpolypen. Z. vergl. Physiol., *3*: 737–775.

Bigelow, R. P. 1891. Notes on *Caravella maxima* Haeckel (*Physalia caravella* Eschscholtz). Johns Hopkins Univ. Circ., *10*: 90–93.

Blanquet, R. S. and H. M. Lenhoff. 1968. Tyrosine enteroreceptor of *Hydra.* Its function in eliciting a behaviour modification. Science, *195*: 633–634.

Burnett, A. L., R. Davidson, and P. Wiernik. 1963. On the presence of a feeding hormone in the nematocyst of *Hydra pirardi.* Biol. Bull., *125*: 226–233.

———, W. Sindelar, and N. Diehl. 1967. An examination of polymorphism in the hydroid, *Hydractinia echinata.* Jour. Mar. Biol. Ass. U.K., *47*: 645–658.

Cazaux, C. 1958. Facteurs de la morphogenèse chez un Hydraire polymorphe, *Hydractinia echinata* Flem. Comptes rend. Acad. Sci., *247*: 2195–2197.

Donaldson, S. 1971. Factors mediating the commensal relationship of *Proboscidactyla flavicirrata* (Hydrozoa) and its sabellid hosts. Abstracts of papers presented at Western Society of Naturalists Meetings, Sacramento, Calif. Dec. 1971.

Ebbecke, U. 1957. Reflexuntersuchungen an Coelenteraten. Publ. Staz. Zool. Napoli, *30*: 149–161.

Fields, W. G., and G. O. Mackie. 1971. Evolution of the Chondrophora: Evidence from behavioural studies on *Velella.* Jour. Fisheries Res. Bd. Canada, *28;* 1595–1602.

Foyn, B. 1927. Studien über Geschlecht und Geschlechtzellen bei Hydroiden. I. Ist *Clava squamata* (Müller) eine gonochoristiche oder hermaphrodite Art? Arch. Entw.-Mech. Org., *109*: 513–534.

Fulton, C. 1963. Rhythmic movements in *Cordylophora.* Jour. Cell. Comp. Physiol., *61*:39–51.

Jacobs, W. 1937. Beobachtungen über das Schweben der Siphonophoren. Z. vergl. Physiol., *24;* 583–601.

Josephson, R. k. 1961a. Colonial responses of hydroid polyps. Jour. Exp. Biol., *38*: 559–577.

———. 1961b. Repetitive potentials following brief electrical stimuli in a hydroid. Jour. Exp. Biol., *38*: 579–593.

———. 1965. Three parallel conducting systems in the stalk of a hydroid. Jour. Exp. Biol., *42*: 139–152.

———, and G. O. Mackie. 1965. Multiple pacemakers and the behaviour of the hydroid *Tubularia*. Jour. Exp. Biol., *43*: 293–332.

———, and J. Uhrich. 1969. Inhibition of pacemaker systems in the hydroid *Tubularia*. Jour. Exp. Biol., *50*: 1–14.

Mackie, G. O. 1960. Studies on *Physalia physalis* (L.) (by A. K. Totton and G. O. Mackie) p. 2, Behaviour and histology. Discovery Repts., *30*: 369–408.

———. 1963. Siphonophores, bud colonies and superorganisms, *in* E. C. Dougherty, ed., The Lower Metazoa. Univ. Calif. Press, Berkeley, p. 329–337.

———. 1964. Analysis of locomotion in a siphonophore colony. Roy. Soc. London Proc., ser. B, *159*: 366–391.

———. 1965. Conduction in the nerve-free epithelia of siphonophores. Am. Zool., *5*: 439–453.

———. 1968. Electrical activity in the hydroid *Cordylophora*. Jour. Exp. Biol., *49*: 387–400.

———. 1970. Neuroid conduction and the evolution of conducting tissues. Quart. Rev. Biol., *45*: 319–332.

———, and D. A. Boag. 1963. Fishing, feeding and digestion in siphonophores. Pubbl. Staz. Zool. Napoli, *33*: 178–196.

———, and G. V. Mackie. 1967. Mesogleal ultrastructure and reversible opacity in a transparent siphonophore. Vie et Milieu, ser. A, Biol. Mar., *28*: 47–71.

Morin, J. G., and J. M. Cooke. 1971a. Behavioural physiology of the colonial hydroid *Obelia*, I, Spontaneous movements and correlated electrical activity. Jour. Exp. Biol., *54*: 689–706.

———, and J. M. Cooke. 1971b. Behavioural physiology of the colonial hydroid *Obelia*, II, Stimulus-initiated electrical activity and bioluminescence. Jour. Exp. Biol., *54*: 707–721.

———, and J. M. Cooke. 1971c. Behavioural physiology of the colonial hydroid *Obelia*, III, Characteristics of the bioluminescent system. Jour. Exp. Biol., *54*: 723–735.

Naumov, D. V. 1969. Hydroids and hydromedusae of the U.S.S.R. Israel Program for Scientific Translations Ltd., Jerusalem.

Rose, S. M. 1967. Polarized inhibitory control of regional differentiation in *Tubularia*, III, The effects of grafts across sea water-agar bridges in electric fields. Growth, *31*: 149–164.

*Spencer, A. N. 1971. Behaviour and electrical activity in *Proboscidactyla flavicirrata* (Hydrozoa). Ph.D. Thesis, Univ. Victoria.

Stokes, D. 1972. In preparation. Functional organization of conducting systems in the colonial hydroid *Hydractinia echinata* (Flemming). Ph.D. Thesis, Univ. Hawaii.

Strickland, D. L. 1971. Differentiation and commensalism in the hydroid *Proboscidactyla flavicirrata*. Pacific Sci., *25*: 88–90.

Totton, A. K. 1960. Studies on *Physalia physalis* (L.) (by A. K. Totton and G. O. Mackie) p. I, Natural history and morphology. Discovery Repts., *30*: 303–367.

Wilson, D. P. 1947. The Portuguese man-of-war. *Physalia physalis* (L.) in British and adjacent seas. Jour. Mar. Biol. Ass. U.K., *27*: 139–172.

Wyttenbach, C. R. 1968. The dynamics of stolon elongation in the hydroid *Campanularia flexuosa*. Jour. Exp. Zool., *167*: 333–352.

Zoja, R. 1891. Sulla transmissibilita degli stimoli nelle colonie di Idroidi R. C. Ist. Lombardo, ser. 2, *24:* 1225–1234.

Evolution of Holopelagic Cnidaria: Colonial and Noncolonial Strategies

Philip J. Phillips   Texas A & M University

ABSTRACT

Coloniality in holopelagic Cnidaria has evolved independently a minimum of seven times in the Hydrozoa and at least once in the Scyphozoa. Ecologically, colonial holoplanktonic organisms are equivalent to noncolonial organisms such as the Trachymedusae. In the evolution of a holopelagic condition a sessile neritic polyp stage is bypassed by direct development, neotenic development, larval parasitism on other or the parent medusa (e.g., Narcomedusae), viviparity (e.g., the scyphozoan *Stygiomedusa fabulosa),* or direct attachment to the water surface by float formation (e.g., Chondrophora). Coloniality, which is here defined to include any polypoid or medusoid homologue which supports developing juveniles of budded or parasitic origin, is very distinct with respect to ontogenetic origin for each group considered. As an adaptive strategy it is extremely successful, allowing invasion of oceanic water masses by removing the necessity of a continental shelf habitat for the attached stages in the life cycle. The direction of evolution in the pelagic Cnidaria is from a meroplanktonic condition to a holoplanktonic situation. The single most important environmental factor selecting for the holoplanktonic condition is tectonic processes in coastal regions which favor survival of organisms that can complete the life cycle in the more environmentally stable oceanic waters. The low species diversity in all oceanic cnidarian groups is correlated with the relative slowness of major tectonic changes in geological history and less frequent establishment of allopatric populations necessary for speciation.

Author's address: 100 Belmont Place, Staten Island, N.Y. 10301.

INTRODUCTION

It is impossible to discuss coloniality in any group of organisms without considering its ecological significance with respect to survival and reproduction. It is my intention, therefore, to discuss the functional significance of various colonial situations in holopelagic Cnidaria in comparison with noncolonial, oceanic Cnidaria. "Coloniality" in pelagic Cnidaria has evolved, independently, several times. The only holopelagic Cnidaria which lack a nurse carrier are the Trachymedusae, in which the polyp is bypassed by the direct development of an actinuloid larva and at least one scyphozoan, *Pelagia noctiluca,* in which the planula bypasses the scyphistoma stage and develops directly into an ephyra. In this article the term "coloniality" is defined, with respect to pelagic Cnidaria, as characterizing a medusoid or hydroid analogue that supports maturing juveniles of intrinsic (budded) or extrinsic (parasitic) origin. This definition of the term "colonial" therefore embraces the Narcomedusae, in which the larva is either attached to the parent medusa or parasitizes another medusa.

The object of this paper is threefold. I intend to demonstrate that where direct development is absent, the nurse carrier is a necessity for successfully adapting to the oceanic environment and that ancestral forms to all these holopelagic groups were neritic species with a distinct sessile stage. Additionally I will try to relate the origin of and speciation in holopelagic Cnidaria to continental tectonic processes.

EVOLUTION OF THE SIPHONOPHORA AND CHONDROPHORA

Garstang (1946) has very adequately elucidated the morphology and interrelations of siphonophores and chondrophores. Although Garstang pointed out the significant differences in float formation in the two groups, he nonetheless retained both in the same order with the rationale that although sufficiently distinct morphologically and embryologically, both evolved from gymnoblastic hydroids. Totton (1954) separated the two groups and assigned each ordinal rank interpreting the chondrophorans to be floating tubularoid hydroids which budded off the free-living medusae *Chrysomitra* and *Discomitra.* Although the siphonanths (siphonophores) and disconanths (chondrophores) are both evolved from gymnoblastic hydroid groups, each evolved independently from distinct hydroid stocks. In any event the affinity of both these orders with gymnoblastic hydroids is well established: the disconanths have tubularoid affinities and the siphonanths have corynoid affinities (Totton, 1965). The basic problem, then, is the functional mechanism whereby these holopelagic superorganisms evolved from sessile neritic progenitors.

Evolution in the siphonophores has progressed from the more primitive cystonect stage through the intermediate physonect condition to the floatless calycophoran condition (Garstang, 1946; Totton, 1965). Garstang interpreted the original ancestor of the Siphonophora to be an overgrown, asexual polyp which never attained the adult hydrozoan form (the sexual medusa) but gave rise to it by budding. Totton also pointed out that "the long axis of a fully grown physonect . . . carries at one end the original larval mouth of the oozooid and its invaginated aboral float at the other." The actinula described by Garstang is certainly an oozooid quite different from the trachyline actinuloid larva.

The unifying theme of evolution in both the chondrophores and siphonophores is the convergent evolution of a float. The origin of both groups resulted from dissociation with a neritic habitat by the development of a float. Within the Siphonophora, however, the float has become secondarily lost in the calycophorans.

It is well known that all holopelagic or holoplanktonic groups have a relatively small number of species or low species diversity when compared to related sessile neritic or meroplanktonic groups. With this low species diversity we also find a very wide oceanic distribution for most species and, indeed, a large number of species which may be almost cosmopolitan. In the case of the siphonophores the evolutionary patterns are very clearly delineated within the extant species, *Velella velella* and *Porpita porpita*. The Narcomedusae and Trachymedusae each have approximately 50 species. This low level of species diversity within each group is indicative of a low rate of speciation. Greater species diversity in neritic waters is a function of greater probability of geographic isolation in neritic areas (Valentine, 1971).

The neritic zone, of all marine habitats, is subject to the greatest amount of environmental stress as a result of tectonic and climatic changes and sea-level fluctuation, thus inducing a greater variety of stress on the organisms inhabiting this region than on those in the more stable oceanic environment. Holoplanktonic Cnidaria most likely evolved at times of drastic change in the continental shelf habitat. Precocious or neotenic development not dependent on the occurrence of a suitable neritic substrate would have considerable survival value in regions of neritic instability. Siphonophores and chondrophores, in essence, are gymnoblastic hydroids which have attached themselves to the oceanic water layers. Indeed, as Mackie (1959) has pointed out, the surface of the water replaces the ancestral neritic substrate of the chondrophores. In both groups the initial development of an aboral float, convergently evolved, serves as a substitute for an ancestral neritc or continental shelf surface for polyp attachment. Float formation is not unique to the chondrophores and siphonophores. The pelagic anemone *Minyas* traps gas bubbles in a fibrous meshwork on its pedal disc (Hedgpeth, 1954), and hydras are known to occasionally trap gas bubbles on the aboral surface and temporarily become surface floaters. Flotation, by enabling Hydroida to become independent of an ancestral substrate, allows greater potential for range expansion and consequent colonization of regions of greater environmental stability.

ORIGINS OF NARCOMEDUSAE AND TRACHYMEDUSAE

The Trachymedusae, at least one species of scyphozoan (*Pelagia noctiluca*), and possibly some pelagic anemones constitute the only holopelagic Cnidaria that are noncolonial. The Trachymedusae have successfully invaded the oceanic habitat by the extreme reduction of the polyp to a directly developing actinuloid larva. Morphologically the Trachymedusae have affinities with the Leptomedusae with respect to bell structure, gonad location, sense organs, and oral structure. The Narcomedusae, in contrast, have a radically different adult morphology and a very distinct type of parasitic larval development. The phyletic affinities of the Narcomedusae are most certainly obscure and have been little investigated. Larval development in Narcomedusae as delineated by Rees (1963), Kramp

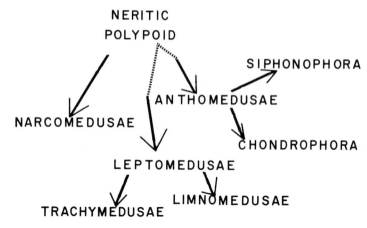

Figure 1

Evolution in the pelagic Hydrozoa. A broken line indicates uncertain relationship or origin.

(1957, 1959), and Russell (1953) involves an attached larva either in association with the parent or parasitic on another medusa. The attached parasitic larva then either grows into an actinuloid type of neotenic larva which frees itself from the host or produces a proliferative stolon or parasitic hydroid homologue which buds off medusae. Rees (1963) interprets the proliferative parasitic stolon of the Narcomedusae to represent the beginnings of "true metagenesis" of the type found in the Anthomedusae and in the Leptomedusae. Rees assumes that the actinuloid larva of some Narcomedusae evolved from a trachymedusan ancestor and, as a corollary of this, implies that both the Anthomedusae and Leptomedusae evolved from a parasitic hydroid. I cannot envision the origin of free-living, sessile hydroids from highly modified parasitic polypoids. Additionally, the fact that both the Narcomedusae and Trachymedusae are holoplanktonic is, by itself, insufficient for demonstrating common ancestry. My interpretation of the relationships of these two groups within the pelagic Hydrozoa is shown in Fig. 1.

In Narcomedusae the polyp stage, by parasitic attachment to a medusa, has bypassed the ancestral neritic or continental shelf substrate. The Narcomedusae must be of very ancient lineage, in that no group of extant Cnidaria has a comparable ontogeny or adult morphology. In the Narcomedusae the medusoid parent or parasitised medusa may be considered the nurse carrier or surrogate for the neritic substrate.

HOLOPLANKTONIC LIMNOMEDUSAE, ANTHOMEDUSAE, AND LEPTOMEDUSAE

The limnomedusan *Proboscidactyla ornata,* commonly taken in the epipelagic waters of the Gulf of Mexico, usually has polypoid structures growing out of the stomach walls and radial canals (Fig. 2) (Phillips, 1972). Several specimens have medusa buds on the polypoid outgrowth. Kramp (1962, as cited by Totton, 1965) has also noted these polypoid structures on *Proboscidactyla ornata.* In essence this is a case of a medusa

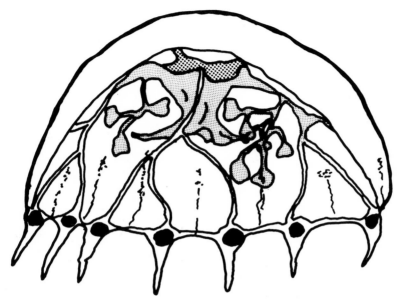

Figure 2

Limnomedusa *Proboscidactyla ornata* (bell diameter 6 mm). Note the hydroid-like outgrowths on the radial canals. The polypoid structures produce medusae by asexual budding.

carrying its own polyp. Reproductive cycles in the Limnomedusae are poorly known. The polyps of *Proboscidactyla,* called *Lar,* when not attached to the medusa are found in association with sabellid worm tubes. This limnomedusan is evidently able to bypass the sessile polyp by asexual budding of polypoids and via this asexual reproductive mechanism is widely dispersed in the upper strata of warm and tropical seas.

The occurrence of polypoid structures growing out of various portions of the medusa is not unique to *Proboscidactyla*. Among the anthomedusae I have captured one specimen of *Zanclea costata* in the Gulf of Mexico which bore fully developed hydroids on its radial canals. Medusa buds are also produced on various other Anthomedusae via attached stolons, which may also be interpreted to be polyp homologues. The asexual production of medusae by medusae is widespread in the Anthomedusae and has evolved independently in various families. *Cytaeis tetrastyla* commonly has medusa buds attached to its stomach walls (Fig. 3) and is often found in epipelagic oceanic waters. Although several species of sessile polyps are known for *Cytaeis*, only one valid species of medusa is known. The medusa of *Cytaeis* commonly bypasses the polyp by asexual medusa formation, thus accounting for wide oceanic distribution. The anthomedusa *Bougainvillia platygaster* has polypoid structures growing out of its stomach walls and, like *Cytaeis tetrastyla*, asexually buds medusae. The polypoid structures on *B. platygaster* have been discussed by Kramp (1959). This type of asexual medusa formation by medusae also occurs in the anthomedusan families Tubulariidae, Clavidae, and Margelopsidae. A similar situation also exists in the Leptomedusae in the family Campanulariidae.

In the family Margelopsidae, the hydroid, where known, is pelagic and both the

Figure 3

Anthomedusa *Cytaeis tetrastyla* (bell height 5 mm), with asexually budded juvenile medusae attached to the base of the stomach.

medusa and polyp may exist separately in the plankton. The margelopsid *Pelagohydra mirabilis* is especially interesting in that the polyp is very large, pelagic, and produces reduced medusae which are not known to occur free in the plankton.

The ability of medusae to complete the life cycle asexually without a sessile neritic stage would obviously be of great survival value in unstable continental shelf areas. In addition, asexual holoplanktonic reproduction would be conducive to oceanic distribution in species where the adult medusa has a short life span. Quite possibly asexual reproduction would take place indefinitely until a swarm of medusae reached a suitable coastal area where sexually produced larvae could settle and produce polyps. Ecologically, the asexual budding of medusae by medusae by whatever the mechanism and the budding of medusae by pelagic hydroids are equivalent, permitting a wide oceanic distribution. Functionally, in terms of relationships with the environment, the conditions in afore-mentioned groups of medusae, Chondrophora and Siphonophora, are convergently evolved responses to the same environmental stress.

HOLOPLANKTONIC SCYPHOZOA

Evolution in the Scyphozoa has been from the neritic condition to the oceanic. In *Pelagia noctiluca* the scyphistoma (polyp) stage has been completely suppressed, and released planulae grow directly into ephyrae (juvenile medusae). *Pelagia* is most closely related to the pelagiid genus *Chrysaora,* which has a sessile scyphistoma. *Chrysaora* passes through an eight-tentacle 16-lappet stage which is identical to the adult meristic condition in *Pelagia.* The juvenile stage of *Chrysaora* is appropriately called the "pelagia" stage. The meristic similarities indicate that *Pelagia* evolved from *Chrysaora* by suppres-sion of the scyphistoma and neotenic development of the ephyra.

True viviparity is known in the scyphozoan *Stygiomedusa fabulosa* (a bathypelagic semaestome). In *Stygiomedusa* the scyphistoma, as described by Russell and Rees (1960), has its base embedded in the tissues of the stomach wall of the parent and obtains its nourishment from surrounding tissues. As opposed to releasing ephyral stages, the attached scyphistoma releases well-developed juveniles. Russell and Rees relate the occur-rence of viviparity to the absence of a suitable substrate for the scyphistoma in the bathypelagic habitat, and the adult medusa serves as a surrogate substrate. As pointed out by Russell and Rees, this is analogous to the reproductive situations in salps and other holoplanktonic forms which have evolved from neritic relatives.

Within the Scyphozoa evolution of holoplanktonic species has happened twice by completely different mechanisms. In both cases they have evolved from a neritic ancestor. With respect to *Stygiomedusa* there is a nurse carrier condition reminiscent of that found in some Narcomedusae.

EVOLUTIONARY PATTERNS IN HOLOPLANKTONIC CNIDARIA

As has been noted, the holoplanktonic condition has evolved in a minimum of seven orders of Hydrozoa and at least twice in the semaestome Scyphozoa. Of the seven orders of Hydrozoa considered, all but the Anthomedusae, Leptomedusae, and the Lim-nomedusae are exclusively holoplanktonic. Of the exclusively holoplanktonic groups only the Trachymedusae and the scyphozoan *Pelagia* lack a nurse carrier for a developing polypoid or hydroid and are therefore noncolonial. The evolutionary relationships of the pelagic Hydrozoa may be envisioned in the manner depicted in Fig. 1. This scheme differs from the phylogeny favored by Rees (1963) with respect to the following:

1. It does not consider the parasitic polypoid condition in some Narcomedusae to be the beginnings of true metagenesis; rather this polypoid is a highly degenerate (or modified) polyp which has not given rise to any leptolinid type hydroid.

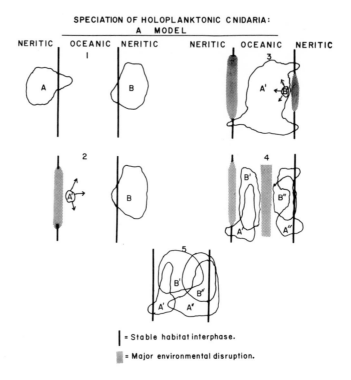

SPECIATION OF HOLOPLANKTONIC CNIDARIA:
A MODEL

NERITIC   OCEANIC   NERITIC   NERITIC   OCEANIC   NERITIC

| = Stable habitat interphase.

▓ = Major environmental disruption.

Figure 4

Model for speciation in holoplanktonic Cnidaria. A and B represent holoplanktonic populations of two different species. A complete explanation is given in the text.

2. Evolution in the hydromedusae has proceeded from the neritic to the oceanic condition; i.e., the Trachymedusae have evolved from an ancestral leptolinid by neoteny and suppression of the polyp.

3. The Narcomedusae and Trachymedusae evolved independently from different groups, the Narcomedusae possibly having a more ancient ancestry than the Trachymedusae.

4. The ancestral hydrozoan which gave rise to extant groups was a neritic polypoid homologue not of trachyline ancestry.

THE ORIGIN OF DIVERSITY WITHIN HOLOPLANKTONIC CNIDARIA

No adequate theory has yet been advanced to account for speciation within holoplanktonic groups because of the difficulty of devining or elucidating mechanisms of geographic isolation in the oceanic habitat which would be comparable to those operating with respect to continental organisms. Geographic variation in coastal populations of neritic species has been extensively documented by Ekman (1953) and many others. Within the Scyphozoa, geographic variation is such as to inspire Mayer (1910) to comment that the Linnean concept of the species does not apply to the majority of semaestome medusae. In these scyphozoans there are numerous geographic varieties and intergrades between species. For example, the semaestome medusa *Chrysaora quinquecirrha* (the common sea nettle) in the Gulf of Mexico differs ecologically and morphologically from those of Carolinian waters (Phillips et al., 1969), and this can be related to geographic isolation of the northern Gulf of Mexico by emergence of the Floridian peninsula during the Pleistocene (Phillips, 1972). When considering the oceanic Cnidaria we are confronted with very widely distributed populations wherein distinct geographic variants are extremely difficult to define by standard methods. It is, however, not unusual to encounter individual swarms of a particular species which may be phenotypically distinct when compared to nearby swarms. In the case of the almost ubiquitous trachymedusan *Liriope tetraphylla,* numerous nominal species were erected on the basis of very minor morphological variation in different medusa swarms. According to Kramp (1959) and Russell (1953) there is only one valid species of *Liriope* and all other species are invalid. Although variation occurs in oceanic populations, it has not yet been quantitatively defined for any one species.

Assuming that speciation in holoplanktonic groups is a result of geographic isolation and the establishment of allopatry, its occurrence in geologic time is a relatively uncommon event, as witnessed by the low species diversity within the major taxa when compared to coastal taxa of similar rank. A hypothetical model for geographic isolation and speciation in the holoplankton is as follows (Fig. 4):

1. Oceanic holoplankton evolved from neritic or continental shelf ancestors (Fig. 4.1).

2. Environmental stress in continental shelf regions resulting from tectonic processes selected for those members of continental shelf populations that could survive and reproduce in oceanic conditions (Fig. 4.2).

3. This newly established race or population invaded extensive areas of the epipelagic zone and became established in oceanic waters and overlapped continental shelf regions (Fig. 4.3).

4. Allopatric populations occurred as a result of extinction within the central region of epipelagic distribution, thus effectively isolating segments of the original population into areas of distribution which may have had continental shelf overlap (Fig. 4.4). Factors responsible for extinction in the central area of the range which resulted from tectonic processes could be any one or more of the following: (a) Drastic temperature changes in the epipelagic realm caused by massive upwelling of cold water, (b) mountain-range emergence or any type of land-mass emergence which would effectively change the characteristics of the upper water strata or (c) climatic changes which would serve to drastically lower temperatures in the upper water strata except in limited regions of shallow tropical water in continental shelf regions.

5. Reestablishment of broad belts of tropical surface waters allowed for the reinvasion by these previously isolated populations, which now may or may not be distinct species (Fig. 4.5).

6. Greatest species diversity would occur in tropical waters, not because speciation occurred in those waters, but because the environmental stability of the tropical habitat allows for the establishment of a large number of species. The mechanism for the establish-

ment of high species diversity in low latitudes has been examined by Valentine (1971) and he concludes the following:

> In environments wherein trophic resources fluctuate, however, a flexible life style is favored. When resources fall to low levels, the populations that live in the greatest variety of habitats and that eat the largest range of food items will have the best chance of surviving. And when resource levels rise, the population that responds most rapidly and extensively by increasing its population size is apt to obtain a larger share of the resources and . . . endure the next unfavorable period. When new species appear they are not easily accommodated, for the accent on selection for high reproductive potential results in a strong competition for resources. Furthermore, resources cannot easily be partitioned, for the resulting reduction in the sizes of the populations involved would greatly lower their chances of survival during unfavorable periods.
>
> Therefore in energetically stable environments, efficient populations can be accommodated within an ecosystem, together with generalists as well, and population sizes may be small on the average. In energetically unstable environments, by contrast, relatively few populations can coexist, for they are chiefly of broadly-adapted, inefficient species that require large populations during favorable periods. Trophic resources are thus especially important because they form a variable, density-dependent limiting factor. The increase in the numbers of species within ecosystems at low latitudes can thus in this view be attributed chiefly to the increasing stability of trophic energy.

The factor inducing extinction in the center of the oceanic range not only aided in the establishment of various allopatric populations with tropical continental shelf overlap, but may well have served as a selective factor for the evolution of cold-water and bathypelagic species. Assuming drastic temperature reduction in the upper water strata and simultaneous extinction of neritic and continental shelf populations that connect with high and low latitudinal segments of the species, there could conceivably be a sufficient number of individuals in higher latitudes that would be able to survive and reproduce under a much lowered temperature regime. Thus there will be eventually a shift in the temperature tolerance of a given population. Hence the origin of the present-day bathypelagic Cnidaria was most probably in high-latitude, cold-water continental shelf areas. The phenomenon of polar emergence of bathypelagic Cnidaria is well known in arctic waters. An excellent example is the almost ubiquitous bathypelagic coronate medusa *Periphylla periphylla,* which exhibits the phenomenon of equatorial emergence where there are regions of strong upwelling. Additionally, *Periphylla* and other bathypelagic medusae occur in the Gulf of Mexico (Phillips, 1972), evidently having successfully crossed over the relatively shallow sills separating the Gulf from the Caribbean. The environmental factor limiting many Cnidaria to the bathypelagic and intermediate water strata is the inability of these medusae and siphonophores to tolerate higher surface temperatures characteristic of tropical and temperate seas. The thermal stability of the bathypelagic water masses would allow for the establishment of broad species diversity analogous to that of shallow tropical seas by the same mechanism, i.e., trophic stability versus trophic instability at higher latitudes coupled with major tectonic changes. The comparable situations with respect to trophic stability and species diversity in the ocean depths as in the tropics have been noted by Valentine (1971).

CONCLUSIONS

A predominant view among marine biologists is that the greatest rate of speciation is found in tropical waters, as judged by the high species diversity in those regions. Zoogeographically, these low-latitudinal areas, particularly the tropical Indo-Pacific, are centers of dispersal, rather than evolution, and functionally are analogous to terrestrial refugia of the Pleistocene, which acted as centers of dispersal once surrounding areas became suitable for range expansion. With respect to the Cnidaria, the oceanic holoplankton, whether colonial or not, can be viewed as a depauperate offshoot of the neritic or continental shelf fauna. Evolution in the oceanic environment, although much slower than in meroplanktonic, continental slope species, occurs via establishment of allopatry by major tectonic changes. The relative slowness of evolution within oceanic waters is a function of the relative slowness of major tectonic changes, as compared with short-term, relatively rapid topographical changes in neritic areas and the greater opportunity for geographic isolation in coastal waters. Coloniality (convergently evolved in at least seven orders of pelagic Hydrozoa and at least one time in the Scyphozoa) and direct development without an attached polyp stage in noncolonial pelagic Cnidaria are equivalent evolutionary strategies enabling successful colonization in the oceanic environment. The functional significance of coloniality (whatever its guise) in pelagic Cnidaria is considerably different than that for sessile taxa. To understand the functional significance of coloniality or for that matter any morphologic feature, one must first analyze its significance in the life history of the organism. In the case of the holoplanktonic Cnidaria a "colonial" or nurse-carrier situation allows for life-cycle completion in oceanic waters without the necessity of a continental shelf substrate for polyp attachment. As has been demonstrated, elimination of the ancestral neritic substrate as a necessity for life-cycle completion, by whatever the ontogenetic mechanism, is a necessity for a holoplanktonic life style. The direction of evolution, for all holoplanktonic organisms, has been from the neritic to the oceanic, with elimination of a sessile, bottom-dwelling stage. In the Hydrozoa the Trachymedusae are not ancestral to the predominantly continental shelf Anthomedusae, Leptomedusae, and Limnomedusae, but have probably evolved from a leptolinid ancestor by direct development of the larva, analogous to the case of the scyphomedusa *Pelagia noctiluca,* wherein the scyphistoma has been lost and the planulae grow directly into ephyrae.

REFERENCES

Ekman, S. 1953. Zoogeography of the Sea. London. 417 p.

Garstang, W. 1946. The morphology and relations of the Siphonophora. Quart. Journ. Micros. Sci., *87*: 103–193.

Hedgpeth, J. G. 1954. Anthozoa: the anemones. U.S. Fish. Bull., *55*: 285–290.

Kramp, P. 1957. Hydromedusae from the Discovery Collections. Discovery Repts. *29*: 1–128.

———. 1959. The hydromedusae of the Atlantic Ocean and adjacent waters. Dana Rept., *46*: 1–283.

Mackie, G. O. 1959. The evolution of the Chondrophora: new evidence from behavioural studies. Roy. Soc. Canada, Trans., *53*: 7–20.

Mayer, A. G. 1910. Medusae of the World, v. I, II and III. Carnegie Foundation, Washington, D.C., 735 p.

Phillips, P. J. 1972. The pelagic Cnidaria of the Gulf of Mexico: zoogeography, ecology and systematics. Unpublished doctoral dissertation, Texas A & M Univ., College Station, Texas, 212 p.

———, W. D. Burke, and E. J. Keener. 1969. Observations on the trophic significance of jellyfishes in Mississippi Sound with quantitative data on the associative behavior of small fishes with medusae. Am. Fish. Soc., Trans., *98*: 703–712.

Rees, W. J. 1963. The evolution of the Hydrozoa, *in* Zoological Society of London Symposium no. 16, p. 199–222.

Russell, F. S. 1953. The Medusae of the British Isles: Anthomedusae, Leptomedusae, Limnomedusae, Trachymedusae and Narcomedusae. Cambridge Univ. Press, England. 530 p.

———, and W. J. Rees. 1960. The viviparous scyphomedusa *Stygiomedusa fabulosa* Russell. Jour. Mar. Biol. Ass. U.K., *47*: 469–473.

Totton, A. K. 1954. Siphonophora of the Indian Ocean together with systematic and biological notes on related species from other oceans. Discovery Repts. *27*: 1–161.

———. 1965. A synopsis of the Siphonophora. British Museum, London, 230 p.

Valentine, J. W. 1971. Plate tectonics and shallow marine diversity and endemism, an actualistic model. Syst. Zool., *20*: 253–264.

Introduction to Bryozoans

Bryozoans are wholly colonial animals in which the coelomate organization allows a wide variety of levels of integration of component zooids and extrazooidal parts. Because of the intimate relations between cuticle, calcified layers, and epidermis, skeletal walls, where developed, are parts of the body wall, and their morphology and mode of growth closely reflect the degree of integration of the colony. Astogenetic changes, polymorphism, and interzooidal communication also reflect varying degrees of integration in the phylum.

The phylum is commonly divided into three major groups: Stenolaemata, Gymnolaemata, and Phylactolaemata. Stenolaemates and gymnolaemates are predominantly marine, most commonly have calcified walls, and have extensive fossil records. Stenolaemates are characterized by conical or tubular zooids, most of which continue to lengthen into later ontogenetic stages. This group, known from Ordovician to Recent, comprised the great majority of Bryozoa until the proliferation of gymnolaemates in the Cretaceous. Most gymnolaemates are characterized by box- or sac-shaped zooids, the length of which is established at early ontogenetic stages. This group is diversified and abundant in modern seas and is known from at least Jurassic to Recent. Phylactolaemates are freshwater Bryozoa lacking a calcareous skeleton and having repeated feeding members suspended in a colony-wide coelom and therefore representing the highest degree of zooid communication in the phylum. Fossil colonies of this group are unknown.

Integration in stenolaemates and gymnolaemates is reviewed by **Boardman and Cheetham** on the basis of the morphology of salient modern and fossil genera and distribution of levels of integration through time. **Kaufmann** proposes a model, based on assumed energy distribution, for the relationship of colony growth form to environment in the gymnolaemates but with application to other colonial groups. **Abbott** infers a relationship between astogeny and polymorphism from a detailed analysis of astogenetic changes in living populations of three species of gymnolaemates, concluding that such changes represent functional levels in the development of the colony. A theoretical basis for the development of polymorphism in gymnolaemates is presented by **Schopf** from the relation between number of polymorphs and environmental stability as implied by the geographic distribution of modern species. **Banta** discusses evolution of avicularia in gymnolaemates and notes the changes in levels of integration implied by the inferred evolutionary steps. Both continuity and discontinuity in modern gymnolaemate skeletons are demonstrated by **Sandberg** from detailed microstructural analysis. Among the papers on stenolaemates, **Utgaard** and **Blake** make the first major attempts to interpret mode of growth and functional morphology of two extinct groups by comparison with some living analogues. Inter- and intracolony variation in a fossil stenolaemate species is treated statistically by **Farmer and Rowell.**

Mode of growth and colony structure in phylactolaemates are reviewed by **Wood,** who presents data on zooid and colony reproduction relative to colony age in three species. **Bishop and Bahr** provide data on feeding rate and colony size in another phylactolaemate species.

Degrees of Colony Dominance in Stenolaemate and Gymnolaemate Bryozoa

Richard S. Boardman and Alan H. Cheetham   Smithsonian Institution

ABSTRACT

All Bryozoa form colonies consisting of one or more kinds of zooids, and some include extrazooidal parts. The zooid is assumed to correspond to the individual in solitary animals as the basic morphologic and functional unit. This correspondence seems clearest for feeding zooids. Nonfeeding polymorphs are so varied in stenolaemate and gymnolaemate Bryozoa that the only feature that all zooids possess is a body cavity, generally regarded as a coelom, contained by a body wall. Extrazooidal parts are connective and/or supportive structures occurring outside zooidal boundaries. The variety of extrazooidal parts known in stenolaemates and gymnolaemates forms a transitional series approaching that of polymorphic zooids.

The first step in analyzing colonies is a description of morphology that reflects degrees of physiologic or structural dependence of a zooid on other parts of the colony beyond that shown by a solitary individual on other individuals. In this comparison the two theoretical extremes are complete autonomy of the individual and complete dominance of the colony.

Degrees of colony dominance in Bryozoa can be predicted from interpretation of morphology based on assumptions of physiologic and functional properties of body

   Authors' address: Department of Paleobiology, National Museum of Natural History, Smithsonian Institution, Washington, D.C. 20560.

wall and zooid organs. Six morphologic integration series, each consisting of states expressing low to high colony dominance, can be interpreted in stenolaemates and gymnolaemates: (1) zooid walls, (2) interzooidal connections by zooidal soft tissue, (3) extrazooidal hard and soft parts, (4) astogeny, (5) morphologic differences between polymorphic zooids, and (6) positional characteristics of polymorphs.

Stenolaemate bryozoans are characterized by elongate zooids that are cone- or tube-shaped and have terminal apertures. Zooids are budded in unpartitioned, multizooidal budding zones and increase in length during most or all of their ontogenetic development. The four orders are known from rocks of Middle Ordovician age; three are thought to have become extinct near the end of the Paleozoic Era; the fourth, the Cyclostomata, are still living in marine waters. Evolution and classification within and among orders are too poorly known to support studies of colony dominance within an order. Living stenolaemates have two general types of zooidal walls, a double-walled arrangement that allows interzooidal communication (colony-dependent character) around ends of interior (colony-dependent structure) walls and a single-walled structure which adds exterior walls to the ends of interior walls to close off distal communication. The great majority of Paleozoic Bryozoa were double-walled. A few single-walled forms which apparently lacked means for interzooidal communication after calcification of interior walls evolved in Ordovician times and continued through the Paleozoic Era. In the post-Paleozoic eras stenolaemates developed communication pores (colony-dependent structures) in interior walls that apparently provided interzooidal communication, and both single- and double-walled forms flourished. Extrazooidal structures take several forms and occur throughout the history of stenolaemates. Paleozoic fenestratid bryozoans commonly display a secondary skeletal covering following the formation of zooids, which in extreme development covered zooids completely and formed massive supporting structures for the colonies. Extrazooidal skeleton also forms commonly in later ontogenetic stages between adjacent zooids in erect growth habits. Generally unpartitioned coelomic space between zooids is known only in a few forms of Paleozoic age. Primary astogenetic differences are difficult to understand in stenolaemates because subsequent zooids commonly cover earliest generations in a colony. Colony growth by superimposed layers of zooids is especially common in double-walled forms throughout stenolaemate history, and apparently small patches of subsequent growth over dead zooids were a normal method of ensuring even growth of colony surfaces. Study of polymorphism is in early descriptive stages and in fossils is handicapped by the simplicity of tubular skeletons characteristic of stenolaemate polymorphs. Intrazooidal polymorphism during the ontogeny of a zooid is suggested by different skeletal structures in segments of the same zooecium in some Paleozoic forms. The general pattern of morphology related to colony dominance through time suggests an early development of most colony-controlled structures and minor extinctions and later independent repetitive development in more viable combinations.

Gymnolaemate Bryozoa comprise two orders, the soft-bodied Ctenostomata and the calcified, far more diversified Cheilostomata. Zooids in gymnolaemates are generally box- or sac-shaped, with dimensions of their principal body cavity determined early in ontogeny. Budding is commonly at distal ends of lineal series separated by exterior walls and consists of outpocketing usually followed by upgrowth of an interior wall to separate the distal portion of the bud. Some cheilostomes seemingly lack lineal series, their zooids being budded in multizooidal zones similar to those in stenolaemates. Communication pores, plugged with soft tissues in all living gymnolaemates investigated, are present in both interior and exterior walls between contiguous zooids. Extrazooidal parts are apparently present in some cheilostomes, being formed either concurrently with budding of zooids or subsequently by coalescence of zooidal tissues. Astogenetic differences between zooids are generally conspicuous and may be restricted to the proximal zone of the colony, including the single primary zooid or a group of primary zooids plus one or more generations of asexually produced zooids, or may occur also in one or more distal zones. Polymorphism of zooids characterizes nearly all gymnolaemates. Morphologic differences between zooids suggest a wide range of degrees of interdependence associated with functions, including feeding, sexual reproduction, support, and connection. Polymorphs can be intercalated in the budding pattern seemingly at random or in repeated groups, or they may be structurally parts of other zooids.

The wide range of degrees of colony dominance inferred from the morphology of gymnolaemates and the stratigraphic ranges of genera of cheilostomes, insofar as known, suggest a pattern from late Jurassic time to the present of increasing integration in the rapidly proliferating cheilostome stocks. Evolution of integration in this group appears to have followed a mosaic pattern rather than a simple progression, and ancestral stocks, retaining the least integrated states known, have persisted to the present. Genera having similar colony forms, regarded as expressing similar adaptations, appear also to have similar levels of integration in particular sets of integrative characters.

INTRODUCTION

Nature of Bryozoan Colonies

All Bryozoa form colonies consisting of one or more kinds of zooids, and some include extrazooidal parts. Zooids are physically connected, asexually replicated morphologic units which separately perform major physiologic or structural functions such as feeding, reproduction, or support. The physical wholeness and implied genetic uniformity of the colony make it the unit of natural selection in Bryozoa (see discussion by Schopf, this volume). An understanding of the evolution of Bryozoa therefore depends on recognition of how natural selection has worked on the morphology and physiology of the colony and its components, both zooids and extrazooidal parts.

The zooid is assumed to correspond to the individual in solitary animals as the basic morphologic and functional unit. This comparison seems most appropriate for feeding zooids, especially in the gymnolaemate genus *Monobryozoon* Remane, in which a colony, unless dividing, has only one feeding zooid among nonfeeding polymorphs (stolons). That feeding zooid is also capable of both sexual and asexual reproduction. Furthermore, the typical primary zooid in stenolaemates and gymnolaemates is capable of feeding before other zooids in the colony which bud from it. Feeding zooids in Bryozoa in general vary widely in morphology, in some species even within a colony. Morphologic variation of zooids continues transitionally to forms in which a markedly different function is performed. In the extremes of polymorphism of zooids, all organs associated with the functions of ordinary feeding zooids may be missing.

The fundamental question in analyzing the nature of animal colonies is how the zooids in colonies differ morphologically and physiologically from individuals in solitary animals. The first step in analysis is description of morphology that reflects degrees of physiologic or structural dependence of a zooid on other parts of the colony beyond that shown by a solitary individual on other individuals.

In the morphologic comparison between solitary and colonial animals, the two theoretical extremes are represented by complete autonomy of the solitary individual at one end of a series and complete colony dominance at the other. Among colonial animals, a high degree of zooid autonomy might be expected at one end of a series and a high degree of colony dominance at the other. Because of the nature of bryozoan colonies, colony dominance can be expected to be expressed by zooid dependence and presence of extrazooidal parts. The position within the series from zooid autonomy to colony dominance is indicated by the degree of integration of zooids and extrazooidal parts, with increasing integration expressing increasing colony dominance.

The potential morphology and physiology by which integration is expressed within a major colonial group are dependent upon phylum level of organization. All Bryozoa possess a body cavity that has been regarded as a coelom. The colony coelom and body wall expand and, by partitioning in various ways, produce the component zooids. In some Bryozoa portions of the colony coelom and body wall may not be assignable to zooids (extrazooidal). This organization allows both the expansion and partitioning of the colony to take a wide variety of forms and therefore to respond to many different aquatic environments. The adaptability of the Bryozoa is evidenced by their abundance from Ordovician times to the present, a span of some 500 million years.

The ordinary feeding zooids are minute and relatively constant in size throughout the phylum, but a wide range of colony size and shape is achieved by development of different numbers and arrangements of zooids, ranging from a few in colonies less than 1 cm in size to enormous numbers in colonies more than 1 m in size. Within a species, shape as well as size of the ordinary feeding zooids is less variable than colony size and shape. This colony variation makes a species adaptable to different environments requiring different growth habits.

Some or all of the zooids in any colony are feeding autozooids possessing at some ontogenetic stage or stages a complete alimentary canal with associated lophophore and musculature. Various polymorphic zooids may lack parts or all of these structures. Sexual structures may be present in all feeding autozooids or may be restricted to sexual polymorphs. Some zooids (kenozooids) lack these kinds of structures entirely and may serve supportive or connective functions only.

Zooids in stenolaemate and gymnolaemate Bryozoa are so varied in morphology and function, even within a colony, that the only feature that all possess is a coelomic space contained by a zooidal body wall. The body wall throughout the phylum consists of cellular and noncellular layers that may or may not be reinforced by calcareous skeleton. The skeletal portion of the body wall is secreted on the side of the epidermis away from the coelom and therefore is exoskeletal throughout, even though in some places it is deposited by epidermis which is infolded into existing coelomic space.

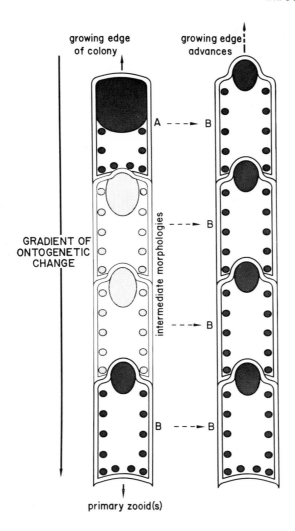

Figure 1

Pattern of ontogenetic difference in zooid morphology in hypothetical bryozoan colony. In series shown on left, zooid morphology increases in complexity from morphology A to morphology B through intermediate morphologies on a gradient directed proximally from colony's growing edge. With further growth of the colony, as indicated on right, zooids of morphology A and intermediate morphologies all change to morphology B, beyond which there is no ontogenetic change shown.

Sources of Variation in Bryozoan Zooids

Because Bryozoa are wholly colonial, bryozoan zooids can be expected to differ morphologically and physiologically from individuals in solitary animals.

The physically connected zooids in a colony are a clone. Individuals in solitary animals usually are physically disconnected and have different genotypes, although they can be disconnected members of clones or connected individuals of different genotypes. Because of their assumed genotypic uniformity, zooids in a colony might be expected to be morphologically identical. Morphologic variation among zooids in a bryozoan colony is common, however (see Farmer and Rowell, this volume), and follows patterns of distribution that have been attributed to four sources (Boardman et al., 1970). These sources of variation within a colony are (1) ontogeny of the zooids, (2) astogeny of the colony, (3) polymorphism, and (4) microenvironment.

1. Ontogenetic variation arises from changes in a zooid (or any extrazooidal part of the colony) during the course of its development, which may or may not continue throughout the life of a zooid. These changes are recognizable among zooids of a colony along a gradient of generally increasing complexity extending proximally from growing extremities, illustrated on the left of Fig. 1. Further development of the colony (right side of Fig. 1) transforms younger, less complex zooids (morphology A) to older, more complex ones (morphology B). Thus zooids and extrazooidal parts of colonies form a sequential record in a proximally directed series of the ontogenetic stages through which the proximal members of the series have progressed.

2. Astogeny is the course of postlarval development of a colony. A bryozoan colony is developed from a primary zooid or group of primary zooids resulting from metamorphosis of the larva. The process of colony founding involves morphologic differences between generations of zooids defining a primary stage of astogenetic change during which a pattern capable of endless repetition of zooids is developed. The primary zone of astogenetic change comprises the zooids, usually belonging to a few generations, which show morphologic differences in more or less uniform progression distally from generation to generation, as shown in Fig. 2. In other words, the zooids in each generation in a distally directed series from the primary zooid or zooids express morphologic characteristics unique to that generation. The primary stage of astogenetic change is followed distally by a primary stage of astogenetic repetition in which large numbers of zooids, usually belonging to many generations, of repeated morphologies are proliferated.

3. Polymorphism is discontinuous variation in the morphology of zooids arising at the same astogenetic level. In zones of astogenetic change polymorphic differences are expressed within the same generation, as shown in Fig. 3. In zones of astogenetic repetition, polymorphic differences can be observed throughout the zone. It is assumed that the morphologic differences defining polymorphs indicate differences in function.

4. Within a colony, variation which cannot be inferred to be an expression of ontogeny, astogeny, or polymorphism is assumed to be microenvironmental. This variation may occur in one or more regions of a colony or in scattered zooids. Microenvironmental variation can be recognized as gradational differences between groups of zooids either belonging to the same generation in a zone of astogenetic change (Fig. 4) or belonging to the same or different generations (at the same ontogenetic stage) in a zone of astogenetic repetition. An environmental change affecting the entire colony or more than one colony is more widespread than those considered to be microenvironmental. In some cases, microenvironmental differences may have been obviously caused by such factors as irregularity in the substrate, crowding, or injury. Temperature-related size differences in clones of zooids have been demonstrated in the laboratory by Menon (1972).

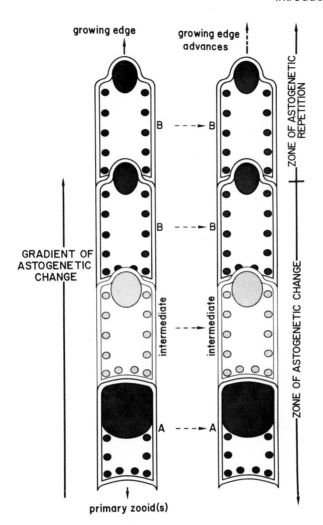

Figure 2

Pattern of astogenetic difference in zooid morphology in hypothetical bryozoan colony. Zooid morphology increases in complexity from morphology A to morphology B through intermediate morphology on a gradient directed distally from colony's primary zooid or zooids. With further growth of the colony, zooids shown do not change in morphology.

As shown in Fig. 5, both ontogenetic and astogenetic differences in morphology are expressed in a colony among zooids parallel to the direction of budding; ontogenetic differences are expressed by generally increasing complexity proximally and astogenetic differences by generally increasing complexity distally. Because of the sequential nature of budding, these differences have relative time significance. Intracolony differences produced by polymorphism and microenvironment are not necessarily sequential.

Morphologic variation in solitary animals obviously includes ontogenetic and environ-

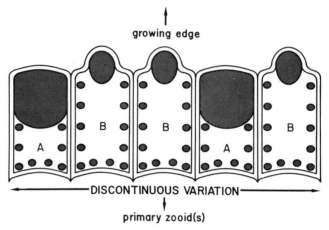

Figure 3

Polymorphic difference in zooid morphology in hypothetical bryozoan colony. Zooids belonging to same generation have morphology A or morphology B, intermediate morphologies being absent.

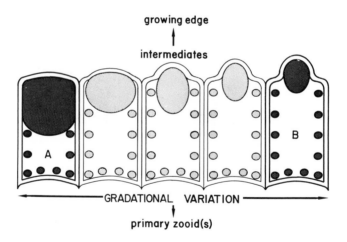

Figure 4

Microenvironmental difference in zooid morphology in hypothetical bryozoan colony. Zooids belonging to same generation have morphologies grading from A to B through series of intermediates. Spatial arrangement of zooid morphologies is for convenience of comparison; in actual colonies, morphologies are commonly irregularly intermixed.

mental differences intermixed with genetic differences between individuals. Polymorphic differences in most solitary animals are due to differences in genotypes, unlike polymorphic differences between zooids in a colony. In a few groups of animals which are not colonial by the definition used here, such as ants and other social insects, some polymorphism is determined independently of genetic differences (Wilson, 1953) and in that respect is comparable to polymorphism in colonial animals. Astogenetic differences

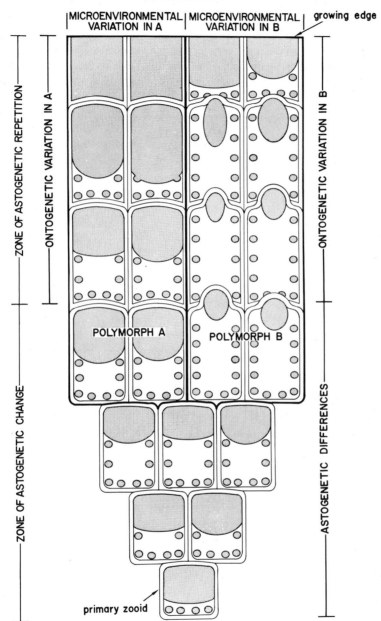

Figure 5

Patterns of ontogenetic, astogenetic, polymorphic, and microenvironmental difference in zooid morphology in hypothetical bryozoan colony. Spatial arrangement of polymorphs A and B is for convenience of comparison (although known in some groups of cheilostomes); in actual colonies, polymorphs are commonly intermixed in budding pattern.

have no relevance to most solitary animals. The change in caste ratios during founding of insect societies (Wilson, 1968) may be comparable to the sequential order of zooid types progressing to repeatable morphology during the stage of astogenetic change in a bryozoan colony. The zooids in the zone of astogenetic change in a bryozoan colony, however, differ from the founding individuals of an insect society in being members of a clone and generally in remaining a necessary structural part of the fully developed colony because of physical connection.

The differences between generations of zooids in a zone of astogenetic change in a bryozoan colony are empirically similar to those that have been described in some graptolites (see discussion by Urbanek, this volume). Urbanek has related astogenetic changes in graptolites to a morphophysiological gradient under monarchic control of the primary zooid. A morphogenetic substance produced by a primary zooid is inferred to have been diffused distally during the growth of the colony, resulting in differentiation of morphology between generations throughout the colony. In Bryozoa, however, the primary zone of astogenetic change comprises a small number of asexual generations of the total in the colony, and thus any morphogenetic substance produced by primary zooids appears not to have been continuously diffused throughout colony development. Also, in some stenolaemate Bryozoa, soft-tissue connections of the primary zooids to asexually produced zooids are apparently interrupted during calcification, again indicating lack of continuing monarchic effect (see p. 144). Further, the common occurrence in Bryozoa of subsequent zones of astogenetic change (Boardman et al., 1970), i.e., zones of change developed distal to zones of repetition without primary zooids, indicates that sequential morphologic differences between generations is not under control of the primary zooids, which, indeed, may no longer have been alive when subsequent zones of change developed.

Understanding morphogenetic gradients in Bryozoa is complicated by their wide variety of simple to complex budding patterns in which morphologic differences can appear on lateral or oblique gradients as well as distally trending ones (Brood, 1972, p. 59–64). The understanding of these gradients requires a knowledge of the relation of isochronous growth lines to budding pattern, a relationship which is known for few Bryozoa.

Morphologic Interpretation of Integration in Bryozoa

The degrees of integration in a bryozoan colony may be predicted from interpretation of morphology based on the following assumptions:

1. The body cavity (possibly excluding the portion contained by the tentacles) is separated from the external environment by body wall having cuticle as its outermost layer.

2. Cuticle is impervious to sufficient physiologic exchange to sustain zooid growth and function.

3. Coelomic connection between zooids or between zooids and extrazooidal parts of a colony permits freer physiologic exchange than does cellular connection.

4. Secretion of calcareous wall requires a source of nutrients.

5. An imperforate calcareous wall is impervious to sufficient physiologic exchange to permit further zooid growth.

6. Morphologic difference between zooids implies physiologic difference between them (e.g., in polymorphism and astogenetic differences).

7. Some zooids in a colony must be feeding autozooids. This and assumptions 2 and 5 require that a zooid lacking feeding organs be connected by tissues directly or indirectly to feeding autozooids.

8. Extrazooidal parts grow under colony control of local or colony-wide extent.

9. Correlated cyclic growth within a group of zooids is not necessarily a result of colony control but can be simultaneous separate responses to a cyclic environment.

An important step in understanding degrees of integration in Bryozoa was Silén's (1944a, 1944b) recognition of two methods by which the noncalcified layers of body walls grow. The distinction between these two kinds of walls, interior and exterior, has been further clarified by Banta (1969b). Interior walls partition preexisting body cavity, growing by edgewise extension (apposition) from other walls, which may be either interior or exterior. Exterior walls extend the body cavity of the primary zooid or zooids to become the outer wall of the colony, growing by simultaneous expansion and intercalation (intussusception). Walls of both types may be calcified.

The distribution through the colony of exterior and interior walls and the relative times of development of zooid walls and the contained lophophore and gut suggested to Silén (1944a) a hypothesis of evolutionary relationships among the three major groups of Bryozoa: Phylactolaemata, Stenolaemata, and Gymnolaemata.

According to Silén's hypothesis, the integration of zooids changed during bryozoan evolution. He assumed the ancestor of Bryozoa to have been a solitary lophophorate in which the feeding organs periodically degenerated and regenerated, as they do in Bryozoa. Silén inferred that the first Bryozoa had become colonial by regenerating more than one set of feeding organs within a common coelom which simultaneously expanded to make room for the increasing number of feeding units. Expansion was brought about by intussusceptive growth of the exterior walls of the primary zooid. This mode of growth corresponds to that in some phylactolaemates. The next stage visualized by Silén is one in which the common coelom came to be partitioned by interior walls so that each alimentary canal occupied a separate compartment, cut off from the common coelom only after the lophophore and gut had formed, as suggested for phylactolaemates and stenolaemates. The vertical walls of zooids in these two groups should thus be predominantly interior walls. The last stage of evolution was inferred by Silén to have been reached through localization of swellings in the coelom of the primary zooid to zooid-sized outpocketings and a delay in the formation of feeding organs until the zooid compartment was cut off from its parent by an interior wall. This is the situation in most gymnolaemates, in which the vertical walls of the zooid should thus comprise exterior lateral walls and interior transverse walls.

According to Silén's hypothesis, the earliest Bryozoa should have had the most highly integrated zooids, and in later Bryozoa the structural autonomy of the zooids should have been progressively stressed.

Whether or not interior and exterior walls evolved in Bryozoa as Silén hypothesized, their mode of growth suggests differing degrees of integration. The walls between living chambers of zooids may be either interior or exterior walls. Exterior walls are similar in mode of growth and morphology to those bounding a solitary individual. These walls potentially separate the zooid from the environment, thus expressing zooid autonomy. Interior walls, in their mode of growth and morphology, express colony dominance in that they do not have the potentiality for separating the zooid from the environment. The combinations of wall types listed in Series A occur in Bryozoa and are described in more detail in the separate sections on Stenolaemata and Gymnolaemata.

Series A: ZOOID WALLS. States arranged in order of increasing integration.
  1. Walls exterior.
  2. Vertical walls partly exterior, partly interior.
  3. Vertical walls wholly interior.
  4. Vertical walls incomplete.
  5. Vertical walls lacking.

Interzooidal connection by zooidal soft tissues is generally lacking in solitary animals, and its presence in colonial animals, therefore, expresses colony dominance. Connected coeloms of zooids or extrazooidal parts can exchange physiologic substances, and, on the basis of assumption 3, the nature of the connection expresses the different states of colony dominance, as listed in Series B. All states except 2, in which zooidal soft tissues are in contact but not shared, are known in Bryozoa.

Series B: INTERZOOIDAL CONNECTIONS BY ZOOIDAL SOFT TISSUE. States arranged in order of increasing integration.
  1. Soft tissues separated in space.
  2. Soft tissues in apposition.
  3. Connections by cells or tissues through mural pores.
  4. Connections by confluent body cavity around ends of vertical walls.
  5. Connections by confluent body cavity through vertical walls (or vertical walls absent).

The presence of extrazooidal soft and hard parts in a colony is an indication of a degree of colony control of growth (Series C), because extrazooidal parts are unique to colonial animals, on the assumption that a solitary individual is comparable to a zooid. Development of extrazooidal parts results in a further loss of zooid autonomy from the condition in solitary animals. In Bryozoa, extrazooidal parts are connective and/or supportive structures occurring outside zooidal boundaries. The variety of extrazooidal parts known in stenolaemate and gymnolaemate Bryozoa forms a transitional morphologic series approaching that of polymorphic zooids. In Bryozoa, extrazooidal parts can develop in three ways:
  1. These parts may be produced in later ontogeny by coalescence and/or resorption of tissues which were parts of zooids (see section on Gymnolaemata). After coalescence the physically continuous structures develop across zooidal boundaries, indicating increased colony control of growth in later ontogenetic stages.
  2. Extrazooidal parts may be entirely extrazooidal in origin, but formed subsequently to budding of zooids (see section on Stenolaemata), indicating colony control throughout growth of the extrazooidal parts and therefore a higher degree of integration.
  3. Extrazooidal parts may develop concurrently with budding of zooids as part of

the budding pattern (see sections on Stenolaemata and Gymnolaemata), indicating colony control of growth throughout development of the colony and therefore a still higher degree of integration.

Series C: COLONY DOMINANCE THROUGH EXTRAZOOIDAL HARD AND SOFT PARTS. States arranged in order of increasing integration.
1. Absent.
2. Formed subsequently to budding of zooids through coalescence and/or resorption of zooidal tissue.
3. Formed subsequently to budding of zooids, extrazooidal in origin.
4. Formed concurrently with budding of zooids as part of budding pattern.

It is assumed (see assumption 6 above) that morphologic difference between zooids implies physiologic difference between them. The astogenetic differences between zooid generations in a zone of change then can be assumed to have been developing toward a repeatable physiology as well as morphology. The similarity of patterns of astogenetic change in different conspecific colonies indicates that the general pattern is largely the result of a genetically controlled sequence of physiologic changes. The inferred sequence of physiologic changes is an expression of colony dominance, and generally is absent in solitary animals. In Bryozoa, states 1 and 5, Series D, are not known, and state 2 has not been recognized with certainty.

Series D: MORPHOLOGIC DIFFERENCES BETWEEN GENERATIONS OF ZOOIDS (ASTOGENY). States arranged in order of increasing integration.
1. All zooids of constant morphology.
2. First zooid or zooids different, all others without generational differences (primary zone of astogenetic change = first zooid(s) + first generation of asexually produced zooids; remainder = primary zone of repetition).
3. Generational differences between asexually produced zooids limited to proximal region of colony (primary zones of astogenetic change and repetition only).
4. Generational differences between asexually produced zooids present in at least one distal region of colony (subsequent zones of astogenetic change and repetition).
5. Generational differences between asexually produced zooids on gradient throughout colony.

Also on the basis of assumptions 6 and 7, polymorphic differences are an expression of colony dominance. Polymorphism is one kind of environmental response by the colony. In feeding, reproduction, and other basic processes, monomorphic zooids in a colony can respond virtually as individuals. The response of a polymorph, however, is through its contribution to the colony as a whole, in direct proportion to its functional specialization. All the states shown in Series E are known in Bryozoa.

Series E: MORPHOLOGIC DIFFERENCES WITHIN GENERATIONS OF ZOOIDS (POLYMORPHISM). States arranged in order of increasing integration.
1. All zooids of same generation of constant morphology.

2. Asexually produced zooids polymorphic, all having feeding and sexually reproductive ability.

3. Asexually produced zooids polymorphic, some lacking feeding or sexually reproductive ability.

4. Asexually produced zooids polymorphic, some lacking both feeding and sexually reproductive ability.

As suggested by Banta (this volume), another measure of dependence, in addition to functional or physiologic difference, is involved; this is a positional and structural dependence of polymorphs on other zooids. Polymorphs intercalated randomly in the colony budding pattern probably contribute their specialized functions as separate operating units. Those assembled in repeated groups of one or more kinds of polymorphs can carry out their specialized functions jointly. Intrazooidal polymorphs (alternating in morphology and corresponding function during the life of a zooid) and adventitious polymorphs (generally developed as appendage-like zooids adding functions to those of the supporting zooid) indicate a higher degree of structural dependence on the supporting zooid than polymorphs intercalated in the budding pattern. (See discussion by Schopf, this volume, for contrasting interpretation.) All the states shown in Series F are known in Bryozoa.

Series F: POSITIONAL DIFFERENCES WITHIN GENERATIONS OF ZOOIDS (POLYMORPHISM). States arranged in order of increasing integration.

1. All zooids of same generation of constant morphology.

2. Asexually produced zooids polymorphic, intercalated in the budding pattern randomly.

3. Asexually produced zooids polymorphic, intercalated in the budding pattern regularly or in single or repeated groups.

4. Asexually produced zooids polymorphic, intrazooidal or adventitious.

In the generalizations expressed above, a wide variety of states of integrative characters has been inferred to occur in Bryozoa as a whole. To illustrate these inferred states and their broad evolutionary history, it is convenient to present evidence in separate discussions divided by principal modes of growth within the phylum. These growth modes generally parallel the major taxonomic groups: Stenolaemata, Gymnolaemata, and Phylactolaemata. Because our experience has been with stenolaemate (tubular) and gymnolaemate (box-like) bryozoans, the following discussions are limited to these groups. (See Wood, this volume, for a review of phylactolaemate bryozoans.)

STENOLAEMATA

Stenolaemate bryozoans are characterized by cone- or tube-shaped zooids with terminal skeletal apertures which are protected by terminal–vestibular membranes in modern species. Elongation of zooids is a feature of ontogenetic development during which calcified body wall is added progressively at apertures. The functioning organs of feeding zooids of many stenolaemates maintain a relatively constant apertural position by means of degeneration–regeneration saltations. The outward growth of zooid wall and correspond-

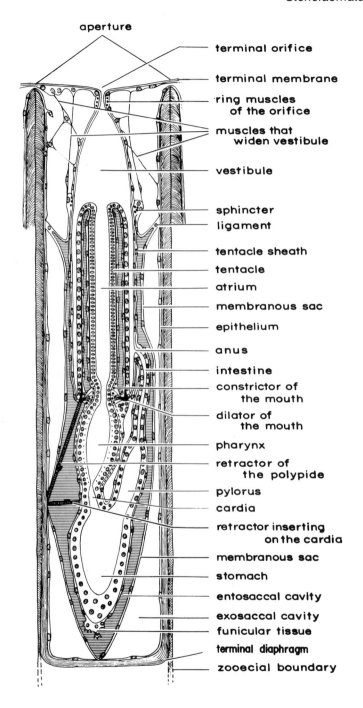

aperture

terminal orifice

terminal membrane

ring muscles
of the orifice

muscles that
widen vestibule

vestibule

sphincter
ligament

tentacle sheath

tentacle

atrium

membranous sac

epithelium

anus

intestine

constrictor of
the mouth

dilator of
the mouth

pharynx

retractor of
the polypide

pylorus

cardia

retractor inserting
on the cardia

membranous sac

stomach

entosaccal cavity

exosaccal cavity

funicular tissue

terminal diaphragm

zooecial boundary

Figure 6

Model of a trepostome autozooid based on organs of a Recent cyclostome (after Neilsen, 1970, fig. 13) and generalized autozooecial living chamber of Paleozoic trepostome.

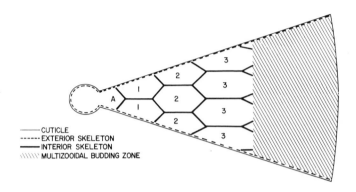

Figure 7

Idealized diagram of section parallel to basal layer of colony (after Borg, 1926, fig. 36) showing the position and extent of a confluent, multizooidal budding zone around the basal margin of a colony. The budding zone is not considered extrazooidal because its position will be occupied by zooids of the fourth and fifth generations as growth proceeds. A is the initial zooid or ancestrula.

ing advancement of functional organs is generally enough to vacate the inner part of the body cavity, which may contain the remains of the degeneration–regeneration process (brown bodies) and may be partitioned into abandoned segments by a series of transverse skeletal diaphragms (Figs. 6 and 13A). The last diaphragm formed functions as the floor of the living chamber. In at least a few genera of modern cyclostomes, the feeding and reproductive organs do not maintain a constant apertural distance outward into long, exterior-walled peristomes. The retracted position of these organs remains inward from the peristome in the body cavity and the vestibular membrane grows outward with the peristome to connect the feeding and reproductive organs with the terminal membrane at the peristomial aperture. In a few of these genera, and in some fossil forms such as most of the fenestratids, living space is so short that organs must have been retracted to constant positions throughout ontogeny. In modern species a membranous sac contains the feeding and reproductive organs of a zooid and is suspended from the skeleton inwardly by ligaments (Fig. 6) or other attachment organ (Boardman, in press). Remnants of membrane in fossils suggest the presence of analogous sacs in stenolaemates of widely different age and classification (see Utgaard, this volume; McKinney, 1969, p. 286).

In stenolaemate bryozoans, asexual reproduction of zooids occurs by the budding of interior zooidal walls from either exterior or interior supporting walls. Budding from exterior basal walls of colonies takes place in distal budding zones which are undivided at right angles to the growth direction and at any one time can extend distally for a distance sufficient to accommodate more than one generation of zooids (Fig. 7). A basal budding zone then can be considered multizooidal and is confluent throughout. Budding of zooids above basal incrusting walls in the erect parts of colonies most commonly takes place interzooidally from points formed by the junctions of established contiguous zooids at growing margins (Fig. 8). The coelomic space above the basal layer into which interzooidal buds project is not considered multizooidal because it is commonly referable to the individual zooids supporting the buds even though the space is laterally confluent. Erect zooidal budding also occurs from erect multizooidal budding structures

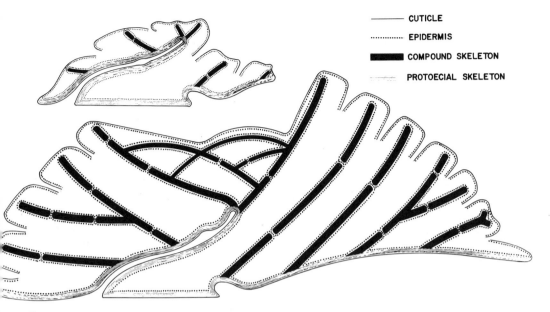

CUTICLE

EPIDERMIS

COMPOUND SKELETON

PROTOECIAL SKELETON

Figure 8

Idealized diagrams of longitudinal sections through center of young lichenoporid colony after Borg (1926, fig. 55). The central zooid with flattened basal disc (protoecium) is the ancestrula. The smaller diagram is a hypothetical earlier growth stage with no attempt made to establish actual relative rates of growth of different zooids. It shows early growth of interior compound walls from the basal protoecial skeleton by an infolding of epidermis. The fold back on itself of the protoecial skeleton provides the basal wall for autozooids to left of ancestrula. Notice potential for interzooidal connections through communication pores in compound walls and coelomic space around ends of zooidal skeletons under outer flexible wall.

of either exterior or interior origin in unilaminar or bilaminar growth habits (Figs. 9, 29, 30, and 35).

As presently understood (Ryland, 1970) the stenolaemates are divided into four orders, the Cyclostomata, Cystoporata, Trepostomata, and Cryptostomata. All the orders are known first from rocks of Middle Ordovician age, and all but the Cyclostomata are thought to have become extinct near the end of the Paleozoic Era. The Paleozoic orders provided a varied and abundant fossil record. Attempts to understand their basic biology, however, are just beginning (e.g., Blake, Utgaard, this volume; Cumings and Galloway, 1915; Tavener-Smith, 1969; Brood, 1970; Boardman and Cheetham, 1969; Boardman, 1971, in press; Gautier, in press).

The Cyclostomata were apparently a relatively unimportant order of few genera during the Paleozoic Era, became the dominant order during the Mesozoic Era after the Paleozoic orders declined, and have since been on the decline as the gymnolaemates became progressively more important. Studies of modern cyclostomes thus provide the major source of knowledge of soft anatomy and mode of growth for possible extrapolation to the three Paleozoic orders of stenolaemates.

Zooid Walls and Interzooidal Connections by Zooidal Soft Tissue

Kinds of zooid walls and means of interzooidal connections are so interrelated in modern cyclostomes and presumably in other stenolaemates as well that the integrative Series A and B are discussed together here.

The basic mode of growth in modern cyclostomes and inferentially in fossil stenolaemates is the double-walled growth concept of Borg (1926, fig. 55; 1933, fig. 26). Borg's double walls refer to the two-part division of body walls of a colony that enclose and partition the body cavity. The exterior body wall separates the colony from its environment and begins with the protoecial or proximal-most wall (Fig. 8) of the first zooid, the ancestrula. The protoecial wall may fold back on itself to the substrate as in the lichenoporids (Fig. 8) and grow outward to form the multizooidal basal wall of the colony (Boardman, 1971, pl. 3, figs. 1–4). The protoecial wall consists of an outer cuticle, calcified layer, epidermis, and peritoneum. The outer or bounding cuticle (Tavener-Smith and Williams, 1972, p. 107) apparently grows from within itself (intussusception) and is reinforced by calcified skeletal layer grown by apposition. The calcified layer is simple, that is, grown by the epidermis from the growing edge and inner side only. The exterior wall of the colony completes the enclosing of body cavity by extending beyond the margins of the protoecial or basal wall and over the remainder of the surface of the colony and consists of bounding cuticle, epidermis, and peritoneum that generally remains uncalcified (Fig. 28E, F). This flexible exterior wall is continuous with the vestibular walls of zooids (Fig. 6) so that it is generally attached to the skeleton of the colony in double-walled forms only around the rim of the basal wall and within the zooids. The sparseness of the skeletal attachment of the outer flexible wall allows for possible interzooidal communication around the ends of zooecia (Fig. 8) by confluent body cavity (state 4 of Series B).

The second wall of the double-walled cyclostome comprises the vertical walls of zooids, which function mainly to partition the existing body cavity of the colony into zooidal and extrazooidal spaces. The vertical walls are interior body walls (compound skeleton of Fig. 8; state 3 of Series A), which are entirely within the exterior wall of the colony. They are generally calcified and compound, that is, calcified from both sides by infolded epidermis. Available evidence indicates that epidermal infolding is done on inner sides of primary or secondary (Fig. 9) calcareous layers of exterior budding walls without involving the outer or bounding, cuticle so that this cuticle is not incorporated into compound interior walls of stenolaemates. The lack of a bounding cuticle is assumed (assumption 1 above) to indicate that vertical interior walls of zooids in cyclostomes and in other stenolaemates are unsuitable for protection against the environment, so that a single zooid removed from a double-walled colony would not be viable. The interior walls of zooids are without hindrance to continued elongation if the depositing epidermis of the interior compound walls does not join with the epidermis of the exterior flexible wall. This separated relationship of the two walls throughout the ontogeny of zooids of double-walled colonies produces some of the more extreme zooidal elongation typical of many stenolaemates.

Other kinds of double-walled modern cyclostomes, such as the hornerids (Borg, 1944, p. 175, text fig. 23), have poorly understood differences in arrangement of the earliest

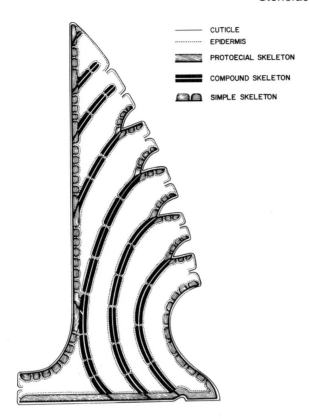

CUTICLE
EPIDERMIS
PROTOECIAL SKELETON
COMPOUND SKELETON
SIMPLE SKELETON

Figure 9

Idealized diagram of longitudinal section through center of hypothetical single-walled cyclostome colony showing basal disc (protoecium) of initial zooid (ancestrula at lower right). The protoecial skeleton extends to left to form the basal skeletal layer of colony. The protoecial skeleton and erect budding wall (left side of figure) are calcified from edges and inside only (simple skeleton) and provide surfaces for the budding of compound interior walls which partition the colony into zooids. At distal terminations of compound walls (to right) a simple zooidal skeleton is added which combines with the erect budding wall to provide a reinforcement to the cuticle around the outside of the colony. Notice that interzooidal connections around the ends of the zooids are shut off by the addition of the simple zooidal skeleton. The zooid at lower left is a basal polymorph (Fig. 29B). The heavier black line on outer side of protoecial and simple skeleton, and the centered white line of the compound skeleton, mark the positions of the primary layer of the skeleton. The diagrammatic pores in the simple skeleton are pseudopores, some of which may penetrate the primary skeletal layer but not the cuticle. The cuticle is contiguous on outer surface of both protoecial and simple skeleton.

zooids from that of the lichenoporids. These different beginnings, however, all develop interior skeletal walls and an exterior flexible wall over the entire colony.

Pores occur commonly in the compound interior walls between zooids of post-Paleozoic cyclostomes (Figs. 8 and 9). The pores have generally been interpreted as open communications between adjacent zooids that allow the passage of epidermis and peritoneum between zooids (Borg, 1926, p. 199). The openness of these interzooidal pores has been questioned (Brood, 1972, p. 45, 64) and many communication pores are closed by a central calcified septum. Many are open in the same colonies, however, and occupied by large cells (Fig. 28D), as reported by Borg. The frequency and pattern of open and closed inter-zooidal pores has not been determined, and their effectiveness as communication pores is not known. It is assumed here, however, that since many are open, they function as communication pores at some time during the life of adjacent zooids. The potential for interzooidal communication through pores and around the ends of zooecia is the rule in post-Paleozoic double-walled cyclostomes (combination of states 3 and 4 of Series B).

A second growth pattern among modern cyclostomes (Borg, 1926, p. 191, fig. 1) has been termed "single-walled." The single wall refers to the development by each zooid of an outermost segment of exterior zooidal wall consisting of outer cuticle, simple skeletal layer (the simple zooidal skeleton of Fig. 9), epidermis, and peritoneum. The exterior zooidal wall is attached to, or is a continuation of, part of the compound, interior zooidal wall (Brood, 1972, p. 44, 45). Single-walled colonies apparently begin growth in the same way that double-walled colonies do with the two skeletal elements, the exterior basal (protoecial) and interior compound walls, recognizable in varying configurations in both kinds of cyclostomes (Borg, 1926, p. 197). The growth of a simple zooidal skeleton (grown from one side only by one zooid) follows ontogenetically the development of compound interior walls, as indicated by the general lack of simple walls in distal budding zones and their progressive development on the ends of compound walls proximally in a colony (Figs. 9 and 28C).

The termination of compound skeletal growth and the beginning of simple skeletal growth is reportedly triggered by the joining of the epidermis of the compound wall with that of the outer flexible wall (Borg, 1926, p. 258). After contact, the bounding cuticle of the outer wall apparently grows outward by intussusception from the ends of the compound walls, followed by the contiguous growth of the simple skeletal layer by apposition. This combined growth gains coelomic space for the zooid and therefore forms a segment of exterior wall that is attached to a vertical interior wall (state 2 of Series A). The cuticle of the terminal–vestibular wall of a zooid is laterally continuous with the bounding cuticle of the exterior zooidal wall segment so is suspended from the simple zooidal skeleton, shutting off confluent zooidal connections around the ends of the zooecia. This limits the possibility of interzooidal connection in single-walled post-Paleozoic cyclostomes to the communication pores in compound walls (state 3 of Series B).

The microstructure and methods of joining of the parts of the skeleton of a single-walled cyclostome reveal much about distribution of soft parts, mode of growth, and inferred function. Some unilaminar single-walled colonies develop an erect, exterior budding wall from which the interior, zooidal compound walls are budded (Figs. 9 and 29B–E). The microstructure of the erect budding wall and simple zooidal walls is comparable and the two walls join to form a continuous skeletal reinforcement of the exterior cuticle except over zooidal apertures and distal budding zones.

In many cyclostomes the skeletal wall can be divided into two major layers based upon differing microstructure and position (Söderqvist, 1969, p. 116). The exterior erect budding walls and simple zooidal skeletons contain a thin outer layer that can appear hyaline, granular, or laminated, and vary from colorless through shades of brown pigmentation. This layer is traceable in some cyclostomes from budding wall into a similar-appearing median layer in the compound interior walls, and from there into the simple zooidal skeleton (Figs. 29C–E and 32D). Because of its outer and medial positions the thin layer can conveniently be considered the primary skeletal layer (Tavener-Smith and Williams, 1972, p. 124, 129). Lining the body cavities defined by the primary layer is the secondary layer that is commonly laminated, colorless, and thicker. In the compound interior walls, the primary layer is in the boundary position between adjacent zooids, and the secondary layer is generally referable to the zooid that it lines. Communication pores in interior compound walls and pseudopores in exterior simple walls generally penetrate the secondary skeletal layer. Pseudopores (Borg, 1926, p. 198) are commonly closed at outer ends by the cuticle or primary layer (Fig. 32E) of the exterior wall, and calcareous septa that close communication pores are in the position of the medial primary layer.

An extreme development of the brown pigmentation of the primary layer is seen in species of two genera, a tubuliporid (Fig. 32D) and *Diplosolen* (Fig. 30D, E), from the same collection from the Kara Sea area. These unusually dark primary layers are only partly soluble in hydrochloric acid so presumably contain a high percentage of organic material, suggesting that primary layers of varying shades are admixtures of calcite and high proportions of organic material. Nonlaminated structures in Bryozoa of Paleozoic age, such as acanthopore cores (see Blake, this volume), lunaria (see Utgaard, this volume), and granular (primary) walls of the Fenestrata (Tavener-Smith, 1969, text-fig. 3), typically contain pigmented particles which might well be modified remains of a high organic content in life.

Interior compound walls give rise to exterior simple walls by dividing into halves along zooidal boundaries in many cyclostomes (Fig. 32D). Part of the primary layer and one side of the secondary layer diverge sharply from the other half of the compound skeleton on all sides of each zooid to form the simple skeleton of the exterior zooidal wall and peristome. In the tubuliporid illustrated, the only microstructural change observed in the transition is the expected change from communication pores in the compound zooidal skeleton to pseudopores in the simple zooidal skeleton. Both types of pores are about equally scattered.

In another common microstructural pattern of change from compound to simple zooidal skeleton, the compound wall seems to end abruptly and the simple skeleton of the zooid is fastened with a visible structural break onto the end of the supporting compound wall related to that zooid (Figs. 29D and 32C). This type of simple zooidal wall forms the exterior zooidal wall and a peristome if present, and commonly contains closely spaced pseudopores. Transitional concentrations of pseudopores among species and both continuous and discontinuous methods of developing simple zooidal walls within a single colony (Fig. 30D, E) suggest that these two common types of microstructure in modern cyclostomes might be extremes of one highly variable microstructure. A majority of the described genera still have not been studied internally.

A measure of flexibility in both mode of growth of simple zooidal skeleton and its taxonomic significance is suggested by a more complex microstructure in the species *Heteropora pacifica* Borg, 1933. Modern cyclostomes have been divided into five equal divisions above the family level (Borg, 1944, p. 19), two divisions of single-walled forms and three divisions of double-walled forms. Borg used the genus *Heteropora* de Blainville, 1830, as the nominate genus for one of his double-walled divisions, the Heteroporina (1944, p. 208).

Borg (1933, p. 284) used the long peristome development in *Heteropora pacifica* (Fig. 31A, C) as a diagnostic character of the species and apparently assumed that the peristomes were grown under an outer coelom and membranous wall as apparently happens in hornerids and other double-walled forms. Sections of a topotype zoarium of *H. pacifica* from the same Middleton Island, Alaska, collection that produced two of Borg's primary types indicate that the long peristome development actually constitutes a single-walled growth.

The topotype zoarium has terminal diaphragms (Nye, 1969, p. 112) which close the smaller polymorphic zooecia and some zooecia of autozooidal size just below distal budding zones (Fig. 31D). The diaphragms seem to have grown by edgewise growth (Boardman and Cheetham, 1969, p. 210, text-fig. 2B), the calcification of the diaphragm developing along its edge and inner side which gradually seals off the aperture. Laminae of the diaphragms are either fastened directly to the ends of compound zooidal walls with structural discontinuity, or are flexed inwardly at junctions with walls, suggesting growth both from diaphragm edges and inner surfaces. There is no indication of growth on outer surfaces of terminal diaphragms, where thin contiguous cuticle can be seen in this and other specimens of cyclostomes.

The nature of terminal diaphragms is further suggested by the common occurrence in them of pseudopores in both single- and double-walled cyclostomes (Fig. 32A, B). Pseudopores widen and open on inner sides of terminal diaphragms or walls and narrow outwardly and are closed on outer sides by either skeleton and contiguous cuticle or cuticle alone (Tavener-Smith and Williams, 1972, figs. 96, 104). Borg reported no pseudopores anywhere in double-walled forms (1926, p. 199). Without recognition of pseudopores in terminal diaphragms of double-walled forms he logically suggested that terminal diaphragms formed under the double walls as part of the system of interior walls and could therefore grow from both sides (Borg, 1933, text-fig. 26).

The microstructure, pseudopores, and contiguous cuticle in terminal diaphragms compare with exterior simple walls of single-walled cyclostomes. An alternative growth model then for double-walled cyclostomes suggests that the double membranous walls and outer coelomic space are discontinued immediately over terminal diaphragms and other simple exterior walls which contain pseudopores such as brood-chamber covers in many double-walled cyclostomes. Growth of the simple skeletal layers is inferred to be on developing edges and inner sides of those layers only. Several membranes that might be multiple layers of cuticle can be developed outside of subterminal diaphragms in zooids that are in a degenerate stage, suggesting considerable flexibility in cuticle formation (Fig. 32B). Also, two terminal diaphragms can develop within the same zooid, the inner one later than the outer one, suggesting further flexibility (Fig. 32A). There seems no reason to assume that a contiguous cuticle could not have occurred on the outer surface of the inner diaphragm.

In the topotype of *H. pacifica,* the terminal diaphragm can modify its function and become a segment of outer zooidal wall that is skeletally comparable with the simple zooidal wall of a single-walled cyclostome (Fig. 31B). The only evident control in the change from a terminal diaphragm to simple zooidal wall is the angle which the zooid makes with the surface of the colony. If the compound walls of the zooid intersect the colony surface at nearly right angles a terminal diaphragm can be formed, the common condition in double-walled cyclostomes. If the compound walls of the zooid are more nearly parallel to the colony surface where they intersect, a segment of simple zooidal wall is needed to maintain living chamber shape, the common condition in single-walled cyclostomes.

Many of the autozooids of the topotype colony developed long peristomes which are distal extensions of the zooecial linings of the compound walls (Fig. 31A). If the peristome wall grew in the same way that the zooecial lining grew, growth was on outer edges and inner surfaces only. This is consistent with the scattering of pseudopores in the peristome wall and the continuation of exterior cuticle from the outer surfaces of terminal diaphragms in adjacent zooids to the outer surface of the peristome. Further flexibility of skeletal growth is indicated by the short segment of terminal diaphragm followed distally by zooecial lining which in combination form an autozooecial peristome (Fig. 31C). Such apparently opportunistic skeletal combinations suggest that the epidermis of the interior walls of the zooids and the epidermis of the colony-wide outer membrane can combine to grow different kinds of exterior wall as needed from zooid to zooid.

The development of zooidal wall segments and peristomes as simple exterior walls makes *H. pacifica* a single-walled species both morphologically (state 2 of Series A) and functionally (state 3 of Series B). Communication around ends of autozooecia with peristomes is assumed to be ruled out in the species. If terminal diaphragms are simple exterior walls with contiguous bounding cuticle, species such as *Heteropora pelliculata* (see Borg, 1926, pl. 3, fig. 4, pl. 4, fig. 2), in which functioning autozooids are isolated from each other by other zooids covered by terminal diaphragms, are also functioning as single-walled cyclostomes. The assignment of these species to a double-walled genus can certainly be questioned, but the alternative question should also be considered in future classifications: in a phylogenetically based classification is there evidence that single- and doubled-walled species can be considered to be congeneric?

Biologic understanding of fossil Bryozoa is based primarily on comparisons of skeletal morphology and microstructure with modern Bryozoa (Boardman and Cheetham, 1969, p. 210, 218). The amount of detail and confidence in inferences concerning fossils is generally in proportion to similarity with modern forms. In less comparable fossil forms, growth surfaces are assumed to parallel skeletal surfaces and, less commonly, skeletal laminae. Dissimilarities are many, confusing, and expected early in the fossil record of the Bryozoa, which started nearly 500 million years ago. Combinations of morphology and microstructure can also be strikingly comparable with those of modern forms. These similarities apparently reflect restrictions placed on colony integration and evolution by the stage of development achieved by the Bryozoa within the animal kingdom.

Some of the simplest, least integrated zoaria which have been placed in the stenolaemates are the corynotrypids (Bassler, 1911). They reportedly ranged from the Ordovician to the Cretaceous periods, but specimens of post-Devonian age are apparently rare and need to be restudied. The corynotrypids are incrusting and uniserial. Zooecia are club-shaped, narrow, or stoloniferous at proximal ends and inflated at the distal, apertural

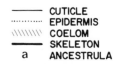

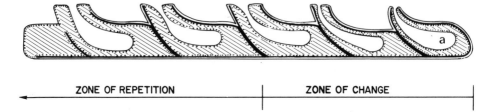

Figure 10

Idealized diagram of longitudinal section through center of young colony of the stomatoporid growth habit. Parts of zooidal walls which separate living chambers from each other are interior compound walls. Parts of zooidal walls which separate living chambers from the environment are exterior simple walls that correspond with simple protoecial and zooidal layers in Fig. 9.

ends (Fig. 33A, B). The zooecial wall is apparently simple and exterior throughout (state 1 of Series A).

A mural pore connects adjacent zooecia, permitting parent–daughter asexual reproduction and presumably communication between zooids (state 3 of Series B). Corynotrypids are so generalized and different from modern stenolaemates that perhaps the best evidence of their bryozoan affinities is through known cheilostomes of similar skeletal arrangement (Fig. 15B), although there are several major morphologic differences between the two groups.

The stomatoporids are an informal grouping of single-layered, incrusting, uniserial to multiserial cyclostomes which range from the Ordovician to the present. Zooecia consist largely of simple exterior walls but developed vertical compound interior walls (Figs. 10 and 33E, G; state 2 of Series A) that partition living chambers. The generalized growth habit can develop complications in patterns of budding in the interior walls (Illies, 1963, 1971) or different microstructure in skeletons. Ordovician forms developed frontal walls with or without pores (Fig. 33D, F, G). It is assumed that the pores in the Ordovician species were pseudopores even though any closing cuticle has been removed during fossilization. The skeletal walls containing the pores appear simple rather than compound and the pores open inwardly with increasing diameters, so that exterior skeletal walls compare structurally with those of single-walled cyclostomes in post-Paleozoic species which have been called stomatoporids (Fig. 33C, E) and which have pseudopores. Communication pores have not been found in the few Paleozoic specimens studied, however, and no communication is assumed around outer ends of zooecia because of the inferred simple zooidal walls. Apparently no provision was made, therefore, for interzooidal connection by soft tissue after calcification in Paleozoic stomatoporids,

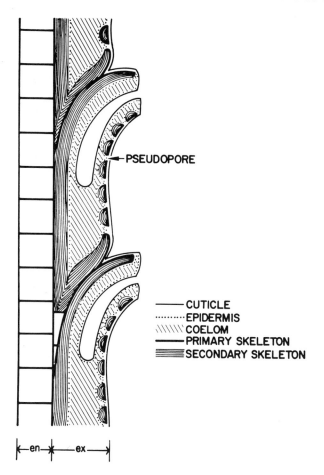

Figure 11

Idealized diagram of longitudinal section through two autozooids of *Crownopora*. Living chambers are restricted to the outer, thick-walled exozonal (ex) part of colony. The thin endozonal (en) skeleton is interior, dark, and granular, and can be traced into the exozone along zooecial boundaries to outermost calcareous layers of simple exterior walls (Fig. 34B). This dark granular layer then is comparable in position to primary skeletal layers of post-Paleozoic cyclostomes (Fig. 32D) and is considered primary in *Crownopora* also. The secondary skeleton in *Crownopora* is grown entirely on inner surfaces of autozooidal living chambers in the exozone. The parallel orientation of laminae in the secondary skeleton suggests that calcification was simultaneous over the entire inner surface of the living chamber, so that continued calcification would progressively reduce the volume of the living chambers. Inferred pseudopores are indicated in simple exterior zooidal walls.

indicating no colony integration for physiologic exchange among adult autozooids (state 1 of Series B). Communication pores do occur in interior compound walls of post-Paleozoic species studied, and the skeletal evidence for interzooidal connections is comparable with other single-walled cyclostomes (state 3 of Series B).

——— CUTICLE
············· EPIDERMIS
\\\\\\\ AUTOZOOIDAL COELOM
//////// EXTRAZOOIDAL COELOM
ⅢⅢⅢⅢ COLUMNAR SKELETON
▬▬▬ LAMINATED SKELETON

Figure 12

Idealized diagram of longitudinal section through autozooids on one side of a bifoliate branch of *Diploclema*. The corresponding skeletal connection between the autozooid in middle of diagram and the supporting median wall is either above or below the plane of the section for the lowermost autozooid in the diagram (see Fig. 35D of transverse section for clarification). The columnar skeleton is an exterior wall; the laminated walls are interior walls. The lowermost autozooid shows a terminal diaphragm without pores, interpreted from poorly preserved diaphragms seen in several autozooecia.

The Ordovician genus *Crownopora* Ross, 1967, is interpreted as another single-walled cyclostome of Ordovician age that appears to have developed adult autozooids that were skeletally isolated from each other (Fig. 11) and had no physiologic communication after calcification (state 1 of Series B). Communication pores could not be found in interior walls of either the endozone or exozone of the type species (Fig. 34). The compound interior walls in the exozone end outwardly by parting along zooecial boundaries to form simple, exterior zooecial walls as in single-walled cyclostomes (state 2 of Series A). The thick exterior skeletal layers have pores, and the laminae of the walls (Fig. 34F, G) indicate growth on the inner surfaces only with pores opening inwardly as pseudopores do in modern cyclostomes. Compared with modern single-walled cyclostomes, then, an outer cuticle is inferred that closed the pores at their outer ends making them pseudopores. The cuticle is inferred to have continued without break in both directions along the outside of the short peristomes (Fig. 34B) to the autozooidal apertures.

The diploclemids (Fig. 12) are a small group of bifoliate species that have been placed in the phylum (Ulrich, 1890, p. 368; Ross, 1964, p. 28; Kopayevich, 1972), although their morphology suggests a mode of growth different from other Bryozoa. They occur near the beginning of the bryozoan record and are known only for a short interval (Middle Ordovician to Middle Silurian age). Diploclemids are unique among bifoliate stenolaemates in that proximal ends of their apparent autozooecia (Figs. 12 and 35) arise from the simple, exterior skeletal wall through which autozooecial apertures open distally, rather than from median walls in the center of the zoarium. The exterior skeletal wall is continuous except for autozooecial apertures and covers both zooecial chambers and small portions of a colony-wide extrazooecial space (Fig. 35A, B). The exterior wall lacks pseudopores and has a columnar microstructure (Fig. 35E) that is unknown in that position in other stenolaemates.

The locus of budding and the sequence of development of the different parts of the skeleton is not known. If the autozooids budded at their proximal ends they could have grown distally in the same direction that autozooids of other tubular Bryozoa grow, but this hypothesis would require budding from an exterior wall already in place. Interpretations of zooidal development in cyclostomes suggest that exterior walls in a corresponding position are zooidal and are developed after a zooid is budded and has grown its interior wall. The exterior skeleton of the diploclemids might be interpreted then as an erect budding wall because it covers both zooidal and extrazooidal spaces and might have been the generative wall that produced new zooids.

A more conservative hypothesis suggests the possibility that budding occurred from the median wall along the midline of skeletal contact with the autozooids and that the interior walls grew within the extrazooidal space both proximally and distally out to a developing exterior zooidal wall. Proximal zooidal growth, however, has not been recognized in cyclostomes. Budding from adjacent autozooids at points of contact just above the median walls (Fig. 35H) might have been possible, but is mentioned here only because those are the only other skeletal contacts that an autozooecium makes.

The interior zooecial skeleton is laminated and apparently was grown from both the zooidal and extrazooidal sides. Both interior and exterior vertical walls are present (state 2 of Series A) if the exterior wall is zooidal and not a budding wall. Single mural pores occur on the reverse side of at least some of the autozooecia just above the median wall at both distal and proximal ends (Fig. 35E, F), apparently connecting zooidal and extrazooidal coelom (state 3 of Series B). No connection seems possible around ends of compound walls because of the simple columnar skeleton.

The overwhelming majority of Paleozoic Bryozoa display zooecia that have compound vertical walls without communication pores, and lack simple zooidal skeleton developed

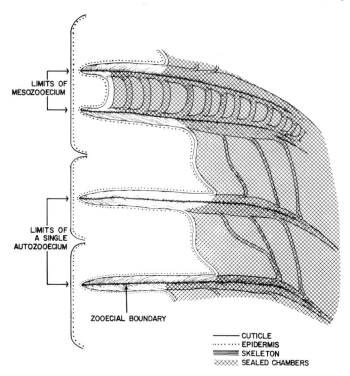

LIMITS OF
MESOZOOECIUM

LIMITS OF
A SINGLE
AUTOZOOECIUM

ZOOECIAL BOUNDARY

——— CUTICLE
········· EPIDERMIS
▓▓▓▓ SKELETON
▒▒▒▒ SEALED CHAMBERS

Figure 13A

Idealized diagram of longitudinal view through exozone of a mesozooid and two autozooids of a generalized trepostome, showing inferred position of outer flexible wall to left consisting of cuticle and epidermis, and position of epidermis lining the living chambers that presumably deposited the skeleton. Proximally, chambers are sealed by combinations of zooecial walls and calcareous diaphragms which both lack communication pores. Growth in chambers subsequent to sealing is thought to have been virtually stopped due to lack of physiologic communication.

distally. The inference is made that vertical walls were wholly interior (state 3 of Series A) and that these forms were double-walled with the possibility of interzooidal connection by confluent body cavity around the ends of the zooecia (state 4 of Series B). (See Blake, Utgaard, this volume; Brood, 1970; Gautier, in press; Boardman and Cheetham, 1969; Boardman, 1971).

The skeletal evidence of double-walled organization of body walls in Paleozoic Bryozoa has been reinforced by the recognition (Boardman, in press) of a bright reddish-brown membrane as the cuticle of the outer wall (Fig. 36A, B) in a dendroid trespostome of Late Ordovician age (see also Utgaard, this volume). The preservation of such a fragile and exposed membrane was presumably made possible by the progressive encroachment of an incrusting monticuliporid trepostome during the life of the supporting dendroid colony. This enveloping membrane appears to be continuous with brown deposits and fragments of membrane within zooecia that have the general shape of a lophophore and gut complex.

Many forms of trepostomes and cystoporates, especially, have one or more skeletal partitions or diaphragms (Fig. 13A), which in combination with a lack of communication pores divide the inner ends of zooecia into skeletally sealed compartments. The last-formed diaphragm is inferred to have been the floor of the living chamber (Boardman, 1971, p. 5). These compartments are considered to be without significant living tissue consistent with assumption 5 above, and evidence of skeletal growth from within a compartment is generally lacking. Living tissue in zooids containing basal diaphragms therefore is inferred to be limited virtually to outermost chambers and surfaces of the colony.

One group of double-walled Bryozoa of Paleozoic age (the ceramoporids of the cystoporates) does have observable communication pores and larger gaps in zooecial walls (Fig. 36C, D), so that interzooidal connections are inferred to have been through vertical walls as well as around ends. Basal diaphragms therefore did not necessarily form closed chambers that were physiologically isolated, and living tissues might well have existed throughout a colony. Consistent with this inference, terminal diaphragms have been discovered in the group that were grown from inner surfaces (Utgaard, this volume) similar to those of younger cyclostomes which have communication pores in vertical walls. The presence of inferred confluent connections around ends and through vertical walls of zooids makes the ceramoporids one of the most highly integrated groups of Bryozoa of any age relative to interzooidal connections (states 3 and 4 of Series A and 3–5 of Series B).

Extrazooidal Hard and Soft Parts

In the Stenolaemata a number of structures connect zooids and are readily interpreted as extrazooidal. As might be expected, however, interpretation of some structures as either zooidal or extrazooidal becomes difficult. Vertical skeletal walls of the zooids themselves are apparently without the precise physical boundaries of ingrowing cuticle so that zooidal boundaries have been recognized in many species (Boardman, 1960, p. 30) by symmetry of skeletal laminae rather than by acutal physical break. Moreover, with the physical continuity of depositing epidermis of interior walls throughout the growing surfaces of a colony, some zooidal structures can be expected to be transitional with extrazooidal structures and distinguishable only on position relative to zooidal chambers.

Some of the more obvious extrazooidal structures include the outermost laminated or fibrous skeletal layers typical of fenestrate bryozoans, which commonly form a secondary backing or covering on the reticulate branches of the zoaria. Early recognition of the outer secondary skeleton as extrazooidal is suggested by the use of the term "sclerenchyma" (Ulrich, 1890, p. 352). The growing edges of the outer extrazooidal layers are generally proximal to the growing tips of the zoaria (Tavener-Smith, 1969, text-fig. 7) where the zooids were budded (state 3 of Series C). After formation of the primary nonlaminated zooidal skeleton and laterally connecting dissepiments, the nonlaminated skeleton continued outward growth in many species in the form of skeletal rods and frontal carinal structures. These were formed concurrently with the enveloping extrazoodial laminated skeleton (Tavener-Smith, 1969, p. 295) by an inferred epidermis on the outer surface of the skeleton. Extreme developments of the extrazooidal skeleton resulted in the complete covering of zooids and massive supporting structures for the colonies (Fig. 36E–G). In the same way, extrazooidal skeleton develops on the reverse sides of many cyclostomes (e.g., hornerids; crisinids, Fig. 38E between pits) and cystoporatids (Fig. 38C, D). Calcification takes place on outer surfaces of skeletons under an outer coelomic space and an outermost flexible wall.

Another kind of structure that seems readily recognizable as extrazooidal is coelomic space within a colony of *Diploclema* (described above, Figs. 12 and 35) that is not assignable to zooids but occupies spaces between them. This extrazooidal space was formed concurrently with the zooids as part of the budding pattern at the growing edges of the colonies (state 4 of Series C). The presence of extrazooidal space between zooids is so rare in stenolaemates, however, that it is one of the characters that casts doubt on the assignment of *Diploclema* to the Bryozoa. The extreme expressions of the irregularity of mural pore size and walls of polymorphic mesozooecia of the ceramoporid genus *Acanthoceramoporella* Utgaard, 1968, produce skeletally open spaces that might be considered extrazooidal (Fig. 36C, D) around relatively well defined autozooids. These spaces do not appear until the exozonal development so are not produced at the budding surface by the budding pattern but by later disruption in wall development of the polymorphic zooids which surround the autozooecia (state 3 of Series C).

Extrazooidal skeletal material is recognizable in similar positions between zooids in their later ontogenetic development in many erect cystoporates, cyclostomes, and bifoliate and dendroid cryptostomes. Generally in these colonies, adjacent zooids become more widely spaced as they increase in length. In many cystoporates (see Utgaard, this volume) the spaces between the tubular zooids are filled with overlapping skeletal cysts (Fig. 37F). The cysts generally extend to outermost zoarial surfaces and are convex outwardly, allowing no recognizable spaces for relating a part of the cystose structures to any one of the adjacent zooids, so the structures are considered to be extrazooidal. In some cystoporates the extrazooidal cysts start at budding surfaces or a zooid may even be set entirely within extrazooidal cysts (Utgaard, this volume) so both states 3 and 4 of Series C can be applied.

Similarly, in some bifoliate cryptostomes, skeletal material in exozones has been differentiated into zooecial walls and range partitions (Karklins, 1969, p. 7, figs. 7, 8) which are considered extrazooidal (Karklins, personal communication). Zooecial boundaries are drawn at the first break or reversal of direction of zooecial laminae outward from the longitudinal axis of the zooecium (Fig. 37C, D). The skeletal material in range boundaries may be cystose in early ontogeny and change to solid skeleton in later stages (Karklins, 1969, figs. 7, 8) with sharp breaks in laminae at zooecial boundaries. In some forms neither the cystose structure nor the sharp zooecial boundaries develop, but there are reversals in laminae orientation between zooecial cavities nevertheless (Karklins, 1969, fig. 6). It seems consistent to consider the skeletal material in the regions between zooecial as extrazooidal, even though the zooecial boundaries are recognized without the benefit of physical breaks in structure. The extrazooidal skeleton in bifoliate cryptostomes may or may not arise as part of the budding pattern (states 3 and 4 of Series C). A comparable development of extrazooidal skeleton occurs between zooids in many double-walled cyclostomes (Fig. 37E). Growth of extrazooidal skeleton on its outer surfaces under an outer coelom and membranous exterior wall in both Paleozoic and modern species seems to be the key to the striking similarity between Paleozoic and modern forms.

Astogenetic Differences

In stenolaemate Bryozoa generational differences are commonly difficult to observe in proximal-most regions of colonies because colony founders grow in confusing three-dimensional masses or are covered by subsequent growth. As a result, little information is available (e.g., Borg, 1926, 1933, for modern cyclostomes; Cumings, 1904, 1905, 1912, for Paleozoic Bryozoa) on this region, which theoretically at least should contain morphologic data important to our understanding of stenolaemate biology, evolution, and classification. As presently understood, colony founders include the primary zooid (ancestrula) (Figs. 8, 9, 28A, B, and 29A) and the group of immediately budded zooids, which show morphologic differences in more or less uniform progression distally from generation to generation (together the zooids of the primary zone of astogenetic change, Boardman et al., 1970). The primary zone of change is followed distally in a colony by the primary zone of astogenetic repetition (Figs. 2 and 10), which includes zooids of endlessly repeatable morphologies and patterns of budding.

The development of subsequent layers of zooid growth on the zooids of primary zones of repetition is typical of double-walled stenolaemates of all ages. These overgrowths can originate from settling of a separate larva on the surface of the supporting colony, or they can originate from within a colony by continued growth of one or several zooids which initiate the overgrowth. Intracolony overgrowth is considered to form a subsequent zone of change (Boardman et al., 1970, fig. 6). The overgrowth does not have an ancestrula but does include the originating zooids from the primary zone of repetition and the first few generations necessary to establish the overgrowth. After the overgrowth is established, a subsequent zone of repetition can form (state 4 of Series D) which can remain incrusting on the supporting surface or can develop erect growth. In intracolony overgrowths it is assumed that the genetic makeup is the same for both the primary and subsequent zones.

Double-walled stenolaemates apparently have used intracolony overgrowths to cover patches of prematurely inactive or dead zooids in order to keep growing surfaces concordant. Also, the normal mode of erect growth of some double-walled colonies (Hillmer, 1971, p. 27) was achieved by developing cyclic intracolony overgrowths (Fig. 37A, B). Many zooecial cavities in the section of Fig. 37A can be traced from one overgrowth to the next.

In double-walled stenolaemates intracolony overgrowths are generally separated from supporting colony surfaces by basal skeletal layers away from their places of origin. It would be of interest to know if the basal layers also included subjacent bounding cuticles, as if the supporting surface were merely providing a foreign substrate, or whether the subsequent incrusting growth and further growth of the supporting colony could all take place under a common outer wall. Single-walled cyclostomes are less well known but intracolony overgrowths seem to be rare, suggesting that the development of simple exterior walls deprived the single-walled colony of a certain amount of flexibility in growth habit and in repairing damage to the colony (state 3 of Series D).

Morphology and Position of Polymorphs

In stenolaemate Bryozoa, some genera are characterized by colonies of monomorphic zooids, some by polymorphic zooids, and some genera include both monomorphism and polymorphism. In fossil stenolaemates not closely related to modern species monomorphic zooecia are assumed to have contained feeding organs for at least a part of their ontogenetic development. In related species containing polymorphic zooecia, the zooecia which compare in morphology with the monomorphic zooecia can be assumed to have been the skeletons of feeding autozooids (Boardman, 1971, p. 2).

Polymorphs are widely variable in morphology and position within stenolaemate colonies. Terms applied to differentiate kinds of polymorphs have been based primarily on soft-part morphology and assumed function in modern cyclostomes (e.g., nanozooid, kenozooid, gonozooid) or skeletal morphology and position within the colony in both modern and fossil stenolaemates (e.g., dactylaethra, firmatopore, nematopore, tergopore, mesozooecium or mesopore, exilazooecium or exilapore). Unfortunately, morphology and function are not as well known for polymorphs in modern cyclostomes as for accompanying feeding autozooids, and most polymorphs are so incompletely known that attempts to use some of the specialized terms seem to cause more confusion than understanding.

Nanozooids were named and their tissues and organs described by Borg (1926, p. 188, 232–239) from the smaller polymorph of the Recent and fossil genus *Diplosolen* Canu, 1918. Borg reported a small operational lophophore with a single tentacle and muscular system (the arrow on the right in Fig. 30A points to a cross section of membranous sac surrounding a tentacle sheath and single tentacle). The alimentary canal was found to be greatly reduced and reproductive structures were not observed, so no suggestion for function was made (presumably state 4 of Series E).

Nanozooids form an outer layer on bifoliate colonies in two collections studied of the genus *Diplosolen* (Fig. 30). Nanozooids are restricted to an outer position in the

---

Figure 13B

Idealized diagram of longitudinal section through autozooids and nanozooids (N) on one side of a bifoliate branch of *Diplosolen*. The autozooids all bud from the median skeletal layer (vertical wall on right). Nanozooids bud from between autozooids at a later growth stage as the early compound walls of the autozooids divide. Each wall is then shared by an autozooid and nanozooid. The heavier black line on outer side of the simple skeleton and the centered white line of the compound skeleton mark positions of the primary layer of the skeleton.

The skeletal arrangement of nanozooids may ultimately be given high taxonomic values if the concepts of single- and double-walled forms remain important in the classification of cyclostomes. *Diplosolen* is a single-walled cyclostome in that an exterior simple wall which reinforces a bounding cuticle and shuts off interzooidal communication around outer ends of zooecia is formed in a later ontogenetic stage. *Diplosolen* differs from the single-walled genera, however, in that the simple exterior walls, in addition to forming peristomes of feeding autozooids (Fig. 9), form the outer walls of nanozooids.

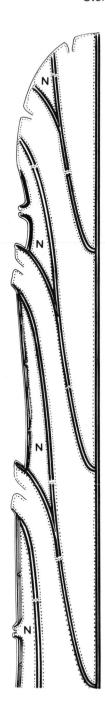

CUTICLE
EPIDERMIS
COMPOUND
SKELETON
SIMPLE
SKELETON

colony because they bud at a later ontogenetic stage between developing autozooids (Fig. 13B). The skeletal walls of the autozooids are compound except for short peristomes and are shared first with adjacent autozooids. A nanozooid is started by the autozooidal wall dividing into two compound walls (just below letter C in Fig. 30E), which separates contiguous autozooids and provides intervening space for the living chamber of the nanozooid (state 3 of Series F).

The living chamber of the nanozooid is closed outwardly by an exterior simple wall which begins its growth at the distal side of an autozooidal aperture (just above letters AZ in Fig. 30D). The compound wall divides into two simple walls; one forms a short peristome on the autozooid and the other grows distally to the proximal side of another autozooidal aperture to form the outer wall of the nanozooid. The living chamber of a nanozooid is long and flattened (Fig. 30A–C) and runs between and around autozooidal apertures (Borg, 1926, fig. 9). A small aperture with a peristome is commonly developed near the midlength of a nanozooid as the simple wall grows distally. This central aperture rather than a terminal one is apparently unique among polymorphs of stenolaemates. After calcification a nanozooid apparently must receive nutrients from adjacent feeding autozooids through communication pores. In return, the calcified simple exterior walls of the nanozooids collectively act as reinforcement to the external cuticle throughout the colony.

The term "kenozooecium" (Levinsen, 1902, p. 3) was defined as a zooid lacking ". . . both polypide and an orifice." The definition was later expressed as, "a chamber in which there is no polypide and as a rule no muscle and no aperture" (Levinsen, 1909, p. v.). Levinsen and later Borg (1926, p. 239–241) applied the term to supporting structures of stenolaemate colonies such as rootlets, rhizoids, or spines of the Crisiidae. Borg tentatively extended the concept of kenozooid to polymorphs occurring regularly above the base of colonies (Borg, 1926, p. 241) and the term has come to be used in stenolaemates for any zooid that is known or assumed to be without polypide, muscular system, and opening in terminal wall (state 4 of Series E) regardless of skeletal character. The smaller polymorphs regularly budded throughout colonies of Recent species placed in the genus *Heteropora* (lower three zooids of Fig. 28E) were considered kenozooids by Borg (1933, p. 368). The basal polymorph (Fig. 29B) could be considered a kenozooid if found to contain only mesenchymatous cells.

Application of Series E, concerned with major functions of organs, is impossible or speculative at best for those polymorphs in modern and fossil stenolaemates which are known only by their skeletal characteristics (Hillmer, 1971, p. 21). Using as starting points the concepts of nanozooid and kenozooid from modern cyclostomes, it can be inferred that a closely tabulated mesozooecium (Figs. 13A and 38F) found commonly in early Paleozoic species is the skeleton of a kenozooid that lacked both feeding and sexually reproductive ability (state 4 of Series E) only because its living chamber (outward from the outermost diaphragm) was too small to house even a smaller polypide such as that of a nanozooid. Exilazooecia, more common in later Paleozoic species, are tubes comparable in size to mesozooecia but lack closely spaced diaphragms. Living chambers of many exilazooecia are large enough to have housed a smaller polypide. Although smaller polypides with feeding or reproductive functions are not known in modern cyclostomes, it does not seem useful to speculate about possible functions of possible organs in Paleozoic forms.

"Dactylethra" is a term proposed by Gregory (1896, p. 12) for ". . . a form of aborted 'zooecia' consisting of short caecal tubes, closed externally." They were reported to occur in *Terebellaria,* a Jurassic genus. Dactylethrae have subsequently been interpreted as kenozooids (e.g., Bassler, 1953, p. G9; Brood, 1972, p. 49). Gregory's drawing of an exterior view (1896, pl. 10, fig. 5), photographs of thin sections from the type suite (Walter, 1969, pl. 8, figs. 8–10), and study of topotypes in USNM collections, all of the type species of *Terebellaria,* indicate that dactylethrae are zooecia similar in morphology to adjacent open zooecia (Tavener-Smith and Williams, 1972, p. 132) except that they are covered by terminal diaphragms containing abundant pseudopores. Neighboring open zooecia must have housed feeding organs. Covered and open zooecia are arranged spirally in zones around the branch; zooecia in the growing margin of any one whirl of the spiral are open; proximally they are all covered. The simplest interpretation is that during earlier growth stages the closed zooecia were open and parts of feeding zooids which were subsequently covered, as commonly happens in modern cyclostomes (Brood, 1972, p. 47).

Various tubular voids occur in extrazooidal skeletons of stenolaemates of all ages. These pits or vacuoles are generally too small to be considered zooids (Borg, 1941, p. 15) and are thought to have functioned only as pores in modern double-walled crisinids (smaller arrow on the left side of the specimen of Fig. 38E) and hornerids (smallest set of cavities on the right side of the specimen of Fig. 28F). The pits were inferred to have made possible necessary exchanges of gases through connecting communication pores with adjacent zooids as thickened secondary (extrazooidal) walls grew (Borg, 1941, p. 15). Larger pits of similar appearance occur in extrazooidal skeleton on the reverse side of the zoarium (Fig. 38C, D) in an undescribed genus of early Paleozoic cystoporatid. The obverse side contains pit-like structures of comparable diameter which would generally be interpreted as exilazooecia because they are long and parallel to the autozooecia. The genus lacks communication pores through interior walls, so apparently any exchange with zooids would have to have been through the coelomic space between the outer surface of the skeleton and the outermost flexible wall.

A larger set of tubes opens between the pits on the reverse side of one of the modern crisinids (larger arrow, Fig. 38E); these are large enough to be considered skeletons of zooids, but their soft parts are unknown. Polymorphs of similar appearance occur on the reverse side of zoaria of the type species of the double-walled lower Paleozoic genus *Pseudohornera* Roemer, 1876 (arrows, Fig. 38A, G). This genus lacks communication pores so that exchange with presumed feeding zooids on the obverse side of the colony would presumably be through the coelomic space between the skeleton and the outermost flexible wall. Again it seems pointless to speculate about possible functions of possible organs.

Examples of intrazooidal polymorphism (state 4 of Series F) have been inferred from skeletal structures of stenolaemate Bryozoa of Paleozoic age (Utgaard, this volume; Boardman, 1971) and are apparently unknown in modern cyclostomes. A three-part polymorphism within a single zooecium is suggested by the skeletal characteristics of the Ordovician–Silurian genus *Calopora* Hall, 1851 (Fig. 38F). Autozooecia characteristically started development as small, polygonal, closely tabulated mesozooecium-like tubes intercalated among larger established autozooecia. These smaller tubes with their smaller living chambers opened at the growing ends of colony branches. After developing several early diaphragms, these thin-walled mesozooecia increased to the full autozooidal size and shape inferred to have housed feeding organs. The third stage of intrazooidal polymorphism is suggested by the development within the autozooidal living chamber of a space-restricting flask-shaped chamber (the outward opening funnels of the chambers are shown in the autozooecia at the top and bottom of Fig. 38F). Functions for the first and last polymorphs in the ontogenetic sequence can only be speculative.

Monticules (or maculae) are clusters of a few polymorphic zooecia which are commonly expressed on zoarial surfaces as prominences or depressions. Monticules are characteristic of one order of stenolaemates, the trepostomes, and are common in most other major groups of Paleozoic Bryozoa. Monticules are fairly evenly spaced in zones of repetition (state 3 of Series F). Monticular polymorphs differ from surrounding intermonticular autozooecia in one or more characters, such as budding pattern, size and shape of cross-sectional area, zooecial wall thickness, concentration of acanthopores within and between walls in exozones, distinctness of zooecial boundaries, and spacing and configuration of intrazooecial structures such as basal diaphragms or cystiphragms lateral to living chambers. Monticules commonly seem to grow as units which disrupt the regular development of intermonticular autozooids immediately around them, causing some differential crowding and irregularity of cross-sectional shapes of adjacent autozooids.

There is little evidence for the function of monticules or their constituent polymorphs. Most monticules contain some large zooecia with ample living chambers for feeding or sexual organs, and many monticules contain concentrations of mesozooecia with no significant living chambers or exilazooecia with extremely narrow ones. Symmetrical surface expression suggests the possibility of monticular control of feeding currents (personal communication, W. C. Banta). Living chambers of presumed feeding autozooids as delineated by cystiphragms in some genera of monticuliporids are arranged around a monticule on nearer or farther sides relative to the center of the nearest monticule, further suggesting some cooperative action among intermonticular autozooids and monticular polymorphs. The constituent zooecia and surface expression of monticules vary so widely that different monticules may well have had different functions.

Monticules empirically fit the concept of cormidia or "colonies within colonies." Cormidia are regularly recurring groups of polymorphs within a colony which fulfill either a certain function or ". . . most of the vital functions of the colony . . . ." (Beklemishev, 1970, p. 490). Monticular polymorphs of extreme expression at least should have had, separately or collectively, a certain function or group of functions different from those of intermonticular zooids (assumption 6 above). The presence of monticules suggests a loss of individual autonomy and corresponding increase of colony dominance.

Colony Dominance through Time in Stenolaemates

Internal structures of the majority of stenolaemate genera are so poorly known that data on integrative characters as outlined in the six series of this paper are generally available only through restudy of specimens. It is not feasible to undertake a comprehensive restudy at this time, so taxa included in this first attempt at colony analysis have been chosen to illustrate some of the salient features of colony dominance through time.

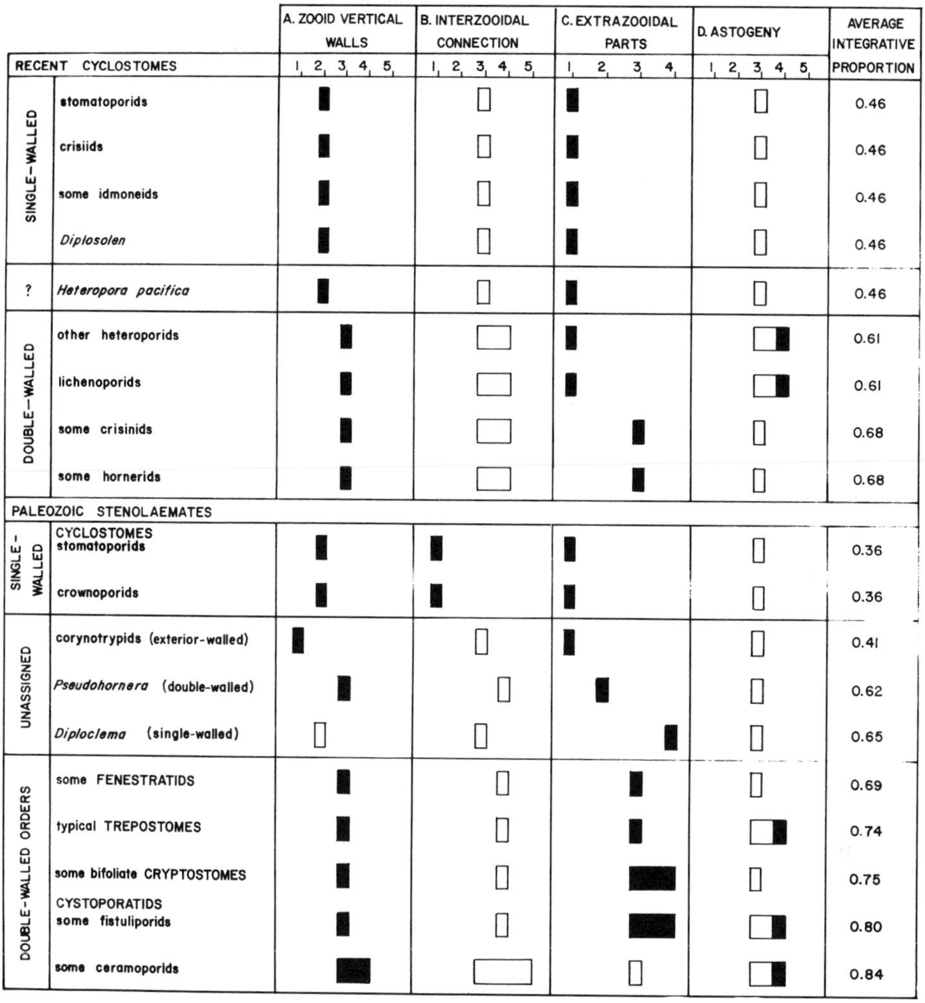

	A. ZOOID VERTICAL WALLS	B. INTERZOOIDAL CONNECTION	C. EXTRAZOOIDAL PARTS	D. ASTOGENY	AVERAGE INTEGRATIVE PROPORTION
RECENT CYCLOSTOMES	1 2 3 4 5	1 2 3 4 5	1 2 3 4	1 2 3 4 5	

RECENT CYCLOSTOMES					
SINGLE-WALLED — stomatoporids					0.46
crisiids					0.46
some idmoneids					0.46
Diplosolen					0.46
? — *Heteropora pacifica*					0.46
DOUBLE-WALLED — other heteroporids					0.61
lichenoporids					0.61
some crisinids					0.68
some hornerids					0.68

PALEOZOIC STENOLAEMATES

SINGLE-WALLED — CYCLOSTOMES stomatoporids					0.36
crownoporids					0.36
UNASSIGNED — corynotrypids (exterior-walled)					0.41
Pseudohornera (double-walled)					0.62
Diploclema (single-walled)					0.65
DOUBLE-WALLED ORDERS — some FENESTRATIDS					0.69
typical TREPOSTOMES					0.74
some bifoliate CRYPTOSTOMES					0.75
CYSTOPORATIDS some fistuliporids					0.80
some ceramoporids					0.84

Figure 13C

Summary of integrative states of stenolaemate Bryozoa discussed in text. States of each character arranged as in Series A–D. States observed in one or more taxa in each group are indicated by solid rectangles; those based only on inference or speculation are indicated by hollow rectangles. The average proportion of complete integration for the four characters is given in column to right. The time period of the Paleozoic Era was approximately 600 to 220 million years ago; the earliest bryozoans known lived approximately 500 million years ago.

Internal morphology and function of polymorphs are so poorly known and apparently so complex that even the salient features are obscure, and Series E and F are not considered in the summary chart (Fig. 13C). There is just enough information on structural relationships and taxonomic distribution of sexual polymorphs, for example, to suggest complex and widely varying patterns which will constitute a major study in themselves.

An understanding of colony dominance in stenolaemates is further complicated by the complex taxonomic diversity represented by four, possibly five, orders distributed through approximately 500 million years, nearly all of Phanerozoic time. Higher taxonomic groups are inadequately known and no attempt has yet been made to recognize broader phylogenetic relationships or homologous structures from Paleozoic to modern faunas. At present, all that can be done in stenolaemates is to recognize similar morphology that occurs widely spaced in time and to speculate about its significance in the interrelated colony development, natural selection, and evolution.

The single- and double-walled modes of growth of stenolaemate colonies and their accompanying exterior and interior body wall arrangements can be interpreted as controlling the morphologic expression of interzooidal connection, extrazooidal parts, and subsequent astogenetic zones. The vertical walls of zooids and extrazooidal skeletal parts are essentially structural features of a colony. Interzooidal connection through or around body walls or through extrazooidal coelom is assumed primarily to express physiologic integration. Astogeny is considered in stenolaemates to be partly structural and partly functional, and polymorphism, if considered in this summary, would primarily express functional integration. It is assumed that the colony functions as a unit in nature and that the different combinations of these integrative characters (Fig. 13C) represented compensating tradeoffs which produced species of varying success through time.

Calcified simple exterior walls have the structural advantage of providing support to the outer cuticle at the colony–environment interface and theoretically would be suitable as outer body walls of solitary animals. Exterior walls have the disadvantage, however, of restricting interzooidal connections where they are attached to the outer ends of compound interior walls. Further, extrazooidal skeleton and subsequent zones of astogeny, by themselves advantageous to colonies and indicative of colony integration, are grown essentially by addition of skeleton to outermost surfaces of established skeletal material. In single-walled forms, the lack of outer depositing epidermis and interconnecting coelomic space apparently makes these kinds of growth difficult if not impossible to achieve.

Interior vertical walls apparently have no capability to form a protective body wall at the colony–environment interface because they lack a bounding cuticle. They would not be suitable for body walls of solitary animals and therefore are an expression of colony integration. Interior skeletal walls have the advantage to the colony, however, of presumably premitting interzooidal communication around their ends and especially in post-Paleozoic forms through communication pores. Interior walls also provide a depositing epidermis on outer skeletal surfaces that is nourished by outer coelom, so that wide areas of extrazooidal skeleton are possible between feeding zooids, thereby adding structural strength to erect colonies. In the same way, interior walls provide remarkable flexibility in growth response to microenvironmental accident or cyclic growth by their ability to develop patches of intracolony overgrowth and subsequent astogenetic zones. Speculation suggests that in double-walled forms microenvironmental accident to the fragile membranous exterior walls has been common enough to require a ready response to injury for survival.

Double-walled stenolaemates were some of the earliest bryozoans in the fossil record and include the overwhelming majority of the wealth of species that lived during the Paleozoic Era. The double-walled organization, with its interior vertical walls and capacity to add growth to outer skeletal surfaces, developed the massive and bushlike stoney bryozoans, which grew up to several feet in dimension, as well as the more delicate fenestrate and bifoliate growth habits of the cryptostomes. In the beginning a small group of genera, the ceramoporids, experimented with communication pores and even larger gaps in zooidal walls. These openings were accompanied by occasional terminal diaphragms grown from within the zooids, which suggests that the openings actually functioned as interzooidal connections for nutrient exchange. Ceramoporids apparently became extinct during the middle of the Paleozoic Era and communication pores are not known again for as long as 170 million years. During Jurassic time communication pores became more abundant and occur generally in Cenozoic and Recent double-walled stenolaemates.

Single-walled stenolaemates were also some of the earliest Bryozoa in the fossil record. Unlike the double-walled forms, they apparently were small, inconspicuous colonies of few taxa during the Paleozoic Era. Paleozoic single-walled stenolaemates are thought to be the only bryozoans without interzooidal connection after calcification of zooidal walls (Fig. 13C) since they apparently lack communication pores in interior walls and connections around the ends of zooecia are cut off by the simple exterior zooidal walls. Lack of neighbor response to injury may have been a major reason for the limited success of the group. The generally shortened dimensions of zooidal body cavity of some of these forms (Figs. 10 and 11) suggest that the feeding and reproductive organs of reasonable proportions maintained much the same retracted position during ontogenetic development. Greatly thickened secondary calcification as seen in the crownoporids (Figs. 11 and 34) would progressively reduce living-chamber volume if continued throughout life, suggesting another growth limitation for these forms. The presence of apparent pseudopores in simple exterior walls of early Paleozoic species and again much later in post-Paleozoic species suggests that pseudopores may have been developed independently and almost certainly have an advantageous function of selective value.

Post-Paleozoic single-walled cyclostomes apparently have communication pores in interior walls. Their distribution needs to be checked further, and single-walled species lacking communication pores would presumably have physiologically isolated zooids. If communication pores provided enough physiologic exchange to support continued growth of nonfeeding neighbors, zooids could make an advantageous tradeoff by adding simple exterior walls which provided skeletal support for the external cuticle of the colony, but shut off interzooidal connection around ends of zooecia, now not needed. The establishment of physiologic exchange seems essential to continued growth of nonfeeding polymorphs such as kenozooids at the stage where they start growth of simple exterior zooidal walls (Fig. 29B may be an example) and thereby seal themselves from connection around ends of zooecial walls.

Whatever the selective improvements were, single-walled cyclostomes underwent a considerable addition of taxa during post-Paleozoic times. The apparent ease with which simple exterior zooidal walls were added to *Heteropora pacifica* (Fig. 31) and the ontogenetic sequence of double- to single-walled in single-walled colonies suggest that an evolutionary transformation from double- to single-walled species could have occurred many times in many different evolving stocks. Modern single-walled groups then may well be polyphyletic in complex patterns and a phylogenetic classification would profoundly modify the present single- and double-walled groupings. The general similarities of several integrative characters in different combinations and widely spaced in time gives promise that characters expressing colony dominance are selective and may have gone through several cycles of convergence and parallelism during the last 500 million years.

A number of integrative states occurring in Paleozoic forms do not seem to have been repeated in later time. These include the lack of interzooidal connection through communication pores in most Paleozoic forms at one integrative extreme and the gaps in the interior walls of the ceramoporids (Fig. 36C, D) at the other extreme. The wholly exterior zooidal walls of the corynotrypids (Fig. 33A, B), with single connecting mural pores, seem to be restricted to early Paleozoic time, but younger species may well have been overlooked. The surrounding of single zooids by extrazooidal skeleton in the budding pattern so that zooids are not contiguous seems to have occurred in only a few cystoporatids and bifoliate cryptostomes. The unique combination of extrazooidal coelomic space, paired communication pores, and interior and exterior wall relationships are so different from other known bryozoans that the diploclemids (Figs. 12 and 35) might well belong to a different phylum. These may be a few of the early integrative characters that either by chance or lack of selective advantage were not reestablished in post-Paleozoic stocks.

In summary, both single- and double-walled stenolaemates have been in existence from the beginning of the fossil record of the phylum. At first appearance they were highly integrated, and, in fact, the highest average integrative proportions belong to Ordovician and Silurian cystoporatids. Single-walled forms of both Paleozoic and Recent ages have the lower average integrative numbers and double-walled forms of the two ages have the higher numbers (Fig. 13C). Several integrative features are common to both ages and some were tried and discarded, apparently permanently, during the Paleozoic Era. Communication pores were apparently discarded during the middle of the Paleozoic Era after only modest success in the ceramoporids and developed independently one or more times during the Mesozoic Era to become the apparent key to the success of single-walled stenolaemates during Cenozoic and Recent times.

GYMNOLAEMATA

Zooids in gymnolaemate bryozoans are highly variable in shape and, in the great majority of known genera, may be polymorphic within a colony. Autozooids generally have a box or sac shape, with the dimensions of their principal body cavity (perigastric coelom) being approximately determined early in ontogeny (Fig. 14). Autozooidal organs undergo cyclic degeneration–regeneration within the same perigastric coelom, the brown bodies thus formed not generally being retained. Structures, commonly with calcified walls, may continue to be elaborated in later ontogenetic stages on the frontal side of the zooid, either from interior walls separating the perigastric and hypostegal coeloms (Fig. 14) or from exterior-walled outpocketings, such as spines. Such structures may include an elongate, tubular peristome, superficially resembling that in some stenolaemates, but with the zooidal orifice and subjacent vestibule lying at its inner end. In living species the orifice is commonly equipped with an operculum (Cheilostomata) or a pleated collar (Ctenostomata). Living species have no membranous sac, the feeding apparatus being protrusible through the orifice by action of muscles directly on the body walls. Traces of opercula and muscle attachments are known in some fossil gymnolaemates. Budding of zooids most commonly is at the distal ends of lineal series separated by exterior walls and consists of outpocketing followed by upgrowth of an interior wall to separate the zooid from the distal portion of the bud, which continues to expand distally as the bud of the next zooid in series (Fig. 14A', A). A bud at the end of a lineal series may be either zooid-sized or multizooidal ("giant bud" of Lutaud, 1961), that is, of sufficient length at any one time to form two or more zooids by ingrowth of transverse walls. Some gymnolaemates appear to lack lineal series, their zooids budding in distal multizooidal zones similar to those in stenolaemates. Some zooids in gymnolaemates either having or apparently lacking lineal series are budded by direct expansion of parts of zooids which are proximal to the distal budding regions, most commonly from the frontal (see Banta, 1972, for detailed description) or basal walls of such zooids. Communication pores, plugged with soft tissues in all living species investigated, are present in both interior and exterior walls between contiguous zooids.

Of the two gymnolaemate orders, the far more diversified Cheilostomata appear to have the greater variety of integrative states, and their morphology is discussed in detail in the following paragraphs. The Ctenostomata have been considered to be much like the simplest cheilostomes in having extensive exterior vertical walls, cellular interzooidal communications, and, in most genera, kenozooids in addition to feeding–reproductive autozooids (Silén, 1942b; 1944a, 1944b; Ryland, 1970). The Recent ctenostome genus *Monobryozoon* Remane has been given special consideration in some discussions of the colonial nature of Bryozoa because of its solitary appearance, resulting from the prominence of its typically single autozooid and the stolonlike morphology of its multiple kenozooids. That the stolonlike structures are kenozooids rather than simple protuberances from the autozooid is indicated by their separation from the autozooid by interior walls with communication organs (Franzen, 1960, p. 140, fig. 9), and *Monobryozoon* is thus colonial, as has been pointed out by Ryland (1970). The kenozooids of *Monobryozoon* also can become autozooids during the process of budding new colonies (Franzen, 1960; Gray, 1971), and thus asexual reproduction in this genus is not limited to production of kenozooids.

Zooid Walls

Inferences based on observed ontogenetic and astogenetic morphologic gradients in colonies of Cheilostomata, coupled with some observations from sectioning and dissection, have resulted in a general model of cheilostome growth (Silén, 1944a, 1944b) that has been confirmed by more recent studies employing extensive sectioning (Lutaud, 1961; Bobin and Prenant, 1968; Banta, 1968b; 1969b), observation of developing colonies (Lutaud, 1961), and scanning electron microscopy of skeletons (Sandberg, 1971; Tavener-Smith and Williams, 1972). At the same time, these techniques have revealed a greater diversity in modes of growth in cheilostomes than previously suspected.

The common mode of budding in Cheilostomata produces zooids in lineal series (Boardman and Cheetham, 1969), each comprising the primary zooid (a in Fig. 14)

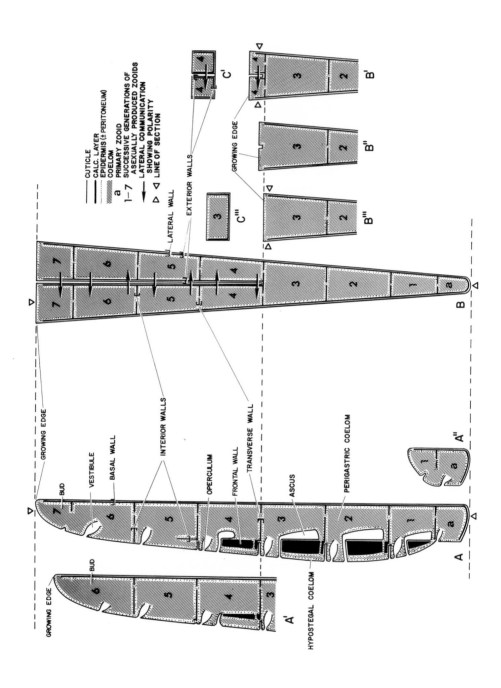

or group of primary zooids and one zooid of each asexually produced generation (1–7 in Fig. 14A, B) to the growing edge of the colony. As shown by Silén (1944b) and Banta (1969b), a series is bounded frontally, laterally, and basally by exterior walls developed by intussusception from the walls of the primary zooid (Fig. 14 A′), whereas zooids within a series are separated by transverse walls developed by apposition from the basal and lateral walls (zooids 6 and 7 in Fig. 14A″, A). Adjacent lineal series are separated by a pair of exterior (lateral) walls that grow by intussusception on two lobes originating when the growing edge coincides in position with the as-yet-unmarked distal end of the axillary zooid (3 in Fig. 14B‴, B″, B′). Transverse walls and those lateral walls adjacent to other zooids are perforated at intervals for passage of communication organs (Fig. 14B), the morphology of which is discussed below.

In numerous cheilostomes with a great variety of growth habits, a colony typically consists of many lineal series increasing in number distally and arranged in lateral contiguity (Fig. 15A). The vertical walls in such cheilostomes, probably representative of the majority of genera, thus include significant interior and exterior components (state 2 of Series A).

Zooid walls appear to be less integrated in genera such as *Pyripora* d'Orbigny and *Rhammatopora* Lang, both of which appeared early in the fossil record (Thomas and Larwood, 1960). In these genera zooids are arranged predominantly uniserially (Fig. 15B), although series branch and in some species branches are laterally contiguous, forming pluriserial portions of colonies with zooids presumably connected by lateral communication organs. A characteristically pyriform zooid of either of these genera is in contact with its predecessor in lineal series only through a narrow proximal cauda. In most species of *Pyripora* and all of *Rhammatopora* the cauda appears not to be significantly wider than a communication organ (Thomas and Larwood, 1960, figs. 3, 4; Frey and Larwood, 1971), and thus the only interior walls present in such species are presumably the pore plates, which are parts of the communication organs. A typical zooid of *Pyripora* or *Rhammatopora* may thus be regarded as surrounded by exterior wall except for its communication organs (state 1 of Series A). In some genera that appeared later in the fossil record, such as *Hippothoa* Lamouroux, there are also species

---

Figure 14

Mode of growth of a colony of an ascophoran cheilostome such as *Schizoporella* (based on Banta, 1969b, 1970, 1971; and other sources). Lophophores, alimentary canals, and associated structures have been omitted for diagrammatic simplicity. A. Sagittal section through a lineal series of zooids, showing ontogenetic and astogenetic gradients of morphologic difference in zooid walls. B. Deep tangential section through two lineal series, the zooids in which are opposite as in *Porina*. Polarity of lateral communications is random. A′. Sagittal section through distal part of colony as it would have appeared at a slightly earlier stage of growth. Note mode of origin of transverse wall between zooids 6 and 7 and of frontal wall of zooid 5 by comparing A′ and A. A″. Sagittal section through colony as it would have appeared at a very early stage of growth to show mode of budding from primary zooid. B′, B″, B‴. Deep tangential sections through distal part of colony as it would have appeared at three successively earlier stages of growth, showing origin of bifurcation of lineal series. C′, C‴. Transverse sections through distal part of colony as it would have appeared at stages represented by B′ and B‴, respectively.

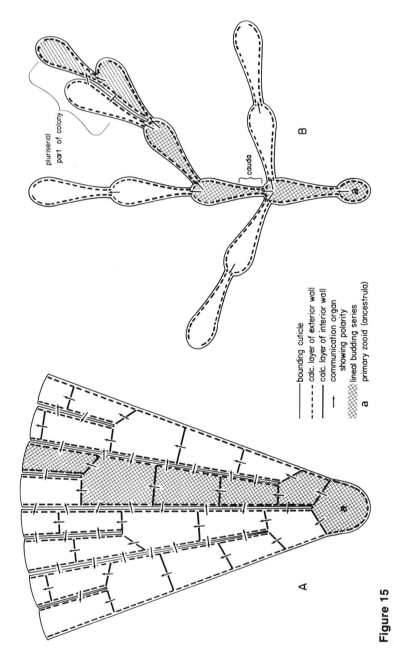

Figure 15

Distribution of exterior and interior vertical walls in *Pyripora* or *Rhammatopora* (B, based on Thomas and Larwood, 1956, 1960; Marcus, 1949) as contrasted with that in a characteristically multiserial cheilostome (A, based on Silén, 1944b; Banta, 1969b). Colonies are shown in deep tangential section.

(e.g., *H. flagellum* Manzoni; see Harmer, 1957; Gautier, 1962) in which zooids are pyriform, narrowly caudate, uniserially arranged, and presumably with no more interior wall than in *Pyripora* and *Rhammatopora*.

Wilbertopora Cheetham, another genus that appeared early in the fossil record, ordinarily displays the typical cheilostome multiserial budding pattern (Cheetham, 1954). Some colonies have been found, however, in which zooids budded uniserially at first and then multiserially (Fig. 39C). These colonies thus appear to link *Pyripora* and *Rhammatopora* with the majority of encrusting cheilostomes. The uniserially arranged zooids in *Wilbertopora* are pyriform, but their caudae are wider than those of many species of *Pyripora* and *Rhammatopora*, suggesting the presence of more extensive interior walls.

Zooids nearly surrounded by exterior walls grow in an apparently different manner in the Eocene to Recent genus *Beania* Johnston. Each zooid is connected to others laterally as well as proximally by narrow tubelike extensions similar to caudae (Silén, 1944b). According to Silén, a zooid is produced by fusion of three buds, one from the proximal zooid and the others from each of the proximolateral zooids and thus is simultaneously a member of three lineal series. Fusion of buds has been demonstrated in other cheilostomes in which zooids are typically segregated in lineal series, as for

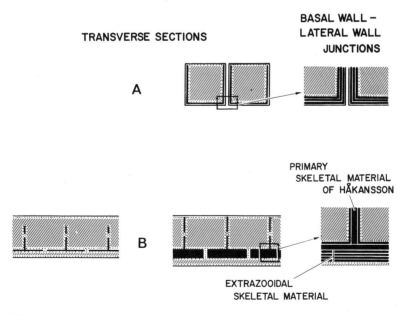

Figure 16

Relationship between lateral walls of contiguous zooids in cupuladriids (B, after Håkansson, in press) as contrasted with that in characteristic cheilostomes (A). Symbols as in Fig. 14. Microstructure of skeleton is represented as lamellar for diagrammatic simplicity (for known structural types see Sandberg, 1971, this volume; and Håkansson, in press).

example in reparative budding (Banta, 1969b) and may also occur in budding of zooids separated by pore chambers in lineal series (Silén, 1944b).

For cheilostomes possessing pore chambers, Silén (1944b; see also Ryland, 1970, fig. 12H) proposed a growth model in which exterior walls bounding a lineal series are constricted just distal to the interior transverse wall to form one or more small chambers at the proximal end of a zooid. Such constrictions, formed according to Silén by intussusceptive folding of the frontal, basal, and lateral walls back upon themselves, might seem to suggest less integration of zooid walls than in typical multiserial cheilostomes. The pore chamber, however, is separated from the proximal zooid by a standard transverse wall with communication organs and is widely confluent with the distal zooidal cavity. Therefore, as pointed out by Silén, zooids within lineal series appear to be separated in the same way as in typical multiserial cheilostomes.

A more highly integrated arrangement than in most cheilostomes is suggested in *Cupuladria* Canu and Bassler and some related genera for which a growth model has been recently proposed by Håkansson (in press) from observations on body wall structure. In *Cupuladria* all walls between contiguous zooids seem to be without bounding cuticles and therefore, on the basis of assumption 1 (p. 130), to be interior walls (Figs. 16B and 17C; state 3 of Series A). Intussusception of the walls of the group of primary zooids (a triad, according to observations by Maturo on metamorphosis of *Cupuladria* and the related genus *Discoporella* d'Orbigny presented at conference of International Bryozoology Association, Durham, England, September 1971) produces the frontal and basal noncalcified walls of the colony within which the vertical walls of zooids grow by apposition. This mode of growth is more like that in some stenolaemates than that in other gymnolaemates. Thus the characteristic cheilostome budding in lineal series appears to be lacking in *Cupuladria*. *Lunulites* Defrance, a genus that appeared earlier in the fossil record than *Cupuladria* or *Discoporella,* appears to grow by a method combining some features of *Cupuladria* with those of more typical cheilostomes (Håkansson, personal communication), and in many species of *Lunulites* zooids are separable in lineal series. As suggested by Sandberg (this volume), a mode of growth similar to that proposed for *Cupuladria* may characterize a number of otherwise dissimilar cheilostomes lacking cuticles bounding vertical walls.

Scanning electron microscopy of cheilostome skeletons (Sandberg, 1971, this volume; Tavener-Smith and Williams, 1972) has revealed a number of ultrastructural types of calcareous layers and varying relationships with cuticular layers. Various ultrastructural types have been found singly or in combination in walls of which the noncalcified layers presumably grew either by intussusception or by apposition. Regardless of ultrastruc-

---

Figure 17

Relationship between transverse and basal walls of zooids in cupuladriids (C, after Håkansson, in press) as contrasted with that in characteristic cheilostomes (A, B). Symbols and representation as in Figs. 14 and 16. Compare A with Figs. 1, 7, and 16 of Sandberg (this volume) and B with description of *Adeona* in the same paper. Transverse wall in each case is an interior wall regardless of skeletal structure. Basal wall in cupuladriids is interior in contrast to that in most other cheilostomes.

SAGITTAL SECTIONS

BASAL WALL–
TRANSVERSE WALL
JUNCTIONS

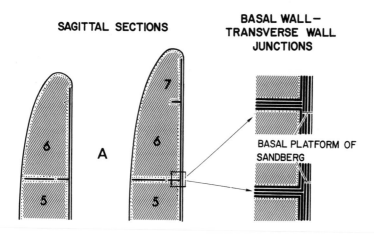

BASAL PLATFORM OF
SANDBERG

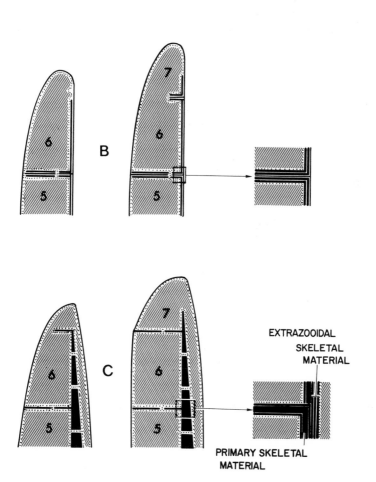

EXTRAZOOIDAL
SKELETAL
MATERIAL

PRIMARY SKELETAL
MATERIAL

ture, calcified layers without bounding cuticles can be assumed to be parts of interior walls. Those having cuticular layers, however, are not necessarily parts of exterior walls. Arrangement of cuticular and calcareous layers would appear to depend on when walls calcified as well as on how they grew. In Fig. 17 are shown some possible types of basal wall–transverse wall junctions (the calcified layers being depicted with lamellar structure for diagrammatic simplicity), based on differences in times of calcification of basal and transverse walls and time of inception of the transverse wall. In all cases shown, the transverse walls have grown by apposition, even though having a cuticle in Fig. 17B, in which growth of the transverse wall is postulated to have begun by infolding of epidermis along a line distal to the forward edge of calcification of the basal wall (and also of the lateral ones).

Interzooidal Soft-Tissue Connections

The soft tissues by which zooids in cheilostomes are interconnected are parts of communication organs that appear remarkably similar in the otherwise morphologically varied species that have been investigated (Levinsen, 1909; Silén, 1944b; Bobin and Prenant, 1968; Banta, 1968b, 1969b). The structure of communication organs has been worked out in detail by Banta (1969b). Each consists of a thin calcareous or cuticular plate with one or more perforations traversed by fibrillar tubes filled with one or more cells providing the actual soft-tissue connection (state 3 of Series B). The cells have polarity such that their nuclei generally lie on the side of the pore plate within the more proximal of the two zooids connected by the organ (see arrows in Fig. 15A and 16). These cells are in contact with epithelial cells lining the body cavities of the connected zooids and with cells grading into the funicular strands that traverse the cavities.

There are some variations on this theme (Banta, 1969b). Organs connecting zooids belonging to different lineal series possess a thickened cuticular ring on the side of the more distal zooid, marking the margin of the interrupted bounding cuticle between series, actually dissolved during formation of the organ. The number of communication organs in a wall, the number of groups of cells passing through a plate, and the number of cells in a group also vary.

As discussed above, the presence of pore chambers is not thought to alter the nature of communications between zooids within lineal series (Silén, 1944b). Recent work by Gordon (1971) on the development of pore chambers suggests that those in lateral walls are formed similarly to lateral communication organs, but the opening dissolved in the cuticle is separated from the pore plate by the width of the chamber (see Silén, 1944b, fig. 42; Ryland, 1970, fig. 12G). Gordon (1971), however, regards the pore chambers themselves as discrete zooids.

Extrazooidal Hard and Soft Parts

In most cheilostomes all hard and soft parts of a colony are referable to zooids. These parts include structures developed by buds before their separation as zooids, such as continuous skeletal layers deposited in the basal wall in structural continuity with those of other zooids in the same lineal series (basal platform of Sandberg, this volume; see also Fig. 17A). Two kinds of tissues and associated calcareous deposits, however, form apparent exceptions that can be regarded as extrazooidal. In *Euthyrisella obtecta* (Hincks), *Cupuladria* and related genera, and sertellids, extrazooidal parts appear to develop concurrently with budding of zooids. In *Metrarabdotos* Canu and some other ascophoran genera, extrazooidal parts develop by coalescence of zooidal tissues.

Harmer (1902) noted the presence in *Euthyrisella obtecta* of a continuous extrazooidal coelom ("colonial body cavity") of perhaps colony-wide extent. This coelom completely invests the single layer of contiguous zooids on both their basal and frontal sides, except for their orifices (Harmer, 1902, fig. 34; redrawn by Blake, this volume, fig. 30). The outer wall of the extrazooidal coelom is a continuous membrane attached to small calcareous papillae apparently deposited on frontal and basal walls of zooecia by the extrazooidal epidermis. The very thin zooecial walls in this species are strengthened on the lateral margins of the erect branches by a system of thick longitudinal calcareous bars (Harmer, 1902, fig. 37), apparently also deposited by extrazooidal epidermis. The zooids and extrazooidal parts of colonies of *E. obtecta* appear to form concurrently at the growing tips of branches (state 4 of Series C).

The basal sides of conical to discoid free-living colonies of *Cupuladria, Discoporella,* and some lunulitid genera have extensive calcareous layers formed by basal apposition (Silén, 1942a, p. 9). The depositing epidermis for these basal layers lines a basal coelom (Marcus and Marcus, 1962, pl. 2, fig. 5) divided by cuticular partitions which arise from the basal cuticle and extend into and end in the calcareous layers (Håkansson, in press, presented at conference of International Bryozoology Association, Durham, England, September 1971). In *Discoporella,* some lunulitids, and some species of *Cupuladria,* narrow basal compartments many times longer than the zooids extend radially from the periphery of the colony toward its center (see, for example, Cook, 1965a, 1965b). In some species of *Cupuladria,* the basal compartments are shorter, being additionally separated by transverse cuticular partitions (Håkansson, in press; Tavener-Smith and Williams, 1972). The basal chambers in *C. biporosa* Canu and Bassler are regularly square and subequal in length to autozooids (see Cook, 1965b, fig. 1j), but those in *C. canariensis* (Busk) form irregular combinations of long and short rectangles up to several times as long as autozooids (Cook, 1965b, fig. 1f). In *C. biporosa* and some other species of *Cupuladria,* series of kenozooids are budded from the basal side of each autozooid (Cook, 1965b; Håkansson, in press; Tavener-Smith and Williams, 1972). These kenozooids have skeletal microstructure comparable to that of the autozooids and the accompanying vibracula but different from that of the basal compartments into which the kenozooidal series open (Håkansson, in press). Some of the kenozooidal series budded from one autozooid and some of those budded from other autozooids are incorporated in the calcareous layers of one basal compartment (Marcus and Marcus, 1962; Cook, 1965b). The basal compartments and associated calcareous layers in these genera thus appear to form a transitional series from extensive, almost colony-wide features to restricted, slightly more than zooid-sized ones. Their size, their variable shape, and the presence in some of series of kenozooids all seem to indicate that these features are extrazooidal. They are formed concurrently with budding of zooids (Håkansson, in press) (state 4 of Series C).

Basal calcareous layers similar to those in *Cupuladria* and allied genera are known in sertellid cheilostomes where they form one side of the erect, usually anastomosing branches of the colony, the other side of which is composed of a single layer of autozooids (Harmer, 1934; Hass, 1948; and others). The depositing epidermis for these basal layers

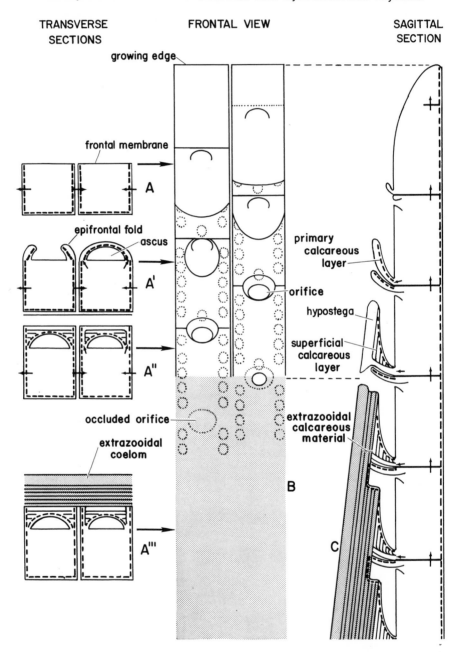

Figure 18

Development of extrazooidal tissue and associated calcareous deposits in an ascophoran cheilostome such as *Metrarabdotos* (based on Cheetham, 1968; Sandberg, this volume). Symbols and representation as in Fig. 15, with extrazooidal parts shaded. During its ontogeny, recorded on proximal gradient from growing edge, a zooid has the following succession of structures on its frontal side: A, apparently lines coelomic compartments of variable size and shape separated by cuticular partitions like those in *Cupuladria* (Hass, 1948; Tavener-Smith and Williams, 1972; Sandberg, this volume, figs. 10, 11). Also as in *Cupuladria,* these compartments include zooids (avicularia). Although the basal compartments of sertellids have been regarded as kenozooids (Marcus, 1926; Harmer, 1934; Hass, 1948), their similarity to those of *Cupuladria* and allied genera suggests that they, too, should be interpreted as extrazooidal. In some sertellids, these compartments develop concurrently with budding of zooids (Harmer, 1934, p. 516) (state 4 of Series C).

A different type of apparently extrazooidal parts occurs in the genus *Metrarabdotos* and some other ascophoran cheilostomes. Frontal calcareous layers extending without interruption across zooidal boundaries succeed deposits, commonly of the same ultrastructure and composition, that are limited to separate zooids (Fig. 18; Sandberg, this volume). These extrazooidal layers are deposited by tissues apparently made continuous by the coalescence of extensions (hypostegal coeloms) of the body cavities of contiguous zooids (state 2 of Series C). Bounding cuticles between lineal series thus end abruptly at the contact between zooidal and apparently extrazooidal layers (Fig. 18; Cheetham, 1968, 1971). Because this type of material is developed late in the ontogeny of zooids, it is restricted to the more proximal parts of colonies. Even after orifices and other parts of subjacent zooids essential to their feeding functions have been occluded, these apparently extrazooidal tissues continue to be depositionally active, presumably through nourishment supplied by more distal feeding zooids.

The tertiary frontal wall in some Late Cretaceous and early Tertiary cribrimorphs is composed in part of calcareous material that apparently spread from localized interzooecial deposits to form a united lattice above the frontal walls of zooids (Larwood, 1962, 1969). This material thus seems to be an extrazooidal deposit developed subsequently to budding of zooids (state 3 of Series C), although the mode of growth of interzooecial deposits in cribrimorphs is obscure (Larwood, 1969, p. 179).

Astogenetic Differences

In some part or parts of a cheilostome colony, there are conspicuous generational differences between zooids. These astogenetic differences may be restricted to the proximal part of the colony, or they may occur in one or more distal parts.

A cheilostome colony is commonly founded by a single primary zooid (ancestrula), the morphology of which may differ slightly or markedly from that of the asexually produced zooids. This difference, however, need not be consistent within a genus (Abbott, this volume) or even within a species (Waters, 1923, p. 558; Powell, 1967, p. 212). At the minimum the ancestrula differs from the asexually produced zooids in possessing an exterior proximal wall (Figs. 15A and 19), but it also commonly differs in size, shape, and frontal structure (Fig. 14A; see also Abbott, this volume). Although the

---

a flexible frontal wall; A', a distomedially advancing frontal fold, on the basal side of which is deposited a primary calcareous layer; A'', a frontally advancing frontal fold, on the basal side of which accretes superficial calcareous material; and A''', a still frontally advancing extrazooidal fold formed by coalescence of frontal folds of contiguous zooids. Superficial material deposited at stage A''' is structurally continuous across zooid boundaries.

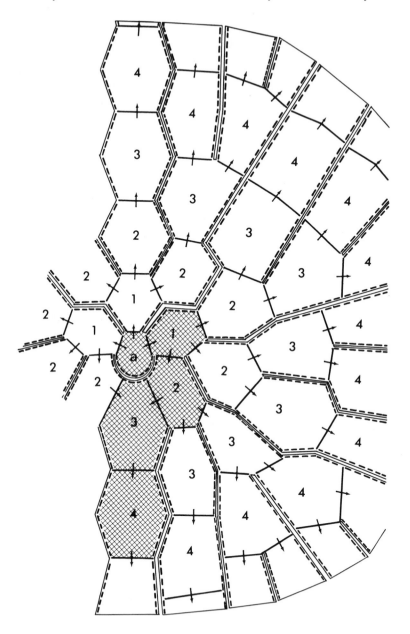

Figure 19

Common astogenetic pattern in encrusting cheilostomes. Colony represented in deep tangential
section with symbols as in Fig. 15A, B, except that lateral communications have been omitted.
Note that zooids in contact with ancestrula belong to more than one generation with consequent
variation in morphology.

difference between ancestrulae and asexually produced zooids has been given much attention (Waters, 1924; and many others), differences between other generations leading to repeatable zooid morphology have been considered in detail in fewer studies (e.g., Illies, 1953; Medd, 1966; Abbott, this volume), perhaps because of the commonly gradational nature of these differences. In well-documented examples, progressive differences in size and other morphologic features have been found to characterize several generations of zooids proximal to those of repeated morphologies (see also Fig. 39A). We do not know of examples in which there are no astogenetic differences between some generations of asexually produced zooids.

In most cheilostomes having a single primary zooid, this zooid buds others only from its distal and lateral margins. Zooids are generally filled in proximally to it by the wrapping of lineal series around it (Fig. 19; see also Waters, 1924; Harmer, 1930; Medd, 1966). The periancestrular zooids in a colony so formed belong to a number of generations with a consequently wide range of morphology (Fig. 19), and thus, as pointed out by Medd, astogenetic gradients cannot generally be recognized just from the distance of zooids from the ancestrula. A slightly different method of periancestrular closure was observed by Stach (1938) in larval metamorphosis and ensuing development of zooids in *Smittina papillifera* (MacGillivray). Five first-generation zooids of similar morphology formed simultaneously around all but the proximal extremity of the ancestrula. These six zooids differentiated within 7 minutes of larval fixation, suggesting to Stach that all six could almost be regarded as primary. In *Membranipora* Blainville metamorphosis is followed by simultaneous development of a pair of primary zooids side by side (Robertson, 1908; "double ancestrula" of Waters, 1925; "twin ancestrulae" of Hastings, 1930; and others). More recent observations on metamorphosis and colony founding (Eitan, 1972; Cook, in press; Maturo, 1971 I.B.A. conference) in a number of cheilostomes including *Cupuladria* have revealed that initial partitioning of the coelom produces three or more primary zooids of virtually indistinguishable morphology, none of which then appears to be properly regarded as an ancestrula. Håkansson (in press) has shown that the zooids of the primary triad in *Cupuladria* have wholly interior vertical walls, and, as discussed above, continuation of this mode of growth results in absence of lineal series throughout the colony. In spite of this difference in mode of growth, astogenetic differences in morphology between generations of asexually produced zooids in *Cupuladria* appear to be similar to those in cheilostomes having an ancestrula (Cook, 1965a, 1965b).

In some cheilostomes, especially in *Cupuladria* and other genera having a free-living growth habit, budding of zooids from colony fragments can produce colonies of approximately the same shape as unfragmented ones, and fragmentation may be so common as to provide an important means of colony reproduction (Marcus and Marcus, 1962). Because fragments from which growth was initiated are more likely to be from the larger zone of astogenetic repetition, especially if they have originated as peripheral "colonial buds" described by Marcus and Marcus (1962), such colonies usually lack a primary zone of astogenetic change. Zooids in such colonies, however, are still the asexually produced lineal descendants of a group of primary zooids necessary for colony founding and thus have the same dependence on primary zooids as those in a colony in which the primary zone is retained.

Apparently the common astogenetic arrangement in cheilostome colonies, then, is a primary zone of change, comprising the primary zooid or zooids and a few generations of asexually produced zooids, followed by a primary zone of repetition, comprising numerous generations of zooids to the growing edge of the colony (state 3 of Series D). More complex astogenetic zonations, with additional (subsequent) zones of change and repetition, have been described for some cheilostomes (Boardman et al., 1970). In these colonies zooid morphologies characteristic of the primary stage of repetition cease to develop, either throughout the colony or in parts of it, and subsequently budded zooids, throughout the colony or in these parts, show another set of astogenetic differences leading to further repeatable morphologies (state 4 of Series D).

Three principal arrangements were recognized by Boardman et al. (1970, figs. 3–5) in cheilostomes in which subsequent zones of astogenetic change and repetition succeed the primary zone of astogenetic repetition either wholly sequentially or in part concurrently:

1. A subsequent zone of change comprising zooids of similar morphologies to those in the primary zone of change (except for that of the primary zooid) is followed by a subsequent zone of repetition comprising zooids of similar morphologies to those in the primary zone of repetition. Some colonies of *Bugula* Oken were given as an example of this arrangement. The subsequent astogenetic changes and repetition described by Abbott (this volume) in "type R" colonies of *Hippoporina porosa* (Verrill) also appear to fit this pattern.

2. The arrangement in 1 is repeated cyclically so that a colony consists of numerous zones of repetition, the zooids in all of which have similar morphologies, separated by zones of change, in all of which the astogenetic differences between zooids are similar. In addition to *Poricellaria* d'Orbigny, the example cited by Boardman et al. (1970), some cellularines and some species of *Tetraplaria* Tenison-Woods (see Cheetham, 1972) appear to show this arrangement.

3. One subsequent zone of change and following subsequent zone of repetition are formed as in 1, but the morphologies of zooids in subsequent zones are different from those in corresponding primary ones. The example cited by Boardman et al. (1970) is a genus, *Kleidionella* Canu and Bassler, in which frontally budded zooids differ from those in basal layers of the colony. Banta (1972) has described in more detail the morphologic differences between zooids in frontally budded and basal layers in *Schizoporella* Hincks. In addition to genera with frontal budding, species of *Cupuladria* having "*doma*-type" colonies (Cook, 1965a, 1965b) appear to represent this kind of astogenetic zonation. In a fully developed *doma*-type colony the primary zone of repetition, comprising autozooids and vibracula, is succeeded by two peripheral generations consisting entirely of larger vibracula and smaller zooids similar in shape to the autozooids but closed frontally and lacking feeding organs and musculature. No more zooids are budded once the apparent subsequent zone of repetition, consisting of a single generation of vibracula and closed zooids, has developed (Cook, 1965b), and the colony thus attains a characteristic maximum size.

Morphology of Polymorphs

It seems impossible to find any genus of cheilostomes in which polymorphism is entirely absent. (Its apparent absence in some cheilostome species listed by Schopf, this volume, refers to the lack of particular zooid types, avicularia and vibracula, considered in his functional analysis.) It is true that some genera seemingly lack the more obvious polymorphs, such as avicularia or ovicelled zooids, but kenozooids seem to be almost universal among cheilostomes, although their occurrence can be either regular or sporadic. Commonly markedly smaller than autozooids in the same species, kenozooids consist of body wall enclosing a coelom that lacks alimentary canal and musculature, although traversed by funicular strands and connected to other zooids by communication organs (state 4 of Series E). The function of kenozooids has been assumed to be supportive or connective (Levinsen, 1909). In some genera zooids of this morphologic type may be present only at obstructions to colony growth, as, for example, at the intersection of opposing growth directions of the same colony or of different colonies (Silén, 1938; Powell and Cook, 1966). Such zooids, if they are invariably terminal members of lineal series, are probably properly regarded as early developmental stages of autozooids whose further development was blocked rather than as polymorphs. If, however, zooids of this morphologic type give rise to daughter zooids, they should be recognized as kenozooids (Fig. 39D).

Ovicelled autozooids, avicularia, and kenozooids are present, together with nonovicelled autozooids, in the great majority of cheilostomes known, and these polymorphs appeared in various genera early in the fossil record. In both living (Marcus, 1949) and fossil (Thomas and Larwood, 1960) representatives of *Pyripora,* kenozooids occur within lineal series as predecessors of autozooids. Apparently similar kenozooids occur in *Rhammatopora* (Thomas and Larwood, 1960) and in *Wilbertopora* (Fig. 39A, C, D), which also has ovicelled autozooids (Fig. 39B) and avicularia (Fig. 40B, C). In *Electra* Lamouroux kenozooids are commonly restricted to regions of opposing budding directions (Marcus, 1926), but they may also be developed within lineal series to form stolons (Bobin and Prenant, 1960). Species of *Membranipora* can have kenozooids within budding series (Levinsen, 1909), and some species have sporadically developed avicularia (Cook, 1968). These genera (including the *Pyripora*-like Jurassic cheilostomes reported by Pohowsky, 1971 I.B.A. conference) apparently represent the least consistently expressed polymorphism in cheilostomes.

Not only do most cheilostomes possess one or more of the conspicuous types of polymorphs (avicularia, vibracula, ovicelled zooids) in addition to their nonovicelled autozooids and, where present, kenozooids, but any of these different types of zooids may in turn occur in more than one form in the same colony. The number of different polymorphs in a cheilostome colony thus can be great, and it can be taken as an indication of the degree of specialization and consequent interdependence of zooids because of the assumed correspondence between morphologic and functional differences. (See discussion by Schopf, this volume, of the theoretical aspects of functional specialization.) The morphologic difference between polymorphs, however, itself is a direct indication of degree of interdependence (see discussion by Banta, this volume). The lack of feeding, digestive, and reproductive apparatus in kenozooids is direct evidence of their dependence upon autozooids morphologically equipped for these functions. Conversely, it must be assumed that, in species having kenozooids, autozooids are less efficient in performing supportive or connective functions and thus in turn are dependent. With similar assumptions, it is possible to infer degrees of interdependence of zooids implied by other principal types of polymorphs known to occur in cheilostomes.

Lack of feeding and reproductive structures apparently gives vibracula and most avicularia as much dependence on autozooids as have kenozooids. Functional feeding organs do occur, however, in avicularia in *Crassimarginatella* Canu as well as in the avicularium-like B-zooids of *Steganoporella* Smitt (see Ryland, 1970). The apparently slighter dependence of avicularian autozooids in these two genera is reflected in a smaller morphologic difference between their skeletal parts and those of their accompanying ordinary autozooids (Fig. 40D). This kind of comparison thus can be made in fossil cheilostomes (Fig. 40B, C), the morphology of which suggests that the earliest avicularia known may have been among the least dependent. The reciprocal dependence of ordinary autozooids on avicularia or vibracula must vary to the same extent.

Polymorphism associated with sexual reproduction in cheilostomes appears to be nearly as varied in the implied interdependence of zooids as it is in morphologic expression. At one extreme are the highly polymorphic zooids of genera such as *Hippothoa* including (Gordon, 1968): (1) ovicelled female zooids, in which the lophophore lacks tentacles and is nonfeeding; (2) male zooids, in which the lophophore has a small number of tentacles but is also nonfeeding; and (3) asexual feeding zooids. Polymorphs in *Hippothoa* differ markedly in skeletal features (Rogick, 1956; and others) as well as in soft parts, and their interdependence appears to be comparable to that between kenozooids and feeding autozooids or between typical avicularia and feeding autozooids.

At the other extreme are such cheilostomes as *Electra* in which the sexes are combined in at least some feeding autozooids, although other feeding autozooids in a colony may be asexual or male (Silén, 1966). Sexual differences in these cheilostomes seem not to be reflected in skeletal or opercular parts of the zooid, and the implied zooid interdependence appears to be minimal. In *Hippopodinella adpressa* (Busk) zooids without apparent skeletal or opercular differences comprise nonfeeding males and feeding females (Gordon, 1968; ovicelled zooids reported by Thornely, 1912, in this species appear not to characterize other populations studied, according to Powell, 1967). If male and female structures alternate in a zooid of *H. adpressa* through intrazooidal degeneration–regeneration, a possibility suggested by Gordon, then the implied zooid interdependence seems not significantly greater than that among hermaphroditic autozooids of *Electra*.

The principal morphologic expression of sexual differences in the majority of cheilostomes is the possession or lack of brood chambers (either ovicells or intrazooecial chambers of various sorts) by two sets of feeding autozooids that either are otherwise morphologically alike in skeletal features (Fig. 39B) or differ in some skeletal features in correlation with their possession or lack of brood chambers (Fig. 40E). An arrangement that appears to characterize many cheilostomes was described by Silén (1944c) in *Callopora* Gray and *Securiflustra* Silén. In these genera, the ovicell is a fold that develops by outpocketing from the distal end of the zooid which produces ova. Embryos are then deposited by this zooid in the external cavity enclosed, except for a proximal opening, by the ovicell. Parts of the walls of the ovicell are calcified, and the coelomic lumen between the walls remains confluent with the zooidal coelom. Because this connection is a pore, Silén (1944c) regarded ovicells as possibly derived from zooid buds. Beklemishev (1970, p. 463) regarded ovicells of scrupocellariids as zooids which, together with autozooids, avicularia, and vibracula, are assembled in repeated groups (cormidia), but he did not present evidence for this interpretation. Recently, Woollacott and Zimmer

(1972) have shown that the fold forming the ovicell in *Bugula neritina* (Linné) is a kenozooid budded from the distal autozooid, to which it is connected by a communication organ. The presence of pore plates between ovicells and distal autozooids in some other cellularines (Hastings, 1943) suggests that these, too, may be regarded as kenozooids. Ovicells may thus represent a wider range of degrees of zooid interdependence than now is generally recognized.

Finally, autozooidal polymorphism in several genera appears to be only partly, or perhaps not at all, associated with sexual reproduction. In *Steganoporella* B-zooids (mentioned above in discussion of avicularia) do not produce ova (Cook, 1964), although possessing functional feeding organs. In *Euthyrisella* Bassler, the structure of B-zooids suggested to Harmer (1902) that ovaries might be limited to these zooids, although direct evidence is lacking. The difference between A- and B-zooids in *Uscia* Banta was considered by Banta (1969a) as likely to be nonsexual. In species of *Poricellaria* d'Orbigny and *Vincularia* Defrance possessing both ovicells and dimorphic autozooids, ovicells are limited to one set of dimorphs, although zooids of that set do not invariably bear ovicells and ovicells are more frequent in the less strongly dimorphic species (Cheetham, 1968; in press). In all these genera, the morphologic difference between autozooids suggests generally less interdependence than the other forms of polymorphism discussed here.

Position of Polymorphs

Polymorphic zooids in cheilostomes may be intercalated in the budding pattern seemingly at random or in repeated groups, or they may be structurally parts of other zooids. The implications of these patterns of zooid interdependence in cheilostomes having avicularia are discussed by Banta (this volume). The more general relationships, including all kinds of polymorphs, are discussed here.

Even among genera in which polymorphs seem to be randomly distributed (state 2 of Series F), some tendency toward regularity of certain polymorphs is apparent. In *Wilbertopora* and related forms (Fig. 40B, C), avicularia are commonly distolateral members of a bifurcation of lineal series, but kenozooids and ovicelled autozooids appear to be sporadically distributed. In *Steganoporella* the avicularium-like B-zooids are commonly, but not invariably axillary in position at a bifurcation (Fig. 40D). In species of *Thalamoporella* Hincks, avicularia are almost invariably distolateral members of a bifurcation, but zooids in that position commonly are of other types (Fig. 40A; see also Powell and Cook, 1966). In such genera as *Coscinopleura* Marsson (see Voigt, 1956) and *Adeonella* Busk (see Harmer, 1957), certain polymorphs are restricted to lateral margins of bilaminate colonies, but again zooids in those positions are not invariably of that polymorphic type. In *Dionella* Medd a small avicularium is commonly, but not invariably, intercalated in the budding pattern distal to each autozooecium (Medd, 1965).

In a number of cheilostome genera polymorphic zooids are arranged regularly enough to be considered repeated groups (state 3 of Series F) generally fitting the cormidium concept of Beklemishev (1970, p. 463). This concept seems to apply best to genera such as *Cupuladria,* in which each autozooid in the zone of astogenetic repetition is associated with a distal vibraculum and, in some species, with a group of basal kenozooids (Cook, 1965a). Also, a segment (except the most proximal ones) of a jointed colony of Oligocene to Recent species of *Poricellaria* can be regarded as a cormidium, each of which consists of a zone of astogenetic change followed by a zone of repetition comprising dimorphic autozooids segregated in different lineal series and bearing adventitious avicularia (Cheetham, 1968; in press). Slightly less regularity is shown by vittaticellids, in which ovicelled autozooids are regularly members of colony segments consisting of a given number of zooids arranged in a particular pattern, differing from genus to genus (Stach, 1935); ovicell-bearing segments, however, are less regularly interspersed among nonovicelled ones. A similar degree of regularity is shown in some species of *Metrarabdotos* in which brooding autozooids are irregularly scattered throughout the colony, but, where they occur, such zooids are regularly accompanied by nonbrooding zooids having special forms of adventitious avicularia (Fig. 40E). In *Woodipora* Jullien and some other genera, irregularly scattered avicularia are each surrounded by a group of characteristically oriented autozooids (Lagaaij, 1952), and in *Lunulites* vibracula are segregated in lineal series between those comprising autozooids, although the patterns by which rows of each polymorph originate differ from species to species (Cheetham, 1966).

Adventitious polymorphs are common among cheilostomes. These are zooids occupying positions on the wall of another zooid with which their sole communication is formed (state 4 of Series F), rather than being intercalated in the budding pattern and communicating with more than one other zooid. Adventitious polymorphs are most commonly avicularia, and they are most generally positioned on the frontal wall of the supporting zooid, although they include, for example, the basal rootlike kenozooids of *Petraliella* Canu and Bassler (see Harmer, 1957). They are present in anascans such as *Ramphonotus* Norman, *Callopora* Gray, and most cellularines, and they characterize the great majority of ascophorans. Their presence is taken as direct evidence of a high degree of zooid interdependence because of their structural integration with the supporting zooid (see p. 134, and discussion by Banta, this volume).

Intrazooidal polymorphism has not been described in cheilostomes, although possible alternation of male and female structures in autozooids of *Hippopodinella adpressa* has been suggested (see discussion above). Male zooids within zooecial cavities of asexual and female zooids in *Hippothoa* (Rogick, 1956; Powell, 1967) appear, from Powell's description of their morphology, to have originated by reparative budding, and thus the position of these zooids appears not to differ from those intercalated in the budding pattern.

Evolution of Colony Dominance in Cheilostomes

As described above and summarized in Fig. 20, a wide range of levels of integration appears to characterize the cheilostomes in all but two of the integrative characters, interzooidal communication (B) and the maximum morphologic difference between polymorphs (E). Although this variation suggests that integration may have played a role in cheilostome evolution, the level of integration of genera and higher taxa seems not to have evolved in a simple progression toward more highly integrated types, for

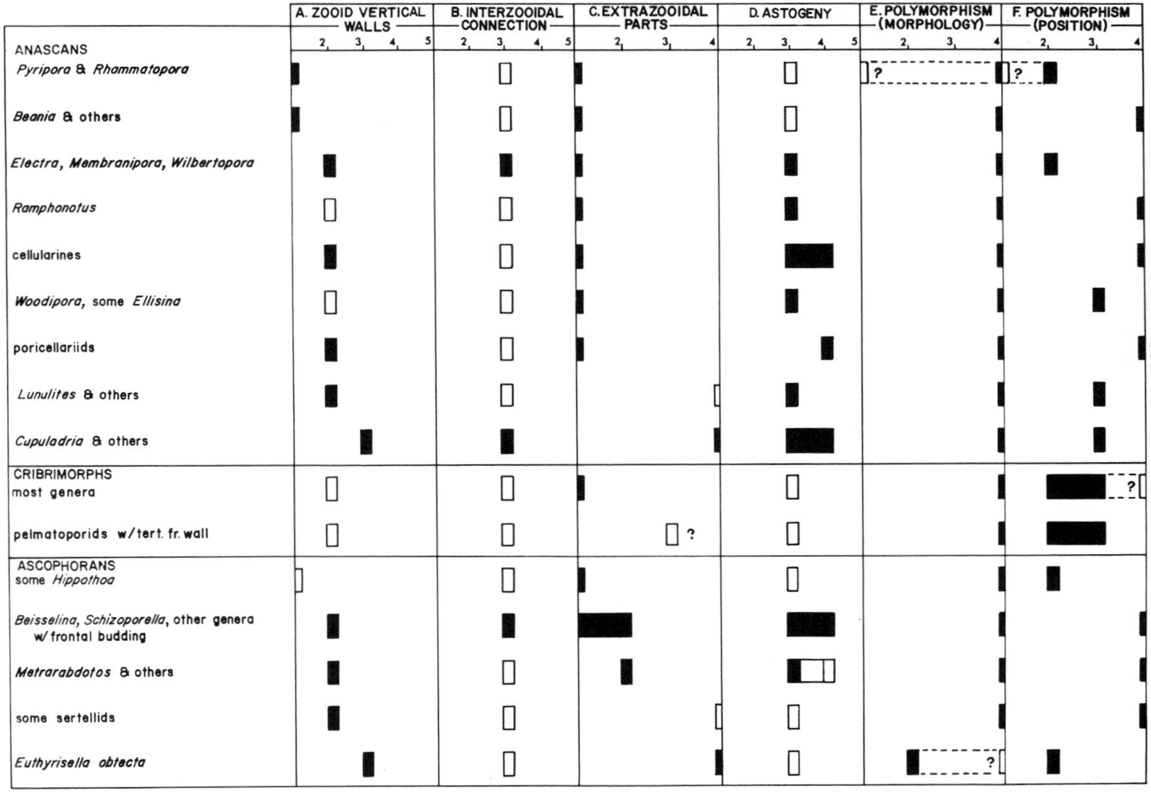

Figure 20

Summary of integrative states in groups of cheilostome genera discussed in text. States of each character arranged as in Series A–F. States observed in one or more taxa in each group are indicated by solid rectangles; those based only on inference are indicated by hollow rectangles, with queries marking most speculative inferences.

two reasons. First, within each of the major groups into which cheilostomes have conventionally been divided—anascans, cribrimorphs, and ascophorans—there is a broad range of levels of integration apparent. The range within the anascans, the group that appeared first in the fossil record, indeed encompasses those of the other groups for each of the integrative characters, if the genera listed in Fig. 20 are consistent samples of the three groups. Second, integration tends to vary within a genus, being high for some characters and low for others. Cellularine genera, for example, are low in C and high in D, E, and F, whereas *Euthyrisella* is high in A and C and low in F and possibly in E.

As integration in cheilostomes appears not to have evolved in a simple progression, its pattern in time will be difficult to recognize. The pattern might best be revealed by plotting the average level of integration for all genera known, from Late Jurassic time (Pohowsky, 1971 I.B.A. conference) to the present. Morphologic and stratigraphic data, however, seem insufficient to estimate average integration with precision. Instead, the genera discussed above and listed in Fig. 20, which were deliberately selected as representatives of extreme integrative states, can give at least a rough estimate of minimum and maximum levels during this interval of time. Although these estimates are likely to change as morphology and stratigraphic ranges become better known, they may have some stability because of being based on recognition of the most conspicuous states of the integrative characters.

For all four characters (A, C, D, and F) that show a significant amount of difference within the cheilostomes, the patterns in time (Figs. 21–24) of minimum and maximum levels appear to be similar. For each character, the oldest cheilostomes known (Pohowsky, 1971 I.B.A. conference) are inferred to express the lowest levels of colony dominance, and this apparent minimum has persisted while the maximum level progressively increased in one or more steps. Although similar in pattern, these stepwise increases involve different genera at the higher levels of different characters, a further expression of the mosaic taxonomic distribution of integrative states mentioned above.

If the ranges of all the integrative characters (Series A–F) can be assumed to represent comparable scales of colony dominance, then the values for different characters might be combined as a measure of total integration. On the assumption of such comparability (see discussion p. 131–134), total integration was calculated as the weighted average of characters A–F for each genus in Fig. 20. This measure of integration shows a pattern in time (Fig. 25) similar to but smoother than those for the separate characters. The steady divergence between its minimum and maximum levels implies an increasing diversity of degrees of colony dominance within successive faunas. Even though this pattern must be regarded as only a tentative indication of the evolution of integration in cheilostomes, it invites comparison with the generally increasing taxonomic diversity of cheilostomes through their history as presently understood (Larwood et al., 1967).

The maximum level of integration in cheilostomes appears to be generally correlated with the total number of cheilostome families extant at any given time (Fig. 26A). Slight deviations in pattern are apparent, the most notable of which is that resulting from a diversity decrease in earliest Tertiary time. Such deviations could be expected to be resolved by plotting average levels of integration, when data become sufficient for that purpose.

The maximum level of integration agrees even more strikingly with the number of anascan families (Fig. 26B), in keeping with the significant advancements in integration

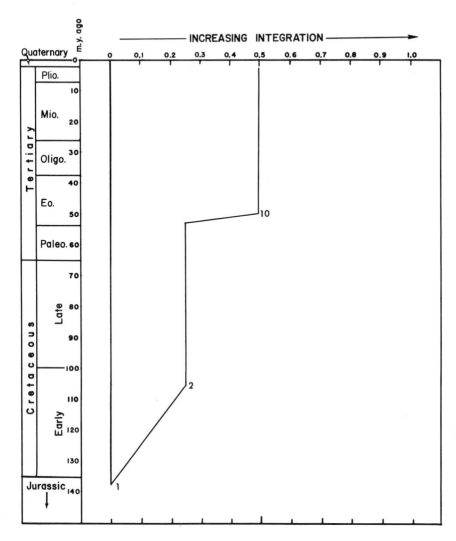

Figure 21

Inferred evolution of integration of zooid vertical walls in cheilostomes. Integration scale proportional to states in Series A. Minimum and maximum integration levels are from genera in Fig. 20 and the following stratigraphic ranges (for genera shown in this figure or in Figs. 22–25): 1, *Pyripora* and other pyriporimorphs, Portlandian (Pohowsky, 1971 I.B.A. conference) to Recent; 2, *Wilbertopora* and other membranimorphs, Albian (Cheetham, 1954; unpublished) to Recent; 3, *Ramphonotus* and other anascans with adventitious avicularia, Cenomanian (Brydone, 1929) to Recent; 4, pelmatoporids with tertiary frontal walls, Santonian to Danian (Larwood, 1962); 5, *Ellisina* with regularly distal avicularia, Campanian (Brydone, 1929) to Recent; 6, lunulitids with secondary basal thickening. Campanian to Recent (Håkansson, in preparation); 7, cellularines, Maastrichtian (Cheetham, 1968) to Recent; 8, poricellariids, Maastrichtian (Cheetham, 1968) to Recent; 9, *Beisselina* and other genera with frontal budding, Maastrichtian (Wiesemann, 1963) to Recent; 10, *Cupuladria,* Ypresian (Gorodiski and Balavoine, 1962, according to Buge and Debourle, 1971) to Recent; 11, *Cupuladria* with *doma*-type colonies, Miocene–Recent (Cook, 1965a).

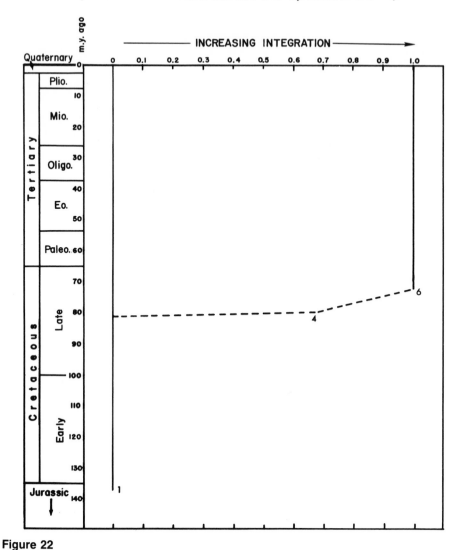

Figure 22

Inferred evolution of integration of extrazooidal parts in cheilostomes. Integration scale proportional to states in Series C. Minimum and maximum integration levels are from genera in Fig. 20 and stratigraphic ranges listed for Fig. 21.

made by anascans during Cretaceous and early Tertiary time. These include: (1) substitution of interior for exterior vertical walls, a trend beginning in the Early Cretaceous and culminating in the appearance of cupuladriids in the Eocene (Fig. 21); (2) development of basal extrazooidal parts with the appearance of lunulitids in the late Cretaceous (Fig. 22); (3) addition of subsequent zones of astogenetic change and repetition with the appearance of cellularines and poricellariids in the Late Cretaceous (Fig. 23); and (4) appearance of adventitious avicularia in addition to or replacing interzooidal and vicarious avicularia in diverse anascan groups, also in the Late Cretaceous (Fig. 24).

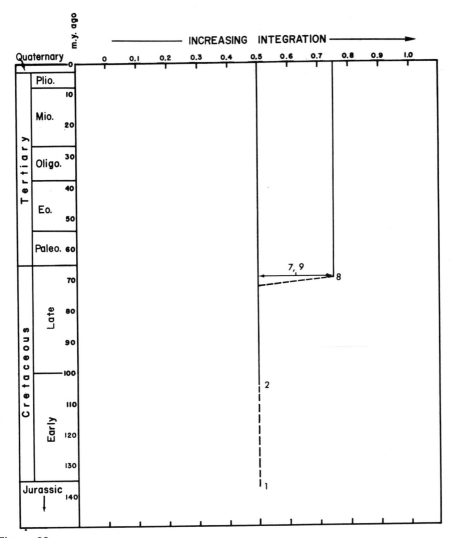

Figure 23

Inferred evolution of integration of astogeny in cheilostomes. Integration scale proportional to states in Series D. Minimum and maximum integration levels are from genera in Fig. 20 and stratigraphic ranges listed for Fig. 21.

The cribrimorphs seem to have participated little in integrative advancements, except for development of tertiary frontal walls, presumably an extrazooidal feature in at least some genera, in mid-Late Cretaceous time (Fig. 22). The multiserial growth form of most cribrimorphs and its implied vertical wall integration appear to have originated with the presumed anascan ancestors of the group, although uniserial colonies with caudate zooecia are known in a few cribrimorph genera (Larwood, 1962). Avicularia of adventitious position may be present in cribrimorphs (Larwood, 1962) possibly as an independent evolutionary development. Lack of further major integrative advancements

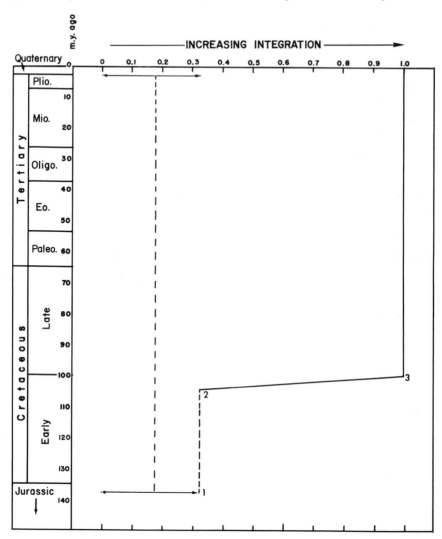

Figure 24

Inferred evolution of integration of position of polymorphs in cheilostomes. Integration scale proportional to states in Series F. Minimum and maximum integration levels are from genera in Fig. 20 and stratigraphic ranges listed for Fig. 21.

in this group seems empirically correlated with its decline in diversity in later Late Cretaceous and earliest Tertiary time (Fig. 26C).

Diversity in the ascophorans (Fig. 26C) seems to be less directly correlated with increasing integration levels than is that in other groups. The majority of ascophoran families appeared in post-Cretaceous time, after many of the important advances in integration had been made by their presumed anascan and possibly cribrimorph ancestors. Extensive interior vertical walls and ubiquitous adventitious avicularia in ascophorans

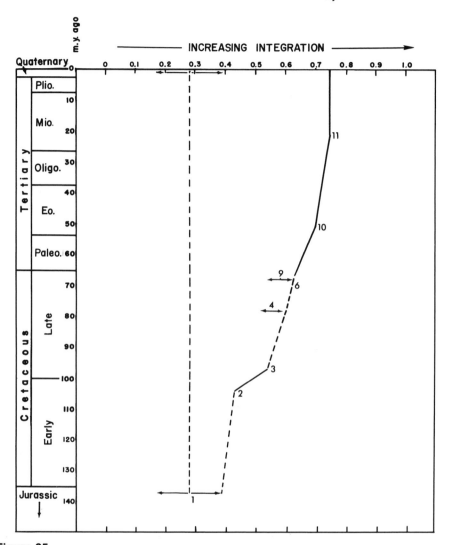

Figure 25

Inferred evolution of total integration in cheilostomes. Integration scale proportional to weighted averages for Series A–F. Minimum and maximum integration levels are from genera in Fig. 20 and stratigraphic ranges listed for Fig. 21.

are probably traceable to the ancestral stocks. Frontal budding and extrazooidal parts developed by coalescence of zooidal structures, however, appear to be significant integrative advancements that may be correlated directly with increased diversity in ascophorans. To a large extent diversity in this group is linked to intrazooidal characters, such as the structure and mode of development of the frontal wall and associated ascus (Harmer, 1902; Silén, 1942b; Banta, 1968b, 1970, 1971; Tavener-Smith and Williams, 1970),

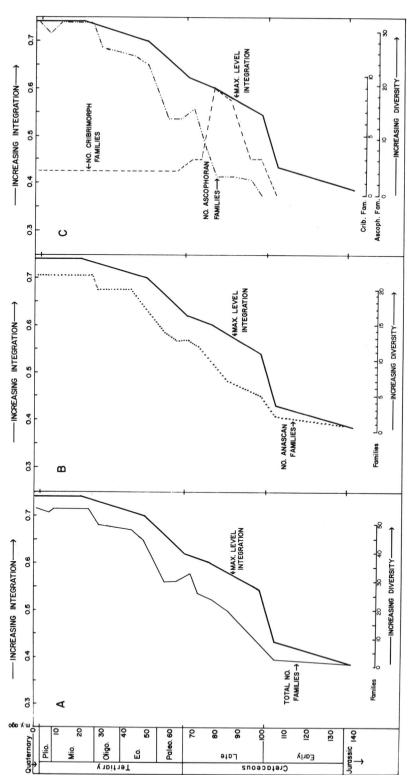

Figure 26

Evolutionary relationship between level of integration and taxonomic diversity in cheilostomes. Maximum level of integration (from Fig. 25) is compared with A, total number of cheilostome families; B, number of anascan families; and C, numbers of cribrimorph and ascophoran families. [Numbers of families are based on data presented by Larwood et al. (1967) slightly modified by Cheetham (1971).]

the form of the orifice and its relation to the opening of the ascus, and the position on the supporting zooid of adventitious avicularia. These characters seem not to be directly related to integration, although there may be indirect relationships between, for example, the mode of growth of the zooidal frontal wall and the strength of the colony to resist bending (Cheetham, 1971). Also, some of them, such as the position of an adventitious avicularium, may involve integrative rearrangements, as in their nearly cormidial groupings with respect to brooding and nonbrooding zooids in *Metrarabdotos* (Fig. 40E) and other genera. For the most part, however, such trends seem to have duplicated those undergone earlier by anascan stocks such as the cellularines.

The mosaic distribution of more and less integrated states of different characters in many cheilostome genera further suggests that selection has not acted upon all aspects of integration in concert; that is, the patterns in time of maximum levels of integration of different characters might represent more than one adaptive trend in cheilostomes. A rigorous test of how many and what kinds of adaptive factors underlie these morphologic expressions of colony dominance must also await the availability of sufficient morphologic and stratigraphic data, but some possible relationships can be discussed here.

The importance of the whole colony, rather than the zooid, as the unit of selection has been stressed by Schopf (this volume), and any consideration of adaptive trends in integration should properly begin with the relationships between integrative characters and the colony as a unit, that is, those characters which "intensify the individuality of the colony" (Beklemishev, 1970). Among the trends suggested by Beklemishev, three appear to have affected cheilostome evolution: (1) structural integration of the colony through development of structures shared by zooids or outside their boundaries, (2) physiologic integration of the colony through increased communication among its parts, and (3) functional integration of the colony through specialization of zooids and their grouping in functional assemblages. Only two of these appear to have changed within the group, as interzooidal communication seems similar throughout, although future studies may show differences in number of communication organs to be significant. Other processes discussed by Beklemishev, such as increasing regularity of budding patterns and increasing complexity of astogeny, appear to be at least partly related to these three, as discussed below.

1. A relation between the nature of zooid vertical walls (A) and the presence or absence of extrazooidal parts (C) as expressions of structural integration is suggested by an apparent correlation in their states. Most of the genera listed in Fig. 20 lie between the extremes represented by *Pyripora*-like forms with extensive exterior vertical walls and no extrazooidal parts and cupuladriids with exclusively interior vertical walls and extensive extrazooidal parts. More integrated states of these characters thus would appear to indicate greater rigidity of the colony through reduction in the proportion of cuticle between calcified walls (except in such genera as *Adeona* discussed by Sandberg, this volume) and binding together of zooid skeletons by structures continuous across zooid boundaries.

2. The most direct expression of potential physiologic integration (see p. 132), the nature of interzooidal communications (B), is apparently invariant in cheilostomes and thus not correlated with other characters. Apparently the level of integration represented by cheilostome interzooidal communication organs is sufficient for maintenance in colonies

of nonfeeding polymorphs and extrazooidal parts, thus permitting evolutionary advances in other aspects of integration.

3. The three remaining characters, astogenetic differences (D) and morphologic and positional differences between polymorphs (E and F), appear in cheilostomes to be at least partly related as an expression of functional integration. Polymorphs can be observed or inferred to be specialized for different functions, and their morphologic differences are partially correlated with their different positions in the budding pattern. Adventitious avicularia, for example, generally differ more from autozooids than do avicularia intercalated among the autozooids (contrast Fig. 40A–D with Fig. 40E; see also discussion by Banta, this volume). Kenozooids, however, can occupy almost any position seemingly

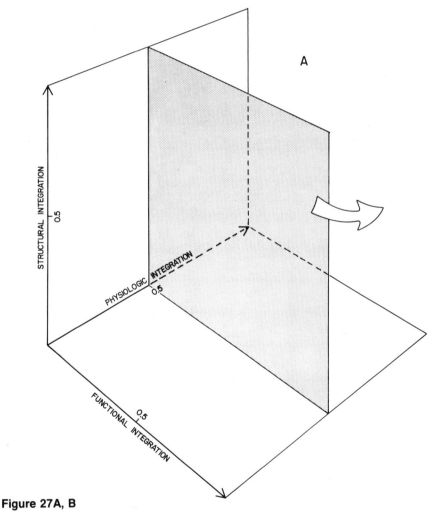

Figure 27A, B

Possible adaptive significance of levels of integration in cheilostomes. Three aspects of integration (discussed in text) are represented in A by mutually perpendicular axes, each increasing in the

with little variation in their distinguishing morphologic features. Astogenetic differences in cheilostomes commonly include the introduction in one or more generations of several kinds of polymorphs that characterize zones of astogenetic repetition (shown diagrammatically in Fig. 5 and discussed by Abbott, this volume). The generation in which particular polymorphs first appear may differ in conspecific colonies, however, with genetics or environment, so that the correlation between astogenetic differences and polymorphism is at best only partial. Furthermore, astogenetic differences are apparent in colonies having monomorphic zooids. Even a partial correlation suggests that the functions performed by zooids in zones of astogenetic change may also change toward repeatable assemblages of functions performed by zooids in zones of astogenetic repetition (see

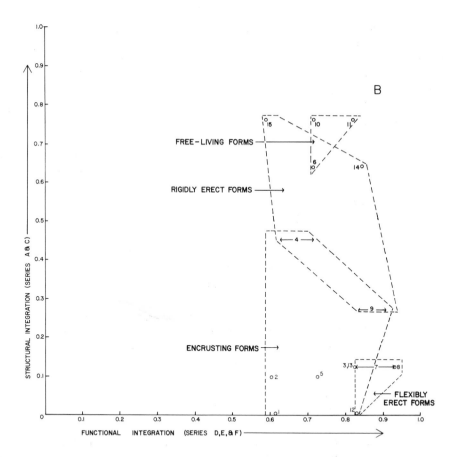

direction of the arrow. Cheilostomes, as far as known, lie in the shaded plane determined by the 0.5 level of physiologic integration. Distribution in this plane, relative to functional and structural integration, is shown in B for the genera named in Fig. 20. Dashed lines in B enclose genera of similar colony form and *do not* indicate total ranges of integration for the colony forms named. Groups of genera are numbered as explained in Fig. 21, with the following additions: 12, *Beania*; 13, most cribrimorphs; 14, sertellids; 15, *Euthyrisella*.

the levels of "functional coloniality" discussed by Abbott, this volume). The zooids in a primary zone of change, for example, would appear to have performed a different set of functions, in establishment of the colony, than those in a zone of repetition, in the growth, maintenance, and reproduction of the colony.

It should be emphasized that these suggested aspects of integration in cheilostomes are not to be regarded as independent factors in the morphologic expression of colony dominance. Indeed, as suggested by Banta (this volume), if adventitious avicularia evolved from autozooids (with vicarious and interzooidal avicularia as intermediate forms), the resulting increase in functional integration would imply increased physiologic and structural integration as well.

Even though the extent of correlation among these three aspects of integration cannot now be examined with precision, their levels in the genera discussed above suggest a possible adaptive explanation of integration in cheilostomes that might merit further investigation. The three aspects of integration can be represented as mutually perpendicular axes (Fig. 27A), with cheilostomes, insofar as known, lying entirely in one plane determined by their invariant interzooidal communication (physiologic integration). With respect to structural and functional integration (shaded plane in Fig. 27A; Fig. 27B), colony forms, a character that has been recognized as a general expression of adaptation in cheilostomes (Stach, 1936; Lagaaij and Gautier, 1965; Cook, 1968; also see Schopf, 1969, for a review), appear to fall in four groups:

1. Genera with encrusting colonies appear, on the average, to be the least integrated. This group of genera includes the oldest cheilostomes known and can be interpreted as relatively unspecialized, although some genera in other groups include typical encrusting colonies. Encrusting colonies are tolerant of a wide variety of substrates and of water movement, but their size is approximately dependent on the substrate.

2. Genera with rigidly erect colonies can be interpreted as specialized for increased size relative to the amount of substrate occupied. The several forms included in this group (subcylindrical-branched, bilaminate-branched, anastomosing, and others) are susceptible to breakage as, for example, in moving water. The risk of breakage would appear to decrease with their increasing structural integration, by reduction of planes of weakness or binding together with extrazooidal deposits.

3. Genera with free-living colonies can be interpreted as specialized both for increased size relative to the substrate, which is generally incorporated into the fully developed colony, and for the instability of the fine-grained sediment on which they most commonly live. Adaptation in this group would appear to involve even more structural integration, on the average, than in the rigidly erect one.

4. Genera with flexibly erect colonies can be interpreted as specialized for increased size, instability of substrate, and water movement, but with a generally different set of adaptations than genera with rigidly erect and free-living forms. Many genera in this group are more lightly calcified than those having rigidly erect colonies, and most have uncalcified joints regularly intercalated in the budding pattern. The great regularity of changes in budding pattern seemingly required for jointing is achieved by cyclic occurrence of similar subsequent zones of astogenetic change and repetition throughout the colony and by regular placement of repeated assemblages of polymorphs. There are a number of genera, such as *Cellaria* Ellis and Solander, that appear to be intermediate,

in both colony form and integration, between this group and the one having rigidly erect colonies. These genera are less regularly jointed and more heavily calcified, and there is evidence, at least in *Cellaria,* that zooid vertical walls include more extensive interior components (Banta, 1968a; Sandberg, this volume).

In summary, the morphology and stratigraphic ranges of genera of cheilostomes, although inadequately known for precise recognition of evolutionary trends, suggest a pattern of increasing structural and functional integration in proliferating evolutionary stocks while ancestral stocks, retaining the least integrated states, also persisted. These apparent trends are in keeping with the importance of the colony as the unit upon which selection has acted (Schopf, this volume), albeit in a mosaic pattern, and with colony form as a general expression of adaptation in cheilostomes (Stach, 1936; and others). Advances in structural integration seem to have been most significant in genera with rigidly erect or free-living forms, in which the risk of breakage may thus be diminished. In some free-living forms, however, the ability to break remains an important character. Advances in functional integration appear to characterize many genera with flexibly erect forms, in which jointing seems to require a more regular budding pattern incorporating subsequent astogenetic zones and repeated groups of polymorphs. Genera having encrusting forms, including stocks that appeared early in the fossil record, seem to show a wide range of relatively low levels of structural and functional integration. Increased colony dominance with concomitantly reduced zooid autonomy would thus appear to have played an important role in broadening the range of adaptation in cheilostomes and thus permitting their striking taxonomic proliferation from Late Jurassic time to the present.

ACKNOWLEDGMENTS

We are grateful for helpful discussion and technical criticism by members of the coloniality seminar at the National Museum of Natural History (see the Introduction to this volume) and by Marie B. Abbott, Patricia L. Cook, Anna B. Hastings, Valdar Jaanusson, Douglas M. Lorenz, Geneviève Lutaud, Frank K. McKinney, and Philip A. Sandberg. F. M. Bayer, A. G. Coates, W. A. Oliver, Jr., and Adam Urbanek helped develop the integration series used in this paper. Helen Tappan Loeblich generously provided locality data for Cretaceous cheilostomes in the National Museum of Natural History collected by her and Alfred Loeblich, Jr. The Cretaceous cheilostome from the Louisiana State University collection, figured in this paper, was loaned to us by H. V. Andersen, J. G. Harmelin and P. L. Cook loaned us specimens of Recent cyclostomes, some of which are figured in this paper. Technical assistance was rendered by Donald A. Dean, JoAnn Sanner, and Larry B. Isham. Research for this paper was supported in part by a Smithsonian Research Foundation Grant (Fund 427206).

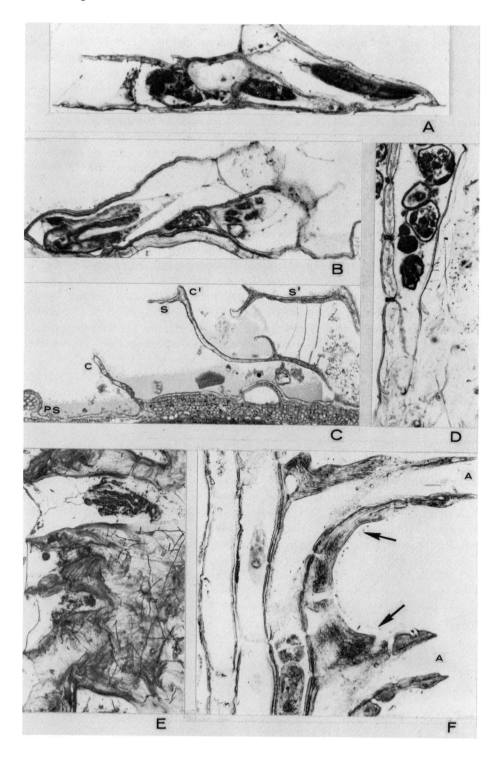

Figure 28A, B

Unidentified single-walled cyclostomes. A. Longitudinal section, USNM 186541, young incrusting colony showing ancestrula (on right) and parts of two generations to growing tip (on left). Preserved gut of ancestrula is retracted into proximal-most basal disc (protoecium). B. Longitudinal section, USNM 186542, through young incrusting colony showing ancestrula (on left). ×150, Dall Collection 925, Popoff Strait, Alaska.

Figure 28C

Diaperoecia sp. Longitudinal section, ×100, of growing end of colony showing sequence of skeletal development: the distal-most edge of protoecial skeleton (PS); incomplete compound interior skeleton that partitions zooids (C); incomplete simple, exterior, zooidal skeleton (S) at termination of compound skeleton (C′); completed simple zooidal skeleton (S′). Collected by J. G. Harmelin at 100 m at station 71.25, Banc du Magaud, Iles d'Hyeres, France.

Figure 28D

Heteroporid cyclostome. Longitudinal section of compound wall, ×100, showing communication pores occupied by large cells, the upper one of the three with median calcareous partition. From Pacific area, loaned by British Museum (Natural History).

Figure 28E

Heteroporid cyclostome. Longitudinal section of preserved double-walled cyclostome, ×100, showing cuticle of outer flexible wall across ends of zooids. From Pacific coast of North America, loaned by British Museum (Natural History).

Figure 28F

Crisina watersi Borg, 1942, a preserved double-wall cyclostome loaned and identified by J. G. Harmelin. Longitudinal section, ×150, showing position of outer flexible wall indicated by concentration of foreign particles on right (arrows). Compound skeletal walls extend distally to right to form peristomes around apertures (A), outer surfaces greatly thickened and inflected by pores that open outward into the outer coelomic space under the outer flexible wall. Collected by M. Vivien at 50 m, Tulear, Madagascar.

Figure 29A

Single-walled cyclostome. Longitudinal section through dried colony USNM 186543, ×50, ancestrula (zooecium at extreme right), basal protoecial wall lacking pseudopores, basal expansion of recumbent zooecia (left), and base of erect growth upward in figure. The wall of the ancestrula (S), rising above the basal level of the zoarium to form the top of the basal disc (protoecium) and the outer skeletal wall of the zooecium, is a thickened, simple exterior wall with pseudopores that is similar in appearance to other exterior zooecial walls in zoarium. Albatross station 2407, Gulf of Mexico.

Figure 29B

Idmonea milneana d'Orbigny. Longitudinal section at base of zoarium, USNM 186544, ×50, showing basal protoecial skeleton (PS) without pseudopores, a basal polymorph opening to left with simple exterior zooecial wall (S) with pseudopores, the erect, simple, exterior budding wall with pseudopores (arrow), and the zooecia opening to right with interior compound walls and exterior simple zooidal walls. Lutetian (Eocene), Chaussy, Seine et Oise, France.

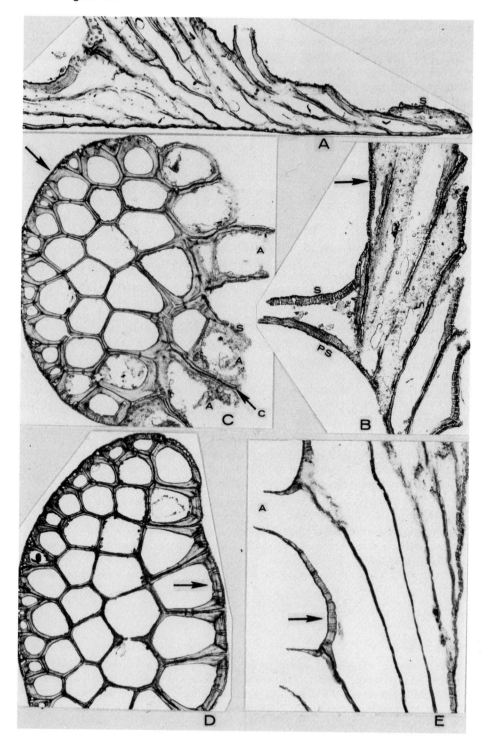

Figure 29C–E

Idmonea californiensis d'Orbigny. C. Transverse section of zoarial branch, USNM 186545, ×50, showing erect budding wall (arrow) on reverse side of branch pitted with combination of pseudopores and irregular solution cavities, compound interior zooecial walls within branch (dark line of primary layer centered along zooecial boundaries) arising from budding wall and opening to right (A). Zooecial walls are compound (C) distally to apertures where zooids are contiguous, and simple (S) where wall separates living chamber from environment. D. Transverse section of same specimen, ×50, cut at level corresponding approximately to that of arrow in E. Budding wall to left and simple zooecial wall to right (arrow) showing pseudopores and abrupt nature of contact with compound walls. Pleistocene age from Dead Man Island, San Pedro, California. E. Longitudinal section, USNM 186546, ×50, showing erect budding wall to right with pseudopores, compound walls ending distally at simple zooecial walls (arrow), apertures (A) opening to left, and primary (darker) and secondary walls (laminated) throughout. Dried Recent specimen from the Pacific at La Jolla, California.

Figure 30A–C

Diplosolen intricaria (Smitt). A. Transverse section of preserved specimen with skeletal and soft parts in place, ×100, showing autozooids (AZ) bounded by compound interior walls except for short peristome, and smaller flattened polymorphs (arrows) limited to outer regions of colony. Body walls of polymorphs, termed "nannozooids," include compound interior walls shared with contiguous autozooecia on their inner sides and simple exterior walls containing apertures on outer sides. B, C. Longitudinal sections, ×50, from same colony showing sequence of development from growing tips of inner compound walls of autozooids (AZ) followed proximally by simple exterior walls of elongated nannozooids (arrow). Growing tip of nannozooid is apparently not its aperture. Collected at 110–130 fathoms from 60 miles north of North Cape, Norway, in Barents Sea, loaned by British Museum (Natural History).

Figure 30D, E

Diplosolen sp. Longitudinal sections of bifoliate branches of dried specimens, USNM 186547, 186548, ×100, showing median wall (M) from which the compound walls of autozooids (AZ) budded in opposite directions. Nannozooecia (arrows) consists of compound walls (C) shared with autozooecia on inner sides of their living chambers and exterior simple walls (S) containing small apertures on outer sides. Dark primary layer is brown organic-rich material. Pseudopores occur in terminal diaphragms and exterior walls. Collected from the Kara Sea, U.S.S.R.

Figure 31A–D

Heteropora pacifica Borg. A. Longitudinal section of dried topotype, USNM 186549, ×150, showing thick compound zooecial walls with communication pores and laminated zooecial linings coming from left, terminal diaphragm (TD) with closely spaced pseudopores covering upper zooecium and zooecial linings extending distally to form thin peristome (P) in lower zooecium. An extremely thin membraneous layer occurs at zooecial boundaries and may be analogous to primary skeletal layers in other genera. B. Longitudinal section from same specimen, ×100, showing segment of outer zooidal wall (W) with same microstructure as terminal diaphragms in zooecium that is more nearly paralleling direction of branch (vertical in picture). C. Longitudinal section of same specimen, ×100, showing peristome consisting of short inner segment of terminal diaphragm (TD) and long outer segment similar in appearance to zooecial lining. D. Longitudinal section of same specimen, ×100, showing incomplete terminal diaphragms that were growing distally. Concentration of foreign particles indicates position of dried cuticle (C). Topotype collected at 12–14 fathoms, from vicinity of Middleton Island, Southern Alaska.

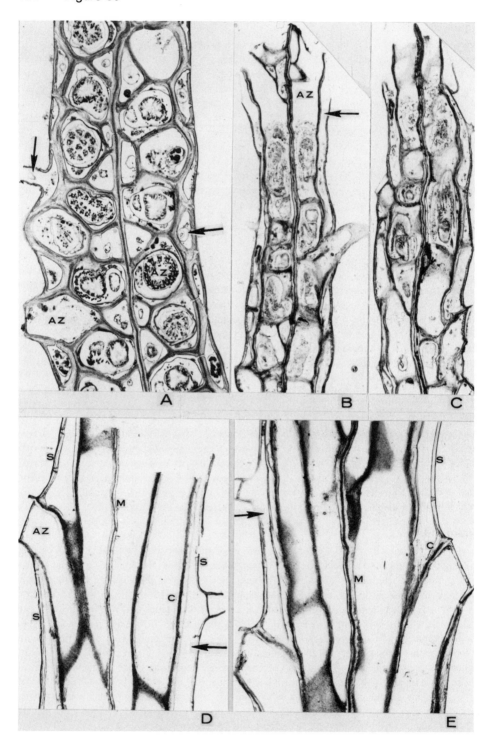

Figure 31   197

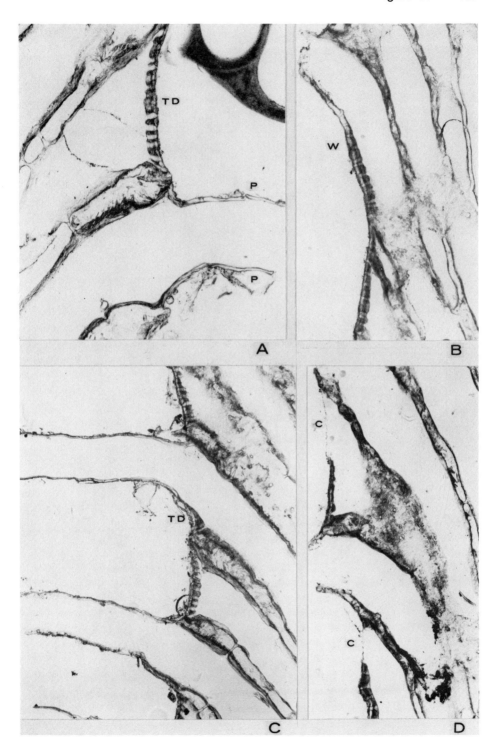

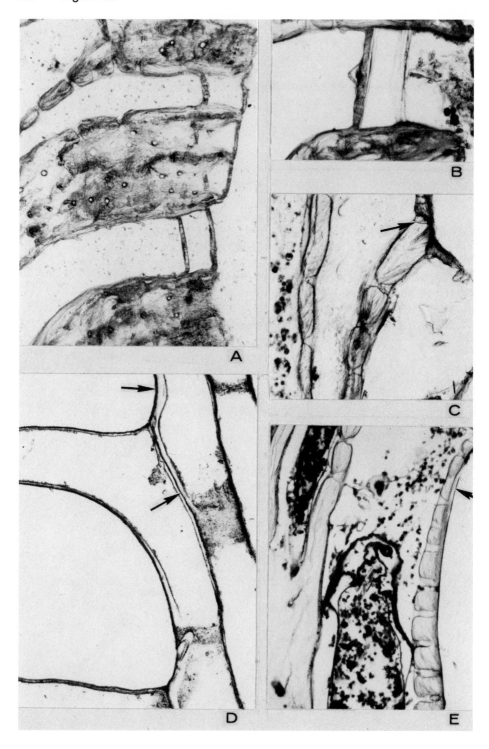

Figure 32A

Heteropora pelliculata Waters. Longitudinal section of dried specimen USNM 186550, ×100, shows paired terminal diaphragms with laminae of inner diaphragms superimposed on those of outer diaphragms proximally along zooecial walls to indicate calcification from within living chambers and formation of inner diaphragms subsequent to outer diaphragms. There is no evidence to suggest that inner diaphragms could not have developed contiguous cuticle on outer surfaces. Collected from Neah Bay, Washington.

Figure 32B

Heteroporid cyclostome sp. Longitudinal section of preserved specimen, ×150, showing a terminal calcified diaphragm followed distally (to right) by a number of parallel membranes. Zooids are in a degenerated state. Specimen from Pacific Area, loaned by British Museum (Natural History).

Figure 32C, E

Single-walled cyclostome. C. Longitudinal section USNM 186551, showing compound wall with communication pores terminating with break in microstructure (arrow), followed distally (to right) by exterior zooidal wall. E. Longitudinal section of same specimen showing pseudopores in thick exterior zooidal wall and disrupted outer cuticle (arrow), both sections ×150. Collected from 156 fathoms at 51°22.52′S. 73°8.64′E. Kerguelan Ridge. S. Indian Ocean.

Figure 32D

Single-walled tubuliporid cyclostome. Longitudinal section, USNM 186552, ×100, of dried specimen showing compound interior walls coming from lower right, separating without break into simple exterior zooidal walls which extend into peristomes to left. Dark primary layer is organic-rich (arrows). Kara Sea, U.S.S.R.

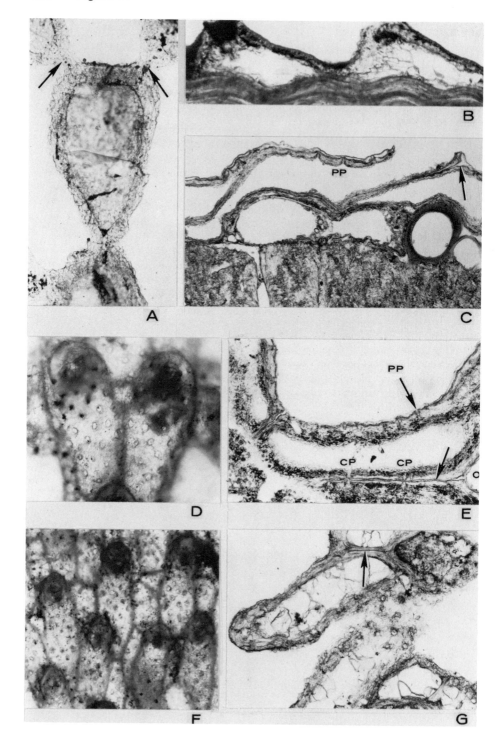

Figure 33A, B

Corynotrypa inflata (Hall). A. Deep tangential section, USNM 186553, ×100, cut through living chamber of autozooecium just above base showing mural pores (arrows) connecting autozooecia. Bellevue Member of McMillan Formation (upper Ordovician), Cincinnati, Ohio. B. Longitudinal section, USNM 186554, ×100, showing simple shapes of autozooecia with apertures at top. Waynesville Formation (upper Ordovician), Oregonia, Ohio.

Figure 33C, E

Stomatoporid cyclostome. C. Longitudinal section, USNM 186555, ×100, showing interior compound walls with hyaline primary layer along zooecial boundary (arrow) and exterior zooecial walls with primary layer and pseudopores (PP). E. Deep tangential section parallel to basal layer of same specimen, ×100, showing vertical compound walls with hyaline primary layer along zooecial boundaries (arrow), communication pores (CP), and exterior zooecial walls with pseudopores (PP). Opening (O) is a gap in colony between zooecia. Pliocene, 15 miles N.E. Myrtle Beach, South Carolina.

Figure 33D, F, G

Stomatoporid cyclostomes. D. External apertural view, USNM 186556, ×100, using transmitted light showing pseudopores in exterior zooecial wall. Bellevue Member of McMillan Formation (Upper Ordovician), Cincinnati, Ohio. F. External apertural view, USNM 186557, ×50, showing patterns of interior compound walls separating living chambers of autozooecia. Corryville Member of the McMillan Formation (Upper Ordovician), Cincinnati, Ohio. G. Deep tangential section USNM 186558, ×100, parallel to basal layer showing compound interior wall with dark zooecial boundary (arrow) and pseudopores in simple exterior walls. Waynesville Formation (upper Ordovician), Oregonia, Ohio.

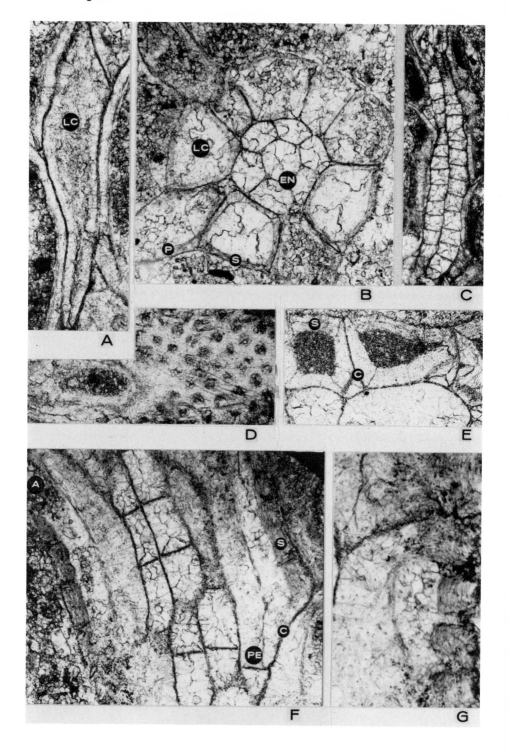

Figure 34A–G

Crownopora singularis Ross. A. Deep tangential section of paratype Y.P.M. 25179, × 100. showing vertical compound interior zooecial walls with dark primary layer along zooecial boundaries and general shape of living chamber (LC). B. Transverse section of holotype, Y.P.M. 25173, × 100, showing thin compound zooecial walls of endozone (EN) in center of branch, compound interior walls with thick secondary, poorly laminated layers lining living chambers (LC) in exozone, and simple exterior zooecial skeleton (S) consisting of outermost, dark primary layer and well-laminated secondary layer. Both layers extend distally into peristomes (P). C. Longitudinal section of paratype Y.P.M. 25179, × 30, showing general relationships between diaphragmed thin-walled endozone and thick-walled exozone. D. Tangential section of paratype Y.P.M. 25180, × 100, through exterior skeletal layer showing rhombic arrangement of pseudopores relative to longitudinal axis of zooecium, aperture at left. E. Transverse section of paratype Y.P.M. 25174, × 100, showing details of microstructure of compound (C) and simple (S) zooecial walls of exozone. Darker regions are mud-filled living chambers. F. Longitudinal section of holotype, × 100, showing distal end of living chamber of zooecium with aperture (A) on left and proximal end (PE) of living chamber on right. Notice continuity of dark primary layer from poorly laminated compound walls (C) to well-laminated simple walls (S), and pseudopores in the simple skeleton. G. Longitudinal section of holotype, × 200, showing microstructure of simple exterior skeleton between pseudopores. Pseudopores open to left on living chamber side of skeleton and are inferred to have been closed by cuticle to right on outside of skeleton. Shoreham Formation (Middle Ordovician) at Crown Point and Manny Corners, New York.

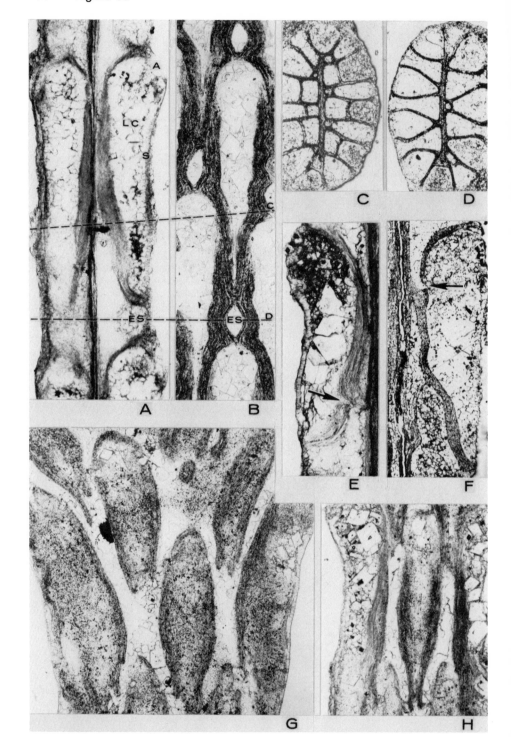

Figure 35A, B, E, G, H

Diploclema sparsum (Hall) A. Longitudinal section, USNM 186559, ×100, showing autozooecia with enclosed living chambers (LC) in profile attached back to back to central median wall. Thick, laminated interior walls extend distally (upward) and outwardly to apertures (A), and proximally and outwardly to proximal ends of the autozooecia where the interior walls abut against simple exterior walls (S). The colony-wide extrazooecial space (ES) projects out between the autozooecia to the simple exterior wall without apertures. B. Tangential section, USNM 186560, ×100, cut just below exterior wall. B aligned with the structures in A according to dashed lines C and D. Notice termination of wall laminae in apertural end of autozooecium, suggesting that the skeleton consists of two parts. E. Longitudinal section of autozooecium, USNM 186561, ×100, showing columnar structure of simple exterior wall and proximal communication pore (arrow) opening into extrazooecial space below. G. Deep tangential section cut just above median wall, USNM 186562, ×100, showing continuous extrazooecial space surrounding autozooecia. H. Deep tangential section, USNM 186574, ×100, cut just above that of G, showing connections between autozooecia in adjacent ranges. Rochester Shale Member of Clinton Formation (Middle Silurian), Lockport, New York.

Figure 35C, D, F

Diploclema sp. C. Transverse acetate peel of USNM 186563. ×50, cut at level corresponding approximately to dashed cross-section line C in photos A and B. Inner areas adjacent to median wall are necessarily part of the extrazooecial space, outer chambers are the autozooecia. D. Transverse peel of same specimen, ×50, cut at level D in photos A and B. Triangular chambers widening outwardly are autozooecial living chambers, flask-shaped areas are extrazooecial. F. Longitudinal peel of same specimen, ×100, showing distal communication pore (arrow) opening into extrazooecial space above. Wenlock Formation (Middle Silurian), Dudley, England.

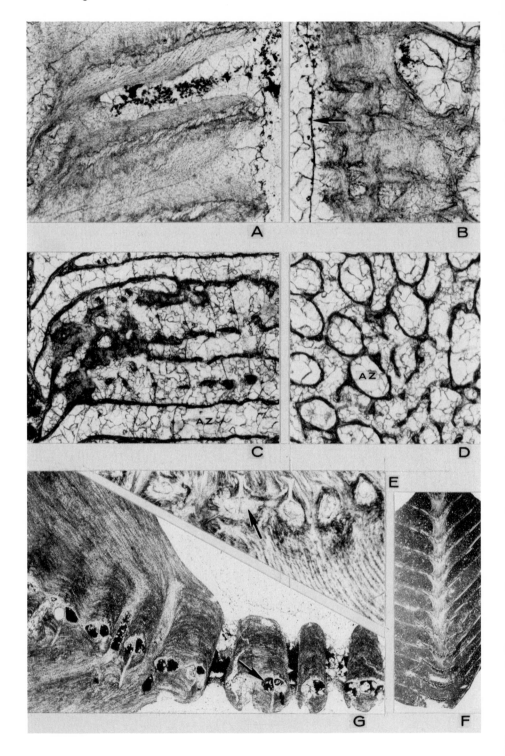

Figure 36A, B

Dendroid trepostome. A. Longitudinal section, USNM 179006, ×100, showing granular iron oxide deposit interpreted to be indications of approximate position and shape of soft parts of lophophore and gut of feeding zooid connecting to fragments of membranous-appearing deposit on right interpreted as the cuticle of outer wall. B. Longitudinal section of same specimen, ×100, showing membranous sheet (arrow) in section inferred to have been the outer cuticle of zoarium at right. Between membrane and skeleton would be outer coelomic space of double-walled colony. Waynesville Formation, Richmond Group (Upper Ordovician), Hanover, Ohio.

Figure 36C, D

Acanthoceramoporella sp. C. Longitudinal section, USNM 159794, ×30, showing well-defined tube of autozooecium (AZ) at bottom and another at top of photo with confused region of disrupted mesozooecia between, which in extreme disruption appear to be extrazooidal. D. Tangential section of same specimen, ×30, showing autozooecial tubes (AZ) cut transversely and extrazooidal space between. Arline Formation (Middle Ordovician), Friendsville, Tennessee.

Figure 36E, F

Archimedes sp. E. Transverse section USNM 182789, ×30, through paired zooecia (arrow) of fenestrate frond with carinal structures (thickened spines in section bifurcating at ends and pointing upward in figure) pointing distally in colony, and many long thin rods pointing proximally in colony. These nonlaminated primary structures are covered by secondary, extrazooidal, laminated skeleton. F. Section cut vertically through center of spiral extrazooidal supporting structure of same specimen, ×2, showing relationship with fronds of zoarium. Chester Group (Upper Mississippian), near Fox Trap Creek, 12 miles south of West Lighton, Alabama.

Figure 36G

Lyropora quincuncialis Hall. Section cut transversely through zooecia (arrow), American Museum Natural History 7873, ×30, showing increasing development of secondary skeleton to left to form massive supporting structure. Chester Group (Upper Mississippian), Chester, Illinois.

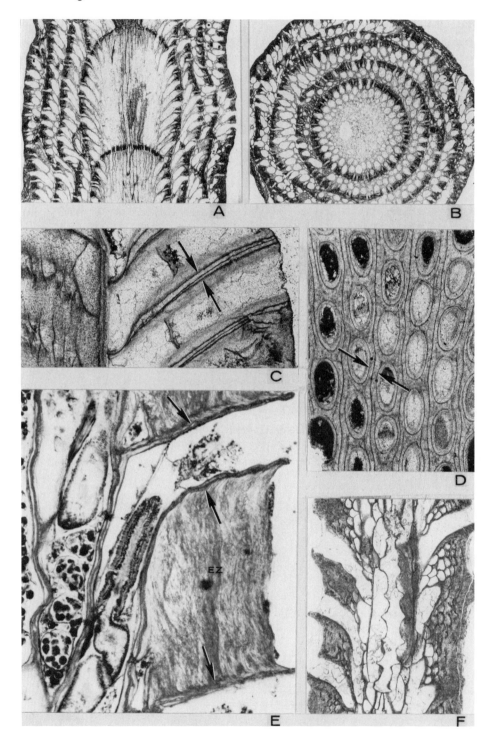

Figure 37A, B

Atagma macroporum (Hamm, 1881). A. Longitudinal sections, USNM 186564, ×7, showing apparent mode of branch growth by development of cyclic overgrowths. Many zooecia in axial regions of branch have uninterrupted living chambers from one cycle to the next. B. Transverse section of same specimen, ×7, Maastrichtian (Cretaceous), Ciply, south of Mons, Belgium.

Figure 37C

Astreptodictya fenestelliformis (Nicholson, 1875). Longitudinal section, USNM 55146, ×50, showing zooecial boundaries at dark lines (arrows) and two layers of extrazooecial wall between arrows. Maquoketa Formation, Richmond Group (Upper Ordovician), Wilmington, Illinois.

Figure 37D

Astreptodictya acuta (Hall, 1847). Tangential section of holotype, AMNH 666/1, ×30, showing zooecial boundaries (arrows) and extrazooecial skeleton between. Trenton Limestone (Middle Ordovician), Middleville, New York.

Figure 37E

Hornerid cyclostome. Longitudinal section, ×150, feeding autozooid in profile, with both soft and skeletal parts in place, showing zooid boundaries (arrows) and extrazooidal skeletal material (EZ) between zooids. Notice that further growth distally (to right) must be on outer skeletal surfaces under an outer membranous wall. Recent specimen from Arctic Ocean, British Museum (Natural History).

Figure 37F

Cheilotrypa hispida Ulrich, 1884. Longitudinal section, syntype, USNM 43277, ×30, showing extrazooidal skeleton and zooids in similar arrangement to that of the hornerid in E. Extrazooidal skeleton begins on budding surface as cysts, which in turn grow by addition to their outer surfaces. Glen Dean Limestone, Chester Group (Upper Mississippian), Sloans Valley, Kentucky.

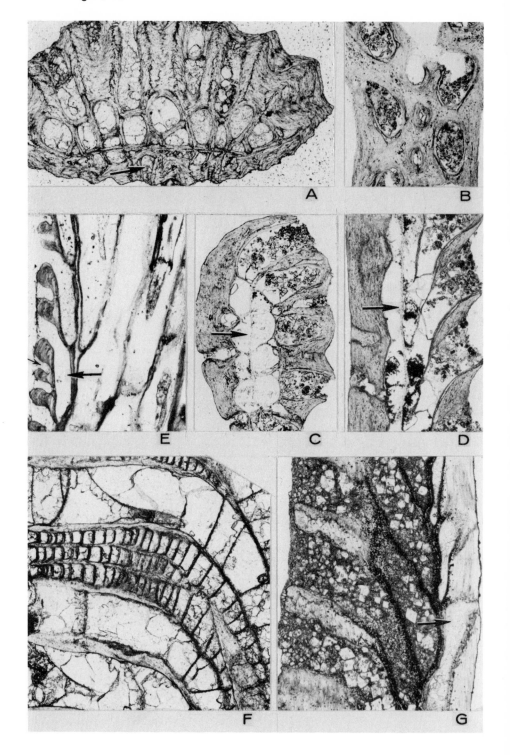

Figure 38A, G

Pseudohornera diffusa (Hall, 1852). A. Transverse section of topotype of type species of the genus, USNM 158341, ×50, showing smaller polymorphs (arrow) budding from reverse side of median layer. Zooecia, inferred to have housed feeding organs, open toward top of figure. G. Longitudinal section of another topotype, USNM 186565, ×50, showing autozooecia opening to left and smaller polymorph (arrow) opening to right. Rochester Shale Member of the Clinton Formation (Middle Silurian), Lockport, New York.

Figure 38B, C, D

Undescribed cystoporatid. B. Tangential section USNM 99061, ×50, showing larger living chambers of presumed feeding autozooids and smaller polymorphs (exilazooecia). C. Transverse section of same specimen, ×50, showing pores opening to left of median wall (arrow) and autozooecia and exilazooecia opening to right. D. Longitudinal section of same specimen, ×50, showing pits to left of median wall (arrow) and autozooecia to right. Kuckers Shale (Ordovician), Kothla, Estonia.

Figure 38E

Crisinid cyclostome. Longitudinal section, USNM 186566, ×150, showing set of smallest pores (smaller arrow) and polymorphs (larger arrow) on left side of specimen, feeding autozooecia opening to right. Philippines expedition of the Albatross, locality D5559, collected 1909.

Figure 38F

Calopora elegantula (Hall, 1852). Longitudinal section of topotype of type species, USNM 186567, ×50, showing cluster of small, closely tabulated mesozooecia in exozone opening to left. The larger tubes are autozooecia; the funnel-shaped skeletal partitions in the autozooecia at the top and bottom of figure indicate inferred polymorphic living chambers. Proximal ends of autozooecia to right in figure are smaller and more closely tabulated. Rochester Shale Member of the Clinton Formation (Middle Silurian), Lockport, New York.

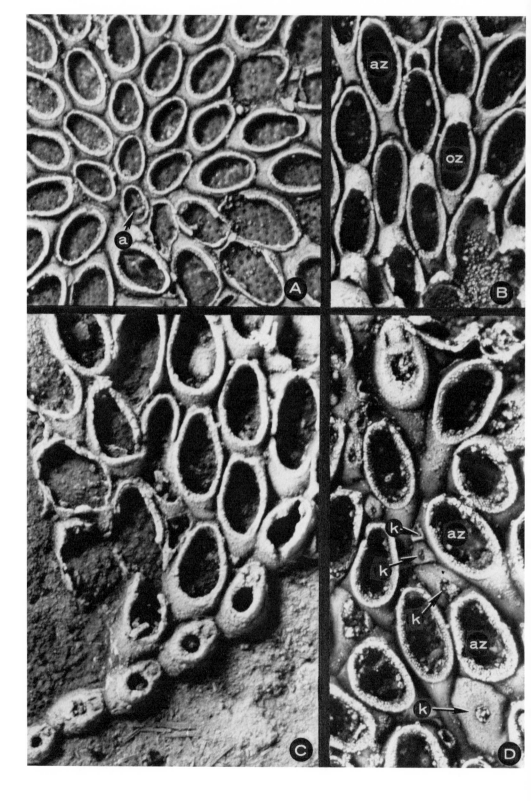

Figure 39

Astogeny and polymorphism in *Wilbertopora mutabilis* Cheetham, Lower Cretaceous (Albian), Texas. A. Characteristic budding pattern and morphologic variation among autozooecia in zone of astogenetic change, those placed proximally to ancestrula (a) belonging to later generations than those distal to it; holotype, LSU 4600, Fort Worth Limestone, Krum, Texas. B. Ovicelled (oz) and nonovicelled (az) autozooecia, zone of astogenetic repetition; USNM 651282, topotype. C. Uniserial zooecia succeeded by characteristically multiserial zooecia, zone of astogenetic change and proximal part of zone of repetition; frontal membranes and opercula of proximal zoids were calcified during life of colony; note small distal spine bases on some zooecia; USNM 186568, Kiamichi Formation, Patton, Texas (Loeblich loc. HTL/235, sample 810). D. Autozooecia (az) and kenozooecia (k), some of which gave rise to daughter zooids; note distal spine bases on some autozooecia as in C; USNM 186569, topotype. All ×50.

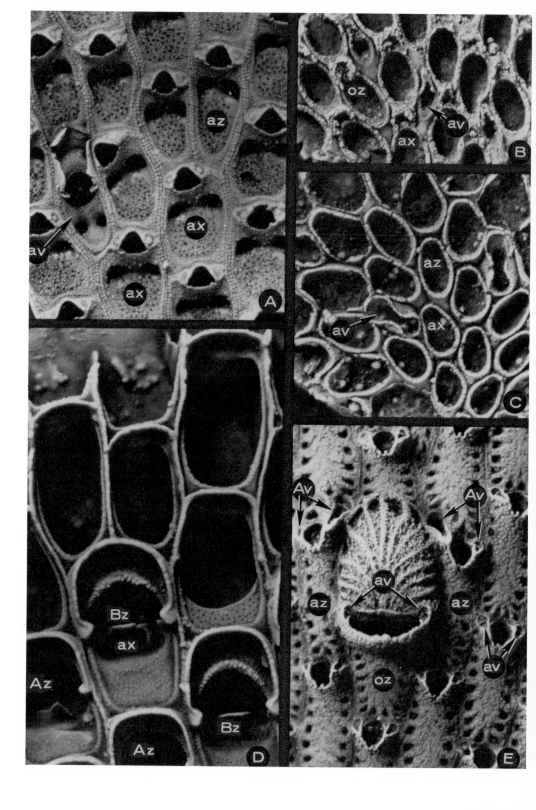

Figure 40

Partly regular arrangement of polymorphs in some cheilostomes. A. Avicularium (av) succeeds axillary autozooecium (ax) distolaterally; note that distolateral successor of another axillary autozooecium is an autozooecium (az); *Thalamoporella biperforata* Canu and Bassler, USNM 186570, Miocene, Cercado de Mao, Dominican Republic. B. Pointed avicularium (av) and ovicelled autozooecium (oz) both succeed axillary autozooecium (ax) distolaterally; *Wilbertopora* sp., USNM 186571, Upper Cretaceous (Cenomanian), Grayson Formation, near Austin, Texas (Loeblich loc. HTL-94, sample 387). C. Spatulate avicularium (av) and nonovicelled autozooecium (az) both succeed axillary autozooecium (ax) as in B; note small distal spine bases on some autozooecia; *Wilbertopora mutabilis* Cheetham, USNM 186572, Lower Cretaceous (Albian), Fort Worth Limestone, Fort Worth, Texas (Loeblich loc. HTL-43, sample 245). D. Axillary and nonaxillary B-zooecia (Bz); note that zooecia nearer growing margin had not developed morphologic features distinguishing A-zooecia (Az) and B-zooecia (Bz); *Steganoporella magnilabris* (Busk), USNM 186573, Recent, Atlantic north of Puerto Rico, Johnson-Smithsonian deep-sea expedition sta. 26, 60–73 m. E. Grouping of brooding autozooecium (oz) having similar avicularia (av), laterally adjacent nonbrooding autozooecia (az) having differentiated avicularia (Av), and other nonbrooding autozooecia having similar avicularia; *Metrarabdotos (Rhabdotometra) micropora* (Gabb and Horn) s.s., USNM 650791, Eocene, lower Red Bluff Formation, St. Stephens quarry, Alabama. All ×50.

REFERENCES

Banta, W. C. 1968a. The body wall of cheilostome Bryozoa, I, The ectocyst of *Watersipora nigra* (Canu and Bassler). Jour. Morph., *125*: 1–10, 1 pl.

——. 1968b. The body wall of the encrusting cheilostome *Watersipora nigra* (Canu and Bassler), p. 93–96, *in* E. Annoscia, ed., Proceedings of the First International Conference on Bryozoa. Atti Soc. Ital. Sci. Nat., *108*: 1–377, pls. 1–13.

——. 1969a. *Uscia mexicana,* new genus, new species, a watersiporid bryozoan with dimorphic autozoids. S. Calif. Acad. Sci. Bull., *68*, pt. 1: 30–35, 4 text-figs.

——. 1969b. The body wall of cheilostome Bryozoa, II, Interzoidal communication organs. Jour. Morph., *129*: 149–170, pl. 1, 26 text-figs.

——. 1970. The body wall of cheilostome Bryozoa, III, The frontal wall of *Watersipora arcuata* Banta, with a revision of the Cryptocystidea. Jour. Morph., *131*: 37–56, 26 text-figs.

——. 1971. The body wall of cheilostome Bryozoa, IV, The frontal wall of *Schizoporella unicornis* (Johnston). Jour. Morph., *135*: 165–184, 5 pls.

——. 1972. The body wall of cheilostome Bryozoa, V, Frontal budding in *Schizoporella unicornis floridana.* Mar. Biol., *14*: 63–71, 3 text-figs.

Bassler, R. S. 1911. *Corynotrypa,* a new genus of tubuliporoid Bryozoa: U.S. Nat. Mus. Proc., *39*: 497–527, 27 text-figs.

——. 1953. Bryozoa, *in* R. C. Moore, ed., Treatise on Invertebrate Paleontology, pt. G, xiii + 253 p., 175 text-figs.

Beklemishev, W. N. 1970. Principles of comparative anatomy of invertebrates (Transl. 3d Russian ed., Moscow, 1964, by J. M. MacLennan; Z. Kabata, ed.). v. 1, Promorphology, xxx + 490 p.; v. 2, Organology, viii + 532 p., Oliver & Boyd, Edinburgh, and Univ. Chicago Press, Chicago.

Boardman, R. S. 1960. Trepostomatous Bryozoa of the Hamilton Group of New York State. U. S. Geol. Survey, Prof. Paper, *340*: 1–87, pls. 1–22, 27 text-figs.

——. 1971. Mode of growth and functional morphology of autozooids in some Recent and Paleozoic tubular Bryozoa. Smithsonian Contr. to Paleobiol., *8*: 51 p., 11 pls., 6 text-figs.

——. Body walls and attachment organs in some Recent cyclostomes and Paleozoic trepostomes, *in* .G P. Larwood, ed., Proceedings of the Second International Conference on Bryozoa, I.B.A., Durham, England, 1971. In press.

——, and A. H. Cheetham. 1969. Skeletal growth, intracolony variation and evolution in Bryozoa: a review. Jour. Paleont., *43*; 205–233, pls. 27–30, 8 text-figs.

——, A. H. Cheetham, and P. L. Cook. 1970. Intracolony variation and the genus concept in Bryozoa. N. Am. Paleont. Conv. Proc., p. 294–320, 12 figs.

Bobin, G., and M. Prenant. 1960. *Electra verticillata* (Ellis & Solander 1786) Lamouroux 1816, (Bryozoaire Chilostome). Cah. Biol. Mar., *1*: 121–156, 1 pl., 11 text-figs.

——, and M. Prenant. 1968. Sur le calcaire des parois autozoéciales d'*Electra verticillata* (Ell. et Sol.), Bryozoaire Chilostome, Anasca, notions préliminaires. Arch. Zool. Expér. Gén. Paris, *109* (2): 157–191, 9 text-figures.

Borg, F. 1926. Studies on recent cyclostomatous Bryozoa. Zool. Bidrag. Uppsala, *10*: 181–507, pls. 1–14, 109 text-figs.

——. 1933. A revision of the recent Heteroporidae (Bryozoa). Zool. Bidrag. Uppsala, *14*: 253–394, pls. 1–14, 29 text-figs.

——. 1941. On the structure and relationships of *Crisina* (Bryozoa Stenolaemata). Arkiv för Zoologi, *33A*: 1–44, pls. 1–4, 16 text-figs.

——. 1944. The stenolaematous Bryozoa. The Swedish Antarctic Expedition, 1901–1903, 271 p., 16 pls., 26 text-figs.

Brood, Krister. 1970. On two species of *Saffordotaxis* (Bryozoa) from the Silurian of Gotland, Sweden. Stockholm Contr. in Geol., *21*(5): 58–68, 9 pls., 4 text-figs.

——. 1972. Cyclostomatous Bryozoa from the Upper Cretaceous and Danian in Scandinavia. Stockholm Contr. in Geol., *26*: 464 p., 78 pls., 148 text-figs.

Brydone, R. M. 1929. Further Notes on New or Imperfectly Known Chalk Polyzoa. Dulau & Co., London. pt. 1, p. 1–40, pls. 1–14.

Buge, E., and A. DeBourle. 1971. Présence dans le Bartonien d'Aquitaine occidentale de *Cupuladria boulangeri* n. sp. (Bryozoaire Cheilostome). Bull. Centre Rech. Pau-SNPA, *5*: 35–47, 2 pls., 2 text-figs.

Cheetham, A. H. 1954. A new early Cretaceous cheilostome bryozoan from Texas. Jour. Paleont., *28*: 177–184, pl. 20, 5 text-figs.

———. 1966. Cheilostomatous Polyzoa from the Upper Bracklesham Beds (Eocene) of Sussex. Brit. Mus. (Nat. Hist.), Bull., Geol. *13*: 1–115, 81 text-figs.

———. 1968. Morphology and systematics of the bryozoan genus *Metrarabdotos*. Smithsonian Misc. Coll., *153*(1): i–vii, 1–121, pls. 1–18, 24 text-figs.

———. 1971. Functional morphology and biofacies distribution of cheilostome Bryozoa in the Danian Stage (Paleocene) of southern Scandinavia. Smithsonian Contr. to Paleobiol. *6*: 87 p., 17 pls. 29 text-figs.

———. 1972. Cheilostome Bryozoa of late Eocene age from Eua, Tonga. U.S. Geol. Survey, Prof. Paper, *640-E*: 1–26, pls. 1–7, 7 text-figs.

———, Study of cheilostome polymorphism using principal components analysis *in* G. P. Larwood. Proceedings of the Second International Conference on Bryozoa, I.B.A., Durham, England. 1971. In press.

Cook, P. L. 1964. Polyzoa from West Africa, I, Notes on the Steganoporellidae, Thalamoporellidae, and Onychocellidae (Anasca, Coilostega). Résultats scientifiques des campagnes de la "Calypso", Iles du Cap Vert. Ann. Inst. Océan., n. ser., *41*(6): 43–78. 1 pl., 13 figs.

———. 1965a. Notes on the Cupuladriidae (Polyzoa, Anasca). Brit. Mus. (Nat. Hist.), Bull., Zool., *13*: 151–187, pls. 1–3, 6 text-figs.

———. 1965b. Polyzoa from West Africa. The Cupuladriidae (Cheilostomata, Anasca). Brit. Mus. (Nat. Hist.), Bull., Zool., *13*: 189–227, pls. 1–3, 4 text-figs.

———. 1968. Polyzoa from West Africa, the Malacostega, pt. I. Brit. Mus. (Nat. Hist.), Bull., Zool., *16*(3): 115–160, pls. 1–3, 20 text-figs.

———. Settlement and early colony development in some species of Cheilostomata, *in* G. P. Larwood, ed., Proceedings of the Second International Conference on Bryozoa, I.B.A., Durham, England, 1971. In press.

Cumings, E. R. 1904. Development of some Paleozoic Bryozoa. Am. Jour. Sci., ser. 4, *17*: 49–78, 83 text-figs.

———. 1905. Development of *Fenestella*. Am. Jour. Sci., ser. 4, *20*: 169–177, pls. 5–7.

———. 1912. Development and systematic position of the monticuliporoids. Geol. Soc. America Bull., *23*: 357–370, pls. 19–22.

———, and J. J. Galloway. 1915. Studies of the morphology and histology of the Trepostomata or monticuliporoids. Geol. Soc. America, Bull., *26*: 349–374, pls. 10–15.

Eitan, G. 1972. Types of metamorphosis and early astogeny in *Hippopodina feegeensis* (Busk) (Bryozoa:—Ascophora). Jour. Exp. Mar. Biol. Ecol., *8*: 27–30, 1 text-fig.

Franzen, A. 1960. *Monobryozoon limicola* n. sp., a ctenostomatous bryozoan from the detritus layer on soft sediments. Zoologiska bidrag från Uppsala, *33*: 138–148.

Frey, R. W., and G. P. Larwood. 1971. *Pyripora shawi*; new bryozoan from the Upper Cretaceous of Kansas (Niobrara Chalk) and Arkansas (Brownstown Marl). Jour. Paleont., *45*: 969–976, pls. 115–116, 2 text-figs.

Gautier, T. G. Growth of bryozoans of the order Fenestrata, *in* G. P. Larwood, ed., Proceedings of the Second International Conference on Bryozoa, I.B.A., Durham, England, 1971. In press.

Gautier, Y. V. 1962. Recherches écologiques sur les Bryozoaires chilostomes en Mediterranée occidentale. Rec. Trav. Stn. Mar. Endoume, Fr., *24*(38): 5–434, 91 text-figs.

Gorodiski, A., and P. Balavoine. 1962. Bryozoaires crétacés et éocènes du Sénégal. Bull. Bur. Rech. Géol. Minières, Paris, *4*: 1–10, pls. 1–2.

Gordon D. P. 1968. Zooidal dimorphism in the polyzoan *Hippopodinella adpressa* (Busk). Nature, *219*: 633–634, 1 text-fig.

———. 1971. Zooidal budding in the cheilostomatous bryozoan *Fenestrulina malusii* var. *thyreophora*. New Zealand Jour. Mar. and Freshwater Res., *5*: 453–460, 5 text-figs.

Gray, J. S. 1971. The Meiobenthic Bryozoa, p. 37–39, 2 text-figs, *in* N. C. Hulings, ed., Proceedings of the First International Conference on Meiofauna. Smithsonian Contr. to Zool., *76*: 205 p.

Gregory, J. W. 1896. The Jurassic Bryozoa. Catalogue of the fossil bryozoa in the Department of Geology, British Museum (Natural History), 239 p., pls. 1–11, 22 text-figs.

Håkansson, E. Mode of growth of the Cupuladriidae (Bryozoa, Cheilostomata) *in* G. P. Larwood, ed., Proceedings of the Second International Conference on Bryozoa, I.B.A., Durham, England, 1971. In press.

Harmer, S. F. 1902. On the morphology of the Cheilostomata. Quart. Jour. Micr. Sci., London, n. s., *46*: 263–350, pls. 15–18.

———. 1930. Polyzoa, Presidential address delivered at the anniversary meeting of the Linnean Society, Session 141, 1930. Linn. Soc. London, Proc., *141*: 68–118, pl. 1.

———. 1934. The Polyzoa of the Siboga Expedition, pt. 3, Cheilostomata Ascophora I, Family Reteporidae. Siboga Exped. Repts., Leyden, *28*c: 503–640, pls. 35–41.

———. 1957. The Polyzoa of the Siboga Expedition, pt. 4, Cheilostomata Ascophora II. Siboga Exped. Repts., Leyden, *28d*: 641–1147, pls. 42–74, text-figs. 49–118.

Hass, H. 1948. Beitrag zur Kenntnis der Reteporiden mit besonderer Berucksichtigung der Formbildungsgesetze ihrer Zoarien und einem Bericht über die dabei angewandts neue Methode für Untersuchungen auf dem Meeresgrund. Zoologica, Stuttgart, *37*: 1–138, 10 pls., 62 text-figs.

Hastings, A. B. 1930. Cheilostomatous Polyzoa from the vicinity of the Panama Canal collected by Dr. C. Crossland on the cruise of the S.Y. "St. George". Zool. Soc. London, Proc., *1929*: 697–740, 17 pls.

———. 1943. Polyzoa I. Scrupocellariidae, Epistomiidae, Farciminariidae, Bicellariellidae, Aeteidae, Scrupariidae. Discovery Repts., Cambridge Univ. Press, 22: 301–510, 9 pls., 66 text-figs.

Hillmer, G. 1971. Bryozoen (Cyclostomata) aus dem Unter-Hauterive von Nordwestdeutschland. Mitt. Geol.-Paläont. Inst. Univ. Hamburg, 40: 5–106, pls. 1–22, 30 text-figs.

Illies, G. 1953. Variationsstatistische Untersuchungen an *Rhiniopora cacus* (Bryd.) (Bryos. Cheil.) aus der Oberkreide von Hemmoor/Niederelbe. Geol. Staatsinst., Hamburg, Mitt. 22: 76–101, pls. 15–16, 8 text-figs.

———. 1963. Über *Stomatopora dichotoma* (LAMX.) und *St. dichotomoides* (d'ORB) [Bryoz. Cycl.] aus dem Dogger des Oberrheingebietes. Oberrheinische Geologische Abhandlungen, *12*: 45–80, taf. 3–9, 22 abb.

———. 1971. Drei Arten der Gattung *Stomatopora* (Bryoz. Cycl.) aus dem mittleren Lias bei Goslar und deren verschiedene Knospungsmuster. Oberrheinische Geologische Abhandlungen, 20: 125–146, 3 taf., 27 abb.

Karklins, O. L. 1969. The cryptostome Bryozoa from the Middle Ordovician Decorah Shale, Minnesota. Minnesota Geol. Survey, SP-6, spec. publ. ser., 78 p., 18 pls. 11 text-figs.

Kopayevich, G. V. 1972. Polymorphism in the family Ptilodictyidae (Bryozoa) Paleont. Jour., *1*: 57–63, 2 pls., transl. (First publ. 1971).

Lagaaij, R. 1952. The Pliocene Bryozoa of the Low Countries. Meded. Geol. Stichting, ser. C, *5*(5): 233 p., 26 pls. 29 text-figs.

———, and Y. V. Gautier. 1965. Bryozoan assemblages from marine sediments of the Rhone delta, France. Micropaleontology, *11*: 39–58, 34 text-figs., 1 chart.

Larwood, G. P. 1962. The morphology and systematics of some Cretaceous cribrimorph Polyzoa (Pelmatoporinae). Brit. Mus. (Nat. Hist.), Bull., Geol., *6*: 285 p., pls. 1–23, 132 text-figs.

———. 1969. Frontal calcification and its function in some Cretaceous and Recent cribrimorph and other cheilostome Bryozoa. Brit. Mus. (Nat. Hist.), Bull., Zool., *18*: 171–182, 10 text-figs.

———. A. W. Medd, D. E. Owens, and R. Tavener-Smith. 1967. Bryozoa, *in* The fossil record. Geol. Soc. London, p. 379–395, text-figs. 11.1–11.3.

Levinsen, G. M. R. 1902. Studies on Bryozoa. Vidensk. Medd. fra den Naturh. Foren. i Kjobenhavn, p. 1–31.

———. 1909. Morphological and systematic studies on the cheilostomatous Bryozoa. Nat. Fortfatt. Forlag, Copenhagen, vii + 431 p., 27 pls., 6 text-figs.

Lutaud, G. 1961. Contribution à l'étude du bourgeonnement et de la croissance des colonies chez *Membranipora membranacea* (Linné), Bryozoaire chilostome. Ann. Soc. Roy. Zool. Belg., *91*: 157–300, pls. 1–8, 28 text-figs.

Marcus, Ernst. 1926. Bryozoa, *in* G. Grimpe and E. Wagler, Die Tierwelt der Nord- und Ostsee; pt. 7 C-1. Leipzig, Akad. Verlagsgesellschaft, p. 1–100, 168 text-figs.

———. 1949. Some Bryozoa from the Brazilian coast. Commun. Zool. Mus. Historia Nat., Montevideo, *3*: 1–33, pls. 1–7.

Marcus, Eveline, and Ernst Marcus. 1962. On some lunulitiform Bryozoa. Bol. Fac. Fil., Cien., Letr., Univ. São Paulo, Zoologia, *24*, p. 281–324, pls. 1–5.

McKinney, F. K. 1969. Organic structures in a Late Mississippian trepostomatous ectoproct (bryozoan). Jour. Paleont., *43*: 285–288, pl. 50, 1 text-fig.

Medd, A. W. 1965. *Dionella* gen. nov. (Superfamily Membraniporacea) from the Upper Cretaceous of Europe. Palaeontology, *8*: 492–517, pls. 67–71, 9 figs.

———. 1966. The zoarial development of some membranimorph Polyzoa. Ann. Mag. Nat. Hist., ser. 13, *9*: 11–22, 3 text-figs.

Menon, N. R. 1972. Heat tolerance, growth and regeneration in three North Sea bryozoans exposed to different constant temperatures. Mar. Biol., *15*: 1–11, 14 figs.

Nielsen, C. 1970. On metamorphosis and ancestrula formation in cyclostomatous bryozoans. Ophelia, *7*(2): 217–256, 41 figs.

Nye, O. B. 1969. Aspects of microstructure in post-Paleozoic Cyclostomata. Atti. Soc. Ital. Sci. Nat. e Museo Civ. St. Nat. Milano, p. 111–114.

Powell, N. A. 1967. Polyzoa (Bryozoa)—Ascophora—from north New Zealand. Discovery Repts., *34*: 199–394, pls. 1–17, 106 text-figs.

———, and P. L. Cook, 1966. Conditions inducing polymorphism in *Thalamoporella rozieri* (Audouin) (Polyzoa, Anasca): Cah. Biol. Mar., *7*: 53–59, 1 text-fig.

Robertson, Alice. 1908. The incrusting chilostomatous Bryozoa of the west coast of North America. Univ. Calif. Pub. Zool., *4*: 253–344, pls. 14–24.

Rogick, M. D. 1956. Studies on marine Bryozoa. vii. *Hippothoa*. Ohio Jour. Sci., *56*: p. 183–191, pls. 1, 2.

Ross, J. P. 1964. Champlainian cryptostome Bryozoa from New York State. Jour. Paleont., *38*(1): 1–32, pls. 1–8, 11 text-figs.

———. 1967. Champlainian Ectoprocta (Bryozoa) New York State. Jour. Paleont., *41*(3): 632–648, pls. 67–74, 4 text-figs.

Ryland, J. S. 1970. Bryozoans. Hutchinson Univ. Lib., London. 175 p., 21 text figs.

Sandberg, P. A. 1971. Scanning electron microscopy of cheilostome bryozoan skeletons; techniques and preliminary observations. Micropaleontology, *17*: 129–151, pls. 1–4, 1 text-fig.

Schopf, T. J. M. 1969. Paleoecology of Ectoprocts (bryozoans). Jour. Paleont., *43*: 234–244, 5 text-figs.

Silén, L. 1938. Zur Kenntnis des Polymorphismus der Bryozoen. Die Avicularien der Cheilostomata Anasca. Zool. Bidrag. Uppsala, *17*: 149–366, pls. 1–18, 80 text-figs.

———. 1942a. On spiral growth of the zoaria of certain Bryozoa. Ark. f. Zool., *34A*, (n:o2): 22p., pls. 1–5, 12 text-figs.

———. 1942b. Origin and development of the cheilo-ctenostomatous stem of Bryozoa. Zool. Bidrag. Uppsala, *22*: 1–59, 64 text-figs.

———. 1944a. The anatomy of *Labiostomella gisleni* Silén (Bryozoa Protocheilostomata) with special regard to the embryo chambers of the different groups of Bryozoa and to the origin and development of the bryozoan zoarium. K. Svenska Vetensk Akad., Handl. (3) *21*: 1–111, pls. 1–5, 85 text-figs.

———. 1944b. On the formation of the interzoidal communications of the Bryozoa. Zool. Bidrag. Uppsala, *22*: 433–488, pl. 1, 59 text-figs.

———. 1944c. The main features of the development of the ovum, embryo, and ooecium in the ooeciferous Bryozoa Gymnolaemata. Ark. Zool. Stockholm, *35A*(17): 1–34, 23 text-figs.

———. 1966. On the fertilization problem in the gymnolaematous Bryozoa. Ophelia, *3*: 113–140.

Söderqvist, S. T. 1969. Observations on extracellular body wall structures in *Crisia eburnea* L. Atti Soc. Ital. Sci. Nat. e Museo Civ. St. Nat. Milano, p. 115–118, 1 pl.

Stach, L. W. 1935. The genera of Catenicellidae. Roy. Soc. Victoria, Proc. n. s., *47*: 389–396.

———. 1936. Correlation of zoarial form with habitat. Jour. Geol. *44*: 60–65, 1 text-fig.

———. 1938. Colony-formation in *Smittina papillifera* (MacGillivray, 1869) (Bryozoa). Zool. Soc. London, Proc., ser. B, *108*, pt. 3: 401–415, 1 pl., 4 text-figs.

Tavener-Smith, R. 1969. Skeletal structure and growth in the Fenestellidae (Bryozoa). Palaeontology, *12*, pt. 2: 281–309, pl. 52–56, 9 text-figs.

———, and A. Williams. 1970. Structure of the compensation sac in two ascophoran bryozoans. Roy. Soc. London, Proc., *175B*: 235–254, pls. 42–48, 35 text-figs.

————, and A. Williams. 1972. The secretion and structure of the skeleton of living and fossil Bryozoa. Roy. Soc. London Phil. Trans., ser. B. Biol. Sci., *264*: 97–159, pls. 6–30.

Thomas, H. D., and G. P. Larwood. 1956. Some "uniserial" membraniporine polyzoan genera and a new American Albian species. Geol. Mag., *93*: 369–376, 3 text-figs.

————. 1960. The Cretaceous species of *Pyripora* d'Orbigny and *Rhammatopora* Lang. Palaeontology, *3*: 370–386, pls. 60–62, 4 text-figs.

Thornely, L. R. 1912. The marine Polyzoa of the Indian Ocean, from H.M.S. Sealark. Linn. Soc. London, Trans., *15*: 137–157, pl. 8.

Ulrich, E. O. 1890. Paleozoic Bryozoa. Illinois Geol. Survey, *8*: 283–688, pls. 29–78.

Voigt, E. 1956. Untersuchungen über *Coscinopleura* Marsson (Bryoz. foss.) und verwandte Gattungen. Geol. Staatsinst., Hamburg, Mitt., *25*: 26–75, pls. 1–12, 7 text-figs.

Walter, B. 1969. Les bryozoaires jurassiques en France. Docum. Lab. Geol. Fac. Sci. Lyon, *35*: 328 p., 20 pls., 16 text-figs.

Waters, A. W. 1923. Mediterranean and other Cribrilinidae. Ann. Mag. Nat. Hist. *12*(9): 545–573, pls. 17, 18.

————. 1924. The ancestrula of *Membranipora pilosa* Linné, and of other cheilostomatous Bryozoa. Ann. Mag. Nat. Hist., ser. 9, *14*: 594–612, pls. 18, 19.

————. 1925. Ancestrulae of cheilostomatous Bryozoa, pt. II. Ann. Mag. Nat. Hist., ser. 9, *15*: 341–352, pls. 21, 22.

Wiesemann, G. 1963. Untersuchungen an der Gattung *Beisselina* Canu 1913 und ähnlichen Bryozoen (Maastrichtien, Danien, Montien). Geol. Staatsinst. Hamburg, *32*: 5–70, pls. 1–12, 22 text-figs.

Wilson, E. O. 1953. The origin and evolution of polymorphism in ants. Quart. Rev. Biol., *28*: 136–156, 10 text-figs.

————. 1968. The ergonomics of caste in the social insects. Am. Naturalist, *102*: 41–66, 9 text-figs.

Woollacott, R. M., and R. L. Zimmer. 1972. Origin and structure of the brood chamber in *Bugula neritina* (Bryozoa). Mar. Biol., *16*: 165–170, 4 figs.

The Effect of Colony Morphology on the Life-History Parameters of Colonial Animals*

Karl W. Kaufmann   University of Chicago

ABSTRACT

The colony morphology of many bryozoans and other colonial animals may be divided into three types: vine-like, encrusting, and upright bushy forms. The colony growth rate for these forms can be expressed by linear, quadratic, and exponential functions of time, respectively. The birth schedule may then be expressed as integrals of these equations.

The limiting factor of larva production is the rate of energy consumption. This is proportional to the number of feeding zoids. The energy must be apportioned among budding new zoids, forming nonfeeding polymorphic zoids and extra calcification, and the production of larvae. These factors are fixed for each species, so that one can determine the effect of colony morphology on the intrinsic rate of natural increase (r) for individual colonies.

Computer modeling of birth schedules and consideration of relative availability of substrata suggest that vine-like and lightly calcified encrusting forms potentially have the highest r and would be selected for under conditions of ample substrate availability. Encrusting and bushy forms with heavy calcification and many polymorphic zoids

Author's address: Department of Geophysical Sciences, University of Chicago, Chicago, Illinois 60637.

*Reprinted from Geological Society of America, Abstracts with Programs (1971 Annual Meeting, Washington), v. 3, no. 7, p. 618.

would be at an advantage under crowded conditions, where substrate availability depends on the death of established colonies and other sessile organisms.

Thus the effect of colony morphology on the life history can be separated from the physical strategy of adaptation.

Intra- and Intercolony Variation in Populations of *Hippoporina* Neviani (Bryozoa–Cheilostomata)

Marie B. Abbott

Marine Biological Laboratory
and The University of Connecticut

ABSTRACT

A partly quantitative analysis of about 250 colonies from three entirely sympatric and synchronous living populations (taken in a Petersen grab at a depth of 41 m, salinity 31.34⁰/₀₀, temperature 12.7°C, on Nov. 17, 1970, in northwestern Block Island Sound, New York) documents the nature, range, and limits of intercolony variation in selected autozooid variates and establishes the existence of morphologic gaps between the populations.

The populations studied represent three species tentatively assigned to the genus *Hippoporina* Neviani: *H. porosa* (Verrill, 1879), *H. americana* (Verrill, 1875), and *H. cf. H. verrilli* Maturo and Schopf, 1968. A polyphyletic origin of the genus is suggested by the ancestrula, which closely resembles the first few generations in *H. porosa* but is a "tata" in the other two species.

The analysis is made within an astogenetic framework (modified from Boardman, Cheetham, and Cook, 1970) which is also a classification of colonies by growth pattern and an index of the level of functional coloniality. The principal innovative features are: (1) The ancestrula is recognized as a separate growth phase; (2) Type A colonies, in which growth has been continuous, are distinguished from Type R colonies, in which growth has been interrupted and renewed (regenerated) one or more times; and (3) colonies are classified by the highest level of functional coloniality

Author's address: Gray Museum, Marine Biological Laboratory, Woods Hole, Massachusetts 02543.

attained. In Level I colonies, represented by the ancestrula and dormant colonies, coloniality is only potential and not yet expressed. In Level II colonies, those undergoing prepolymorphic change and/or repetition, coloniality is only partially or incompletely expressed. In Level III colonies, those in which polymorphs have appeared, coloniality is fully functional.

INTRODUCTION

In marine bryozoa the colony originates with an ancestrula (commonly single, but twinned or multiple in some species) metamorphosed from a sexually produced larva and grows by vegetative budding into a theoretically infinite series of successive generations of intercommunicating zooids, all sharing and expressing the same genotype. This clonal nature of the colony has not, however, been sufficiently recognized in bryozoan systematic and taxonomic studies; furthermore, these have been based largely on the typologic rather than the population concept of the species.

This paper is intended as a pilot study, the purpose of which is to document the nature and limits of variation in selected surficial skeletal morphologic characters within and between living colonies of three apparently related populations provisionally assigned to the encrusting ascophoran genus *Hippoporina* Neviani, 1895. The analysis is designed within an astogenetic framework which stresses the functional and systematic importance of the condition of colonial polymorphism.

PREVIOUS COLONY-VARIATION STUDIES

As far as the writer knows, no study has yet been made in which intercolony variation in a bryozoan species is analyzed quantitatively in colonies drawn from demonstrably sympatric and synchronous living populations rather than in isolated or type specimens or in collections from diverse and perhaps inadequately documented sources.

Existing studies actually treating colony variation have been based on fossil material, principally: Illies (1953) on the cribrimorph *Rhiniopora cacus* from the Upper Cretaceous of Germany; David and Mongereau (1961) on *Cellaria fistulosa* from the Miocene of France; Cheetham (1966) on cheilostomes from the Eocene of southern England; Cuffey (1967) on *Tabulipora carbonaria* from the midcontinent; and Farmer and Rowell (1971) on the Pennsylvanian *Fistulipora decora*.

The most recent discussions on the taxonomic significance and measurement of quantitative characters in bryozoa are in Cheetham (1966, 1968). Although some authors include statistically processed measurements of type or other specimens in their descriptions, they rarely indicate where in the colony the measured zooids are located or whether they were selected at random or in a predetermined pattern or sequence. Their data thus do not necessarily reflect the full range of intracolony variation of even that specimen and tell nothing of intercolony variability in populations of the species represented. Other workers in systematic bryozoology have reported measurements without giving the total number of observations or their dispersion, or without noting whether they represent zooids in the same or different colonies or a mixture of both. The value of such quantitative data is at best very limited.

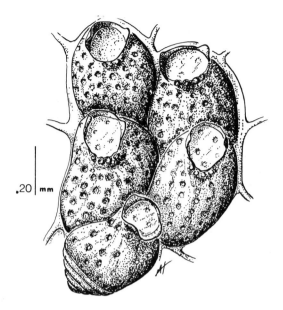

.20 | mm

Figure 1

Hippoporina porosa (Verrill, 1879). Colony 47b (USNM 180746). Station A, Block Island Sound, N.Y., Nov. 17, 1970. Growth phase A2 (ancestrula plus generations I and II). Distinguishable from both associated *Hippoporina* species by nontata ancestrula with transversely wrinkled proximal tip and by significantly and consistently larger size of zooids in all growth phases (Table 5, Figs. 5 and 6). The suboral beaded rim is more pronounced in ontogenetically older zooids.

MATERIALS AND METHODS

The *Hippoporina* population samples studied were taken by Petersen grab at a previously described and located permanent oceanographic station in Block Island Sound, New York (Abbott, 1971, in press) at a depth of 41 m, salinity 31.43 $^0/_{00}$, water column isothermal at 12.7°C, on Nov. 17, 1970. These particular population samples, out of a suite of over 40 such samples from the same site, each with full ecologic data, were selected as the source of materials for this pilot study of intercolony (population) and interspecies variation mainly because all three species apparently assignable to the genus *Hippoporina* were reproducing sexually at this same time and place.

The genus *Hippoporina* itself was selected for four reasons: (1) Its three species are dominants in a bryozoan epifaunal community comprising about 30 species in 24 genera. They are therefore present in large numbers (almost always >15 and commonly >50 colonies) at all temperatures and in all samples in the study suite, although in varying proportions. (2) All ontogenetic and astogenetic stages of each species occur within the entire sample suite, although not necessarily within any one sample. (3) Its external morphologic characters, those most readily accessible in dried or fossil material, are relatively few and simple. (4) The availability of ecologic control data makes possible future studies on the possible effects of known and measured environmental factors on colony variation in this genus and other associated genera as well.

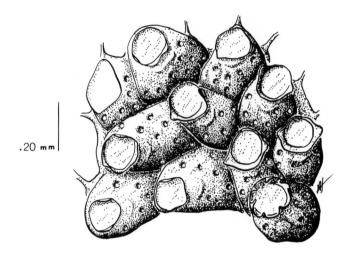

.20 mm

Figure 2

Hippoporina americana (Verrill, 1875). Colony 50a (USNM 180747). Station A, Block Island Sound, N.Y., Nov. 17, 1970. Growth phase A2 (tata ancestrula plus generations I–III). Distinguishable from other *Hippoporina* species by smooth, low, imperforate area proximal to orifice, by small number (usually <15) and large size of pores in proximal half to two-thirds of frontal surface, and by horseshoe-shaped secondary orifice. Ancestrula differentiated from that of *H.* cf. *H. verrilli* (Fig. 3) by short cryptocyst (not visible) and three pairs of lateral spines.

The three *Hippoporina* species are relatively easy to distinguish from each other even if colonies are very young, infertile, or poorly preserved (Figs. 1–3). They are *Hippoporina porosa* (Verrill, 1879), type locality: Fishers Island Sound, N.Y.; *Hippoporina americana* (Verrill, 1875), type locality: off Noank, Conn.; and *Hippoporina* cf. *H. verrilli* Maturo and Schopf, 1968, from northwestern Block Island Sound. About 250 *Hippoporina* colonies were examined and parts or all of 78 randomly selected colonies measured. Specimens figured or measured have been deposited in the Division of Invertebrate Paleontology of the U.S. National Museum of Natural History, Washington, D.C.

Although large numbers of dead colonies of *Hippoporina* and other species are associated with living colonies in this bryozoan community, only the latter were treated in this study. A colony was considered alive if it could be seen or presumed to include feeding zooids when taken, but the entire colony need not necessarily consist of living zooids. If, however, a colony contained only dead or presumably dead zooids, then the colony as a functioning unit was considered dead. Among the principal criteria weighed and evaluated together for classifying a colony as living are the presence of a more-or-less intact epifrontal membrane, which usually appears as a transparent, lustrous, or hyaline sheath over the entire upper surface of the colony and which extends beyond the calcified portion of it at the growing edge; pigmentation beneath the frontal walls and in the ovicells; an operculum fitted into condyles; and the absence of characteristics obviously associated with dead and worn colonies, such as chalky appearance, empty zooids, sealed opercula, and broken walls or spines, clearly not caused by accidents in collection or preparation.

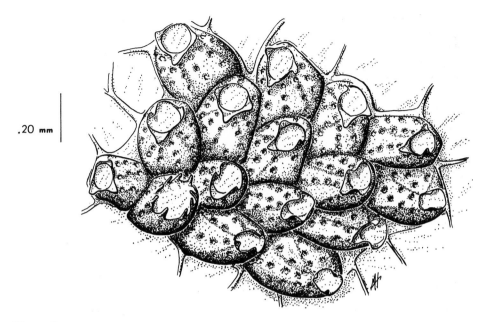

.20 mm

Figure 3

Hippoporina cf. *H. verrilli* Maturo and Schopf, 1968. Colony 47 (USNM 180748). Station A, Block Island Sound, N.Y., Nov. 17, 1970. Growth phase A3 (ancestrula plus generations I–IV). Distinguishable from both associated *Hippoporina* species by significantly smaller zooid size, by a distinct trilobate secondary orifice, and by the presence of occasional avicularia. Ancestrula with four pairs of lateral spines and no cryptocyst.

Types of Colonies

The living zooids in a colony may exhibit either of two types of growth patterns; the zooids may be budded in an uninterrupted, continuous growth sequence from a recognizable ancestrula (or clearly identifiable ancestrular site), or they may be budded in an interrupted, discontinuous sequence from existing apparently dead but actually merely dormant zooids, in one or more cycles of colony regeneration. A colony in which growth has been continuous may be called a Type A colony, one showing regeneration a Type R colony. If it is impossible to distinguish the origin of the living zooids, usually because the colony is too fragmented or broken even though alive when taken, that colony may be termed a Type X colony. Like any other colony, it represents a genotype and can therefore be listed for ecological diversity and density counts, but it is of little value in colony-variation study because of the lack of an astogenetic base line. On the premise, based on the scanning of several hundred colonies of *Hippoporina* and other species in this and other associated population samples, that the patterns of variation in skeletal morphology may be affected or possibly even determined by the regenerative process, Type A and Type R colonies were analyzed separately. Quantitative study was concentrated on small young Type A colonies, which constitute a high proportion of the population sample in all three species.

A few problems arose in grouping colonies by type. In large Type R colonies with many confluent and overlapping new growth sites it is sometimes difficult to determine whether the new growth represents one or more colonies. The nature of the contact between regenerating areas may be a useful criterion, at least in these *Hippoporina* species: if the growing edges flare up in a pseudobilaminate fold, the lobes belong to two incompatible genotypes and therefore by definition to different colonies; if they appear to merge or become fully confluent, to the same colony. It is also not always easy to distinguish between Type R colonies, in which growth is discontinuous, and Type A colonies, in which growth of the colony as a whole appears to be uninterrupted but in which some or all of the older zooids in the earlier generations are moribund or dead. Ordinarily, revitalization of the cystid and growth of a new bud is abrupt and easily recognizable, but occasionally new growth (especially that initiating frontal budding) begins with an apparent gradual renaissance of the epifrontal membrane of an existing older zooid; it is these latter cases that are ambiguous.

Character Variates

The zooid character variates selected for measurement or count are those most accessible in encrusting bryozoa, those of ventral exoskeletal morphology, the specific choice of variates depending on the particular astogenetic growth phase and species of the colony under study (Table 1). In all cases, five standard variates—length and width of the zooid, length and width of the primary orifice, and the number of frontal pores—were measured or counted. Measurements were taken in frontal projection on the specimens themselves, at $100\times$ magnification, to the nearest 0.01 mm.

Processing of tabulated data was confined to the simplest methods commensurate with yielding the information desired at this stage of study. For each set of ungrouped raw data the following statistics were calculated: arithmetic mean, variance, standard deviation, standard error of the mean, and coefficient of variation (corrected if $N<15$).

The statistical data are reported according to the following sequential format, extended from that of Cheetham (1966), in which both its position and the presence or absence of parentheses automatically identify a given statistic.

(Number of Zooids, Number of Colonies)   Mean $\pm$ Standard Error
(Standard Deviation)   Range   (Coefficient of Variation).

A suggested abbreviated formula, illustrated for the variates Lz and lz in Tables 3–5, would be $(N, N\text{col})$ $\bar{x} \pm$ S.E. (s) Range (V).

For this study, only those statistics pertinent to the demonstration or testing of the proposed astogenetic framework (Lz, lz, V) are presented, the others being considered more appropriate to the actual systematic description and revisions of *Hippoporina* and related genera (manuscripts in preparation).

INTERCOLONY-VARIATION ANALYSIS

A bryozoan colony is a group of structurally and/or functionally integrated individuals vegetatively reproduced from one zygote. It is, therefore, a clone, and variations in appearance and behavior within it must all be reflections of the phenotypic range of

Table 1

Zooid Character Variates Analyzed Quantitatively in *Hippoporina* Species[a]

Growth Phase[b]	Variates						
			H. porosa				
A1	La	la	Lo	lo	Ltrem	Lgym	Nfp
A2	Lz	lz	Lo	lo			Nfp
A3	Lz	lz	Lo	lo			Nfp
R1	Lz	lz	Lo	lo			Nfp
R2	Lz	lz	Lo	lo			Nfp
R3	Lgz	lgz	Lo	lo	Lov	lov	Nfp
			H. americana				
A1	La	la	Lfm	lfm	Lgym	Lcryp	
A2	Lz	lz	Lo	lo	Lolo		Nfp
A3	Lz	lz	Lo	lo	Lolo		Nfp
A4	Lz	lz	Lo	lo	Lolo		Nfp
			H. cf. *H. verrilli*				
A1	La	la	Lfm	lfm	Lgym		
A2	Lz	lz	Lo	lo			Nfp
A3	Lz	lz	Lo	lo			Nfp
A4	La	lz	Lo	lo	Lav	lav	Nfp

[a]Abbreviations for zooid character variates:

L, length	o, primary orifice	gym, gymnocyst
l, width	ov, ovicell	olo, olocyst
a, ancestrula	av, avicularium	cryp, cryptocyst
z, zooid	fm, frontal membrane	trem, tremocyst
gz, gonozooid	Nfp, number of frontal pores	

The last four terms abbreviated are used only in a descriptive and locative sense.

[b]See text for description.

expression possible to one genotype, if rare somatic mutations are disregarded. Since such variations are clearly nongenetic in origin, they must be attributed to one or more other sources (Boardman, Cheetham, and Cook, 1970, p. 300).

General Requirements for Intercolony-Variation Analysis

For valid comparison of zooids from different colonies representing different genotypes (intercolony variation) there appear to be two general requirements. First, the sources of (intracolony) nongenetic variation must be recognized and each source then substracted (or otherwise discounted) from the total apparent intercolony variation in order to isolate and identify the nature and magnitude of the true intercolony variation ascribable to genetic differences alone. Second, the relation in time and space of the populations from which the compared colonies are drawn—whether allopatric or sympatric—must be either documented or inferred to determine whether they can or cannot represent one or more species.

The most obvious source of intracolony, interzooid variation is polymorphism, the mechanism for division of labor evolved most elaborately in the two perhaps most success-ful groups of colonial animals, the coelenterates and the bryozoa. In the latter, polymorphism is commonly but not necessarily always associated with the sexual-reproductive function of the colony, the gonozooid being the most conspicuous example in cheilostomes and virtually the only expression of polymorphism in the three *Hippoporina* species of this study.

The chronologic age of the zooid, which determines the stage of ontogenetic develop-ment it has reached, and its astogenetic position, the point or chronologic stage in the life of the colony at which the zooid appears, also affect morphology and must be distinguished from each other. Both ontogenetic and astogenetic stages form a sequential record parallel to the direction of budding but in opposing gradients, the former directed proximally away from the growing edge, the latter distally away from the ancestrula (Boardman, Cheetham, and Cook, 1970, p. 301). The effects of ontogenetic and astogene-tic changes and of polymorphism may be eliminated or at least reduced markedly by restricting observations to corresponding phases or stages of these three extragenetic factors. The chief problem arising in this regard lies in the definition and delimitation of astogenetic stages (see below).

The effects of the environment on a bryozoan population may be considered on two levels. The first comprises the major, usually abiotic, limiting factors such as temperature, salinity, turbulence, etc., which operate not only on an entire colony but on the population as a whole to determine its distribution and viability. These may be called macroenviron-mental, to differentiate them clearly from other minor, transient, incidental, local factors, both abiotic and biotic (such as crowding, overgrowth, or grazing by another epibenthic organism, substrate irregularities, exposure to light) which operate on another level and affect only a given colony or part of it. These are usually termed microenvironmental. In this study, in which hundreds of entire colonies as well as fragments were examined and which is limited to colonies taken alive and with their associated living community, the cause of a specific microenvironmentally induced variation could usually be identified. Any macroenvironmental effects are automatically eliminated from consideration by the restriction of this pilot study to one population sample, assuming that no significant environmental change has occurred in the life of the colony.

The fully sympatric and synchronous nature of the *Hippoporina* population sampled for this study is implicit in the sampling method. Since the three *Hippoporina* species are reproducing sexually at the same time and place, it would be physically possible for them to interbreed if their gene pools were compatible. If any gaps between them can be inferred qualitatively and/or documented quantitatively, the taxa must represent three distinct, noncompatible gene pools, i.e., three species, and not spatially, temporally, or reproductively isolated subspecies.

Astogenetic Growth-Phase Framework

Any framework or classification within which variation in populations of colonial animals—whether morphologic, physiologic, or ecologic—is observed should theoreti-cally serve two functions simultaneously.

1. It should clearly recognize the coloniality of the animal and the functional as well as morphologic significance of the condition of colonial polymorphism. Only when

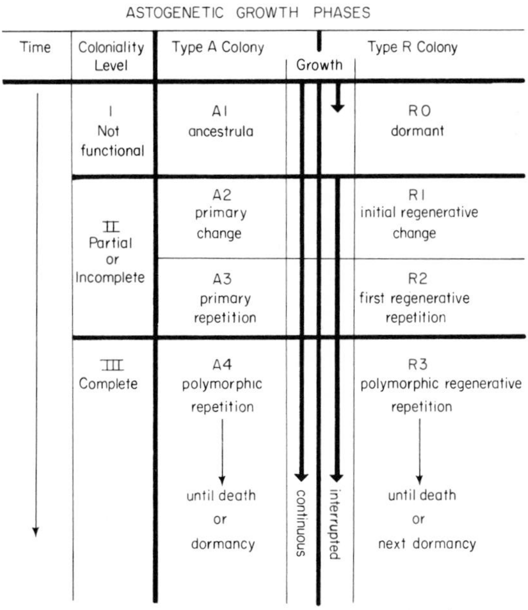

ASTOGENETIC GROWTH PHASES

Time	Coloniality Level	Type A Colony	Growth	Type R Colony
	I Not functional	A1 ancestrula	↓	R0 dormant
	II Partial or Incomplete	A2 primary change		R1 initial regenerative change
		A3 primary repetition		R2 first regenerative repetition
	III Complete	A4 polymorphic repetition		R3 polymorphic regenerative repetition
		↓ until death or dormancy	continuous ⬇ interrupted	↓ until death or next dormancy

Figure 4

Chart showing astogenetic growth phases to or through which Type A colonies (growth continuous from the ancestrula) and Type R colonies (growth interrupted and renewed) pass with time, correlated by the level of functional coloniality represented. In the ancestrula (A1) and in dormant (R0) colonies, coloniality is potential but not yet functional (Level I). In colonies undergoing pre-polymorphic primary change (A2) or initial regenerative change (R1), and primary repetition (A3) or the first regenerative repetition (R2), coloniality is only partial or incomplete (Level II). Only in colonies in which polymorphs have appeared, and division of labor with differential use of energy and materials has begun, is coloniality expressed as completely (Level III) as is possible to the species.

the first polymorphic zooids appear and division of labor in the colony begins, can the colony be considered a fully developed, integrated colonial individual expressing phenotypically the full functional morphologic potential of its specific genotype.

2. It should be so structured as to operate as a template, if perhaps only a pragmatic one, for identifying, isolating, and if necessary measuring extragenetic, intracolony sources of variation from other genetically induced intercolony (population) and interpopulation (interspecies) variations which carry a different order of systematic or phyletic significance.

A classification of bryozoan colonies which attempts to meet the above criteria—devised primarily for living encrusting Cheilostomata but potentially applicable with modifications to other growth forms and possibly to fossil forms—is herewith proposed. In this classification, summarized in Fig. 4, the concepts and terminology of Boardman, Cheetham, and Cook (1970) are modified and extended in a number of significant directions to become functionally as well as morphologically and topographically descriptive. The principal modifying features of this classification are:

1. Type A colonies, in which growth has been continuous, are distinguished from

Type R colonies, in which growth has been interrupted and renewed (regenerated) one or more times.

2. The level or degree of functional coloniality to which a given colony has developed, usually as a function of its chronologic age, is the principal basis for differentiating astogenetic growth phases. In the ancestrula and in dormant colonies coloniality is only potential (Level I), not expressed or functional to any degree. At the other extreme, in colonies in which polymorphic zooids have appeared and division of labor with differential use of energy and materials by individual components of the colonial organism has begun, coloniality is expressed as completely (Level III) as is possible to the species to which the colony belongs. Coloniality continues to function at this level until death or dormancy. At intermediate stages of development, in colonies in which zooids have budded from either the ancestrula or from a previously dormant parental zooid and are undergoing only prepolymorphic change and/or repetitive growth, coloniality may be considered intermediate, partial, or incomplete (Level II).

3. A given colony is classified by the highest level of functional colonality which it exhibits as belonging to or representing a particular (abstract) astogenetic *growth phase*. The concrete expression of this abstraction lies in the number and development of visible, tangible astogenetic *zones* exhibited by the colony, each corresponding to an astogenetic growth phase *to or through* which the colony has progressed. A zooid, on the other hand, is characterized or classified by the astogenetic *zone* in which it is located; e.g., an A2 zooid must be located in the A2 zone of a colony, but that colony itself may be an A2, A3, or A4 colony.

4. The ancestrula is recognized as a separate and distinct astogenetic growth phase, partly because of the level of colonial function it represents and partly because the intercolony variations it exhibits must be at least in part genetic in origin. The dormancy growth phase in Type R colonies, whether expressed in single or multiple regenerative cycles, may be correlated functionally with the ancestrula only in the level of coloniality (Level I) it represents.

Application and Test of Astogenetic Framework

An obvious prerequisite for the comparison of zooids in corresponding growth phases of two or more colonies is recognition of the boundaries within a colony of these growth phases. In Type A colonies the distinction between the ancestrula (A1) and its next succeeding series of zooids (A2) is usually self-evident. The first appearance of polymorphs and the entry into the fully colonial A4 phase is also clearly recognizable. The transition from growth phase A2 to A3, analogous to but not necessarily equivalent to the transition from PC (primary change) to the PR (primary repetition) zones of Boardman, Cheetham, and Cook (1970), may be much more gradational, and the rate of change—and therefore the number of zooid generations required for the transition to be completed—may not be the same for each character (Boardman and Cheetham, 1969, p. 255). The problem is further compounded if the species is an encrusting ascophoran, such as *Hippoporina,* in which only the calcified ventral exoskeleton is exposed.

In an attempt to assess the value of statistical parameters of characters as a discriminatory tool for recognition of growth-phase boundaries, generations I, II, and III of each species

Table 2

Coefficients of Variation for Principal Zooid Character Variates in *Hippoporina* Species, Type A Colonies, by Growth Phase[a]

Lz see: lz		H. porosa Table 5	H. americana Table 3	H. cf. H. verrilli Table 4
Lo	AI	6.10	—	—
	A2,I	11.06	5.71	6.75
	A2,II	11.25	9.69	6.91
	A2,III	10.64	9.26	6.02
	A3	6.63	11.77	7.47
	A4	—	3.23	11.46
lo	AI	6.03	—	—
	A2,I	7.73	8.03	9.15
	A2,II	5.16	8.57	6.62
	A2,III	4.59	9.61	7.69
	A3	4.70	11.41	2.29
	A4	—	0.00	7.36
Nfp	AI	14.40	—	—
	A2,I	21.40	29.49	23.17
	A2,II	17.13	37.21	19.45
	A2,III	14.18	27.76	21.65
	A3	—	26.82	14.67
	A4	—	24.78	17.93

[a]Based on same numbers of colonies as in Tables 3–5.

were studied separately. The five principal character variates (Lz, lz, Lo, lo, Nfp) were measured or counted on every visible, entire zooid in each of these generations in a total of 78 Type A colonies.

A priori, it would seem that in order to serve as a growth-phase boundary indicator a character should possess at least these three properties: (1) It should be relatively insensitive to ontogenetic change, i.e., it should reflect the anticipated astogenetic changes modified only by known or unknown microenvironmental factors; (2) it should be of a size range great enough to be adequately revealed by a 0.01-mm measurement interval; and (3) it should be relatively stable or at least consistent, this stability being expressed in low (<10) value of its coefficient of variation (V).

In Table 2, V values for the principal zooid characters are given by species. It is immediately apparent that frontal pore counts (Nfp), although meristic and therefore theoretically preferable to the continuous variates, have unacceptably high V's, possibly because their count is even more subject to human error than linear measurements, possibly because of their high sensitivity to ontogenetic changes—especially those involving progressive calcification of the frontal wall. The dimensions of the internal orifice (Lo, lo) appear very stable, V for all but the A4 zooids of *Hippoporina* cf. *H. verrilli* being <10, but the actual size of the orifice itself (0.10–0.16 mm depending on species and growth phase) is very small compared to the unit of measurement.

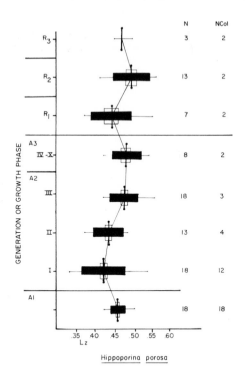

Hippoporina porosa

Figure 5

Range diagram showing change in Lz (mm) in *Hippoporina porosa* with change in growth phase or generation by colony type. Zooids apparently cease to grow in length after generation III in Type A colonies; the ancestrula is consistently larger than the zooids of generations I and II (A2); and Type R zooids are appreciably shorter in the R1 zone of initial regenerative change. On this and succeeding range diagrams, horizontal lines show the observed range, vertical lines the mean, solid black horizontal bar the standard deviation, outlined black vertical rectangle the standard error of the mean.

The length and width of the autozooid (Lz, lz) remain as the characters which might best test the growth-phase discriminatory power of sequential generation analysis. Accordingly, these variates and their central tendencies were plotted graphically in gradient or range diagrams similar to those used for demonstrating clines (Simpson, Roe, and Lewontin, 1960, p. 354ff.) against applicable growth phases for which processed data were available. They are shown in Figs. 5 and 6 for *Hippoporina porosa*, Figs. 7 and 8 for *H. americana*, and Figs. 9 and 10 for *H. cf. H. verrilli*.

Inspection of the range diagrams indicates that zooid length in *Hippoporina porosa*, width in *H. americana*, and both length and width in *H. cf. H. verrilli* cease to increase after generation III. Data for zooid width in *H. porosa* suggest, rather less distinctly, a similar decrease. In *H. americana*, however, zooid length shows no such consistent cessation or deceleration of growth increment at this phase of colony development.

It does then seem that breaks in slope (i.e., in the rate of change) in size of zooid,

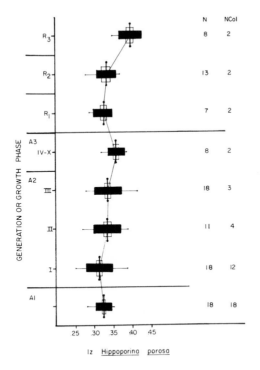

Figure 6

Range diagram showing change in lz (mm) in *Hippoporina porosa* with change in growth phase or generation, by colony type. The growth gradients are less pronounced than for Lz, but follow the same general pattern.

in either length or width or both, between generation III and the next succeeding generations may be used as a criterion for marking the transition from the A2 to the A3 growth phase. This criterion should be applied with caution until data from a larger number of zoids, from a larger number of colonies and different populations, and from sequential generation analysis on A3 zoids become available from work in progress on this and associated population samples.

Data for Type R colonies are currently available only for *Hippoporina porosa*. They appear to support the validity of distinguishing between Type A and Type R colonies. The R1 growth-phase zoids are more nearly comparable to A2 zoids, at least in length, than to either A3 or R2 zoids.

A major anomaly is presented by the several autozooids immediately adjoining the single fully developed gonozooid in the one large A4 *Hippoporina americana* colony present in the population sample. They are significantly larger than the zoids of all earlier generations of other colonies of that species. To test quantitatively whether this change is gradational or the result of chance variation, the length and width of 18 contiguous zoids, one from each successive generation of that colony, were measured and plotted (Fig. 11). It is clear that the variation is not gradational but random, or

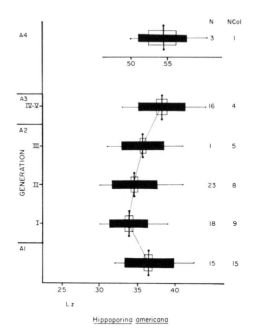

Hippoporina americana

Figure 7

Range diagram showing change in Lz (mm) in Type A colonies of *Hippoporina americana* with change in growth phase or generation. Zooid length shows no consistent cessation or deceleration of growth increment at the transition between A2 and A3; other growth-phase boundaries appear similarly gradational. The autozooids representing A4 are significantly longer than zooids of earlier generations. This difference is not gradational but random (cf. Fig. 11).

perhaps cyclic, and also that zooid width within as well as between colonies is more stable than zooid length. Growth rate in the latter, at least in this species, may be triggered by an unidentified microenvironmental factor, possibly a positive or negative photo- or geotropism. A very large colony such as this one may also have been affected in the course of its lifetime by macroenvironmental factors such as temperature changes.

VARIATION IN *HIPPOPORINA* POPULATIONS: OVERVIEW

The astogenetic framework, which is also a classification of bryozoan colonies and an index of coloniality level, was further tested by employing it in the study of morphologic variations, growth phase by growth phase, in each of the three *Hippoporina* populations of the study sample.

Coloniality Level I: Growth Phases A1 and R0

If or while the *Hippoporina* or any other bryozoan colony consists of only an ancestrula (growth phase A1) or only of apparently dead but actually merely dormant zooids as evidenced by later regeneration of the same colony (growth phase R0—not R1 because

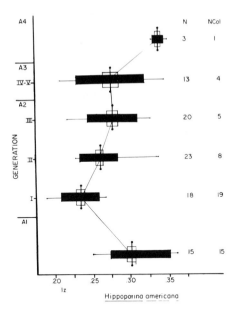

Figure 8

Range diagrams showing change in lz (mm) in Type A colonies of *Hippoporina americana* with change in growth phase or generation. Width of zooid appears to stabilize after generation III. The zooids representing A4 are the same as those shown in Fig. 7.

the zooids are not functionally alive at the same time as the later regenerating zooids), it is not operating as a colony. Its coloniality is potential but not yet expressed. At this level, however, zooid skeletal characters are particularly easy to discern—in the ancestrula because it is visible in its entirety, and in the dormant zooids because of the absence of the obscuring epifrontal membrane.

Even when seen alone without later zooids the three ancestrulae are rather easily distinguished from each other. Those of *Hippoporina americana* and *H.* cf. *H. verrilli* are both "tatas" (Figs. 2 and 3), quite unlike their own succeeding zooids but comparable to the ancestrula and the vegetatively produced zooids of typical calloporids or to the ancestrulae of cribrilinid species. They are similar in size (Tables 3 and 4) and shape, although *H. americana* frequently appears more trapezoidal or cuneiform in outline, a difference not revealed by the properties of its variates. They can be distinguished from each other by the presence of a short cryptocyst in *H. americana* and the highly consistent number of paired lateral spines (three pairs in *H. americana* and four pairs in *H.* cf. *H. verrilli*). The significantly larger ancestrula of *H. porosa* (Table 5) differs qualitatively from succeeding zooids only by the presence of a crescentic nonperforated gymnocyst-like proximal border (Fig. 1) of uncertain origin or microstructure.

All variation in ancestrulae is implicitly intercolonial. The interspecies gaps are clear. If the three species currently placed in *Hippoporina* are correctly assigned, the genus is probably polyphyletic in origin in view of the two types of ancestrulae.

Regeneration, and therefore Type R colonies and R0 growth phases, occurs in all three species but is uncommon in *Hippoporina* cf. *H. verrilli*, not only in this but

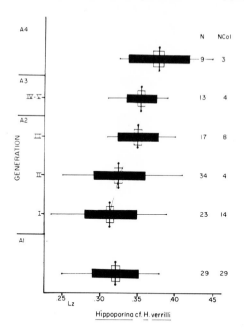

Figure 9

Range diagram showing change in Lz (mm) in Type A colonies of *Hippoporina* cf. *H. verrilli* with change in growth phase or generation. Zooids apparently cease to increase in length after generation II or do so much more slowly.

Table 3

Intercolony Variation in *Hippoporina americana*[a] Populations at Station A, Block Island Sound, Nov. 17, 1970: Lz and lz

Variate	Growth Phase	(N, Ncol)	$\bar{x}$ ± S.E.	(s)	Range	(V)
Lz	A1	(15, 15)	0.365 ± 0.008	(0.032)	0.32–0.42	(8.74)
	A2,I	(18, 9)	0.338 ± 0.006	(0.025)	0.30–0.39	(7.34)
	A2,II	(23, 8)	0.347 ± 0.006	(0.031)	0.30–0.41	(8.87)
	A2,III	(19, 5)	0.358 ± 0.006	(0.027)	0.31–0.41	(7.55)
	A3,IV and V	(16, 4)	0.382 ± 0.008	(0.031)	0.33–0.44	(8.01)
	A4	(3, 1)	0.543 ± 0.022	(0.038)	0.50–0.57	(7.51)
lz	A1	(14, 14)	0.300 ± 0.007	(0.027)	0.25–0.36	(9.13)
	A2,I	(17, 9)	0.240 ± 0.006	(0.026)	0.19–0.27	(10.94)
	A2,II	(23, 8)	0.263 ± 0.005	(0.027)	0.23–0.34	(9.96)
	A2,III	(20, 5)	0.268 ± 0.007	(0.033)	0.22–0.33	(12.29)
	A3,IV and V	(15, 4)	0.278 ± 0.011	(0.044)	0.21–0.35	(16.00)
	A4	(3, 1)	0.337 ± 0.006	(0.011)	0.33–0.35	(3.52)

[a]Parts or all of 18 randomly selected colonies measured.

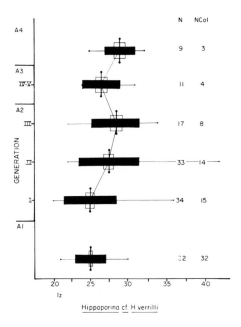

Figure 10

Range diagram showing change in lz (mm) in Type A colonies of *Hippoporina* cf. *H. verrilli* with change in growth phase or generation. Zooids apparently cease to increase in width after generation III.

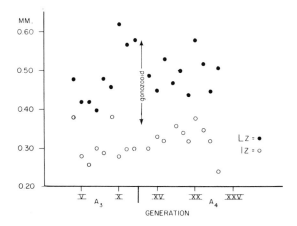

Figure 11

Scatter diagram showing distribution of values for Lz and lz in 19 contiguous zooids, one from each successive generation in one *Hippoporina americana* colony. Generations XII and XIV represent A4 in Figs. 7 and 8. Variation appears not gradational but random or perhaps cyclic, and zooid width appears more stable than zooid length.

Table 4

Intercolony Variation in *Hippoporina* cf. *H. verrilli*[a] Populations at Station A, Block Island Sound, Nov. 17, 1970 : Lz and Iz

Variate	Growth Phase	(N, Ncol)	$\bar{x} \pm$ S.E.	(s)	Range	(V)
Lz	A1	(29, 29)	0.326 ± 0.006	(0.032)	0.25–0.38	(9.86)
	A2,I	(33, 14)	0.313 ± 0.006	(0.034)	0.24–0.39	(10.93)
	A2,II	(34, 14)	0.326 ± 0.006	(0.034)	0.25–0.41	(10.37)
	A2,III	(17, 8)	0.351 ± 0.007	(0.027)	0.31–0.40	(7.80)
	A3,IV and V	(13, 4)	0.354 ± 0.006	(0.020)	0.31–0.39	(5.89)
	A4	(9, 3)	0.378 ± 0.014	(0.041)	0.33–0.45	(11.19)
Iz	A1	(32, 32)	0.253 ± 0.004	(0.029)	0.21–0.30	(9.00)
	A2,I	(34, 15)	0.250 ± 0.006	(0.034)	0.20–0.36	(13.68)
	A2,II	(33, 14)	0.274 ± 0.007	(0.039)	0.22–0.42	(14.25)
	A2,III	(17, 8)	0.284 ± 0.008	(0.032)	0.22–0.34	(11.19)
	A3,IV and V	(11, 4)	0.264 ± 0.008	(0.025)	0.24–0.31	(9.78)
	A4	(9, 3)	0.291 ± 0.009	(0.026)	0.25–0.32	(9.20)

[a]Parts or all of 36 randomly selected colonies measured.

Table 5

Intercolony Variation in *Hippoporina porosa*[a] Populations at Station A, Block Island Sound, Nov. 17, 1970: Lz and Iz

Variate	Growth Phase	(N, Ncol)	$x \pm$ S.E.	(s)	Range	(V)
Lz	A1	(18, 18)	0.046 ± 0.005	(0.020)	0.43–0.51	(4.40)
	A2,I	(18, 12)	0.431 ± 0.014	(0.059)	0.34–0.55	(3.71)
	A2,II	(13, 4)	0.443 ± 0.011	(0.040)	0.38–0.50	(9.20)
	A2,III	(18, 3)	0.484 ± 0.009	(0.039)	0.43–0.57	(7.99)
	A3,IV–X	(8, 2)	0.490 ± 0.014	(0.039)	0.43–0.55	(8.15)
	R1	(7, 2)	0.451 ± 0.021	(0.056)	0.38–0.55	(12.86)
	R2	(13, 2)	0.505 ± 0.013	(0.048)	0.42–0.57	(9.61)
	R3	(3, 2)	0.483 —	—	0.46–0.51	—
Iz	AI	(18, 18)	0.323 ± 0.005	(0.020)	0.28–0.35	(6.10)
	A2,I	(18, 12)	0.313 ± 0.008	(0.035)	0.25–0.39	(11.06)
	A2,II	(11, 4)	0.334 ± 0.011	(0.037)	0.27–0.39	(11.25)
	A2,III	(18, 3)	0.338 ± 0.008	(0.036)	0.28–0.42	(10.64)
	A3,IV–X	(8, 2)	0.361 ± 0.008	(0.023)	0.32–0.39	(6.63)
	R1	(7, 2)	0.327 ± 0.008	(0.021)	0.29–0.35	(6.72)
	R2	(13, 2)	0.335 ± 0.007	(0.027)	0.28–0.37	(8.15)
	R3	(8, 2)	0.404 ± 0.011	(0.032)	0.35–0.43	(8.16)

[a]Parts or all of 24 randomly selected colonies measured.

also in other associated population samples. If they had been alive, most R0 phase colonies would be seen to have progressed to full coloniality (Level III), with gonozooids showing all stages of ovicell development. Although no quantitative study in R0 colonies has yet been made, it would probably be highly desirable to include the properties of the zooids of the last generation of the R0 phase in a modern comparative systematic study.

Coloniality Level II: Growth Phases A2 and R1, A3 and R2

In all Type A *Hippoporina* colonies the first generation consists of a triad, one distal and two distolateral buds produced by the ancestrula. At about the time this original triad has produced the zooids of the fourth or fifth generation, these zooids and also those of the preceding (II and III) generations begin proximolaterally directed periancestral budding. In *H. porosa* and *H. americana* the true generation number of the proximolaterally directed growth can be identified fairly easily, but in *H.* cf. *H. verrilli* this growth is so closely appressed to the ancestrula that it is impossible to determine whether the contact is merely physical or functional as well without dissection and destruction of the colony. A similar problem and possible source of sampling error was also encountered by Illies, whose term "Zirkulus" (1953, p. 82ff.) appears to carry a locative or topographic rather than chronologic or developmental connotation.

A typical A2 *Hippoporina* colony consists of the ancestrula, which functions as a feeding autozooid, plus the first three generations. The primary astogenetic changes are a relatively rapid increase in size (Tables 3–5, Figs. 5–10) and marked increase in regularity of outline. Colonies are fan-shaped, and zooids are frontally inflated and commonly irregularly pentagonal or hexagonal in outline. Distolateral budding is usually bilaterally symmetrical but in *H. americana* particularly may be suppressed on one side for no apparent reason.

In a colony which has progressed to the A3 growth phase, zooids are progressively more advanced ontogenetically toward the ancestrula, the most visible signs of ontogenetic aging being the development of a secondary orifice, especially in *Hippoporina americana* and *H.* cf. *H. verrilli,* and the sealing or coalescing of frontal pores, most pronounced in *H. americana*. Colonies are now oval or round, the growing edge usually forms a semicircle or arc $>200°$ or even a complete circle as proximolateral as well as distal and distolateral budding progresses. Zooids either cease to grow in size or, as in *H. americana,* do so at a slower and less regular rate; they are less inflated and commonly rectangular in outline. There is no evidence of polymorphism, but the colony has preempted substrate space and carries the potential for further zooid differentiation and full coloniality. The zone of primary repetition comprises generations IV through at least VIII in all three species. The ancestrula usually becomes nonfunctional at some stage in this growth phase.

Regenerated colonies operating at a comparable prepolymorphic level of coloniality differ from colonies in which growth has been uninterrupted in that growth phases R1 and R2 are very commonly telescoped in terms of the number of generations required to progress through them. The R1 zone of initial regenerative change is usually limited to the first (or, rarely, the first two) generations after the parental cystid is reactivated. Zooids from this zone are characterized by significantly smaller size (Figs. 5 and 6) and great irregularity of form, especially if they are budded frontally. By far the most common type of regenerative budding is terminal. When frontal budding does occur, it is limited to the R1 zone, to the first generation of initially regenerated zooids, which almost invariably produce the next generation by terminal budding. Zooids in the R2 zone (the zone of first regenerative repetition) are very similar to those in the A3 zone in size, regularity, and lack of gradients in variates. A colony may require only one or two generations to effect the transition to full polymorphic coloniality. It may be significant that Type A colonies of *Hippoporina americana* and *H. porosa,* in which regeneration is most common, may require 15 or 20 generations of continuous growth before the first polymorphs appear; in contrast, Type A colonies of *H.* cf. *H. verrilli,* in which regeneration is much more rarely seen, may produce polymorphs as early as generation IX and usually always by generation XII. Regeneration may possibly be looked on as a "quick recovery" mechanism for rapid exploitation of any advantageous environmental changes, whether fortuitous or cyclic. Type R colonies, once reactivated, achieve full coloniality much more quickly than Type A colonies. They seem to retain even in dormancy the level of coloniality attained in life.

Because of the absence of polymorphism and primary astogenetic change, variation within growth phases A3 and R2 would theoretically be expected to be less than in either of the contiguous phases. The A3 phase is rather poorly represented in all the *Hippoporina* species in the environment of Nov. 17, 1970, but particularly poorly in the cases of *H. porosa* (only six colonies) and *H. americana* (four colonies). Some of the few available population statistics for this phase are included in the range diagrams and Tables 2–5.

Coloniality Level III: Growth Phases A4 and R3

With the development of polymorphic heterozooids, the colony achieves full colonial development and the ability not only to survive and maintain itself but also to reproduce and disperse its species sexually. All the *Hippoporina* species brood their larva in ovicelled gonozooids, but only one (*H.* cf. *H. verrilli*) produces adventitious avicularia as well, and these only rarely. When present, they are small, narrow, directed proximolaterally or laterally, and located beside the orifice. The ovicells in *H. porosa* and *H.* cf. *H. verrilli* are similar, both globose and finely punctate, the latter smaller in size; that of *H. americana* consists of a flattened frontal area, perforated by a few large irregular pores, surrounded by a smooth vertical wall or "ectocyst."

Although growth phases A4 and R3 are termed phases of polymorphic repetition, the first generation containing polymorphs may properly be considered a zone of polymorphic change, only succeeding generations being truly zones of polymorphic repetition. To add to the semantic confusion, some large colonies may exhibit alternating bands of polymorphic and nonpolymorphic repetition within the major zone of polymorphic repetition. Nevertheless, it remains true that the colony, once it has entered this growth phase and remains alive, is operating at its highest level of coloniality.

In this study sample, the polymorphic phase is absent entirely in *H. porosa* and is represented by only two colonies of *H. americana* and four of *H.* cf. *H. verrilli.* Some of the available statistics are also included in the range diagrams and in Fig. 11.

SUMMARY AND CONCLUSIONS

The colonial mode of the bryozoa is both an asset and a disadvantage in their systematic study, an asset because it permits comparison within a colony of phenetically variable but genetically identical individuals, a disadvantage because in intercolony comparisons the sources of nongenetic variation must be assessed and isolated to identify the nature and magnitude of the true intercolony variation ascribable to genetic differences alone.

If differences due to polymorphism can be eliminated by recognition of an astogenetic growth phase or zone (A4 or R3 in this study) when it appears and by restriction of comparisons to autozooids within a given polymorphic phase, and if the effects of astogenetic changes can be eliminated or reduced by further restriction of comparisons to zooids from corresponding growth phases of different colonies, as has been attempted in this study, then any remaining intracolony variation ought to be either ontogenetic or microenvironmental in origin.

The effects of ontogenetic variation may be greatly reduced in either or both of two ways: (1) by selecting zooid character variates logically presumed or actually observed to be relatively insensitive to ontogenetic change, e.g., the variates Lz and lz used to test the astogenetic framework; and (2) by grouping compared zooids not only by generation number *from* the ancestrula, but *to* it, i.e., by generation number proximal to the growing edge, and comparing the sum of variations thus secured with those for zooids grouped by generation from the ancestrula.

If ontogenetic effects were minimized by such groupings, microenvironmental factors would remain as the sole as-yet-undetected source of variation within a colony. To aid in identifying such factors, the following model or structural framework for quantitative intracolony variation, applicable primarily but not exclusively to encrusting species, is suggested by the results of the intercolony-variation study:

I. Zooids belonging to the same astogenetic zone
   A. Contiguous
      1. In the same generation proximal *from* the growing edge
      2. In adjacent generations proximal *to* the growing edge
   B. Not contiguous
      1. But not randomly selected
         a. In different generations
         b. In both the same and different generations
      2. Randomly selected
II. Zooids belonging to different astogenetic zones (interior classification as in A and B)

A comparison of zooids from different colonies, representing not only corresponding growth phases but parallel positions within the growth-phase zones, would presumably yield the fullest possible information on the range of phenotypic variation possible to the respective genomes. The information derived from data on contiguous zooids would also bear on the problem of somatic inheritance: the transfer of some substance, enzyme, or substrate along a physiologic gradient and its possible attenuation in two or three generations. This might be considered a special case of microenvironmental control of variation.

Polymorphism is the particular mechanism evolved by colonial organisms for maximizing efficiency in the division of labor, comparable to the evolution of organ systems in solitary organisms. The existence and phenetic expression of polymorphism should be recognized explicitly in the systematic, which is ultimately also the phylogenetic, study of colonial organisms. The concept of a species in bryozoa should be governed and defined by the full range of both intra- and intercolony variation in characters, by the individuality of the entire fully developed colony and not on the individuality of the separate components. It is on bryozoan colonies and not on zooids that selection pressures ultimately operate. The whole is indeed greater than the sum of its parts.

ACKNOWLEDGMENTS

The writer is grateful to A. H. Cheetham, U.S. National Museum, Smithsonian Institution, and to P. L. Cook, British Museum (Natural History), for stimulating discussions on bryozoan systematics and coloniality that have assisted the preparation of this paper. This expression of appreciation does not of course imply their endorsement of any of the views expressed in it.

The study samples were collected under U.S. Navy Contract 00140-69-C-0031 to the Marine Research Laboratory of the University of Connecticut through the courtesy of A. J. Nalwalk, Chief Scientist. This paper is Contribution No. 85 of the Marine Research Laboratory.

The illustrations were prepared by or under the direction of M. N. Hubbard of the University of Connecticut; the manuscript was typed by E. S. Montiero of the Systematics-Ecology Program, Marine Biological Laboratory, Woods Hole. The research was supported in part by a grant from the Research Foundation of the University of Connecticut. All these sources of assistance and support are acknowledged with warm thanks.

REFERENCES

Abbott, M. B. 1971. Intra- and intercolony variation in populations of *Hippoporina* Neviani (abst.). Geol. Soc. America, Abstracts with Programs, *3*(7): 487.

————. Seasonal diversity and density in bryozoan populations from Block Island Sound, New York, U.S.A. Proceedings of the Second International Conference on Bryozoa, Durham, England, September 1971. In press.

Boardman, R. S., and A. H. Cheetham. 1969. Skeletal growth, intracolony variation, and evolution in bryozoa: a review. Jour. Paleont. *43*(2): 205–233, pls. 27–30.

————, A. H. Cheetham, and P. L. Cook. 1970. Intracolony variation and the genus concept in bryozoa. N. Am. Paleont. Conv. Proc., Sept. 1969, pt. C, p. 294–320.

Cheetham, A. H. 1966. Cheilostomatous polyzoa from the Upper Bracklesham beds (Eocene) of Sussex. Brit. Mus. (Nat. Hist.), Bull., Geol., *13:* 1–115.

————. 1968. Morphology and systematics of the bryozoan genus *Metrarabdotos*. Smithsonian Misc. Col., *153*(1): 1–121, pls. 1–18.

Cuffey, R. J. 1967. Bryozoan *Tabulipora carbonaria* in Wreford Megacyclothem (Lower Permian) of Kansas. Univ. Kansas Paleont. Contrib., Bryozoa, art. 1, 96 p., 9 pls.

David, L., and N. Mongereau. 1961. Un exemple d'étude statistique en paléontologie: *Cellaria fistulosa* (Bryozoa, Cheilostomata) du Vindobonien de la région lyonnaise. Bull., Bur. Rech. Géol. Minières, *3:* 29–41, pls. 1–3.*

Farmer, J. D., and A. J. Rowell. 1971. Variation in the bryozoan *Fistulipora decora* (Moore and Dudley) from the Bell limestone (Pennsylvanian) of Kansas. (abst.). Geol. Soc. America, Abstracts with Programs, *3*(7): 562.

Illies, Gisela. 1953. Variationsstatistische Untersuchungen an *Rhiniopora cacus* (Bryd.) (Bryoz.-Cheil.) aus der Oberkreide von Henmoor/Niederelbe. Mitteil., Geol. Staatsinst. Hamburg, *22:* 76–101, pls. 15, 16.

Maturo, F. J. S., and T. J. M. Schopf. 1968. Ectoproct and endoproct type material: Re-examination of species from New England and Bermuda named by A. E. Verrill, J. W. Dawson and E. Desor. Postilla (Peabody Mus. Nat. Hist., Yale Univ.), *120:* 2–95.

Simpson, G. G., A. Roe, and R. C. Lewontin. 1960. Quantitative Zoology. Harcourt, Brace & World, New York. 440 p.

*Not seen.

Ergonomics of Polymorphism: Its Relation to the Colony as the Unit of Natural Selection in Species of the Phylum Ectoprocta

Thomas J. M. Schopf   University of Chicago

ABSTRACT

The critical notion in this paper is that natural selection in ectoprocts is operating at the colonial level of organization. Given this, the most fundamental aspects of the development of important types of polymorphs (avicularia, vibracularia) can be understood. Specialization within the colony will be enhanced to the extent that the colony as a whole is the beneficiary in terms of optimization of the production of larvae viable in the next generation.

A model for the development of polymorphism in ectoprocts is derived from the model which E. O. Wilson (1968) used to predict the characteristics of caste formation in the social insects. Predictions from the model include (1) the continual evolutionary advantage for a species to partition the work of a colony into tasks to be performed by specific polymorphs; (2) the enhancement of polymorph development to the extent that the environmental stimulus is repeated; and (3) the counteracting influence of a variable environment, which will make polymorphism unrewarding to the colony, and thus lead to the disappearance of polymorphs.

In response to these predictions, the extremely widespread development of polymorphs in ectoprocts is anticipated: approximately 75 percent of the cheilostome fauna from

Author's address: Department of the Geophysical Sciences, and Committee on Evolutionary Biology, University of Chicago, Chicago, Illinois 60637.

the tropics, the arctic, and the deep sea have at least an avicularium, a vibracularium, or both. The greater degree of specialization in polymorphs in more stable environments is anticipated; the development of two or more types of these polymorphs, or three or more, is significantly higher in the tropics than the arctic. Similarly, the absence of polymorphs in species of the typical estuarine fauna is explicable. Finally, various specialized polymorphs are most extensively developed in more stable environments.

INTRODUCTION

The answer to the question of whether the basic unit of an ectoproct is a single zooid *or* the whole colony, is yes! On the one hand, perhaps the correct place to begin in terms of explaining selection with respect to physiological properties is to consider aspects of individual zooids. However, the full evolutionary panorama seems only comprehensible in terms of selection at the colony level. At this degree of organization, all the main features of an ectoproct as a sexually reproducing and polymorphic unit are fully displayed. Such a colony is apparently an integrated sensory unit. It also is capable of passing food among individuals on a local level. And a colony is a single genotype. This paper is an analysis of biotechnology at the level of organization of the colony, or more precisely in ergonomics: the quantitative study of the distribution of work, performance, and efficiency in a society or colony (Wilson, 1963).

But if selection is at the colony level of organization, how large is a colony? Although much has been written about the potential ability of colonies to increase without bound, this is certainly not the usual field situation. I know of no data which could be used to determine if a ''mature'' colony size exists for various species, as is the case in social insects. My impression is that a mean size exists more often than is commonly realized. As a first judgment, it is perhaps best to consider the unit of selection in the vast majority of cases to be a colony that is sufficiently large that the zone of serial repetition is dominant [the stage of astogenetic repetition (Boardman and Cheetham, 1969)].

The second general point regarding evolution within the ectoprocts is that the development of polymorphic individuals (castes) is essentially a behavioral response. Polymorphs result from the way that members of the colony react to the selection by environmental influences. Since approximately 75 percent of cheilostome ectoprocts exhibit polymorphism, the behavior of these animals appears to be both complicated and plastic.

We do not yet understand the vast majority of the specific behavioral responses leading to polymorphism, although I believe that ectoproct–arthropod interactions will prove to be extremely important (see Kaufmann, 1971). I thus attribute much more biological significance to the existence of polymorphism than is sometimes done. There does appear to be a correlation between the appearance of some types of polymorphism and irregularities or small spaces in the growth pattern of colonies (this is the lack-of-space-for-a-normal-zooid-and-therefore-avicularia-form theory). Yet I believe that this correlation does not explain the causal factor since there is no obvious reason why avicularia are especially suited for simply filling space. Indeed, avicularia are extremely complex zooids.

THE MODEL

I am here concerned with one major question. What are the general selection pressures that determine the kinds and degree of polymorphism? In particular, why is there so much variation in polymorphism among species of ectoprocts? For the purpose of the present analysis, only avicularia and vibracularia (here combined and referred to as defensive polymorphs) are considered as types of polymorphism (see Figs. 2 and 3). These are well-established variations, and they are abundantly represented in ectoprocts. However, at least five other categories of polymorphism can be recognized. These include (with examples 1–4 from the West Atlantic tropical fauna) (1) kenozooecia *(Pasythea tulipifera);* (2) dwarf zooecia without polypides *(Pyripora audens);* (3) "zoécies saillantes" *(Schizoporella trimorpha);* (4) two types of opercula for normal zooids *(Dakaria vaginata* and *Gemelliporidra aculleata);* and (5) sexual dimorphism in male and female zooids *(Hippothoa hyalina* visible in calcified skeleton; *Hippopodinella adpressa* and *Hippoporidra senegambiensis* visible in polypides; see Cook, 1968; Gordon, 1968).

I find it helpful to think of the development of polymorphism as a contingency plan. It is a tradeoff within the colony such that some specialists are evolved at the expense of generalists. And since the environment is always changing over time, some "potential generalists" may, in fact, turn out many generations later to have become "specialists," and vice versa. Add also the fact that the terms "specialist" and "generalist" are relative to each other, and I conclude that a mix of generalists and specialists will always be maintained. The environment is a dynamic entity, and any purely static approach which too narrowly conceives of fixed responses is bound to be inadequate when viewed over a time span of many generations.

The initial step in the consideration of the development of polymorphism (Fig. 1A) is to conceive of need curves for the cases in which two castes exist, such as feeders and defenders. From Figure 1A, the number of individuals of cast 1 (α) needed to perform task 1 (feeding) is fewer than would be required if caste 2 were providing for task 1; and vice versa. (The terms "defending" and "feeding" should not be taken too literally, since one type of defense that feeders could and probably do provide is simply to produce so many individuals that all potential enemies are satiated, with the result that some feeders survive to do the business of the colony. Similarly, one type of feeding that defenders could have is through a functional polypide.)

Begin with the upper part of the line for task 2 and move down that line toward point 0_1. α will move down, and β will move out to the right. (α is the number of individuals of cast 1 and β is the number of individuals of caste 2 for any point in the graph at which both tasks are fulfilled.) You will see that α will decrease in value *faster* than β is increasing. Next begin with point 0_1 and move to the right on the line for task 1. Now α will decrease in value *slower* than β is increasing. Thus point 0_1 represents a nodal point. It is the minimum sum of $\alpha + \beta$ for any point on the parts of the lines we have just traced. (These parts of the lines for task 1 and task 2 are the only parts compatible with the completion of both tasks). If the production of larvae viable in the next generation is fixed, then this production will be highest *per zooid* at point O_1 since the fewest number of zooids is required at that point. This discussion follows the technique of linear programming analysis as used by Wilson (1968) in his study of castes in social insects.

(If we were to replace "number" of individuals of a caste in Fig. 1 by the percentage of the energy of the colony devoted to that caste, we would be closer to obtaining one type of information that we are after. It should be possible to feed colonies isotopically labeled carbon via specific compounds or phytoplankton, and then monitor the relative amounts that go into maintaining the colony, and into forming normal zooids, larvae, and various polymorphs. This would provide an answer in terms of energetics to the question of the relative cost of different castes within a colony.)

Let us now vary the environment. We can conceive of a change in selection pressures which would favor a different optimal mix of castes (Fig. 1B). This considers the case where additional selection results in a colony with additional members of caste 2 (de-

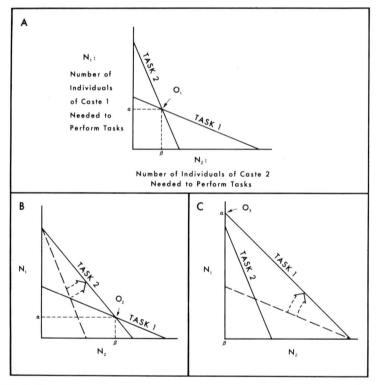

Figure 1A, B, C

Graphic representations of the number of individuals of a given caste required to perform different tasks. O_1, O_2, and O_3 are the equilibrium points toward which evolution will proceed given the specified relationships of castes to tasks. α is the number of individuals of caste 1. β is the number of individuals of caste 2. 1A is the initial condition. B is a later condition at which the importance of task 2 has increased, with a change in the ratio of α to β. C is yet a later condition at which the importance of task 1 has greatly increased, so much so that caste 2 is no longer of selective advantage (see text for further explanation).

fenders). The new optimum mix, point 0_2, has a larger proportion of caste 2 than in the case previously considered. (Compare the sum of $\alpha + \beta$ with that in Fig. 1A.)

Now let us vary the environment in another way (Fig. 1C). This considers the case where there is considerably increased selection for caste 1 (feeders). There is, in fact, no point at which a caste suitable for task 2 (defenders) is able to provide an adequate feeding response (i.e., perform task 1) for this environmental situation. As a result a separate defending caste is no longer of selective advantage. The optimum mix ($\alpha + \beta$) thus consists entirely of feeders.

With this background, we can summarize as follows:

1. As long as the needs for different tasks occur with relatively constant frequencies, it is advantageous for a species to evolve in such a way that each mature colony includes one caste specialized to respond to each kind of task. Limited only by the creative potential of a colony, and the external constraints of the environment, the number of castes will come to equal the number of tasks. Throughout this process, the energy expenditure (here seen as the number of individuals) required to maximize the production of viable larvae will be reduced by selection (Fig. 1A).

2. The countering force which leads to variations in the ratio of castes, and indeed to their disappearance, is an environment in which the selection pressures are continually changing (Fig. 1A, C). Stability of the selection pressure enhances both caste formation and further caste specialization up to the point where selection pressures are relatively variable for the particular task being performed.

3. But from 1, a species will tend to evolve a substitute caste that matches the task and *is* within the optimal mix. This new caste will also be subject to variation in the selecting force, as stated in 2.

4. Thus the more stable the selection forces, the larger the number of species with castes, the larger the average number of castes per species, and the more specialized the castes that can evolve.

5. In the long run, a very specialized caste is likely to be relatively rare and its specific value to the colony is likely to be high because it is dependent upon a more specific environmental stimulus for its formation than is a generalized caste. However, the rarer the selection force, the more likely it is to change. Thus in evolutionary time the turnover rate of very specialized castes is likely to be much higher than is the turnover rate of generalized castes.

THE DATA

Now let us consider what we know of the distribution of castes in the light of these predictions from the model. Diagrammatic zooids with different types and shapes of polymorphs are in Figs. 2 and 3. More realistic, excellent pictorial summaries of various types of avicularia, and different shapes of mandibles, each shown to scale, are presented by Kaufmann (1971, figs. 1, 2).

We have anticipated the widespread extent of polymorphism within colonies of cheilostome ectoprocts, but the apparently low degree of polymorphism in cyclostome ectoprocts remains unexplained. I will return to that problem later. Within the cheilostomes,

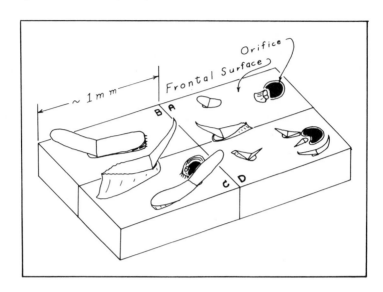

Figure 2

Schematic sketch of four hypothetical zooids from an encrusting ectoproct colony. Drawn to demonstrate a variety of sizes and shapes of avicularia. Zooids A, C, and D have frontal "adventitious" avicularia which rest on normal zooids; zooid B is an immense "vicarious" avicularium which has taken the place of a normal zooid in the budding sequence. Zooid A has a small, suboral rounded avicularium and a larger, frontal "spathulate" avicularium. Zooid C has a small, frontal spathulate avicularium and a large frontal, pointed avicularium. Zooid D has a pointed, suboral avicularium and three pointed, frontal avicularia. A pointed frontal avicularium is between zooids A and D.

polymorphism appears more highly developed in the more stable tropical environment than in the less-stable, high-latitude arctic environment. [To some, the arctic appears stable since the temperature regime probably fluctuates over a very narrow range. However, the seasonal oscillations in nutrients are both more severe and more pronounced in the arctic than in temperate or tropical latitudes. In addition, the case can be argued in another way. To the extent that faunal diversification is causally related to environmental stability (which I believe to be so for much the same reasons that polymorphism is best developed in more-stable environments), we are justified in considering the arctic the least stable of the shallow marine habitats. On a global scale, faunal diversity declines monotonically from low to high latitudes in ectoprocts (Schopf, 1970) and in many other groups of organisms.]

The first comparison is of the presence or absence of polymorphs (Table 1, column 1, for 1 or more polymorphs); 74.2 percent of the Arctic cheilostomes and 77.3 percent of the tropical fauna had at least one type of polymorphism. We would expect to find this type of difference approximately 50 percent of the time ($\chi^2 = 0.450$). And 76.2 percent of the deep-sea cheilostomes (from more than 2000 m) are also polymorphic. The reason why about 25 percent of the cheilostome species should fail to show polymorphism in any of these environments is unknown. Perhaps the existence of an

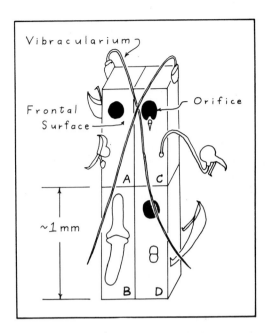

Figure 3

Schematic sketch of four hypothetical zooids from an erect ectoproct colony. Drawn to demonstrate
a variety of sizes and shapes of avicularia, and one type of vibracularium. Zooids A, C, and
D have frontal or lateral "adventitious" avicularia which rest on normal zooids; zooid B is an
immense "vicarious" avicularium which has taken the place of a normal zooid in the budding
sequence. Zooid A has a lateral vibracularium, a lateral pointed avicularium, and a frontal
birds-head "*Bugula*-type" avicularium. Zooid C has a lateral vibracularium, a small suboral pointed
avicularium, and a pedunculate "*Bugula*-type" avicularium. Zooid D has a frontal oval avicularium
and a lateral pointed avicularium.

equivalent high degree of polymorphism in the arctic, tropics, and deep sea indicates
some type of saturation with respect to this type of adaptation.

The evolution of a second and third type of polymorphism is not developed to the
same extent in faunas from the tropics and the arctic. Two or more types of polymorphism
occur in 24.9 percent of the tropical fauna and only 16.6 percent of the arctic fauna
($\chi^2 = 3.26$, $P < 0.10$). Three or more types of polymorphism occur in 8.4 percent
of the tropical fauna and only 2.5 percent of the arctic fauna ($\chi^2 = 3.83$, $P \simeq 0.05$;
Yates's correction for continuity since the total frequency is greater than 40, but one
class has an observed frequency of less than 10 [Simpson, Roe, and Lewontin, 1960,
p. 190]).

Considering deep-sea versus continental shelf faunas, there are no differences which
I consider significant, given the smaller deep-sea fauna (Table 1). The number of species
known from the deep sea is much lower, but this cannot be assumed to be a reliable
index to diversity. Between 1 and 6 km depth, from 20 to 30 percent of the stations
sampled for ectoprocts on the two best documented expeditions yielded them, but few

Table 1

Development of Polymorphism in Four Major Faunas of the World

Variable	Fauna	Number of Usable Cheilostome Species	Percentage of Species in the Fauna Possessing Various Numbers of Polymorphs Among the Zooids		
			1 or more polymorphs	2 or more polymorphs	3 or more polymorphs
Latitude	American Arctic	120	74.2	16.6	2.5
	West Atlantic Tropical	273	77.3	24.9	8.4
Depth	Atlantic Basin (>2000 m)	42	76.2	16.7	4.8
	Estuarine	10	0	0	0

See Appendix for the faunal lists.

samples have been obtained relative to shelf areas (Schopf, 1969). A more reliable guide to diversity would be the number of species per 100 (or other standard number of) individuals collected, but I presently have no information on this.

In any event, the deep-sea and shelf faunas contrast sharply with the apparent absence of polymorphism in species indigenous to estuaries (Table 1). Each of the 10 estuarine species in the Appendix is thought to live and reproduce there.

Specimens of only one of these 10 species, *Cryptosula pallasiana,* has ever been noted to have avicularia. R. C. Osburn, who probably examined material of *C. pallasiana* from tens to hundreds of localities over 40 years, described this species near the end of his career as follows (Osburn, 1952, p. 471): "Avicularia are usually wanting but occasionally there is a small median, suboral one mounted on a small unbonate process; I have found these only rarely on Atlantic specimens and at only two Pacific stations but at one of the latter the avicularia are well distributed over the colony." My much more limited experience of examining material from several reduced salinity embayments along the east coast (especially on Cape Cod) has revealed no specimens with avicularia. Osburn (1952, p. 471) had included two Pacific localities of depths of 29 and 23 fathoms, and it may have been from these offshore areas that the avicularia-bearing Pacific specimens were derived. Although this cannot be determined from published information, no avicularia-bearing *C. pallasiana* from estuarine areas seem to have been reported. As a working hypothesis, it seems more reasonable to me to consider the avicularia-bearing *C. pallasiana* as an aberration from the standard picture, and not to give them more than passing notice.

Perhaps of greater immediate concern is the record of species with avicularia in waters of reduced salinity. Osburn (1944) notes three such species in Chesapeake Bay. These species are *Schizoporella unicornis:* "It does not withstand lower salinity very well, but is found sparingly in the lower part of the Bay, where the salinity is not below 18 (1.8%). A dead specimen was found near the mouth of the Patuxent River (salinity 11) but it may have been carried there on an oyster shell" (Osburn, 1944, p. 44–45); *Bugula turrita:* "In the Chesapeake Bay it occurs only in the lower part toward the mouth, when the salinity is not below 20'' (Osburn, 1944, p. 41); and *Microporella ciliata:* "Mouth of the Chesapeake Bay and inside to a salinity of about 20 (2.0%), not common.'' This report of 1944 by Osburn served to amplify an earlier account of 1932 (Osburn, 1932b) in which the following data were given: *Schizoporella unicornis:* "Sta. 8173, near mouth of bay, 12.8 meters''; *Bugula turrita:* "Sta. 8827, mouth of bay, 18.3 meters, two small colonies attached to shell fragments''; *Micoporella ciliata:* "Sta. 8173, mouth of bay, 18.3 meters''; *Bugula gracilies* var. *uncinata:* "Sta. 8893, not far from the mouth of the bay, at 44.83 meters.'' Osburn notes in 1944 that "In the summers of 1943 and 1944 the author made collections in the shallower waters from the head of the Bay to near its mouth with the result that the earlier list has been extended from 18 to 28 species.'' Possibly all these deeper-water avicularia-bearing species were in fact from the 1932 collections.

All the salinity data appear to have been obtained during the 1944 period of collection, and thus it would seem to represent surface (or very shallow) salinities. Osburn (1944) noted "The salinities given throughout this paper are those recorded at the time of collection.'' Chesapeake Bay is perhaps the world's best-known salt-wedge estuary, wherein fresher, lighter water flows seaward along the surface and saltier, heavier water comes up the estuary along the bottom. Young croker fish have, for example, been found as much as 130 miles up the bay (said to be carried by bottom currents), although the croker spawns in the ocean off the mouth and to the south of Chesapeake Bay (Pritchard, 1951). And in 1932, Osburn (1932b) noted "enormous masses'' of hydroids that "must have been brought in by tides and currents from somewhere near the mouth of the bay, for these hydroids also do not flourish where the salt content is too low.'' Surface water in the middle bay area with a salinity of $16^0/_{00}$ overlies bottom water with a salinity of $23^0/_{00}$ (Pritchard, 1952). Presumably surface salinities of 18 to $20^0/_{00}$ near the mouth of the bay would overlie bottom waters of salinities of 24 to $27^0/_{00}$, and probably higher, since it is closer to the ocean and dilution has been less. Thus, considering the nature of the currents, and the salt wedge, there appears to be good reason for believing that avicularia-bearing, normal-salinity species were not indigenous to lower salinity waters of Chesapeake Bay but were in fact transported to the place where they were collected.

A considerably more detailed report (Maturo, 1959) of larval settlement in the Newport River–Bogue Sound region near Beaufort, North Carolina, discusses 7 species with avicularia. Of the three stations sampled, only one of them (Maturo's station 1) had measured salinities consistently below $30^0/_{00}$, and only avicularia-bearing species from that station will be discussed here. Maturo notes that despite the occurrence of several species with settle at that station, very few species occur there naturally (only *Bugula neritina* appears to be actually reported from this locality). "It appears that conditions in the river are suitable for larval settling and metamorphosis but are not favorable for development to maturity'' (Maturo, 1959, p. 125). Perhaps tidal transport accounts for the movement of larvae into that region. Among the reasons indicated by Maturo for the lack of adult-sized colonies was a lack of "adult tolerances to salinity changes.'' Thus there appears to be no evidence from this study that avicularia-bearing species colonize and reproduce in a region of salinity of less than $30^0/_{00}$.

In addition to these general features, a number of specific cases are consistent with the theory. Lagaaij (1963) noted that *Hippoporella gorgonensis* usually lacks its avicularia when the species occurs in reduced salinity. Smitt (1873) remarked that *Anarthropora minuscula* lacks avicularia in the European arctic but has them in the Gulf of Mexico. An arctic variety of *Dendrobeania murrayana* (its var. *quadridentata*) has avicularia sparingly present (Osburn, 1923, 1932a), whereas the typical boreal form has avicularia well developed. Similarly, the arctic *Bugula simpliciformis* only rarely has avicularia (Osburn, 1932a, 1936) whereas lower latitude species of *Bugula* (except *Bugula neritina*) are characterized by avicularia. *Cribrilina annulata,* an arctic species, lacks avicularia, whereas its close relative *C. punctata,* more typically temperate to boreal than arctic, is characterized by a pair of them.

If this general line of reasoning is correct, the value of the counterexamples ("exceptions") which undoubtedly exist is that they are telling us about other factors which may control the existence of polymorphs. I will return to this later.

Next, the existence of rare and abundant castes in the same colony also becomes explicable. It will now make sense to test whether or not the rarer of two or more polymorphic castes exhibits a more specialized response.

At the first level of analysis, the less-abundant zooids modified for defense are certainly more specialized than normal zooids that are also capable of feeding and reproducing.

On a more specific level, we can compare the occurrence of vicarious defensive polymorphs in different faunas. Defensive zooids which are similar in size and shape to and take the place of normal zooids, i.e., are vicarious, must fit into the regular, sequential geometric patterns typical of colony growth (Figs. 2 and 3). In contrast, defensive zooids which occur either on the frontal surface of normal zooids, or are squeezed into small spaces, can exist without interrupting the regular growth pattern (Figs. 2 and 3). In general, therefore, vicarious polymorphs are judged by me to require more specific genetic and developmental controls during the astogeny of the colony than do adventitious defensive polymorphs. Vicarious polymorphs are anticipated to require a more pervasive selection for their origin, and in this sense are more specialized. This idea is amenable to testing by modern methods of developmental biology, and it would be very useful to do so.

This deductive analysis is in agreement with the relative frequencies of occurrence of vicarious polymorphs. Among species which exhibit polymorphism, vicarious polymorphs occur in 26.1 percent of the tropical fauna, but only 5.6 percent of the arctic fauna. In this same group, adventitious polymorphs exist in more than 85 percent of both the tropical and the arctic faunas.

These theoretical deductions permit additional analyses. By considering changes in shape of a single type of avicularium from the arctic to the tropics, we should be able to deduce a phylogenetic gradient. This way of analyzing a fauna assumes that the fauna of one area (say the arctic) is in a lower grade of evolution than another area (say the tropics). This could be the case if there are lower rates of evolution in the arctic, or if the environment is not "preserving" reproductively isolated local populations from going extinct. Apparently, rates of evolution for species are latitude independent (Stehli et al., 1972). Accordingly, the "brake" on evolutionary diversification would appear to be related to environmental restriction versus enhancement of incipient species.

Table 2

Percent of Encrusting Species with Only One Type of Avicularium Which Exhibit the Following Avicularium Shapes

	Rounded	Pointed
Arctic	60.3	39.7
Tropic	34.8	64.2

Such a "brake" is likely to be most effective where environmental stimuli are most variable and unpredictable, and to be least effective where environmental stimuli are most uniform and predictable.

If it is granted that properties of faunas can be considered as either "less advanced" or "more advanced," then we can examine the distribution of that property in a latitudinal gradient. The subsample I have chosen consists of (1) those species with one adventitious or vicarious avicularium *and* which are primarly encrusting (but may become erect), often well calcified; and (2) those species with two or more types of adventitious or vicarious polymorphs with a similar style of growth. Basically, this is the subsample of the total fauna which appears to have invested most strongly in holding onto space by using extensive calcification and defensive polymorphs; (these would tend to be the K-selection species as used by Kaufmann, this volume). Excluded from the subsample are poorly calcified genera with an erect, bushy growth as well as free-living colonies; both of these types appear to me to be in general adapted for a different method of living (see pp. 266 and 283 for taxa omitted). The results will apply whether the pattern of evolution in adding a second type of polymorph has been from adventitious to vicarious, or vice versa.

Rounded avicularia relative to pointed avicularia are proportionately better represented in the arctic than in the tropics (Table 2). Following the earlier principle that lower-latitude forms will in general evolve a more efficient behavioral response, we would conclude that the rounded avicularia are simply less-modified forms of the pointed avicularia. The fact that the operculum of the normal zooid is always rounded provides a simple transition to the avicularium mandible. If grasping is the functional significance of both shapes of opercula, the pointed forms appear to be more effective, although this remains to be tested with live material.

We might next ask about the shape of this type of avicularium for encrusting species with two or more types of frontal avicularia; i.e., species with more highly evolved behavioral responses (Table 3). For both arctic and tropical faunas, we now find that approximately 75 percent of each fauna has at least one avicularium that is pointed (Figs. 2 and 3). This is consistent with considering this shape a modified and more efficient form of the rounded mandible. One needs to ask, however, why 72 percent of this sample of the tropical fauna has at least one avicularium which is rounded (Figs. 2 and 3). Our theory would suggest that differences in behavioral responses are accentuated in relatively constant environments. By the time a species is so attuned to its environment that it has developed two types of frontal, adventitious avicularia, it may well have passed to a more highly evolved behavioral stage. Under these conditions, the significance

Table 3

Percent of Encrusting Species with Two or More Types of
Avicularia Which Exhibit the Following Avicularium Shapes

	Round	Pointed
Arctic	57.2	71.4
Tropic	72.0	76.0

of rounded versus pointed avicularia may assume greater ecologic significance. There needs to be a systematic attempt to discover the functional significance of rounded and pointed frontal avicularia in encrusting tropical species.

Vibracularia are polymorphic zooids in which the operculum has become yet more accentuated than in an avicularium so that the mandible develops into a long, whip-like structure (Fig. 3). In *Caberea ellisi,* I have watched a vibracularium triggered into action by a probe or other physical stimulus. It sweeps forward and downward across the front of the colony striking a blow against anything on the frontal surface, and triggering the next lowest vibracularium into action, like falling dominoes. This very specialized type of polymorphic structure is much more common in tropical species than in arctic species. Of species with polymorphism, 13.3 percent of the tropical forms have vibracularia, whereas only 2.2 percent of arctic forms have them.

SPECULATION

Let us try to pursue this general analysis of caste development one step further and link it with strategies of life history. The question is: How does the development of polymorphism relate to species which are specialized in obtaining and holding space or other environmental resources (K-selection) as opposed to those species which are specialized in reproducing rapidly and often (r-selection)?

To the extent that polymorphism detracts or contributes to the acquirement and maintenance of a resource, it will be selected against or accentuated. Since polymorphism is, to begin with, a specialized response to particular environmental stimuli, we presume that its development is favored by environmental reinforcement. This would appear more likely to occur in colonies that become large and inhabit a particular locality (K-selection). Hence polymorphism should be accentuated in K-type species.

Selection for reproductive plasticity (r-selection) is likely to be reinforced in variable environments (such as estuaries). Hence the reduction in polymorphism in such areas may be related to type of life history, as well as a paucity of reinforcing stimuli. A reconciliation between life-history phenomena and ergonomic analysis remains. It is possible that many apparent enigmas, such as the absence of polymorphism in a particular species of an otherwise very polymorphic genus (as in the tropical *Bugula neritina*), will become not only understandable but predictable when the life-history characteristics become better known.

Finally, I would like to consider the question of why polymorphism is better developed in cheilostome ectoprocts than in cyclostome ectoprocts or in insects. The reason that we can conceive of polymorphism having an adaptive advantage in colonial organisms is that the other members of the colony appear to share a larger genetic similarity to each other on the average than to the offspring. Thus selection for specialization within a colony (a special type of kin selection) does not violate the precepts of natural selection.

In ectoprocts, all individuals in a colony are usually of the same genotype (some merging of colonies of the same species occasionally occurs). In the social Hymenoptera, in particular the ants, social bees, and social wasps, sisters share three-fourths of the genes in common as a consequence of the haplodiploid mode of sex determination (Hamilton, 1964; Wilson, 1971). All offspring share on the average one-half of the genes in common. From the point of view of kin selection, there is more to be gained in the short run by helping a nearly identical genetic copy than in helping one's immediate offspring. Within the ectoprocts, we should therefore ask why all species in suitable environments do not develop polymorphism. The answer appears to be that not only must the genetic characteristics be of this type, but that starting materials which can be modified by selection also must be present. In this paper, all the polymorphism considered in the cheilostomes consist of modifications of the operculum which covers the orifice. Recent cyclostomes are organized with another type of hydrostatic skeleton and lack an operculum.

The percentage of polymorphic species in tropical ant faunas averages 26 percent (highest 41.2 percent) as compared with 77 percent in the tropical ectoproct fauna surveyed. Two reasons for this difference which are consistent with theory suggest themselves. First, the land is not as well buffered with respect to environmental fluctuations as is the sea, and so this difference may represent an extension of the effect of climatic stability on the development of polymorphism. Second, the zooids of an ectoproct colony share 100 percent of their genes, whereas the sisters of an ant colony only share 75 percent of their genes; thus kin selection would be expected to be higher in ectoprocts than in ants.

TAXONOMIC IMPLICATION

If a taxonomic group occurs over a wide latitudinal distribution, we can conclude that the existence and degree of development of polymorphism may be an unreliable guide to identification. Within a species, polymorphism may be considered as existing in different degrees in different local populations subject to different selection pressures. The genes responsible for its development can be considered to be switched on or off, depending upon environmental factors. A parallel situation is the change in allele frequencies at a biallelic locus (producing an enzyme) from 0.65 to 0.35 over only 20 km (Schopf and Gooch, 1971). Within a genus or family, the existence of polymorphism should be considered as a behavioral trait whose development is closely attuned to the environment. A statement about the polymorphic variability within a genus or family may be no more than a statement about the degree of stability in the region from which the specimens were collected.

CONCLUSION

Heretofore, no systematic analysis of the general features leading to ectoproct polymorphism had been prepared; each case was considered by itself. Using the basic elements of a model for the development of castes in social insects (Wilson, 1968), predictions of the general distribution of castes in ectoprocts were made. The most accessible tests confirmed the predictions from theory and point the way to additional research on the origin and maintenance of polymorphism in ectoprocts. In conclusion, this kind of theory can be used because selection is occurring at the colony level; thus individual fitness can be simplified as long as the loss is more than compensated for by the increase in selection of one's kin through performance of the entire colony.

ACKNOWLEDGMENTS

In discussions at the Marine Biological Laboratory, E. O. Wilson provided the inspiration to consider ectoproct colonies in the same light as insect colonies. I am grateful to him for calling to my attention his paper on ergonomics of caste in social insects (Wilson, 1968). However, even in its most important aspects, the model could only be tested after expending a great deal of effort in compiling reasonably accurate faunal lists. For commenting on my check list for the tropical fauna, I am greatly indebted to F. J. S. Maturo, University of Florida, and Robert Lagaaij, Royal Dutch Shell; for similar remarks on my arctic check-list, I benefited from comments of Karen Bille-Hansen, Copenhagen, and Neil Powell, National Museum of Canada. I thank Drs. Wilson, Maturo, and Lagaaij for extensively commenting upon a draft of the manuscript. Naturally I am solely responsible for the errors that probably remain.

REFERENCES

Barbosa, M. M. 1964. Catálogo das espécies atuais de Bryozoa do Brasil com indicações bibliográficas. Mus. Nac., Rio de Janeiro, 46 p.

Bille-Hansen, K. 1962. Bryozoa. Meddelelser om Grønland., *81*(6): 1–74.

Boardman, R. S., and A. H. Cheetham. 1969. Skeletal growth, intracolony variation, and evolution in Bryozoa: a review. Jour. Paleont., *43:* 205–233.

Braga, L. M. 1968. Notas-sobre alguns Briozoários incrustantes da região de Cabo Frio. Inst. Pesq. Marin., Pub. 025, 23 p.

Busk, George. 1854. Catalogue of Marine Polyzoa in the Collection of the British Museum, I, Cheilostomata (part). Trustees of the British Museum, London. 54 p.

————. 1855. Zoophytology. Quart. Jour. Micros. Sci., *3:* 253–256, pls. 1–2.

————. 1878. Polyzoa, v. 3, p. 283–289, *in* H. W. Feilden, ed., Narrative of a voyage to the Polar Sea during 1875–6 in H. M. Ships *Alert* and *Discovery*. Sampson Low, *et al.,* London.

————. 1880. List of Polyzoa collected by Capt. H. W. Feilden in the North-Polar Expedition; with descriptions of new species. Jour. Linn. Soc., *15:* 231–241, pl. 13.

————. 1884. Report on the Polyzoa . . . pt. I, The Cheilostomata. Report on the Scientific Results of the Voyage of H.M.S. *Challenger*. Zoology, *X*, pt. V: xiv + 216 p., 36 pls.

Calvet, Louis. 1906. Bryozoaires. Expéditions scientifiques du *Travailleur* et du *Talisman*. Masson et Cie, Publishers, Paris. p. 354–495, pls. 26–30.

————. 1931. Bryozoaires provenant des campagnes scientifiques du Prince Albert Ier de Monaco. Résultats des campagnes scientifiques accomplies sur son yacht par Albert Ier. Monaco. *83:* 155 p., 2 pls.

Canu, F. 1928. Trois nouveaux Bryozoaires d'eau douce. Bull. Soc. Hist. Natur. Afrique du Nord., *19:* 262–264, pl. 30.

———, and R. S. Bassler. 1923. North American Later Tertiary and Quaternary Bryozoa. Smithsonian Inst., U.S. Nat. Mus., Bull., *125:* 302 p., 47 pls.

———, and R. S. Bassler. 1928a. Bryozoaires du Brésil. Bull. Soc. Sci. Seine-&-Oise., *9:* 58–100, 9 pls.

———, and R. S. Bassler. 1928b. Fossil and Recent Bryozoa of the Gulf of Mexico. U. S. Nat. Mus., Proc., *72*(14): 1–199, pls. 1-34.

Cheetham, A. H. 1968. Morphology and systematics of the Bryozoan genus *Metrarabdotos*. Smithsonian Misc. Coll., *153*(1): 1–121, 18 pls.

———, and P. A. Sandberg. 1964. Quaternary Bryozoa from Louisiana udlumps. Jour. Paleont., *38:* 1013–1046.

Cook, P. L. 1968. Observations on living Bryozoa. Atti Soc. Ital. Sci.Nat. Mus. Civ. St. Nat. Milano., *108:* 155–160.

Desor, E. 1848. Recent zoological investigations among the shoals of Nantucket. Boston Soc. Nat. Hist., Proc., *3:* 65–68.

Dunbar, M. J. 1953. Arctic and subarctic marine ecology: immediate problems. Arctic, *6:* 75–90.

Ekman, S. 1953. Zoogeography of the Sea. Sidgwick & Jackson, London. 417 p.

Gordon, D. P. 1968. Zooidal dimorphism in the Polyzoan *Hippopodinella adpressa* (Busk). Nature, *219:* 633–634.

Hamilton, W. D. 1964. The genetical evolution of social behavior, I, II. J. Theoretical Biol., *7:* 1–52.

Hastings, A. B. 1943. Polyzoa (Bryozoa) I. Discovery Repts., *22:* 301–510, pls. 5–13.

———. 1963. Notes on Polyzoa (Bryozoa) VI. Some setiform heterozooecia. Ann. Mag. Nat. Hist., ser. 13, *6:* 177–184.

Hazel, J. E. 1970. Atlantic continental shelf and slope of the United States—Ostracode Zoogeography in the Southern Nova Scotian and Northern Virginian Faunal Provinces. U. S. Geol. Survey, Prof. Paper, *529-E:* 29 p., 69 pls.

Hincks, Thomas. 1877. On Polyzoa from Iceland and Labrador. Ann. Mag. Nat. Hist., ser. 4, *19:* 97–112, pls. 10–11.

———. 1880. A History of the British Marine Polyzoa. J. Van Voorst, London. 601 p., 83 pls., in 2 v.

———. 1881. Contributions towards a general history of the marine Polyzoa. Foreign Membraniporina (2nd series). Ann. Mag. Nat. Hist., ser. 5, *7:* 147–161.

———. 1888. The Polyzoa of the St. Lawrence: A study of arctic forms. Ann. Mag. Nat. Hist., ser. 6, *1:* 214–227, pls. 14–15.

———. 1889. The Polyzoa of the St. Lawrence: A study of arctic forms. Ann. Mag. Nat. Hist., ser. 6, *3:* 424–433, pl. 21.

———. 1892. The Polyzoa of the St. Lawrence: A study of Arctic forms. Ann. Mag. Nat. Hist., ser. 6, *9:* 149–157, pl. 8.

Jullien, J. 1882. Dragages du *Travailleur,* Bryozoaires. Bull. Soc. Zool. France, *7:* 497–529, pls. 13–17.

———, and L. Calvet. 1903. Bryozoaires. Résultats des campagnes scientifiques accomplies sur son yacht par Albert Ier. Monaco. *23:* 188 p., 18 pls.

Kaufmann, K. W. 1971. The form and functions of the avicularia of *Bugula* (Phylum Ectoprocta). Postilla. Yale Univ. *151:* 26 p.

Kirkpatrick, R. 1890. Polyzoa, p. 504–506, *in* H. N. Ridley, Notes on the zoology of Fernando Noronha. Jour. Linn. Soc., *20:* 473–592.

Kluge, G. A. 1962. Bryozoa of the Northern Seas of the USSR. Diagnoses of the fauna of the USSR [Opred. Faune SSSR] *76:* 584 p. (In Russian).

Lagaaij, Robert. 1963. New additions to the Bryozoan fauna of the Gulf of Mexico. Pub. Inst. Mar. Sci., Texas., *9:* 162–236, pls. 1–8.

———. 1968. Fossil bryozoa reveal long-distance sand transport along the Dutch coast. Koningkl. Nederl. Adad. Weteñschappen, Amsterdam, Proc., ser. B, *71:* 31–50, pl. 1

Landsborough, D. 1852. A Popular History of British Zoophytes or Corallines. Reeve and Co., London. 404 p., 20 pls.

Levinsen, G. M. R. 1909. Morphological and systematic studies of the Cheilostomatous Bryozoa. Nationale Forfatteres Forlag, Copenhagen. 431 p., 27 pls.

————. 1916. Bryozoa. Danmark-Ekspeditionen til Grønlands Nordøstkyst 1906–1908. Meddelelser om Grønland., *43:* 433–472, pls. 19–24.

Linnaeus, C. 1761. Fauna Suecica. Laurentii Salvii; Holmiae. 576 p. [not seen].

————. 1767. Systema Naturae. 12th ed. Laurentii Salvii, Holmiae. *1:* 1–1327. [not seen].

Marcus, Ernst. 1937. Bryozoarios marinhos Brasileiros I. Bol. Fac. Phil., Sci. Letras, no. 1. Zool. no. 1. Univ. São Paulo. p. 1–224, pls. 1–29.

————. 1938. Bryozoarios marinhos Brasileiros II. Bol. Fac. Phil., Sci. Letras, no. 4. Zool. no. 2, Univ. São Paulo, p. 1–196, pls. 1–29.

————. 1939. Bryozoarios marinhos Brasileiros III. Bol. Fac. Phil. Ciênc. Letras, no. 13. Zool. no. 3., Univ. São Paulo. p. 111–353, pls. 5–31.

————. 1941a. Bryozoarios marinhos do littoral Paranaense. Arq. Mus. Paranaense. *1:* 7–36.

————. 1941b. Sôbre os Briozoa do Brasil. Bol. Fac. Fil., Ciênc. Letras, no. 22, Zool. no. 5, Univ. São Paulo. p. 3–208, pls. 1–18.

————. 1942. Sôbre Bryozoa do Brasil II. Bol. Fac. Fil., Ciênc. Letras, no. 25. Zool. no. 6. Univ. São Paulo. p. 57–106, pls. 1–5.

————. 1944. *Beania cupulariensis* Osb. (Bryozoa Cheilost.), Nova para o Brasil. Comun. Zool. Mus. Hist. Nat. Montevideo, *1*(12): 1–3.

————. 1949. Some Bryozoa from the Brazillian coast. Comun. Zool. Mus. Hist. Nat. Montevideo, *3*(53): 1–33, pls. 1–7.

————. 1953. Notas sôbre Briozoos marinhos Brasileiros. Arq. Mus. Nacional, *42:* 273–342, pls. 1–8.

————, and E. Marcus. 1962. On some lunulitiform Bryozoa. Bol. Fac. Fil., Ciên. Letras, no. 261, Zool. no. 24, Univ. São Paulo. p. 281–324, 5 pls.

Maturo, F. J. S., Jr. 1959. Seasonal distribution and settling rates of estuarine Bryozoa. Ecology, *40:* 116–127.

————. 1966. Bryozoa of the southeast coast of the United States: Bugulidae and Beaniidae (Cheilostomata: Anasca). Bull. Mar. Sci., *16:* 556–583.

————, and T. J. M. Schopf. 1968. Ectoproct and Entoproct type material: reexamination of species from New England and Bermuda named by A. E. Verrill, J. W. Dawson and E. Desor. Postilla, Yale Univ. *120:* 95 p.

Menzies, R. J. 1963. Abyssal Bryozoa collected by expeditions of the Lamont Geological Observatory, I, Bicellariellidae (Bugulidae of Authors), *Kinetoskias.* Novitates, Am. Mus. Nat. Hist. *2130:* 8 p.

Moll, J. P. C. 1803. Eschara zoophytozoorum seu phytozoorum ordine pulcherrima ac notata dignissima genus, etc. Vindobonae. 70 p., 4 pls.

Nordgaard, O. 1906. Bryozoa from the 2nd *Fram* Expedition 1898–1902. Report of the Second Norwegian Arctic Expedition in the *Fram* 1898–1902. *8:* 44 p., 4 pls.

Norman, A. M. 1894. A month on the Trondhjem Fiord, XIII. Ann. Mag. Nat. Hist., ser. 6, *13:* 112–133, pls. 6–7.

Osburn, R. C. 1912. Bryozoa from Labrador, Newfoundland, and Nova Scotia, Collected by Dr. Owen Bryant. U. S. Nat. Mus., Proc., *43:* 275–289, pl. 34.

————. 1914. Bryozoa of the Tortugas Islands, Florida. Papers Tortugas Lab. Carnegie Inst. Washington, *5:* 181–222.

————. 1923. Bryozoa. Rep. Can. Arctic Expedition 1913–18, 8(D): 1D–3D.

————. 1927. The Bryozoa of Curacao. Bijdragen tot de Dierkunde Utgegeven door het Koninklijk Zoologisch Genootschap Natura Artis Magistra, Amsterdam, *25:* 123–132.

————. 1932a. Biological and Oceanographic Conditions in Hudson Bay. 6. Bryozoa from Hudson Bay and Strait. Contrib. Can. Biol. Fisheries, *7*(29): 361–376.

————. 1932b. Bryozoa from Chesapeake Bay. Ohio Jour. Sci., *32:* 441–446.

————. 1933. Bryozoa of the Mount Desert Region, p. 291–385, 15 pls., *in* Biological Survey of the Mount Desert Region, conducted by William Procter. Wistar Inst. Anat. Biol., Phila.

————. 1936. Bryozoa collected in the American Arctic by Captain R. A. Bartlett. Jour. Wash. Acad. Sci., *26:* 538–543.

————. 1940. Bryozoa of Porto Rico with a résumé of the West Indian Bryozoan fauna. Scientific Survey of Porto Rico and the Virgin Islands. New York Acad. Sci., *16*(3): 321–486, 9 pls.

―――. 1944. A Survey of the Bryozoa of Chesapeake Bay. State of Maryland, Board of Natural Resources, Department of Research and Education, Pub. no. 63, 59 p., 5 pls.

―――. 1947. Bryozoa of the Allan Hancock Atlantic Expedition, 1939. Allan Hancock Atlantic Expedition, Rep. no. *5*, 66 p., 6 pls.

―――. 1949. The Genus *Parellisina* (Cheilostomata Anasca, Bryozoa). Allan Hancock Foundation Pub. Occasional Paper no. *10:* 9 p., 1 pl.

―――. 1952. Bryozoa of the Pacific coast of America, part 2, Cheilostomata–Ascophora. Allan Hancock Pacific Expeditions, *14*(2): 271–611, pls. 30–64.

Packard, A. S., Jr., 1863. A list of animals dredged near Caribou Island, Southern Labrador, during July and August, 1860. Can. Nat. Geol., *8:* 401–429, pls. 1–2.

Pallas, P. S. 1766. Elenchus Zoophytorum. Franciscum Varrentrapp; Hagae Comitum. 541. p .

Pourtales, L. F. de. 1867. Contributions to the fauna of the Gulf Stream at great depths. Bull. Mus. Comp. Zool., *1:* 103–120.

Powell, N. A. 1967. Sexual dwarfism in *Cribrilina annulata* (Cribrillinidae-Bryozoa). Jour. Fisheries Res. Bd. Can., *24:* 1905–10.

―――. 1968a. Bryozoa (Polyzoa) of Arctic Canada. Jour. Fisheries Res. Bd. Can., *25:* 2269–2320, 14 pls.

―――. 1968b. Studies on Bryozoa (Polyzoa) of the Bay of Fundy Region, II, Bryozoa from fifty fathoms, Bay of Fundy. Cah. Biol. Mar., *9:* 247–259, pls. 1–5.

―――, and P. L. Cook. 1967. Notes on *Tremogasterina* Canu and *Tremogasterina robusta* (Hincks) (Polyzoa, Ascophora). Cah. Biol. Mar., *8:* 7–20.

―――, and G. D. Crowell. 1967. Studies on Bryozoa (Polyzoa) of the Bay of Fundy region, I, Bryozoa from the intertidal zone of Minas Basin and Bay of Fundy. Can. Biol. Mar., *8:* 331–347, pls. 1–3.

―――. 1968a. Bryozoa (Polyzoa) of Arctic Canada. Jour. Fisheries Res. Bd. Can., *25:* 2269–2320, 14 pls.

―――. 1968b. Studies on Bryozoa (Polyzoa) of the Bay of Fundy Region, II, Bryozoa from fifty fathoms, Bay of Fundy. Cah. Biol. Mar., *9:* 247–259, pls. 1–5.

Pritchard, D. W. 1951. The physical hydrography of estuaries and some applications to biological problems. Sixteenth N. Am. Wildlife Conf. Trans., 1951. Wildlife Management Inst., Washington D.C., p. 368–376.

―――. 1952. Salinity distribution and circulation in the Chesapeake Bay estuarine system. Jour. Mar. Res., *11:* 106–123.

Ridley, S. O. 1881. Polyzoa. Zool. Soc. London, Proc., for 1881. p. 44–61, pl. 6.

Rucker, J. B. 1967. Paleoecological analysis of Cheilstome Bryozoa from Venezuela-British Guiana shelf sediments. Bull. Mar. Sci., *17:* 787–839.

Ryland, J. S. 1969. A nomenclatural index to "A History of the British Marine Polyzoa" by T. Hincks (1880). Brit. Mus. (Nat. Hist.), Bull., Zool., *17*(6): 206–260.

Schopf, T. J. M. 1969. Geographic and depth distribution of the Phylum Ectoprocta from 200 to 6,000 meters. Am. Phil. Soc., Proc., *113:* 464–474.

―――. 1970. Taxonomic diversity gradients of ectoprocts and bivalves and their geologic implications. Geol. Soc. America, Bull., *81:* 3765–3768.

―――, and J. L. Gooch. 1971. Gene frequencies in a marine ectoproct: a cline in natural polulations related to sea temperature. Evolution, *25:* 286–289.

Shier, D. E. 1964. Marine Bryozoa from Northwest Florida. Bull. Mar. Sci., *14:* 603–662.

Silén, Lars. 1951.Bryozoa. Reports of the Swedish Deep-Sea Expedition 1947–1948. Göteborgs ˙Kungl. Vetenskaps- Och Vitterhets-Samhälle. v 2, Zool. Fasc. I, no. 5, p. 61–69.

Simpson, G. G., A. Roe, and R. C. Lewontin. 1960. Quantitative Zoology. Harcourt, Brace & World, Inc., New York. 440 p.

Smitt, F. A. 1868. Kritisk FörteckingÖfver Skandinaviens Hafs-Bryozoer. Öfversigt af Kong. Vetenskaps-Akad. Förhandlingar 1867, Bihang. p. 1–230, pls. 24–28.

―――. 1872. Floridan Bryozoa, pt. I, Kongl. Svenska Vetenskaps-Akad. Handlingar, *10:* 1–20, 5 pls.

―――. 1873. Floridan Bryozoa, pt. II, Kongl. Svenska Vetenskaps-Akad. Handlingar, *11:* 1–83, 13 pls.

Stehli, F. G., R. G. Douglas, and I. A. Kafescioglu. 1972. Models for the evolution of planktonic Fora-minifera, p. 116–128 *in* T. J. M. Schopf, ed., Models in Paleobiology, Freeman Cooper & Company.

Waters, A. W. 1888. Supplementary Report on the Polyzoa. Report on the Scientific Results of the Voyage of the H.M.S. *Challenger*. Zoology, *XXXI*, pt. LXXIX: 1–41, pls. 1–2.

Wilson, E. O. 1963. The social biology of ants. Ann. Rev. Entomol., *8:* 345–368.

———. 1968. The ergonomics of caste in the social insects. Am. Nationalist, *102:* 41–66.

———. 1971. The Insect Societies. Belknap Press of Harvard Univ. Press. x + 548 p.

APPENDIX: FAUNAL LISTS AND SUMMARY OF POLYMORPHIC DATA

General Remarks

Without check lists that are keyed to illustrations, the study of properties of faunas on a worldwide basis is severely handicapped, especially for those not personally familiar with the faunas. And in order to be familiar with all the faunas requires years of study, by which time the questions that originally appeared to be important may have faded in the light of the enormous amount of straight taxonomic work to be done. This may be one reason that taxonomists have not contributed an amount of general information and theory commensurate with their education and dedicated effort.

In order for the reader to reconstruct the data which served as the basis for the preceding summaries, he would need to determine the number and type of avicularia and/or vibracularia for each species in these faunas. To facilitate that, but also to permit an entry into the literature for other potentially interesting general characteristics of these faunas, such as the degree of development of brood chambers, the following faunal lists are keyed to a citation and to the type of polymorphism. The citation is given to (1) a figure and a description (if available), or (2) a description (if available), or (3) a faunal record.

Perhaps inevitably there are a few exasperating difficulties in preparing a faunal list. I will give one example. *Smittina arctica* Norman, 1894, is considered by Kluge (1962) to be the same as *Smittina minuscula* (Smitt, 1868). However, specimens referred to *S. arctica* by Levinsen (1916) are in my view clearly different from those figured by Kluge (1962). This argues in favor of keeping *S. arctica* distinct from *S. minuscula*. The important question for us, however, is the true identification of specimens referred to *S. arctica* by Osburn (1936) and Bille-Hansen (1962), and to *S. minuscula* by Nordgaard (1906), Osburn (1932a), and Bille-Hansen (1962), none of whom, excepting Nordgaard, figured their material. Hence an independent decision about these specimens and the possible synonymy of *S. arctica* and *S. minuscula* as represented in American arctic material is practically impossible without examination of the actual specimens. In this and similar situations, I have usually followed the decision of Bille-Hansen (1962) or Powell (1968a), both of whom have considered the arctic fauna in detail. If still in doubt, I have usually kept apart species that some have synonymized.

Information regarding polymorphism may have come from the cited reference, or from another source on that particular species. This must be so since a single reference may be suitable for an illustration but may omit mention of avicularium shape, etc., and vice versa. The abbreviations are:

A, Adventitious polymorphism
B, Vicarious polymorphism

1, One type of avicularium
2, Two types of avicularia
3, Three types of avicularia
p, pointed or very elongate avicularium
r, rounded or spathulate avicularium
v, vibracularium
0, no avicularia or vibracularia
i, insufficient information

As anyone knows who has tried to classify the types of polymorphism in a large fauna into discrete classes, decisions regarding a few species must be carefully weighed. My general concept of a vicarious avicularium is that of a large (approximately zooid-sized) avicularium which usually (but not necessarily) appears to take the place of a normal zooid in the budding pattern. My concept of an adventitious avicularium is that of a small (much smaller than zooid-sized) avicularium which usually occurs on or between normal zooids but may occur as a discrete entity arising from an anastomosing network of connections (as in *Retevirgula tubulata* or *R. caribbea*). My concept of a vibracularium is that of a very extended mandible significantly longer than a zooid. These concepts are operationally derived from the view that polymorphisms of these types are designed for fundamentally different tasks, which is a different way of considering these polymorphs than is sometimes used (e.g., Hastings, 1963).

I. Check List and Summary of Polymorphism in Tropical West Atlantic Fauna

The tropical West Atlantic fauna is defined as those species which occur shallower than about 125 m in the areas bounded by about 23°S (Cabo Frio, Brazil), and about 25°N (i.e., southern Florida, but also in all of the Gulf of Mexico which extends to about 30°N off Texas). These faunal boundaries correspond to a conservative interpretation of the West Atlantic Tropical Faunal Province, as suggested by Ekman (1953) in his Zoogeography of the Sea. Subtropical provinces extend southward from Cabo Frio, and northward from southern Florida, in the latter case stopping at Cape Hattaras. Neither the fauna of Santos Bay on the south nor Cape Hattaras on the north are included. In addition, I excluded such species as *Palmicellaria skenei, Fenestrulina ampla,* and *Reptadeonella costulata* which have only been reported in the tropical western Atlantic at 128 m, at almost 22°S (in Canu and Bassler, 1928a). These species appear to be from near the shelf break at a depth in the lower part of (or possibly below) the seasonal thermocline. It seems to me that these species are more likely to be representative of a subtropical faunal province than the tropical faunal province, and they are therefore omitted from the check list. In questions of this sort, my guiding principle has been to try to characterize the fauna of a faunal province and not automatically to include every species that happened to have once been found within that area.

The basis for this check list is the list of species mentioned in Lagaaij's 1963 paper on northern Gulf of Mexico forms and generously provided in manuscript form by Lagaaij. It included approximately 160 species of cheilostomes with taxonomic and nomenclatorial emendations from original publications; 278 species are cited in the present list. Lagaaij's list was checked and amplified by reference to Busk (1884), Canu and Bassler (1923, 1928a, 1928b), Cheetham (1968), Cheetham and Sandberg (1964), Hastings (1943), Hincks (1880), Kirkpatrick (1890), Lagaaij (1963, 1968), Levinsen (1909), Marcus (1937, 1938, 1939, 1941a, 1941b, 1942, 1944, 1949, 1953), Marcus and Marcus (1962), Maturo (1966), Osburn (1914, 1927, 1940, 1947, 1949), Pourtales (1867), Powell and Cook (1967), Ridley (1881), Rucker (1967), Shier (1964), Smitt (1872, 1873), Braga (1968), and Barbosa (1964).

The citation to a figure is for a tropical western Atlantic specimen if possible; when this is not possible, the citation to a figure is for a nontropical western Atlantic specimen. For example, several species were described and illustrated by Marcus from south of Cabo Frio (and hence not strictly tropical) and were not refigured by Osburn in his tropical studies; a Marcus reference is given for many of these species. In a few cases (e.g., Busk, 1884) the citation is for a tropical species, the illustrated specimen for which may have come from somewhere else in the world.

There are 278 Cheilostome species in this fauna; 5 of these are insufficiently described or illustrated for a decision to be made regarding the presence or absence of avicularia and/or vibracularia. Hence the sample on which this analysis is based consists of 273 species.

Of these 273 species, 211 (77.3 percent) exhibit at least one type of avicularium and/or vibracularium; 68 species (24.9 percent) exhibit at least two types of avicularia and/or vibracularia, or combination thereof; 23 species (8.4 percent) exhibit at least three types of avicularia and/or vibracularia or combination thereof; and 62 species (22.7 percent) exhibit neither avicularia nor vibracularia.

Of the 211 species with one or both types of polymorphism, 55 species (26.1 percent) have vicarious forms of polymorphism; the remaining 156 species (73.9 percent) have only adventitious forms of polymorphism; 16 species (7.6 percent) have only vibracularia; 12 species (5.7 percent) have a vibracularium and an avicularium, and 183 species (86.8 percent) have only avicularia. Thus of the 211 polymorphic species, 28 species (13.3 percent) have vibracularia, in contrast to 195 species (92.4 percent) which have avicularia.

There are 109 species with a single type of avicularium and a well-calcified, lamellar skeletal growth (primarily encrusting but sometimes becoming erect). (This excludes *Beania, Bicellaria, Bicellariella, Bugula, Caberea, Canda, Caulibugula, Cellaria, Chlidonia, Nellia, Savignyella, Scrupocellaria, Synnotum, Vittaticella, Smittipora abyssicola, Siphonoporella dumonti, Cupuladria,* and *Mamillopora.*) Of the 109 species, 70 (64.2 percent) have a pointed or very elongate and narrow avicularium, and 38 species (34.8 percent) have a rounded, or spatulate type of avicularium; 1 species was indeterminate. For comparison, there are 50 species of similar growth with two or more types of avicularia. Of these 50 species, 38 (76.0 percent) have at least one pointed or very elongate and narrow avicularium, and 36 species (72.0 percent) have at least one rounded or spatulate type of avicularium. Thus in increasing from 1 to more than 1 type of avicularium in a species for this sample of the fauna, the pointed forms increase from 64.2 to 76 percent presence, and rounded forms increase from 34.8 to 72 percent presence. This is an approximate increase of 20 percent for the pointed types, and of 100 percent for the rounded types.

Check List of West Atlantic Tropical Ectoproct Fauna

Poly-morphism	Species	Citation
	abyssicola, Rectonychocella; see *abyssicola, Smittipora* (part B)	
	abyssicola, Smittipora (part A); see *levinseni, Smittipora*	
B1	*abyssicola, Smittipora* (part B)	Smitt, 1873: 6; as *Vincularia abyssicola*
	abyssicola, Vincularia; see *abyssicola, Smittipora* (part B)	
0	*acanthina, Chaperia*	Marcus, 1953: 281
A1p	*aculeata, Gemelliporidra*	Osburn, 1940: 425
B1p	*acutirostris, Smittipora*	Marcus, 1949: 10
	acutirostris, Velumella; see *acutirostris, Smittipora*	
A1p	*advena, Celleporaria*	Smitt, 1873: 69; as *Discopora advena*
	advena, Discopora; see *advena, Celleporaria*	
A2	*aegyptiacum, Synnotum*	Marcus, 1937: 58
A3rrr ⎫ B1p ⎭	*albirostris Celleporaria*	Smitt, 1873: 70; as *Discopora albirostris* forma *typica*
	albirostris, Discopora, forma *pusilla;* see *pusilla, Hippoporella*	
	albirostris, Discopora, forma *typica;* see *albirostris, Celleporaria*	
	albirostris, Holoporella; see *albirostris, Celleporaria*	
A1r	*alipioi, Stephanosella*	Marcus, 1953: 295
0	*ambigua, Scruparia*	Marcus, 1953: 279
A1p	*americana, Cystisella*	Canu and Bassler, 1928b: 113
0	*americana, Coleopora*	Osburn, 1940: 411
	americana, Hippodiplosia; see *verrilli, Hippoporina*	
	americana, Hippoporina; see *verrilli, Hippoporina*	
	americana, Velumella; see *levinseni, Smittipora*	
A3rrr	*anderseni, Parasmittina*	Canu and Bassler, 1928b: 118, as *Mucronella egyptiaca*
0	*anguina, Aetea*	Osburn, 1940: 345
	antiqua, Mollia; see *antiqua, Floridina*	
B1r	*antiqua, Floridina*	Smitt, 1873: 12; as *Mollia antiqua*
	antillea spinosa, Exechonella; see *antillea, Exechonella*	
A1r	*antillea, Exechonella*	Osburn, 1927: 128; as *Lepralia antillea*
	antillea, Hippexechonella; see *antillea, Exechonella*	
	antillea, Lepralia; see *antillea, Exechonella*	
0	*apsata, Pelmatopora*	Shier, 1964: 626
A1p	*aragaoi, Lagenipora*	Marcus, 1953: 308

Check List of West Atlantic Tropical Ectoproct Fauna (Continued)

Poly-morphism	Species	Citation
0	arborescens, Membranipora	Lagaaij, 1963: 166; as Conopeum commensale
0	arborescens, Rhynchozoon (?)	Canu and Bassler, 1928a: 89
A1p } B1r }	areolata, Smittina	Canu and Bassler, 1928a: 87
	armata, Bugula (part); see armata, Caulibugula armata, Bugula (part); see dendrograpta, Caulibugula	
A1	armata, Caulibugula	Levinsen, 1909: 104; as Bugula caraibica
A2rr } B1r }	atlantica, Celleporaria	Marcus, 1953: 310; as Holoporella atlantica
	atlantica, Holoporella; see atlantica, Celleporaria	
0	audens, Pyripora	Marcus, 1949: 5
	audouinii, Escharella; see irregularis, Hippopodina	
	audouinii, Lepralia; see feegeensis, Hippopodina	
0	aurantiacum, Stylopoma	Canu and Bassler, 1928a: 78
	aviculare, Synnotum; see aegyptiacum, Synnotum	
	avicularia, Bugula; see dentata, Bugula	
	avicularis, Cellepora; see dichotoma, Harmerella	
A1p	aviculifera, Phylactella	Osburn, 1914: 213; as Phylactella collaris, var. aviculifera
B1	bassleri, Cellaria	Smitt, 1873: 4; as Cellaria tenuirostris
A1r	bellula, Bellulopora	Lagaaij, 1963: 184
0	bellula, Electra	Marcus, 1937: 37
0	bellula, Electra, var. ramosa	Osburn, 1940: 355
A1 } Av }	bellula, Scrupocellaria	Osburn, 1947: 21
A1p	bernardi, Hippopodina	Lagaaij, 1963: 185
A3 } Av }	bertholletii, Scrupocellaria	Marcus, 1938: 24
	biaperta, Hippothoa; see biaperta, Schizoporella	
A2pr	biaperta, Schizoporella	Smitt, 1873: 46; as Hippothoa biaperta
	biaperta, Stephanosella; see biaperta, Schizoporella	
A1p	bipartita, Adeona	Marcus, 1949: 25
Av } Bv }	biporosa, Cupuladria	Marcus and Marcus, 1962: 285; as Cupuladria canariensis
	bisinuata, Escharella; see bisinuata Petraliella	
A2pr	bisinuata, Petraliella	Smitt, 1873: 59; as Escharella bisinuata

Check List of West Atlantic Tropical Ectoproct Fauna (Continued)

Poly-morphism	Species	Citation
A2pp	*bispinosa, Rhynchopora*	Ridley, 1881: 50 (no fig.)
A2 } Av }	*boryi, Caberea*	Lagaaij, 1963: 182
0	*brasiliensis, Exechonella*	Canu and Bassler, 1928a: 72
A1p	*brongniartii, Chorizopora*	Lagaaij, 1963: 185
A1p} B1r}	*brunnea, Celleporaria*	Canu and Bassler, 1928b: 148; as *Holoporella vagans*
B1r	*buskii, Steganoporella*	Marcus, 1938: 22
0	*buski, Margaretta*	Canu and Bassler, 1928b: 113; as *Tubucellaria cereoides*
Av } Bv }	*canariensis, Cupuladria*	Marcus and Marcus, 1962: 285
	canariensis, Membranipora; see canariensis, *Cupuladria*	
	canui, Schizoporella; see cornuta, *Schizoporella*	
	caraibica, Bugula; see armata Caulibugula	
Av	*caraibica, Canda*	Levinsen, 1909: 142 (no fig.)
	caraibica, Caulibugula; see armata, Caulibugula	
0	*caraibica, Quadricellaria*	Canu and Bassler, 1928b: 17
A1r	*caribbea, Retevirgula*	Osburn, 1947: 15; as *Pyrulella* *caribbea*
Bv	*caribbea, Vibracellina*	Osburn, 1947: 11
	caribbea, Pyrulella; see caribbea, Retevirgula	
	castanea, Mucronella; see castanea, Utinga	
A2pp	*castanea, Utinga*	Marcus, 1949: 21
0	*cecilii, Arthropoma*	Lagaaij, 1963: 186
A1p	*centetica, Parelissina*	Marcus, 1953: 281
	cereoides, Tubucellaria; see buski, Margaretta	
	cervicornis, Cellularia; see regularis, *Scrupocellaria*	
A1p	*cervicornis, Chaperia*	Osburn, 1947: 15 (no fig.)
	cervicornis, Eschara; see cervicornis, Porella	
A1r	*cervicornis, Porella*	Smitt, 1873: 66; as *Eschara* *cervicornis*
	cervicornis, Scrupocellaria; see regularis, *Scrupocellaria*	
	ciliata, Bicellariella; see ciliata, Bicellariella, *forma edentata*	
A1	*ciliata, Bicellariella, forma edentata*	Marcus, 1953: 292
A1p	*ciliata, Microporella*	Marcus, 1937: 110
	ciliata, Porellina; see ciliata, Microporella	
	ciliata personata, Microporella; see personata, *Microporella*	
	cleidostoma, Lepralia; see porcellanum, *Cleidochasma*	
	cleidostoma, Hippoporina; see porcellanum, *Cleidochasma*	
0	*clypeata, Acanthocella*	Canu and Bassler, 1928b: 39

Check List of West Atlantic Tropical Ectoproct Fauna (Continued)

Poly-morphism	Species	Citation

A2pr	coccinella, Anarthropora
	collaris, Phylactella, var. aviculifera;
	  see aviculifera, Phylactella
	commensale, Conopeum; see arborescens,
	  Membranipora
		Rucker, 1967: 831

A1r	compressum, Antropora	Osburn, 1927: 124; as Membrendoecium compressum
	compressum, Membrendoecium; see compressum, Antropora	
A2pp	condylata, Chaperia	Rucker, 1967: 821
	connexa, Steganoporella; see sinuosa, Siphonoporella	
0	contei, Catenicella	Osburn, 1940: 465; as Vittaticella contei
	contei, Vittaticella, see contei, Catenicella	
A1r	contractum, Cleidochasma	Marcus, 1937: 98; as Perigastrella contracta

	contracta, Hippoporina; see contractum,
	  Cleidochasma
	contracta, Lepralia, var. serrata;
	  see contractum, Cleidochasma
	contracta, Perigastrella; see contractum,
	  Cleidochasma	

0	contraria, Figularia	Lagaaij, 1963: 182
A1p	corderoi, Coleopora	Marcus, 1949: 18
A1p	coriacea, Micropora	Marcus, 1949: 16
A2 Av	cornigera, Scrupocellaria (part)	Marcus, 1937: 55
	cornigera, Scrupocellaria (part); see pusilla, Scrupocellaria	
Bv	coronata, Cranosina	Osburn, 1940: 363 (no fig.), see figure in Hincks, 1881: 147
A1p	coronata, Cellepora	Smitt, 1873: 51
A2pr	cornuta, Schizoporella	Canu and Bassler, 1928b: 93; as Schizopodrella incrassata
	costazii, Siniopelta; see hassallii, Celleporina costazii, Costazia; see hassallii, Celleporina costifera, Escharella; see costifera, Escharoides	
A1p	costifer, Escharoides	Osburn, 1914: 203; as Escharella costifera
B1r	crassimarginata, Crassimarginatella cucullata, Lepralia; see subovoidea, Watersipora	Osburn, 1940: 363 (no fig.)
A1	cupula, Mamillopora	Smitt, 1873: 33
A1	cupulariensis, Beania curvirostris, Membranipora; see curvirostris, Parellisina	Osburn, 1914: 190
B1p	curvirostris, Parellisina	Osburn, 1940: 361

Check List of West Atlantic Tropical Ectoproct Fauna (Continued)

Poly-morphism	Species	Citation
A1	*dendrograpta, Caulibugula*	Marcus, 1938: 29; as *Caulibugula armata*
A2	*dentata, Bugula*	Canu and Bassler, 1928b: 41; as *Bugula avicularia*
0	*denticulata, Biflustra*	Osburn, 1940: 353; as *Acanthodesia tenuis*
0	*diaphana, Halysisis*	Lagaaij, 1963: 202
	dichotoma, Arborella; see *dichotoma, Tetraplaira*	
	dichotoma, Cellepora; see *dichotoma, Harmerella*	
A3prr	*dichotoma, Harmerella*	Smitt, 1873: 53; as *Cellepora avicularis*
	dichotoma, Schismopora; see *dichotoma Harmerella*	
0	*dichotoma, Tetraplaria*	Osburn, 1914: 202; as *Arborella dichotoma*
B1r	*dipla, Labioporella*	Marcus, 1949: 14
	distans, Hippothoa; see *flagellum, Hippothoa*	
B1r	*distorta, Thalamoporella*	Osburn, 1940: 380
0	*divae, Monoporella*	Marcus, 1953: 286
0	*divaricata, Hippothoa*	Marcus, 1939: 134
Av	*doma, Cupuladria*	Canu and Bassler, 1928b: 64; as *Cupularia doma*
	doma, Cupularia; see *doma, Cupuladria*	
A2 } Av }	*drachi, Scrupocellaria*	Marcus, 1953: 288
B1	*dumonti, Siphonoporella*	Canu and Bassler, 1928b: 68
	eburnea, Gemellipora (part); see *flagellum, Hippothoa*	
	eburnea, Hippothoa; see *flagellum, Hippothoa*	
A1p	*echinata, Parasmittina*	Shier, 1964: 638; as *Parasmittina trispinosa* var. *echinata*
	echinata, Smittina; see *echinata, Parsmittina*	
A1r } B1p }	*edax, Hippoporidra*	Smitt, 1873: 63; as *Lepralia edax* forma *calcarea*
	edax, Lepralia; see *edax, Hippoporidra*	
	edax, Lepralia, forma *calcarea;* see *edax, Hipporporidra*	
	edax, Lepralia, forma, *janthina;* see *janthina, Hippotrema*	
	edax, Lepralia, forma *typica;* see *edax, Hippoporidra*	
	egyptiaca, Mucronella; see *anderseni, Parasmittina*	
	egyptiaca, Parasmittina; see *anderseni, Parasmittina*	
	egyptiaca, Smittina; see *anderseni, Parasmittina*	

Check List of West Atlantic Tropical Ectoproct Fauna (Continued)

Poly-morphism	Species	Citation
	elegans, Steganoporella; see *magnilabris, Steganoporella*	
	elegans, Vittaticella; see *uberrima, Vittaticella*	
0	*elongata, Mollia*	Marcus, 1949: 16
0	*errans, Aplousina*	Canu and Bassler, 1928a: 60
A1p	*errata, Schizoporella*	Marcus, 1941a: 22; as *Schizoporella unicornis*
A1r	*evelinae, Reteporellina*	Marcus, 1953: 302
B1r	*evelinae, Steganoporella*	Marcus, 1949: 12
B1r	*exilimargo, Crassimarginatella*	Canu and Bassler, 1928a: 61
	falcifera, Schizopodrella; see *"spongites, Stylopoma"*	
B1p	*falcifera, Thalamoporella*	Marcus, 1937: 52
A1p	*feegeensis, Hippopodina*	Osburn, 1940: 412
A1p	*fenestrata, Gigantopora*	Smitt, 1873: 47; as *Hippothoa fenestrata*
	fenestrata, Hippothoa; see *fenestrata, Gigantopora*	
	fenestrata, Stenopsis; see *fenestrata, Gigantopora*	
	figularis, Cribrilina; see *floridana, Reginella*	
	figularis, Figularia; see *floridana, Reginella*	
B1r	*filum, Aplousina*	Canu and Bassler, 1928b: 21; as *Aplousina tuberosa*
A1p	*fissurata, Hoppomenella*	Osburn, 1940: 431
	fissurata, Lepralia; see *fissurata, Hippomenella*	
	flabellata, Bugula; see *simplex, Bugula*	
0	*flagellum, Hippothoa*	Smitt, 1873: 35; as *Gemellipora eburnea*
	flectospinata, Retevirgula; see *caribbea, Retevirgula*	
	floridana, Acanthocella; see *floridana Reginella*	
A1p	*floridana, Aimulosia*	Osburn, 1947: 34
	floridana, Cribrilina (part); see *clypeata, Acanthocella*	
	floridana, Cribrilina (part); see *floridana, Reginella*	
A1r	*floridana, Hippadenella*	Canu and Bassler, 1928b: 105
	floridana, Puellina; see *floridana, Reginella*	
0	*floridana, Reginella*	Smitt, 1873: 23; as *Cribrilina floridana*
A1p ⎫ B1p ⎭	*floridana, Schizoporella*	Osburn, 1914: 205
A1 ⎫ Av ⎭	*frondis, Scrupocellaria*	Hastings, 1943: 361
B1r	*fulgens, Tremoschizodina*	Marcus, 1953: 308
A1r	*gabriellae, Lacerna*	Marcus, 1953: 297
	galeata, Chaperia; see *patula, Chaperia*	

Check List of West Atlantic Tropical Ectoproct Fauna (Continued)

Poly-morphism	Species	Citation
0	*gigantea, Aplousina*	Osburn, 1940: 357
A1	*glabra, Bicellaria*	Busk, 1884: 35
A2rp	*glabra, Gemellipora,* forma *glabra*	Smitt, 1873: 37
	glabra, Gemellipora, forma *striatula;* see *venusta, Trypostega*	
	glabra, Gemelliporina; see *glabra, Gemellipora*	
Av	*goesi, Setosellina*	Lagaaij, 1963: 172
A5ppprp	*gorgonensis, Hippoporella*	Lagaaij, 1963: 191
B1p	*gothica, Thalamoporella,* var. *floridana*	Osburn, 1940: 378
	granulata, Thalamoporella; see *mayori, Thalamoporella*	
	granulata, Tremogasterina; see *mucronata, Tremogasterina*	
A1p	*granulifera, Antropora*	Osburn, 1940: 359 (no fig.)
	granulosa, Siphonoporella; see *granulosa, Labioporella*	
B1r	*granulosa, Labioporella*	Canu and Bassler, 1928b: 69; as *Siphonoporella granulosa*
	gulo, Trigonopora; see *tenue tenue, Metrarabdotos (Biavicularium)*	
A2 } Av }	*harmeri, Scrupocellaria*	Osburn, 1947: 20
A1p } B1p }	*hassallii, Celleporina*	Marcus, 1937: 121; as *Siniopelta costazii*
	hastingsae, Electra; see *monostachys, Electra*	
A1p	*hastingsae, Reptadeonella*	Cheetham and Sandberg, 1964: 1039
	heckeli, Adeona; see *violacea, Reptadeonella*	
	heckeli, Cellepora; see *violacea, Reptadeonella*	
0	*hirtissima, Beania*	Marcus, 1937: 62
	horsti, Schizoporella; see *signatum, Rimulostoma*	
0	*hyalina, Hippothoa*	Marcus, 1937: 79
A1r } B1r }	*ignota, Celleporina*	Osburn, 1940: 461 (no fig.); as *Costazia ignota*
	ignota, Costazia; see *ignota, Celleporina*	
	ignota, Lagenipora; see *ignota, Celleporina*	
A2pr	*imbellis, Cellepora*	Busk, 1884: 195
	imperfecta, Gephyrophora; see *imperfecta, Hippaliosina*	
A1p	*imperfecta, Hippaliosina*	Marcus, 1953: 306
	incrassata, Schizoporella; see *cornuta, Schizoporella*	
	informata, Stylopoma; see "*spongites, Stylopoma*"	
	innominata, Cribrilina; see *radiata, Cribrilaria*	
	innominata, Puellina; see *radiata, Cribrilaria*	

Check List of West Atlantic Tropical Ectoproct Fauna (Continued)

Poly-morphism	Species	Citation

0	*inornata, Lepralia;* see *venusta, Trypostega*
	intermedia, Beania; klugei, Beania
	irregularis, Alderina; see *smitti, Alderina*
	irregularis, Cellaria; see *bassleri, Cellaria*
	irregularis, Hippopodina
	irregularis, Membranipora; see *smitti, Alderina*
	isabelleana, Hippothoa (part); see *trichotoma,*
	  Cribellopora
	isabelleana, Hippothoa (part); see *errata,*
	  Schizoporella
	jacotini, Escharella, var. *spathulata;*
	  see *spathulata, Parasmittina*
	janthina, Hippoporidra; see *janthina,*
	  Hippotrema
		Osburn, 1940: 414

A2pp	*janthina, Hippotrema*	Smitt, 1873: 63; as *Lepralia edax* forma *janthina*
	janthina, Lepralia; see *janthina, Hippotrema* *johnstoniae, Bugula;* see *johnstoniae, Halophila*	
0	*johnstoniae, Halophila*	Smitt, 1872: 17
A1	*klugei, Beania*	Marcus, 1937: 61
	labrosa, Phylactella; see *antillea, Exechonella*	

0	*laciniosa, Electra*
	lacroixii, Biflustra (part); see *exilimargo,*
	  Crassimarginatella
	lacroixii, Biflustra (part); see *filum, Aplousina*
	lacroixii, Biflustra (part); see *leucocypha,*
	  Antropora
	lacroixii, Membranipora; see *leucocypha,*
	  Antropora
		Shier, 1964: 612

A1	*lafontii, Savignyella*	Marcus, 1937: 78
B1p	*lanceolata, Tremogasterina* *landsborovii, Escharella,* var. *minuscula;* see *smittiella, Smittina* *lata, Gemellipora;* see *lata, Tremoschizodina*	Canu and Bassler, 1928b: 48
A1r	*lata, Tremoschizodina*	Smitt, 1873: 36; as *Gemellipora* *lata*
B1r	*latirostris, Parellisina*	Osburn, 1940: 361
A1r	*leucocypha, Antropora*	Marcus, 1937: 46; as *Crassimarginatella leucocypha*
A1	*levinseni, Caulibugula*	Osburn, 1940: 394
B1r	*levinseni, Smittipora*	Marcus, 1953: 282
0	*ligulata, Aetea*	Osburn, 1940: 347
A1p	*lineata, Callopora*	Smitt, 1873: 7; as *Membranipora* *lineata*
	lineata, Membranipora; see *lineata, Callopora*	
A1p	*longicollis, Lekythopora*	Lagaaij, 1963: 199
A1p	*longirostris, Exochella*	Lagaaij, 1963: 194
Av	*longiseta, Crepidacantha* *lowei, Cupularia;* see *umbellata, Discoporella*	Canu and Bassler, 1928b: 135

Check List of West Atlantic Tropical Ectoproct Fauna (Continued)

Poly-morphism	Species	Citation
Av	*lyncoides, Gigantopora*	Ridley, 1881: 47
A2 Av }	*maderensis, Scrupocellaria*	Rucker, 1967: 825
A2rr B1r }	*magnifica, Celleporaria*	Osburn, 1914: 216; as *Holoporella magnifica*
B1r	*magnilabris, Steganoporella*	Marcus, 1953: 284
A1p	*magniporosa, Gemelliporidra*	Canu and Bassler, 1928b: 103
A1p B1p }	*magnirostris, Rhamphostomella*	Canu and Bassler, 1928b: 120
	malleolus, Tremogasterina; see *mucronata, Tremogasterina*	
0	*malusii, Fenestrulina*	Osburn, 1940: 433; as *Fenestrulina malusi*
	mammillata, Cellepora; see *atlantica, Celleporaria*	
	mammillata, Cellepora, var. *atlantica;* see *atlantica, Celleporaria*	
B1	*mandibulata, Cellaria*	Osburn, 1947: 18 (no fig.)
A2pp	*marginata, Hippopetraliella*	Canu and Bassler, 1928b: 80; as *Petraliella marginata*
A2pr	*marginata, Lagenicella*	Rucker, 1967: 830
	marginata, Petraliella; see *marginata, Hippopetraliella*	
	marsupiata, Retepora; see *marsupiata, Sertella*	
	marsupiata, Reteporellina; see *marsupiata Sertella*	
A3rpp	*marsupiata, Sertella*	Smitt, 1873: 67; as *Retepora marsupiata*
A1p	*martae, Escharoides*	Marcus, 1949: 22
B1r	*mayori, Thalamoporella*	Osburn, 1914: 197; as *Thalamoporella granulata*
0	*membranacea, Membranipora*	Osburn, 1940: 349 (no fig.)
0	*mesitis, Pyrulella*	Marcus, 1949: 6
A2 Av }	*micheli, Scrupocellaria*	Marcus, 1953: 289
A1	*microoecia, Bugula*	Osburn, 1914: 187
A1	*minima, Bugula*	Osburn, 1940: 390 (no fig.)
	minuscula, Anarthropora; see *monodon, Anarthropora*	
0	*mirabilis, Beania*	Marcus, 1937: 60
A2pp	*monoceros, Arachnopusia*	Osburn, 1940: 366; (no fig.)
A1p	*monodon, Anarthropora*	Smitt, 1873: 31; as *Anarthropora minuscula*
0	*monostachys, Electra*	Lagaaij, 1963: 168
Av Bv }	*monotrema, Cupuladria*	Busk, 1884: 207; as *Cupularia monotrema*
	monotrema, Cupularia; see *monotrema, Cupuladria*	

Check List of West Atlantic Tropical Ectoproct Fauna (Continued)

Poly-morphism	Species	Citation
A2 pr	*montferrandii, Codonellina*	Lagaaij, 1963: 196
A1r B1r	*mordax, Celleporaria*	Shier, 1964: 643; as *Holoporella mordax*
	mucronata, Escharipora; see *mucronata, Tremogasterina*	
A1p	*mucronata, Hippopleurifera*	Smitt, 1873: 45; as *Hippothoa mucronata*
	mucronata, Hippothoa; see *mucronata, Hippopleurifera*	
A2pr B1r	*mucronata, Tremogasterina*	Smitt, 1873: 24; as *Escharipora* (?) *mucronata*
0	*neritina, Bugula*	Marcus, 1937: 66
	neritina, Bugula, var. *minima;* see *minima, Bugula*	
A2pr B1r	*nitida, Parasmittina*	Marcus, 1937: 104; as *Smittina trispinosa* var. *nitida*
A1r	*numma, Parasmittina*	Marcus, 1949: 22; as *Smittina numma*
	nodosa, Cellaria; see *bassleri, Cellaria* *oculata, Nellia;* see *tenella, Nellia*	
A1p	*opertus, Cauloramphus*	Canu and Bassler, 1928b: 35
A3ppp	*ornatissima, Gemelliporidra*	Canu and Bassler, 1928a: 79
A2 Av	*ornithorhynchus, Scrupocellaria*	Marcus, 1953: 287
A1r B1r	*osburni, Trematooecia*	Marcus, 1953: 311
A1r	*palliolata, Aimulosia*	Canu and Bassler, 1928b: 109; as *Lepralia palliolata*
	palliolata, Lepralia; see *palliolata, Aimulosia*	
B1r	*parvicella, Floridina*	Lagaaij, 1963: 177
Av	*parviseta, Escharina*	Marcus, 1949: 27; as *Mastigophora parviseta*
	parviseta, Mastigophora; see *parviseta, Escharina*	
0	*parvula, Floridinella*	Canu and Bassler, 1928b: 59
A1p	*parva, Antropora*	Canu and Bassler, 1928a: 61; as *Membrendoecium parvus*
i	*patellaria, Mollia*	Smitt, 1873: 12
A1p	*patula, Chaperia*	Canu and Bassler, 1928b: 115 (no fig.); as *Chaperia galeata*
	periporosa, Hincksina; see *tubulata, Retevirgula*	
A1p	*personata, Microporella*	Osburn, 1947: 36 (no fig.); as *Microporella ciliata personata*
A2rr	*pertusa, Cigclisula*	Smitt, 1873: 72; as *Discopora pertusa*
	pertusa, Discopora; see *Cigclisula pertusa* *pertusa, Escharella;* see *pertusa, Hippoporina*	

Check List of West Atlantic Tropical Ectoproct Fauna (Continued)

Poly-morphism	Species	Citation
	pertusa, Hippodiplosia; see *pertusa, Hippoporina*	
A1p	*pertusa, Hippoporina*	Marcus, 1938: 41; as *Hippodiplosia pertusa*
	pertusa, Trematooecia; see *pertusa, Cigclisula*	
A1r	*pesanseris, Escharina*	Marcus, 1939: 142; as *Mastigophora pes-anseris*
	pesanseris, Hippothoa; see *pesanseris, Escharina*	
	pesanseris, Mastigophora; see *pesanseris, Escharina*	
0	*petasus, Membraniporella*	Canu and Bassler, 1928b: 36
A1p ⎫ B1r ⎭	*phrynoglossum, Rhynchozoon*	Marcus, 1937: 115
	plagiopora, Porina (part); see *plagiopora, Reptadeonella*	
	plagiopora, Porina (part); see *hastingsae, Reptadeonella*	
A1p	*plagiopora, Reptadeonella*	Smitt, 1873: 30; as *Porina plagiopora*
Av	*poissoni, Crepidacantha,* var. *teres*	Lagaaij, 1963: 200
	poissoni, Crepidacantha; see *poissoni, Crepidacantha,* var. *teres*	
	porcellana, Hippoporina; see *porcellanum, Cleidochasma*	
	porcellana, Lepralia; see *porcellanum, Cleidochasma*	
A1p	*porcellanum, Cleidochasma*	Marcus, 1937: 96; as *Hippoporina porcellana*
Av	*porosa, Escharina*	Marcus, 1953: 299; as *Mastigophora porosa*
	porosa, Hippothoa; see *porosa, Escharina*	
	porosa, Mastigophora; see *porosa, Escharina*	
	prevailae, Gemelliporella; see *rotundora, Cleidochasma*	
A1r	*prominens, Reteporella*	Canu and Bassler, 1928b: 124
A3prr	*protecta, Parasmittina*	Osburn, 1940: 437 (no fig.); as *Smittina trispinosa* var. *protecta*
A2rr	*protecta, Trematooecia*	Osburn, 1940: 459
A1p	*pulcherrima, Lepralia*	Canu and Bassler, 1928a: 82
0	*pumicosa, Callopora*	Canu and Bassler, 1928b: 33
	pumicosa, Exechonella; see *antillea, Exechonella*	
	pungens, Schizopodrella; see *pungens, Schizoporella*	
A1p	*pungens, Schizoporella*	Canu and Bassler, 1928b: 95; as *Schizopodrella pungens*
A2pp	*pusae, Arachnopusia*	Marcus, 1953: 294
	pusilla, Cellularia; see *pusilla, Scrupocellaria*	

Check List of West Atlantic Tropical Ectoproct Fauna (Continued)

Poly-morphism	Species	Citation
A1r	pussila, Hippoporella	Smitt, 1873: 70; as *Discopora albirostris forma pusilla*
	pusilla, Holoporella; see *pusilla, Hippoporella*	
A1 } Av }	pusilla, Scrupocellaria	Smitt, 1872: 13; as *Cellularia pusilla*
0	pyriformis, Chlidonia	Lagaaij, 1968: 44, 50; as *Cothurnicella pyriformis*
	radiata, Colletosia; see *radiata, Cribrilaria*	
A1p	radiata, Cribrilaria	Marcus, 1937: 73; as *Colletosia radiata*
	radiata, Cribrilina; see *radiata, Cribrilaria*	
	radiata, Puellina; see *radiata, Cribrilaria*	
0	recta, Aetea	Osburn, 1940: 346
A2 } Av }	regularis, Scrupocellaria	Smitt, 1872: 14; as *Cellularia cervicornis*
	reticulata, Buffonellaria; see *reticulata, Stephanosella*	
A1p	reticulata, Smittoidea	Lagaaij, 1963: 196
A1p	reticulata, Stephanosella	Canu and Bassler, 1928b: 89; as *Buffonellaria reticulata*
0	reticulum, Conopeum	Osburn, 1940: 351
	retiformis, Caberea; see *retiformis, Canda*	
Av	retiformis, Canda	Marcus, 1953: 290
	retiformis, Scrupocellaria; see *retiformis, Canda*	
A1p	rosacea, Cycloperiella	Osburn, 1947: 31
A2pp	rostratum, Rhynchozoon	Smitt, 1873: 50; as *Cellepora verruculata*
	rostrigera, Escharella; see *rostrigera, Hippaliosina*	
A1p	rostrigera, Hippaliosina	Smitt, 1873: 57; as *Escharella rostrigera*
	rostrigera, Lepralia; see *rostrigera, Hippaliosina*	
A1p	rotundora, Cleidochasma	Rucker, 1967: 827; as *Gemelliporella prevailae*
	rozierii, Steganoporella; see *gothica, Thalamoporella,* var. *floridana*	
	rozierii, Thalamoporella; see *gothica, Thalamoporella,* var. *floridana*	
A1p	rubra, Gephyrophora	Osburn, 1940: 415
	rubra, Hippomenella; see *mucronata, Hippopleurifera*	
A1p	rugosa, Stephanosella	Osburn, 1940: 423
	sanguinea, Escharella; see *sanguinea, Schizobrachiella*	
	sanguinea, Schizoporella; see *sanguinea, Schizobrachiella*	
A1p	sanguinea, Schizobrachiella	Smitt, 1873: 54; as *Escharella sanguinea*
	savartii, Acanthodesia; see *savartii, Membranipora*	

Check List of West Atlantic Tropical Ectoproct Fauna (Continued)

Poly-morphism	Species	Citation
0	*savartii, Membranipora*	Osburn, 1940: 352; as *Acanthodesia savartii*
	savartii, Membranipora; see savartii, Biflustra	
A2rr	*schubarti, Celleporaria*	Marcus, 1939:159; as *Holoporella schubarti*
0	*scopae, Escharina*	Canu and Bassler, 1928a: 97; as *Mastigophora scopae*
	scopae, Mastigophora; see scopae, Escharina	
	scrupea, Scrupocellaria; see ornithorhynchus, Scrupocellaria	
A2rr } Blr	*serrulata, Spiroporina?*	Smitt, 1873: 27; as *Porina serrulata*
	serrulata, Cigclisula; see serrulata, Spiroporina?	
	serrulata, Porina; see serrulata, Spiroporina?	
Av	*setigera, Crepidacantha*	Smitt, 1873: 58; as *Escharella setigera*
	setigera, Escharella; see setigera, Crepidacantha	
0	*sica, Aetea*	Marcus, 1937: 28
	signata, Lacerna; see signatum, Rimulostoma?	
	signata, Parsmittina; see signatum, Rimulostoma?	
A1p	*signatum, Rimulostoma?*	Lagaaij, 1963: 197; as *Parasmittina signata*
A1	*simplex, Bugula*	Smitt, 1872: 18; as *Bugula flabellata.*
	simplex, Canda; see caraibica, Canda	
A1	*sinuosa, Cellaria*	Osburn, 1940: 383 (no fig.)
	sinuosa, Labioporella; see sinuosa, Siphonoporella	
0	*sinuosa, Siphonoporella*	Osburn, 1940: 377; as *Labioporella sinuosa*
0	*smitti, Alderina*	Smitt, 1873: 8; as *Membranipora irregularis*
A1r	*smittiella, Smittina*	Osburn, 1947: 37
A2pr	*solidum, Rhynchozoon*	Osburn, 1914: 201
A2pr } B1r	*spathulata, Parasmittina*	Smitt, 1873: 60; as *Escharella spathulata*
	spathulata, Smittina; see spathulata, Parasmittina	
	spathulata, Smittina trispinosa; see spathulata, Parasmittina	
	spinosum, Gephyrotes	Canu and Bassler, 1928b: 29
	spongites, Hippothoa; see "spongites, Stylopoma"	
	spongites, Schizoporella; see "spongites, Stylopoma"	

Check List of West Atlantic Tropical Ectoproct Fauna (Continued)

Poly-morphism	Species	Citation
A1p ⎫ B1r ⎭	"*spongites, Stylopoma*"	Osburn, 1940: 424; as *Stylopoma informata*
	stellata, Escharipora; see *stellata, Triporula*	
A1p	*stellata, Triporula*	Smitt, 1873: 26; as *Escharipora stellata*
A1r ⎫ B1r ⎭	*subalba, Holoporella*	Canu and Bassler, 1928b: 146
0	*subovoidea, Watersipora*	Marcus, 1937: 118; as *Watersipora cucullata*
A1p ⎫ B1p ⎭	*subsulcata, Bracebridgia*	Smitt, 1873: 28; as *Porina subsulcata*
	subsulcata, Porina; see *subsulcata, Bracebridgia*	
	tehuelcha, Membranipora; see *tuberculata, Membranipora*	
0	*tenella, Electra*	Marcus, 1937: 38
	tenella, Membranipora; see *tenella, Electra*	
A1	*tenella, Nellia*	Marcus, 1939: 131; as *Nellia oculata*
	tenue, Metrarabdotos; see *tenue tenue, Metrarabdotos (Biavicularium)*	
A2pr	*tenue tenue, Metrarabdotos (Biavicularium)*	Marcus, 1953: 304; as *Trigonopora gulo*
B1p	*tenuirostre, Copidozoum*	Marcus, 1937: 48
	tenuirostris, Callopora; see *tenuirostre, Copidozoum*	
	tenuirostris, Cellaria; see *bassleri, Cellaria*	
	tenuis, Acanthodesia; see *denticulata, Biflustra*	
A1	*tenuis, Nellia*	Osburn, 1940: 400 (no fig.)
	tenuis, Smittia; see *tenue tenue, Metrarabdotos (Biavicularium)*	
	tenuis, Trigonopora; see *tenue tenue, Metrarabdotos (Biavicularium)*	
	tenuissima, Callopora; see *tenuissima, Parellisina*	
0	*tenuissima, Membranipora*	Lagaaij, 1963: 165
B1p	*tenuissima, Parellisina*	Canu and Bassler, 1928b: 34; as *Callopora tenuissima*
A1p	*thrincota, Smittina*	Shier, 1964: 637
B1r	*transversalis, Steganoporella*	Marcus, 1949: 11
A1p	*triangulata, Cribella*	Canu and Bassler, 1928a: 82
0	*trichotoma, Cribellopora*	Lagaaij, 1963: 187
B1r	*trimorpha, Schizoporella*	Canu and Bassler, 1928a; 81
	trispinosa, Parasmittina, var. *echinata;* see *echinata, Parasmittina*	
A1p	*trispinosa, Smittia,* var. *ligulata*	Ridley, 1881: 53
	trispinosa, Smittina (part); see *spathulata, Parasmittina*	

Check List of West Atlantic Tropical Ectoproct Fauna (Continued)

Poly-morphism	Species	Citation
	trispinosa, Smittina (part); see *nitida, Parasmittina*	
A2pp	*trispinosa, Parasmittina,* var. *munita*	Marcus, 1937: 108; as *Smittina trispinosa* var. *munita*
	trispinosa, Smittina, var. *nitida;* see *nitida, Parasmittina*	
	trispinosa, Smittina, var. *protecta;* see *protecta, Parasmittina*	
0	*truncata, Aetea*	Marcus, 1938: 11
A1r	*tuberculata, Drepanophora*	Osburn, 1914: 200; as *Rhynchozoon tuberculatum*
0	*tuberculata, Membranipora*	Marcus, 1937: 33
	tuberculata, Nitscheina; see *tuberculata, Membranipora*	
B1p	*tuberculata, Velumella*	Canu and Bassler, 1928a: 63
	tuberculatum, Rhynchozoon; see *tuberculata, Drepanophora*	
	tuberosa, Aplousina; see *filum, Aplousina*	
	tuberosa, Crassimarginatella; see *filum, Aplousina*	
	tuberosa, Hemiseptella; see *filum, Aplousina*	
	tuberosum, Metrarabdotos; see *tenue tenue, Metrarabdotos (Biavicularium)*	
0	*tubigerum, Conopeum*	Osburn, 1940: 352
A1r	*tubulata, Retevirgula*	Osburn, 1940: 356; as *Hincksina periporosa*
0	*tubulosa, Celleporaria*	Canu and Bassler, 1928b: 147; as *Holoporella(?) tubulosa*
	tubulosa, Holoporella; see *tubulosa, Celleporaria*	
0	*tulipifera, Pasythea*	Osburn, 1940: 462
A1	*turrita, Bugula*	Marcus, 1937: 68
	turrita, Cellepora; see *turrita, Cigclisula*	
A2rr⎫ B1r⎭	*turrita, Cigclisula*	Smitt, 1873: 65, as *Lepralia turrita*
	turrita, Holoporella; see *turrita, Cigclisula*	
	turrita, Lepralia; see *turrita, Cigclisula*	
	turrita, Trematooecia; see *turrita, Cigclisula*	
A1p	*typica, Antropora*	Lagaaij, 1963: 171
0	*typica, Floridinella*	Canu and Bassler, 1928b: 59
A1p	*typica, Gemelliporidra*	Osburn, 1940: 425
A2	*uberrima, Vittaticella*	Lagaaij, 1963: 202
	umbellata, Cupularia (part); see *umbellata, Discoporella*	
	umbellata, Cupularia (part); see *doma, Discoporella*	
Av	*umbellata, Discoporella*	Marcus and Marcus, 1962: 290
A2rp	*undulata, Umbonula*	Canu and Bassler, 1928b: 119

Check List of West Atlantic Tropical Ectoproct Fauna (Continued)

Poly-morphism	Species	Citation
	unguiculata, Trigonopora; see *unguiculatum*	
	unguiculatum, Metrarabdotos	
	(Uniavicularium)	
	unguiculatum, Metrarabdotos; see	
	unguiculatum unguiculatum, Metrarabdotos	
	(Uniavicularium)	
A1p	*unguiculatum unguiculatum, Metrarabdotos*	Marcus, 1953: 304; as
	(Uniavicularium)	*Trigonopora unguiculata*
	unicornis, Schizoporella; see *errata,*	
	Schizoporella	
	unicornis, Schizoporella, var. *isabelleana;*	
	see *errata, Schizoporella*	
	unicornis, Schizoporella, var. *pungens;*	
	see *pungens, Schizoporella*	
	uvulifera, Aimulosia; see *uvulifera, Hippoporella*	
A1p	*uvulifera, Hippoporella*	Osburn, 1914; 210; as *Lepralia*
		uvulifera
	uvulifera, Lepralia; see *uvulifera, Hippoporella*	
	vagans, Holoporella; see *brunnea, Celleporaria*	
0	*vaginata, Dakaria*	Canu and Bassler, 1928a: 77
A1r	*venusta, Trypostega*	Smitt, 1873: 37; as *Gemellipora*
		glabra forma *striatula*
A1p	*verrilli, Hippoporina?*	Lagaaij, 1963: 192; as
		Hippodiplosia americana
A1	*verrucosa, Lagenipora*	Osburn, 1940: 450
	verruculata, Cellepora; see *rostratum,*	
	Rhynchozoon	
	verruculata, Rhynchozoon; see *rostratum,*	
	Rhynchozoon	
	violacea, Adeona; see *violacea, Reptadeonella*	
A1p	*violacea, Aptonella*	Canu and Bassler, 1928a: 83
	violacea, Porina; see *violacea, Reptadeonella*	
A1p	*violacea, Reptadeonella*	Marcus, 1949: 24; as *Adeona*
		violacea
A1p	*vulgaris, Escharina*	Lagaaij, 1963: 188
A1	*zanzibarensis, Caulibugula*	Osburn, 1940: 395 (no fig.)
i	sp., *Canda?*	Ridley, 1881: 45 (no fig.)
i	sp., *Celleporaria*	Rucker, 1967: 833
i	sp., *Margaretta*	Rucker, 1967: 830; as
		Tubucellaria sp.
	sp., *Tubucellaria;* see sp., *Margaretta*	

II. Check List and Summary of Polymorphism in Arctic American Fauna

The arctic fauna is defined as those species which occur shallower than about 150 m in the area bounded by (1) the Atlantic Ocean extension into Davis Strait (between

Canada and Greenland) north of the southern end of Baffin Island (ca. 62°N) on the
west and Disko Bay (ca. 70°N) on the east; and (2) the Arctic Ocean north of Canada
from Davis Strait on the east (ca. 60°W) and the Canadian Archipelago (ca. 125°W)
on the west. This region is close to that indicated as purely arctic by Hazel (1970)
in a very well documented discussion and by Dunbar (1953). The boundaries suggested
by Dunbar had previously been accepted in a paper on arctic Bryozoa (Powell, 1968a).

The major recent analyses of arctic bryozoa are by Bille-Hansen (1962) and Powell
(1968a); Hansen's summary of the Baffin Bay fauna is essentially the arctic fauna for
the area of northern Greenland and Canada as defined above; Powell's records were
culled to extract the arctic forms. These sources were supplemented by arctic records
of species published by Busk (1878, 1880), Nordgaard (1906), and Osburn (1923, 1932a,
1936).

The first preference in literature citations is to *figured* specimens from the *arctic*
or *subarctic* of the western North Atlantic or Arctic Oceans; second preference is to
described specimens from this area; third preference is to a *faunal record* from this
area. Characteristics of avicularia of species unfigured or undescribed from these areas
were obtained from other literature (e.g., Kluge, 1962).

There are 121 Cheilostome species in this fauna; 1 of these is insufficiently decribed
for a decision to be made regarding the presence or absence of polymorphic structures.

Of these 120 species, 89 (74.2 percent) exhibit at least one type of avicularium and/or
vibracularium; 20 species (16.6 percent) exhibit at least two types of avicularia and/or
vibracularia, or combination thereof; 3 species (2.5 percent) exhibit at least three types
of avicularia and/or vibracularia, or combination thereof; and 31 species (25.8 percent)
exhibit neither avicularia nor vibracularia.

Of the 89 species with one or both types of polymorphism, 5 (5.6 percent) have
vicarious forms of polymorphism; the remaining 84 species (94.4 percent) have only
adventitious forms of polymorphism; 2 species (2.2 percent) have a vibracularium *and*
an avicularium, and no species has only a vibracularium. Thus of the 89 polymorphic
species, 2 (2.2 percent) have vibracularia in contrast to 89 species (100 percent) which
have avicularia.

There are 58 species with a single type of avicularium and a well-calcified, lamellar
skeletal growth (primarily encrusting but sometimes erect). (This excludes *Bicellaria,
Bicellariella, Bugula, Caberea, Dendrobeania, Eucratea, Kinetoskias, Microporina
articulata, Notoplites, Pseudoflustra, Securiflustra, Scrupocellaria, Terminoflustra,* and
Tricellaria.) Of these 58 species, 23 (39.7 percent) have a pointed or very elongate
and narrow avicularium, and 35 (60.3 percent) have a rounded or spatulate-shaped
avicularium. For comparison, there are 14 species of similar growth with two or more
types of avicularia. Of these 14 species, 10 (71.4 percent) have at least one pointed
or very elongate and narrow avicularium, and 8 species (57.2 percent) have at least
one rounded or spatulate type of avicularium. Thus in increasing from 1 or more than
1 type of avicularium in a species for this sample of the fauna, the pointed forms
increase from 39.7 to 71.4 percent presence, and rounded forms stay about the same
changing from 60.3 to 57.2 percent presence. This is approximately an 80 percent increase
for the pointed types.

Check List of Atlantic Arctic American Ectoproct Fauna

Poly-morphism	Species	Citation
0	*abyssicola, Escharella*	Osburn, 1933: 340; as *Mucronella abyssicola*
	abyssicola, Mucronella; see *abyssicola, Escharella*	
A1p	*acutirostris, Porella*	Hincks, 1889: 429
	alba, Porella; see *minuta, Porella*	
0	*annulata, Cribrilina*	Powell, 1967: 1905
	annulata, Lepralia; see *annulata, Cribrilina*	
A1r	*aperta, Porella*	Powell, 1968a: 2307
A1	*arborescens, Kinetoskias*	Powell, 1968a: 2291 (no fig.)
	arctica, Callopora; see *arctica, Tegella*	
0	*arctica, Electra*	Powell and Crowell, 1967: 338; as *Electra crustulenta,* var. *arctica*
	arctica, Membranipora; see *arctica, Tegella*	
	arctica, Menipea; see *gracilis, Tricellaria* var. *inermis*	
A2	*arctica, Scrupocellaria*	Nordgaard, 1906: 7 (no. fig.); as *Menipea elongata*
	arctica, Smittina; see *majuscula, Smittina*	
A1p	*arctica, Electra*	Osburn, 1933: 311; as *Tegella arctica*
	arctica, Tegella; see *arctica, Electra*	
A1p	*arctica, Umbonula*	Powell, 1968a: 2294
A2pp	*armifera, Tegella*	Powell, 1968a: 2287
	articulata, Cellaria; see *articulata, Microporina*	
A1	*articulata, Microporina*	Powell, 1968a: 2288
A1r	*auriculata, Schizomavella*	Powell, 1968a: 2299 (no fig.)
	auriculata, Schizoporella; see *auriculata, Schizomavella*	
A1r	*auriculata, Schizoporella,* var. *lineata*	Nordgaard, 1906: 16 (no fig.); as *Schizoporella lineata*
	bella, Porella; see *bella, Smittina*	
A1r	*bella, Smittina*	Osburn, 1933; 337
	biaperta, Schizoporella; see *biaperta, Stephanosella*	
A2rp	*biaperta, Stephanosella*	Nordgaard, 1906: 15, as *Schizoporella biaperta*
	bidenkapi, Schizoporella; see *groenlandica, Smittina*	
	bilaminata, Cellepora; see *bilaminata, Rhamphostomella*	
A1r	*bilaminata, Rhamphostomella*	Hincks, 1877: 111; as *Cellepora bilaminata*
A1p	*bispinosa, Schizoporella*	Nordgaard, 1906: 17
	borealis, Microporina; see *articulata, Microporina*	
	borelis, Salicornaria; see *articulata, Microporina*	
0	*carbasea, Carbasea*	Powell, 1968a: 2283 (no fig.)
	carbasea, Flustra; see *carbasea, Carbasea*	
	catenularia, Electra	Nordgaard, 1906: 10 (no fig.);

Check List of Atlantic Arctic American Ectoproct Fauna (Continued)

Poly-morphism	Species	Citation
	catenularia, Membranipora; see catenularia, Electra	as *Membranipora catenularia*
A1p	cellata, Rh amphostomella	Bille-Hansen, 1962: 39 (no fig.)
	cervicornis, Cellepora; see surcularis, Celleporina	
A1	ciliata, Bicellaria	Osburn, 1923: 7d (no fig.)
A1p	ciliata, Microporella	Powell, 1968a: 2300
A1p	ciliata, Microporella var. arctica	Osburn, 1936: 542 (no fig.)
B1r	coarctata, Myriapora	Powell, 1968a: 2314
	coarctatum, Myriozoum; see coarctata, Myriapora	
A1r	compressa, Porella	Powell, 1968a: 2307
A3rri	concinna, Porella	Hincks, 1889: 428
	concinna, Smittina; see concinna, Porella	
	condylata, Schizoporella; see stylifera, Schizoporella	
0	connectens, Escharella	Powell, 1968a: 2302
	contigua, Cellepora; see contigua, Lepraliella	
A1p	contigua, Lepraliella	Osburn, 1912: 281; as *Cellepora contigua*
	contigua, Rhamphostomella; see contigua, Lepraliella	
A2pp	costata, Rhamphostomella	Hincks, 1889: 426
A2pp	costata, Rhamphostomella, var. cristata	Hincks, 1889: 426
0	crassicosta, Membraniporella	Powell, 1968a: 2291
A2pp	craticula, Callopora	Powell, 1968a: 2286 (no fig.)
	craticula, Membranipora; see craticula, Callopora	
	cruenta, Schizoporella; see crueuta, Stomachetosella	
0	cruenta, Stomachetosella	Powell, 1968a: 2296
A1p	crustacea, Schizoporella	Osburn, 1932a: 376 (no fig.); as *Myriozoella crustacea*
	crustacea, Myriozoella; see crustacea, Schizoporella	
	crustulenta, Electra, var. arctica; see arctica, Electra	
A1	cucullata, Bugula	Nordgaard, 1906: 8; as *Bugula harmsworthi*
A1p	cymbaeformis, Cauloramphus	Hincks, 1888: 217; as *Membranipora cymbiformis*
	cymbaeformis, Membranipora; see cymbaeformis, Cauloramphus	
	digitata, Flustra; see carbasea, Carbasea	
	discreta, Callopora; see smitti, Copidozoum	
0	divaricata, Hippothoa	Nordgaard, 1906: 15 (no fig.)
A2rp	elegantula, Cystisella	Osburn, 1912: 284; as *Porella perpusilla*
A2 } Av }	ellisii, Caberea	Powell, 1968a: 2288
A1p	elmwoodiae, Schizoporella	Nordgaard, 1906: 17; as *Schizoporella stormi*

Check List of Atlantic Arctic American Ectoproct Fauna (Continued)

Poly-morphism	Species	Citation
	elongata, Menipea; see *arctica, Scrupocellaria*	
A1p	*elongata, Phidolopora*	Hincks, 1877: 107; as *Retepora wallichiana*
	emucronata, Discopora; see *polita, Hemicyclopora*	
	erecta, Scrupocellaria; see *arctica, Scrupocellaria*	
0	*expansa, Hippothoa*	Maturo and Schopf, 1968: 56
A1r	*fragilis, Cystisella*	Hincks, 1888: 233; as *Porella elegantula* var. *rostrata*
A2	*fruticosa, Dendrobeania*	Packard, 1863: 409; as *Menipea fruticosa*
	fruticosa, Dendrobeania var. *quadridentata;* see *quadridentata, Dendrobeania*	
	fruticosa, Menipea; see *fruticosa, Dendrobeania*	
	gracilis, Menipea; see *gracilis, Tricellaria* var. *inermis*	
0	*gracilis, Tricellaria* var. *inermis*	Busk, 1855: 254; as *Menipea arctica*
A1r	*groenlandica, Smittina*	Nordgaard, 1906: 19; as *Schizoporella bidenkapi*
	harmsworthi, Bugula; see *cucullata, Bugula*	
A1r	*hincksi, Rhamphostomella*	Nordgaard, 1906: 31
0	*hincksi, Stomachetosella*	Osburn, 1933: 324; as *Stomachetosella producta;* see Powell, 1968a: 2297
	hippopus, Eschara; see *hippopus, Hippoporella*	
	hippopus, Lepralia; see *hippopus, Hippoporella*	
	hippopus, Hippoponella; see *hippopus, Hippoporella*	
A1r	*hippopus, Hippoporella*	Osburn, 1933: 332; as *Hippoponella hippopus.*
0	*hyalina, Hippothoa*	Osburn, 1933: 321
	hyalina, Schizoporella; see *hyalina, Hippothoa.*	
0	*immersa, Escharella*	Osburn, 1933: 339; as *Mucronella immersa*
	immersa, Mucronella; see *immersa, Escharella*	
	incrassata, Cellepora; see *surcularis, Celleporina*	
	indivisa, Escharella; see *connectens, Escharella*	
	indivisa, Mucronella; see *connectens, Escharella*	
A1p	*jacksoni, Escharoides*	Powell, 1968a: 2294
	jacksoni, Peristomella; see *jacksoni, Escharoides*	

Check List of Atlantic Arctic American Ectoproct Fauna (Continued)

Poly-morphism	Species	Citation
	jacksoni, Smittia; see *jacksoni, Escharoides*	
A2pr	*jeffreysi, Parasmittina*	Powell, 1968b: 256
	jeffreysi, Smittina; see *jeffreysi, Parasmittina*	
0	*labiata, Escharella*	Bille-Hansen, 1962: 44 (no fig.)
	labiata, Phylactella; see *labiata, Escharella*	
	laevis, Smittina (Porella); see *aperta, Porella*	
	landsborovii, Smittia, var. *porifera;* see *majuscula, Smittina*	
0	*laqueata, Escharella*	Nordgaard, 1906: 27 (no fig.)
	levinseni, Schizoporella; see *limbata, Stomachetosella*	
	limbata, Schizopoporella; see *limbata, Stomachetosella.*	
0	*limbata, Stomachetosella*	Nordgaard, 1906: 18 (no fig.); as *Schizoporella levinseni*
A2pp	*lineata, Callopora*	Powell, 1968a: 2286 (no fig.)
	lineata, Schizoporella; see *auriculata, Schizoporella,* var. *lineata*	
0	*loricata, Eucratea*	Maturo and Schopf, 1968: 52; as *Gemellaria willisii*
A1r	*majuscula, Smittina*	Powell, 1968 b; 255
B1	*membranaceotruncata, Terminoflustra*	Powell, 1968a: 2284 (no fig.)
	membranaceo-truncata, Flustra; see *membranaceotruncata, Terminoflustra*	
A1r	*minuscula, Smittina*	Nordgaard, 1906: 28
A1r	*minuta, Porella*	Nordgaard, 1906: 25; as *Porella alba*
	murrayana, Bugula; see *murrayana, Dendrobeania*	
A2	*murrayana, Dendrobeania*	Osburn, 1933: 317
	murrayana, Dendrobeania, var. *fruticosa;* see *fruticosa, Dendrobeania*	
	murrayana, Dendrobeania, var. *quadridentata,* see *quadridenta, Dendrobeania*	
	nigrans, Callopora; see *nigrans, Hincksina*	
A1p	*nigrans, Hincksina*	Nordgaard, 1906: 12; as *Membranipora nigrans*
	nigrans, Membranipora; see *nigrans, Hincksina*	
	nigrans, Tegella; see *nigrans, Hincksina*	
	nodulosa, Cellepora; see *nodulosa, Schizmopora*	
A1r	*nodulosa, Schizmopora*	Powell, 1968a: 2302
	nordlandica, Eschara; see *nordlandica, Lepralioides*	

Check List of Atlantic Arctic American Ectoproct Fauna (Continued)

Poly-morphism	Species	Citation
0	*nordlandica, Lepralioides*	Nordgaard, 1906: 22 (no fig.); as *Eschara nordlandica*
	novanglia, Smittina; see rigida, Smittina	
	obesa, Hippodiplosia; see obesa, Schizoporella	
A1r	*obesa, Schizoporella*	Powell, 1968a; 2299
A1r	*ortmani, Schizomavella*	Bille-Hansen, 1962: 31 (no fig.)
	ortmani, Schizoporella; see ortmani, Schizomavella	
	ovata, Cellepora; see ovata, Rhamphostomella	
A1r	*ovata, Rhamphostomella*	Nordgaard, 1906: 32
i	*patula, Porella*	Osburn, 1923: 11D (no fig.)
	pavonella, Discopora; see arctica, Umbonula	
	pavonella, Mucronella; see arctica, Umbonula	
	peachi, Bugulopsis; see peachii, Tricellaria	
	peachii, Cellularia; see peachii, Tricellaria	
	peachii, Mucronella; see immersa, Escharella	
0	*peachii, Tricellaria*	Powell, 1968a: 2290
	perpusilla, Eschara; see elegantula, Cystisella	
A1r	*pertusa, Hippodiplosia*	Hincks, 1892: 154; as *Lepralia pertusa*
	pertusa, Lepralia; see pertusa, Hippodiplosia	
	plana, Lepralia; see plana, Myriozoella	
A1r	*plana, Myriozoella*	Maturo and Schopf, 1968: 62; as *Lepralia plana*
	plana, Palmicellaria; see plana, Porella	
A1r	*plana, Porella*	Hincks, 1888: 221; as *Porella skenei* form *plana*
A1r	*plicata, Rhamphostomella*	Nordgaard, 1906: 30
0	*polita, Hemicyclopora*	Bille-Hansen, 1962: 68 (no fig.)
A1r	*porifera, Schizomavella*	Powell, 1968b: 253
	porifera, Schizoporella; see porifera, Schizomavella	
	porifera, Smittia; see majuscula, Smittina	
	praelucida, Cheilopora; see Cheilopora sincera	
	praelucida, Mucronella; see sincera, Cheilopora	
	princeps, Porella; see producta, Stomachetosella (part 2)	
A1r	*proboscidea, Porella*	Bille-Hansen, 1962: 35 (no fig.)
	producta, Schizoporella; see hincksi, Stomachetosella	
	producta, Stomachetosella (part 1); see	

Check List of Atlantic Arctic American Ectoproct Fauna (Continued)

Poly-morphism	Species	Citation
	hincksi, Stomachetosella	
A1r	*producta, Stomachetosella* (part 2)	Hincks, 1892: 152; as *Monoporella spinulifera var. praeclara*
	propinqua, Hippodiplosia; see *propinqua, Hippoporina*	
A2rr	*propinqua, Hippoporina*	Powell, 1968b: 254
	propinqua, Porella; see *propinqua, Hippoporina*	
A1p	*punctata, Cribrilina*	Osburn, 1923: 9D (no fig.)
A2	*quadridentata, Dendrobeania*	Osburn, 1923: 7D (no fig.); as *Dendrobeania murrayana* var. *quadridentata*
	radiatula, Lepralia; see *radiatula, Rhamphostomella*	
A1p	*radiatula, Rhamphostomella*	Hincks, 1877: 104; as *Lepralia radiatula*
	reticulatopunctata, Hippodiplosia; see *reticulatopunctata, Hippoporina*	
A1r	*reticulatopunctata, Hippoporina*	Powell, 1968b: 2301
	reticulatopunctata, Lepralia; see *reticulatopunctata, Hippoporina*	
	reticulatopunctata, Schizoporella; see *reticulatopunctata, Hippoporina*	
	reticulatopunctata, Smittia; see *reticulatopunctata, Hippoporina*	
	reticulatopunctata, Smittina; see *reticulatopunctata, Hippoporina*	
A1r	*rigida, Smittina*	Osburn, 1933: 338; as *Smittina novanglia*
A1r	*rosacea, Escharopsis*	Bille-Hansen, 1962: 30 (no fig.); as *Ragionula rosacea*
	rosacea, Ragionula; see *rosacea, Escharopsis*	
A1r	*saccata, Cystisella*	Osburn, 1912: 283; as *Porella saccata*
	saccata, Cystisella, var. *rostrata;* see *fragilis Cystisella*	
	sarsii, Eschara; see *sarsii, Posterula*	
	sarsii, Escharoides; see *sarsii, Posterula*	
	sarsii, Escharopsis; see *sarsii, Posterula*	
A1p	*sarsii, Posterula*	Hincks, 1888: 218; as *Escharoides sarsii*
	scabra, Cellepora; see *scabra, Rhamphostomella*	
A1r	*scabra, Rhamphostomella*	Powell, 1968a: 2312 (no fig.)
A2 ⎫ Av ⎭	*scabra, Scrupocellaria*	Osburn, 1933: 314
A2	*scabra, Scrupocellaria* var. *paenulata*	Osburn, 1932a: 371 (no fig.)
	scutulata, Discopora; see *scutulata, Harmeria*	
0	*scutulata, Harmeria*	Osburn, 1936: 541 (no fig.)
	scutulata, Lepralia; see *scutulata, Harmeria*	

Check List of Atlantic Arctic American Ectoproct Fauna (Continued)

Poly-morphism	Species	Citation
	securifrons, Flustra; see *securifrons, Securiflustra*	
A1	*securifrons, Securiflustra*	Powell, 1968a; 2284 (no fig.)
A1p	*septentrionalis, Amphiblestrum*	Bille-Hansen, 1962; 66 (no fig.); as *Amphiblestrum flemingi* var. *septentrionalis*
	septentrionalis, Sertella; see *cellulosa, Retepora*	
	serrulata, Flustra; see *serrulata,. Membranipora*	
0	*serrulata, Membranipora*	Busk, 1880: 234; as *Flustra serrulata*
A1	*simpliciformis, Bugula*	Osburn, 1932a: 369
A1r	*sincera, Cheilopora*	Osburn, 1912: 283; as *Mucronella praelucida*
	sincera, Eschara; see *sincera, Cheilopora*	
	sincera, Hemeschara; see *sincera, Cheilopora*	
	sinuosa, Schizoporella; see *sinuosa, Stomachetosella*	
0	*sinuosa, Stomachetosella*	Osburn, 1933: 324
	skenei, Porella, form *plana;* see *plana, Porella*	
	smitti, Callopora; see *smitti, Copidozoum*	
A1p } B1p	*smitti, Copidozoum*	Powell, 1968a: 2286
	smitti, Menipea; see *smitti, Notoplites*	
A2	*smitti, Notoplites*	Powell, 1968a: 2289
A1r	*smitti, Porella*	Powell, 1968b: 255
A1p	*smitti, Smittina*	Nordgaard, 1906: 28 (no fig.)
	solida, Flustra; see *solida, Pseudoflustra*	
A1	*solida, Pseudoflustra*	Bille-Hansen, 1962: 37 (no fig.)
	solida, Smittina; see *solida, Pseudoflustra*	
	sophiae, Membranipora; see *arctica, Tegella*	
A1p	*spathulifera, Doryporella*	Osburn, 1912: 282; as *Lepralia spathulifera*
	spathulifera, Lepralia; see *spathulifera, Doryporella*	
A1r	*spinigera, Rhamphostomella*	Nordgaard, 1906: 32
	spinulifera, Escharelloides; see *spinulifera, Hincksipora*	
0	*spinulifera, Hincksipora*	Hincks, 1889: 431; as *Mucronella spinulifera*
	spinulifera, Monoporella, var. *praeclara;* see *producta, Stomachetosella* (part 2)	
	spinulifera, Mucronella; see *spinulifera, Hincksipora*	
A1p	*spitzbergensis, Bidenkapia*	Osburn, 1923: 8D (no fig.); as *Callopora spitzbergensis*

Check List of Atlantic Arctic American Ectoproct Fauna (Continued)

Poly-morphism	Species	Citation
	spitzbergensis, Callopora; see *spitzbergensis, Bidenkapia*	
	spitzbergensis, Cribrilina; see *spitzbergensis, Reginella*	
0	*spitzbergensis, Reginella*	Powell, 1968a: 2292
	spitzbergensis, Tegella; see *spitzbergensis, Bidenkapia*	
	stormi, Schizoporella; see *elmwoodiae, Schizoporella*	
A1r	*struma, Porella*	Powell, 1968a: 2309
	stylifera, Emballotheca; see *stylifera, Schizoporella*	
	stylifera, Escharella; see *stylifera, Schizoporella*	
0	*stylifera, Schizoporella*	Nordgaard, 1906: 18; as *Schizoporella condylata*
A1r	*subgracila, Myriapora*	Powell, 1968a: 2314
	subgracile, Myriozoum; see *subgracila, Myriapora*	
	subgracilis, Leieschara; see *subgracila, Myriapora*	
	surcularis, Cellepora; see *surcularis, Celleporina*	
A1r ⎱ B1r ⎰	*surcularis, Celleporina*	Powell, 1968a: 2301
	surcularis, Costazia; see *surcularis, Celleporina*	
	surcularis, Porella; see *surcularis, Celleporina*	
	surcularis, Schizomopora; see *surcularis, Celleporina*	
0	*suturata, Eschara*	Nordgaard, 1906: 21
	ternata, Cellularia; see *ternata, Tricellaria*	
	ternata, Menipea; see *ternata, Tricellaria*	
A2	*ternata, Tricellaria*	Osburn, 1933: 315
A1p	*trifolium, Amphiblestrum*	Osburn, 1933: 314
	trifolium, Membranipora; see *trifolium, Amphiblestrum*	
A2pr	*trispinosa, Smittina*	Osburn, 1932a: 373 (no fig.); as *Smittina trispinosa* var. *nitida*
	trispinosa, Smittina, var. *nitida;* see *trispinosa, Smittina*	
0	*tubulosa, Cylindroporella*	Osburn, 1933: 322
	tubulosa, Porina; see *tubulosa, Cylindroporella*	
	umbonata, Porella; see *proboscidea, Porella*	
	unicornis, Callopora, var. *armifera;* see *armifera, Tegella*	

Check List of Atlantic Arctic American Ectoproct Fauna (Continued)

Poly-morphism	Species	Citation
	unicornis, Membranipora; see *unicornis, Tegella*	
	unicornis, Membranipora, var. *armifera;* see *armifera, Tegella*	
A1p	*unicornis, Tegella*	Busk, 1880: 234 (no fig.); as *Membranipora unicornis*
	variolosa, Mucronella; see *connectens, Escharella*	
	ventricosa, Cellepora; see *ventricosa, Celleporina*	
A1r ⎫ B1r ⎭	*ventricosa, Celleporina*	Powell, 1968a: 2301
	ventricosa, Costazia; see *ventricosa, Celleporina*	
0	*ventricosa, Escharella*	Powell, 1968a: 2303
	ventricosa, Mucronella; see *ventricosa, Escharella*	
	wallichiana, Retepora; see *elongata, Phidolopora*	
0	*whiteavesii, Callopora*	Powell and Crowell, 1967: 340
	willisii, Gemellaria; see *loricata, Eucratea*	

III. Check List and Summary of Polymorphism in Atlantic Deep-Sea Fauna

The Atlantic deep-sea fauna is, I think, best defined as those species which occur deeper than the break between the continental shelf and continental slope (often approximately 200 m), and north of the Antarctic Convergence (between 50° and 60°S). The subsample of the deep-sea fauna used in this paper is those species found deeper than 2000 m, excluding those forms thought to be emplaced by turbidity currents. Such specimens include representatives of *Bugula neritina* (see Calvet, 1906: 379, reported 4060 m, Azores), *Scrupocellaria pusilla* (see Calvet, 1906: 374; reported 2018 m, northwest of Spain), *Cellaria fistulosa* (see Calvet, 1906: 401; reported 3700 m, Canary Islands), *Retepora ramulosa* (see Calvet, 1906: 454, reported 2300 m, Cape Verde Islands). (*Cellepora dichotoma* var. *attenuata* from 2170 m off Monaco may also be turbidity current emplaced; see Calvet, 1931: 122.)

The major sources of data on Atlantic deep-sea cheilostomes are the reports of the expeditions of the *Hirrondelle* (Jullien and Calvet, 1903), the *Challenger* (Busk, 1884; Waters, 1888), the *Travailleur* and the *Talisman* (Jullien, 1882; Calvet, 1906), and of Prince Albert I of Monaco (Calvet, 1931). These reports were supplemented by those of Lagaaij (1963), Silén (1951), and Menzies (1963).

There are 42 species of cheilostomes in this fauna. Of these, 32 (76.2 percent) exhibit at least one type of avicularium and/or vibracularium; 7 species (16.7 percent) exhibit at least 2 types of avicularia and/or vibracularia, or combination thereof; 2 species (4.8 percent) exhibit at least 3 types of avicularia and/or vibracularia, or combination thereof; 10 species (23.8 percent) exhibit neither avicularia nor vibracularia.

Check List of Atlantic Deep-Sea Ectoproct Fauna (>2000 m)

Poly-morphism	Species	Citation
2	alcicornis, Cribilina	Jullien, 1882: 508
1	alternata, Haswellia	Calvet, 1906: 447
2	bicornis, Camptoplites	Busk, 1884: 40; as *Bugula bicornis*
1	biseriata, Cellaria	Calvet, 1906: 401 (no fig.)
1	boreale, Tessaradoma	Busk, 1884: 174
0	cavalieri, Euginoma	Lagaaij, 1963: 179
	clausa, Menipea; see *marsupiata*, Scrupocellaria	
1	crateriformis, Notoplites	Busk, 1884: 16; as *Cellularia crateriformis*
1	cyathus, Kinetoskias	Busk, 1884: 44
0	eburnea, Pasythea	Busk, 1884: 5
1	edwardsi, Fedora	Jullien, 1882: 514
0	edwardsi, Lagenipora	Jullien, 1882: 510
1	enucleata, Jubella	Jullien, 1882: 519
1	evocata, Bicellaria	Jullien, 1882: 508
v	folini, Setosellina	Jullien, 1882: 523
	gracilis, Tessaradoma; see *boreale*, Tessaradoma	
v	implicata, Heliodoma	Calvet, 1906: 396
0	inermis, Palmicellaria	Jullien, 1882: 517
2	jullieni, Retepora	Calvet, 1906: 453
1	leontodon, Bugula	Busk, 1884: 39
v	ligata, Caberea	Calvet, 1906: 378
0	longipes, Crepis	Calvet, 1906: 393
2 } v	macandrei, Scrupocellaria	Busk, 1884: 23
0	magnifica, Cellaria	Busk, 1884: 93, as *Salicornaria magnifica*
1	margaritifera, Himantozoum	Busk, 1884: 41; as *Bugula margaritifera*
2 } v	marsupiata, Scrupocellaria	Jullien, 1882: 506
1	mirabile, Himantozoum	Busk, 1884: 39; as *Bugula mirabilis*
1	minuta, Bifaxaria	Busk, 1884: 81
1	navicularis, Bicellaria	Busk, 1884: 32
v	pandora, Ascosia	Jullien, 1882: 505
1	perrieri, Smittia	Jullien, 1882: 515
1	pocillum, Kinetoskias	Busk, 1884: 45
0	reticulata, Bifaxaria	Busk, 1884: 82
1	reticulata, Camptoplites, var. *unicornis*	Busk, 1884: 40; as *Bugula reticulata* var. *unicornis*

Check List of Atlantic Deep-Sea Ectoproct Fauna (>2000 m) (Continued)

Poly-morphism	Species	Citation
2	*rotundata, Cornucopina*	Silén, 1951: 65 (no fig.)
v	*roulei, Setosellina*	Calvet, 1906: 395
1	*simplex, Canda*	Busk, 1884: 26
0	*simplex, Nellia*	Busk, 1884: 27
2	*strangulatum, Myriozoum*	Calvet, 1906: 427
0	*spectrum, Smittia*	Jullien, 1882: 515
1	*vemae, Kinetoskias*	Menzies, 1963: 5
0	*vermiformis, Euginoma*	Jullien, 1882: 520
v	*vulnerata, Setosella*	Jullien, 1882: 524
1?	sp., *Bicellariella*	Calvet, 1931: 70

IV. Check List of Estuarine Fauna

Poly-morphism	Species
0	*crustulenta, Electra* (Pallas, 1766)
0	*melolontha, Aspidelectra* (Landsborough, 1852)
0	*membranacea, Membranipora* (Linnaeus, 1767)
0	*monostachys, Electra* (Busk, 1854) [*Electra hastingsae* (Marcus, 1938 = *E. monostachys* (Busk, 1854); *fide* Ryland, 1969: 213)]
0	*pallasiana, Cryptosula* (Moll, 1803)
0	*pilosa, Electra* (Linnaeus, 1761)
0	*reticulum, Conopeum* (Linnaeus, 1767)
0	*seurati, Conopeum* (Canu, 1928)
0	*tenuis, Acanthodesia* Desor, 1848
0	*truitti, Conopeum* Osburn, 1944

Evolution of Avicularia in Cheilostome Bryozoa

William C. Banta   The American University

ABSTRACT

The available evidence supports the prevalent assumption that avicularia are modified autozoids, and that they differentiated gradually from monomorphic autozoids. Some living species, including *Steginoporella magnilabris,* possess "ordinary" autozoids ("A-zoids") and polymorphs ("B-zoids") with an augmented apertural apparatus. B-zoids probably represent incipient interzoecial avicularia. If these considerations are valid, we can find a steady increase in development of coloniality. Coloniality first increased by division of labor (development of polymorphism), then physiological interdependence of polymorphs, and finally structural interdependence of polymorphs. The same pattern may prove to be true in other colonial polymorphic animals.

For purposes of this paper, let us define coloniality as the degree to which are developed the features or characters possessed by all colonial animals but lacking in solitary animals. We can then recognize increasing coloniality by noting the extent to which these "colonial characters" are expressed. Beklemishev (1970) attempted to do this by proposing a series of "steps" or "grades" between solitary animals and the most highly evolved colonial forms. He proposed three main trends in increasing coloniality: (1) a weakening of individuality of zoids by increasing organic continuity; (2) a strengthening of the individuality of the colony by specialization of zoids into polymorphs; and (3) formation of "cormidia," clusters of polymorphs forming functional units.

Author's address: Department of Biology, The American University, Washington, D.C. 20016.

AUTOZOID   AVICULARIUM

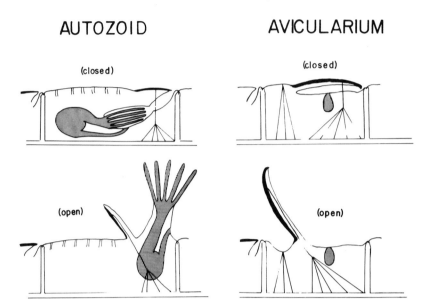

Figure 1

Diagrammatic longitudinal sections of an autozoid and interzoecial avicularium. The polypide and setigerous organ are shaded.

Beklemishev's arguments are extensive and complicated; it is possible to discuss no more than a tiny corner of his work. This paper is an attempt to take one morphological type of polymorph, the cheilostome avicularium, and analyze what we know about it in terms of increasing and decreasing coloniality.

Avicularia are chambers connected to autozoids (feeding zoids) by communication organs. They do not feed; a functional polypide (lophophore plus gut) is absent. There is a cuticular appendage, the mandible, that moves in one plane and is operated by a strong set of muscles (striated muscles, as far as is known) (Fig. 1). Previous authors have invariably homologized avicularia and autozoids. The mandible is considered to represent an operculum; its muscles are augmented opercular muscles. The polypide is supposedly represented in the avicularium by a tiny mass of cells located beneath the mandible; the mass is sometimes partly extrusible and is then called a setigerous organ.

The supposed homology between avicularia and autozoids is further strengthened by the existence of avicularia which occupy the same place in the budding pattern in which one would expect to find autozoids. Such avicularia are termed "interzoecial." The available evidence (Silén, 1938; and others) indicates that they are budded in the same way as autozoids. Interzoecial avicularia are characteristically more autozoid-like than avicularia occurring elsewhere (adventitious avicularia). Interzoecial avicularia are usually larger, the mandibles are more operculum-like, and the muscles originate and insert more like opercular muscles than is the case with adventitious avicularia. Furthermore, there are a few species (mostly in the genus *Steginoporella*) which possess special autozoids

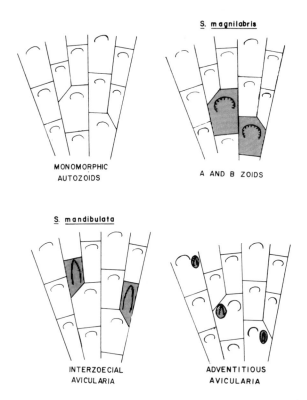

Figure 2

Diagrams of four stages in evolution of avicularia. Heterozoids are shaded.

called B-zoids. B-zoids differ from ordinary A-zoids in possessing augmented opercula and opercular muscles. They are in positions in the colony almost comparable to the usual position of interzoecial avicularia. Interzoecial avicularia in *Steginoporella* and some other genera are usually located just distal to the bifurcation of a longitudinal zoid row; that is, they are the first zoids in a new zoid row (Fig. 2). B-zoids, on the other hand, are most commonly developed just proximal to the bifurcation of a zoid row; they bud two new zoids distal to themselves rather than the usual one zoid (Fig. 2).

The idea that avicularia represent modified autozoids was first proposed by Farre (1837), but it is usually associated with the work of Harmer (1900). Harmer proposed that one species which possesses A- and B-zoids, *Steginoporella magnilabris* (Busk), resembles primitive cheilostomes at an early stage in the evolution of avicularia. He reasoned that B-zoids were evolved by gradual modification of autozoids in a monomorphic colony; that is, a colony composed entirely of autozoids resembling or identical to A-zoids. The implications of this evolutionary process have been discussed by Cheetham (1972). In *Steginoporella*, avicularia occur in some species (*S. mandibulata,* for example), whereas A- and B-zoids occur in others (*S. magnilabris*), but in no case do avicularia occur on colonies with B-zoids.

Silén (1938) agreed in large part with the theory of Harmer, but differed with him over the homology of interzoecial avicularia and B-zoids. Silén reasoned that the difference in location of the two zoid types in the budding pattern (explained above) made homology impossible. Recent work (Banta, 1972) indicates that the same zoid type can be transferred from one budding site to another, so Silén's objection is considerbly weakened. Furthermore, gradual evolution of polymorphism among autozoids has been convincingly demonstrated in a recent paper by Cheetham (1972). For purposes of this paper, it is presumed that B-zoids in *Steginoporella* represent incipient avicularia, and that ancestors of avicularia-bearing cheilostomes went through comparable stages.

If we assume that a monomorphic form gave rise to a dimorphic one possessing A- and B-zoids, does the change from monomorphic to dimorphic represent a change in the degree of coloniality? To answer this question we need criteria for recognizing coloniality. I propose the following criteria: (1) physiological dependence of one zoid on another; (2) structural dependence (organic continuity) between zoids, and (3) division of labor (specialization) among zoids. By this definition an increase in any of these criteria represents an increase in coloniality.

Applying these criteria to the differentiation of A- and B-zoids from monomorphic autozoids, we see that there is some increase in coloniality. There is no evidence for any significant increase in physiological dependence of one zoid on another. In the absence of evidence to the contrary, we should presume that A- and B-zoids are equally efficient in gathering food and exchanging nutrients with neighboring zoids. Nor is there any obvious evidence of increase in structural dependence. The body wall morphology of B-zoids is approximately the same as in A-zoids, and they occupy the same relative position in the colony as do many A-zoids. But there is an apparent expression of division of labor. B-zoids are surely performing some operation that is performed less well by A-zoids, and vice versa. Dimorphism in these forms cannot be called a "sport of nature"; *Steginoporella magnilabris* is almost circumtropical in distribution and always possesses A- and B-zoids. We can only presume that the dimorphism is a feature that is being preserved by selection and the B-zoids do something better than A-zoids, and vice versa. Perhaps B-zoids are specialized to prevent predation from some unknown animal, whereas A-zoids are specialized for more rapid sexual reproduction. Regardless of the function being selected, we may conclude that the development of autozoidal dimorphism represents an increase in coloniality by division of labor.

Now let us presume that the selective forces that caused evolution of autozoid dimorphism continued to act on our hypothetical *Steginoporella*-like ancestor of avicularia-bearing cheilostomes. We see what may be the result of these forces in another species of *Steginoporella*. *S. mandibulata* possesses true avicularia; the polypide has been lost, and the operculum, now called a mandible, has greatly increased in relative size. Two other changes have also taken place: the polymorph with the enlarged operculum-mandible (the "heterozoid") has been transferred from a position as a bifurcating zoid to that of a twinned zoid (Fig. 2), and the heterozoid has decreased in size relative to "normal" A-zoids. Does the presumed evolution of avicularia from B-zoids represent an increase in coloniality? There must be an increase in physiological dependence of the heterozoid on autozoids; the polypide has been lost. Obviously, more nutrients must be transferred from autozoids to avicularia than vice versa. Note that it is the loss of the polypide

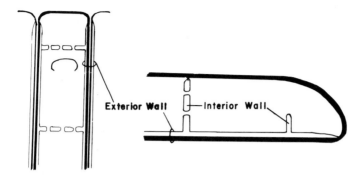

Figure 3

Diagrammatic frontal view (left) and longitudinal section of an autozoid to distinguish between interior and exterior walls.

(and hence loss of autonomy) that signals the development of true avicularia; the lack of autonomy was subjectively recognized as a characteristic of avicularia by bryozoan workers, even though the theoretical justification for it was lacking. With regard to the third criterion, there has surely been an increase in division of labor; the autonomy of the autozoid has been sacrificed to the function of avicularia, whatever that may be. Is there also a change in the structural dependence of the polymorph on the autozoid? This question is more subtle than was the case with the other two criteria.

The avicularia of *S. mandibulata* are interzoecial; that is, they occupy the same position in the colony as autozoids. The walls that separate zoids can be divided into two morphological categories: exterior and interior walls (Silén, 1942). Exterior walls are topologically outside the animal; that is, they separate the body cavity from the outside. Exterior walls can be recognized (sometimes even in fossils) by the presence of a cuticle layer to the outside of the calcareous layer. A solitary animal's body wall is composed almost entirely of exterior walls; an example is the ancestrula of some cheilostome colonies. Interior walls, on the other hand, are partitions which divide up a preexisting coelomic space into compartments (Fig. 3). Interzoidal communication pores are developed only in interior walls. Two zoids separated by an interior wall are more dependent structurally than are zoids separated by a pair of exterior walls (obviously, two zoids cannot be separated by less than two exterior walls). The theoretical basis for distinguishing between interior and exterior walls has been discussed elsewhere (Banta, 1969).

It would seem that one can measure the structural dependence of zoids by measuring the proportion of interior to exterior walls. In *Steginoporella*, the vertical walls are the ones appropriate to measure because these are the walls that separate zoids from one another. An avicularium in *S. mandibulata* is invariably the more proximal of a pair of twinned zoids; that is, it is the proximolateral daughter zoid of a bifurcating mother zoid (Fig. 2). Moreover, they are shorter than autozoids. Both these features make for a smaller proportion of lateral (exterior) to transverse (interior) walls than is the case with autozoids. The change in position of heterozoids in the budding pattern between an ancestor similar to *S. magnilabris* and one like *S. mandibulata*, therefore, would be expected to result in an increase in structural dependence.

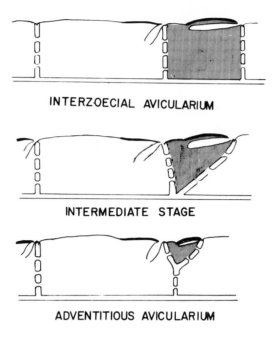

INTERZOECIAL AVICULARIUM

INTERMEDIATE STAGE

ADVENTITIOUS AVICULARIUM

Figure 4

Diagrammatic longitudinal sections through an autozoid and avicularium (shaded) to illustrate a proposed evolutionary sequence between interzoecial and adventitious avicularia.

So far this discussion has included only avicularia produced at the same place in the colony budding pattern as ordinary autozoids. Such zoids are produced from primogenial buds, normally located on vertical walls, and are called interzoecial avicularia. In many cheilostomes, however, zoids are produced from buds located elsewhere, typically on the frontal wall. Such buds are called adventitious buds, and the avicularia produced are adventitious avicularia.

If the theory of evolution of avicularia from polymorphic autozoids is correct, and if adventitious avicularia were evolved from interzoecial avicularia, then it follows that avicularium buds were transferred from the primogenial position to an adventitious position. This transferral may have been sudden or gradual. One population may have begun to produce avicularia in the new (adventitious) position, or, alternatively, avicularia might have been moved slowly through intermediate stages between primogenial and adventitious. We already have a model for a sudden transferral: frontal budding. Here the genetic information necessary for production of an autozoid was transferred suddenly from the primogenial to the adventitious position. We know that the transferral was sudden because (1) the primogenial layer must be produced before frontally budded zoids, and (2) primogenial buds could not have produced frontal zoids in this case because they are already "used" in producing the autozoids in the primogenial layer (see Banta, 1972).

It is also possible to propose a model for gradual transferral. Interzoecial avicularia located just distal to a zoid row bifurcation might have become reduced in size, then

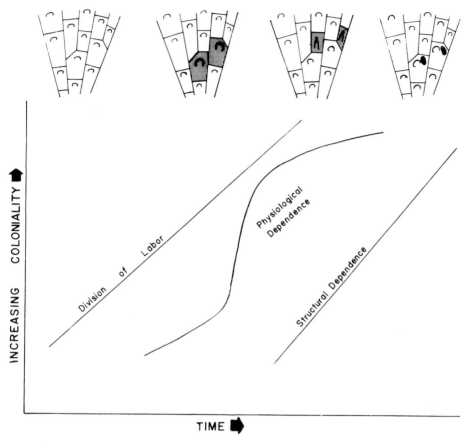

Figure 5

Summary of the evolutionary sequence proposed here. Each of the three criteria for measuring coloniality increase between each of the four stages illustrated (top), but not necessarily at the same times or rates.

moved up the transverse wall of the parent autozoid by a gradual change in the location of the transverse wall of the avicularium (Fig. 4). By this model, the primogenial bud may have divided into two buds: one remaining in the primogenial position, and one becoming an adventitious (frontal) bud. The adventitious avicularium produced by this hypothetical process would be located in the lateral-apertural ("lateral-oral") position. This hypothesis is supported by two lines of evidence: (1) adventitious avicularia occur most commonly in lateral-apertural position, and (2) there is evidence that the physiological processes that cause zoid row bifurcation are similar to or the same as those causing production of lateral-apertural avicularia in *Schizoporella* (Banta and Holden, unpublished).

Present evidence does not allow us to choose between the two hypotheses: transferral could have been sudden or gradual. Perhaps both processes occurred in different evolutionary lines.

Now we should ask: Does the transformation of avicularia from interzoecial to adventitious involve an increase in the degree of coloniality? Although there is no obvious evidence for increase in division of labor or physiological dependence, there seems to be an increase in structural dependence. Virtually all the walls of an adventitious avicularium are interior walls. An adventitious avicularium may become quite small and begin to look like an appendage of the autozoid which produced it. It would seem, therefore, that there has been an increase in coloniality by increase in structural dependence.

If the interpretations of structure made here are correct, it seems probable that evolution of avicularia in cheilostome Bryozoa followed these steps:

1. Autozoids specialized according to function. B-zoids began assuming a function which they performed better than A-zoids, and vice versa.

2. Physiological dependence evolved. The polypide was lost. Specialization continued.

3. Structural dependence evolved. Avicularia began to lose their structural integrity and began to take on the characters of an appendage of an autozoid.

The evolutionary changes involved are summarized in Fig. 5.

The hypothesis presented here is one that could be readily tested by careful study of the fossil record. In fact, much of the material necessary is probably already present in our museums. The theory offers an unusual opportunity to study evolution of polymorphism and evolving coloniality. The chance is especially good in the Bryozoa, because details of body wall structure are preserved in these fossils and because the Bryozoa have such an excellent fossil record. But it also offers a hint that we may profitably look for similar patterns and evolutionary sequences in other polymorphic colonial animals.

ACKNOWLEDGMENTS

I thank Alan H. Cheetham and Richard S. Boardman for helpful discussions and suggestions. I also thank Susan A. Cummings for help in preparing the illustrations and Dennis M. McFaden for reading the manuscript. The research was supported in part by an NSF institutional grant to The American University.

REFERENCES

Banta, W. C. 1969. The body wall of cheilostome Bryozoa, II, Interzoidal communication organs. Jour. Morph. *129*(2): 149–169.

———. 1972. The body wall of cheilostome Bryozoa, V, Frontal budding in *Schizoporella unicornis floridana*. Mar. Biol., *14:* 63–71, 3 figs.

———, and P. M. Holden. Bud width alone does not control zoid row bifurcation in *Schizoporella unicornis floridana* Osburn. In manuscript.

Beklemishev, W. N. 1970. Principles of comparative anatomy of invertebrates, v. I, Promorphology. (Transl. 3rd ed., Moscow, 1964. Oliver & Boyd, Edinburgh.)

Cheetham, A. H. 1972. Study of cheilostome polymorphism using principal components analysis. Proceedings of the Second Conference of the International Bryozoology Association, Durham, England, 1971. In press.

Farre, A. 1837. Observations on the minute structure of some of the higher forms of polypi. Roy. Soc. London, Trans. *1837:* 387–426.

Harmer, S. F. 1900. A revision of the genus *Steganoporella*. Quart. Jour. Micros. Sci., *43:* 225–297.

Silén, L. 1938. Zur Kenntnis des Polymorphismus der Bryozoen. Die Avicularien der Cheilostomata Anasca. Zool. Bidrag Uppsala, *17:* 149–366.

———. 1942. Origin and development of the Cheilo-Ctenostomatous stem of Bryozoa. Zool. Bidrag Uppsala, *22:* 1–59.

Degree of Individuality in Cheilostome Bryozoa: Skeletal Criteria

Philip A. Sandberg   University of Illinois at Urbana-Champaign

ABSTRACT

The degree of individuality or distinctiveness of members in a cheilostome bryozoan colony is reflected in the skeleton in several ways. These include continuity or discontinuity of intervening cuticular layers or of calcified layers in the primary skeleton, the presence of zoarial layers transcending boundaries between individuals, the presence of discrete zooecial linings, and the nature and abundance of interzooecial communications.

In some forms, individuals in a linear series are underlain by a common basal skeletal layer, above which additional basal layers are continuous with layers in the transverse walls, both distally and proximally. In others, no basal layers are shared by any two zooecia, and cuticular layers extend from basal to frontal surfaces along proximal-distal boundaries between individuals. Some cheilostomes, such as *Cellaria,* have walls in which cuticle is absent and growth is inferred to have taken place according to Borg's double-walled cyclostome model. This mode of growth occurs in taxonomically diverse cheilostomes and is regarded as of adaptive significance in development of some particular colonial forms.

INTRODUCTION

Fine structure of the bryozoan skeleton has been an important criterion in recent interpretations of features such as acanthopores, stolonal tubes, mesopores, etc., as

Author's address: Department of Geology, University of Illinois, Urbana, Illinois 61801.

heterozooids or as intra- or interzooidal structures (e.g., Tavener-Smith, 1969a; Blake and Towe, 1971). In the present paper I am also concerned with fine skeletal structure, specifically in cheilostome bryozoans, but from a somewhat different tack. The questions of importance here are: (1) how distinct from one another are the individuals (here primarily the unmodified individuals)?, (2) how, morphologically, are levels of individuality manifested in the skeleton?, and (3) what developmental differences are responsible for observed variations in degree of individuality of members of a cheilostome colony?

Basically, individuality or coloniality of particular skeletons or skeletal parts in cheilostomes are functions of two major factors, ontogenetic stage and mode of skeletal development. The latter is at least partly related to colonial form. Let us consider these two factors and their influence on individuality in cheilostome bryozoan skeletal parts.

ONTOGENETIC DEVELOPMENT

In longitudinal sections of basal-transverse junctions in lamellar calcite walls, layers can be seen curving up into the transverse wall from the basal walls both distal and proximal to it (Figs. 1 and 7). These continuous layers may be further followed into other zooecial walls (Fig. 13). It is readily apparent from low-magnification study of longitudinal sections that these basal layers and their continuations into other zooecial walls represent distinct secretions by individual zooids on either side of the transverse wall. Below those layers there often are longitudinal layers parallel to the basal surface which continue, uninterrupted, straight across the zooecial boundary beneath the transverse wall (Figs. 1, 7, and 16). This group of continuous basal layers may be thick (i.e., composed of many layers) or thin. Followed to the distal growing edge of the colony, those layers can be shown to be the depositional product of a marginal undifferentiated colonial zone distal to the delineation of zooecia in the linear series. In fact, deposition of this sequence of continuous basal layers, which I have called the basal platform, is terminated by the inception of transverse wall development. Subsequently deposited basal wall layers carry on into the transverse walls, and other zooecial walls as well. The basal platform forms a distally tapering wedge which is of very fragile construction and hence is often lost along margins of fossil or dead, transported or impacted modern colonies. The thickness of the basal platform is related to the width distal to, i.e., secretory time prior to, transverse wall development.

The next stage in ontogenetic development of the skeleton is one in which skeletal secretion is effected by well-defined individual zooids, with layers being deposited on the interior of the zooecium (although not necessarily over the entire interior surface) and also on the outer surface. Depending on the particular type of ultrastructure (most evident in lamellar types), growth layers or increments in the basal and transverse walls may be traced into the frontal or subfrontal wall. Note in Fig. 13 the rapid distal advance of the poorly layered initial part of the cryptocyst and the well-laminated later surrounding portions. Those later increments are more evenly distributed but still are thicker on the distal and frontal surfaces of the cryptocyst than on the lower internal surface. This and the zooecial lining discussed below, as well as earlier examples (Sandberg, 1971, pl. 2, fig. 12), are representative of the common situation in which presence of a secretory epithelium means neither continuous active secretion, nor uniformity in rate or amount of deposition at various points along the epithelial surface (see Tavener-Smith and Williams, 1970, p. 242–243).

The phase of highly individualized skeletal secretion is a common characteristic in development of most encrusting and erect cheilostomes, but it is by no means universal. A quite different growth pattern, typical of forms lacking internal cuticle, is discussed later in this paper. When a phase of distinct individual secretion is present it commonly terminates with formation of a partial or total zooecial lining, almost always formed of an ultrastructure distinct from the primary skeleton (Fig. 12; Sandberg, 1971, pl. 2, figs. 7, 8, 9, 11) and sometimes even of a different mineralogy (Sandberg, in preparation). Distribution of mineralogies of skeletal carbonates and the ultrastructural types associated with either $CaCO_3$ polymorph are well shown by combined use of Feigl solution staining (Friedman, 1959; Cheetham, Rucker, and Carver, 1969) and scanning electron microscopy (Schneidermann and Sandberg, 1971). In Fig. 14 the superficial frontal aragonite is clearly delineated by the grainy precipitate and by the brightness resulting from a higher emission of secondary electrons from the high surface area of that precipitate, relative to the smoother calcite and embedding plastic.

Limitation of the zooecial lining by such features as the ascus (Tavener-Smith and Williams, 1970, fig. 12, p. 242) does not entirely explain the frequent nonuniformity of distribution of those deposits. Commonly deposition will commence in one limited region of the zooecial interior (often the basal-distal ''corner'') and spread out subsequently over progressively larger areas.

In later ontogeny, prominently in the lower portions of erect rigid colonies, skeletal layers which transcend zooecial boundaries and may ultimately occlude orifices may be laid down on the frontal wall (Fig. 9; Boardman and Cheetham, 1969, fig. 4). These colonial layers, deposited at the lower margin of the hypostegal coelom (Borg, 1926, p. 196; Banta, 1971, p. 168–169), serve as structural support of various erect rigid colonial forms (Cheetham, 1971, p. 8–9, fig. 4). Similar reinforcing layers, often thicker than the total height of the zooecial cavity, are developed basally in many discoidal or unilamellar-fenestrate cheilostomes (Figs. 8, 10, 11) and fenestellid cryptostomes (Tavener-Smith, 1969b). In the reteporid cheilostomes, these basally thickening layers are usually deposited in irregularly polygonal, cuticle-separated extrazooidal chambers (Harmer, 1934, pls. 36, 37). Similar deposits occur basally in many lunulitiform cheilostomes (see Haakansson, in press). Boundaries between adjacent extrazooidal chambers are typically sutured or interfingering, suggesting alternating dominance of skeletal secretion of one or the other chamber.

The features discussed above may be summarized as an ontogenetic gradient from the colony margin proximally toward older parts of the colony, which includes (1) initial basal colonial layers (basal platform, not present in all forms) developed prior to differentiation of individual zooids, (2) intermediate skeletal secretions recognizable as distinct products of individual zooids, and (3) late ontogenetic frontal or basal thickenings associated with presence of a hypostegal coelom and most common in discoidal, linguiform, or erect bilamellar or unilamellar colonies.

MODE OF COLONIAL GROWTH

The degree of individuality in cheilostomes is strongly influenced by the way in which the colony grows, particularly as reflected by the presence and distribution of cuticle layers. In lateral and basal walls, cuticle layers are usually present (double in lateral walls and in basal walls of bilamellar forms, Banta, 1969, p. 169, figs. 23–25). The double cuticle of such walls is well shown in SEM figures of etched modern material (Fig. 6). Cuticle does not usually occur in transverse walls, except in the region of attachment of the upper edge of the wall and the frontal membrane (Banta, 1971, p. 169) (Fig. 13). In the case of forms with a basal platform, invagination of the cuticle into the transverse wall from below is precluded by those basal layers continuous across the interzooecial boundary (Figs. 1, 7, and 16). In fact, the traditional view, based on light microscopical observations and stated or implied in numerous recent papers on cheilostomes (e.g., Ryland, 1970, p. 87, figs. 12d, 14, 15e; Banta, 1968, p. 499, and 1971, p. 166–167; Cheetham, 1971, figs. 2, 5, 6, 7), holds that transverse walls are single, i.e., that "the proximal wall of any given zooecium is the distal wall of the preceding zooecium" [see, however, Ryland, 1970, fig. 3, and Banta's comment (1969, p. 499) on median subdivision of some transverse walls]. Thus far my SEM studies indicate that this is rarely, if ever, the case. Transverse walls are normally formed of layers secreted on both sides of the transverse wall by the proximal and distal transverse epithelial sheets (Tavener-Smith and Williams, 1970, p. 242) of the two adjacent zooecia. Those transverse wall layers are continuous with basal, lateral, and frontal layers and form parts of two discrete individual zooecial skeletons, secreted in both directions toward the centers of the bounding zooecia. As mentioned earlier, care should be taken not to equate presence of secretory epithelium with uniform or continuous skeletal secretion.

Transverse walls with cuticle do occur (as in *Adeona grisea* Lamouroux) and these are associated with absence of a basal platform (Sandberg, in preparation). The resulting zooecia are clearly separated from adjacent zooecia on all sides by cuticle-lined walls.

Forms with cuticle lacking in transverse walls (Figs. 1, 7, and 16) but present in lateral and basal walls (Fig. 6) are predominant among the cheilostomes examined ultrastructurally thus far. However, there exists a broad spectrum of zooecial individuality, as expressed in ultrastructural discontinuity and cuticular separation at zooecial limits. Cheilostomes with cuticle-bounded zooecia, e.g., *Adeona grisea,* lie at one extreme. At the opposite end are forms in which cuticle, although present on the colony exterior, is either entirely lacking within the walls of the colony or else is restricted to a few particular sites, usually associated with kenozooecial boundaries. The absence of cuticle and usually also of distinct zooecial wall limits has been noticed in a variety of cheilostomes, e.g., *Cellaria* (Banta, 1968, p. 499; Sandberg, 1971, pl. 2, fig. 8), *Myriapora* (Figs. 2, 5, and 17; Banta, 1968, p. 499), *Cupuladria* (Håkansson, in press), *Flabellopora, Mamillopora, Triphyllozoon, Veleroa?,* and *Conescharellina* (Figs. 2–5, 8, 12, 15, and 17. For one of these (*Cupuladria*), Håkansson has shown recently, using soft–hard embedding methods, that growth occurs in the mode of a double-walled cyclostome, with an outward marginal bulging of the hypostegal coelom prior to development of each new zooecium. Consequently, the cuticle is not intercalated into the walls (Banta,

1969) and is limited to the colony exterior (except in association with kenozooecial deposits on the basal surface similar to Fig. 10). The other forms I have studied only as dried material with the SEM, but in them no internal cuticles are apparent. Clear boundaries between skeletal secretions of any two adjacent zooids are usually also lacking (Figs. 2, 4, 5, 12, 16, and 17). The colonies of cheilostomes with noncuticulate walls are lunulitiform, cellariiform, eschariform (with radicel attachment), vinculariiform, or reteporiform. Individuals in such colonies tend to be elongated roughly perpendicular to the frontal surface (Figs. 15 and 17), although in some erect forms elongation parallel to the long axis of the branch occurs (e.g., *Cellaria*; see Harmer, 1926, p. 336, fig. 13). In many, axial kenozooecia (Fig. 15) or tubular connections between zooecia (Figs. 2, 5, and 17) are well developed.

The appearance of the primary skeleton in calcitic forms without internal cuticle varies appreciably among different taxa. Some forms have impersistent laminations parallel to the bounding walls (Fig. 2), but these are generally subdued or absent. Most commonly, the skeleton is a packed but somewhat porous mass of equidimensional or lath-like crystallites of variable orientations, suggestive of fairly rapid deposition (Fig. 12). Accretionary laminations define old positions of the advancing front of skeletal deposition. These lineations may lie nearly transverse to the wall surface and progressively alter their shape and curvature as the various skeletal parts are formed at the calcification front (Fig. 12). Initial growth is in a distal direction, but zooecial linings, often patchy or restricted, are developed later in some forms.

Aragonitic skeletons without internal cuticle have thus far been studied by me only in *Mamillopora cupula* Smitt. In that species, distally oriented crystallite cones or fans are nucleated near the center of the wall along the growing front and radiate toward adjacent cones or fans or toward the wall surfaces (Figs. 3 and 4). Some weak accretionary lineations have been observed.

Exterior zooecial limits in the noncuticulate forms discussed above need not correspond to internal boundaries between individuals. This is in contrast to cuticle-bounded zooecia, in which the cuticle delimits (at least laterally) the zooecium, and emerges at the frontal surface roughly over its position at the basal surface. In species of *Cellaria*, the noncoincidence of internal walls and superficial zooecial boundaries, as defined by margins of the frontal membranes, is particularly well shown (Harmer, 1926, p. 336, fig. 13).

In summary, mode of colonial growth greatly influences degree of individuality in cheilostome colonies. The common pattern in encrusting and many erect cheilostomes is one of distinct individuals with basal and lateral walls cuticle-bounded externally and skeletal layers developing within each zooecium along a growth front advancing perpendicular to the walls, into the zooecial cavity. Basal platforms may be present or absent, with resulting influence on transverse wall cuticle development and increased individuality. Distinctive zooecial linings frequently occur. In some discoidal and erect forms, skeletal features and some few observations on soft tissues indicate a more colonial front of calcification, with cuticle absent except at the outer colony surface. Accretion of walls occurs largely in a distal direction, although later zooecial linings may be formed.

Study of other cheilostomes with the SEM will undoubtedly reveal presence of skeletal growth without cuticles in lateral walls (double-walled cyclostome, see Boardman, 1971,

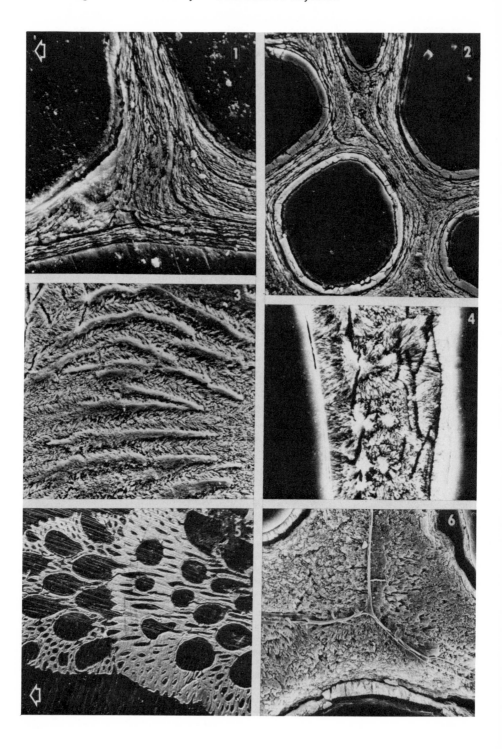

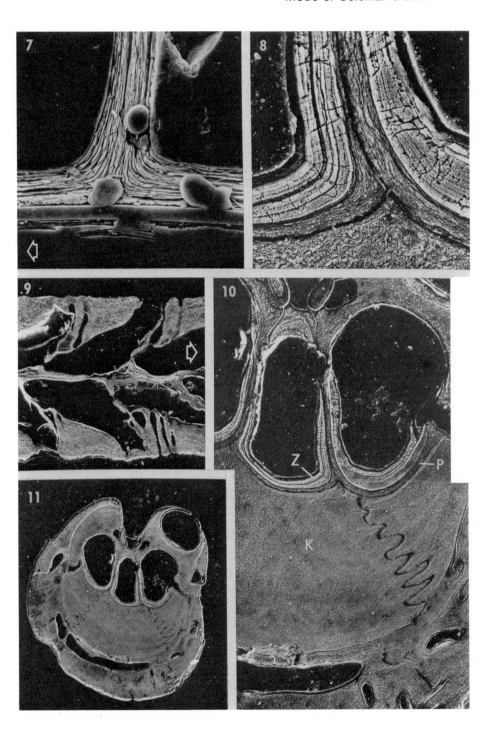

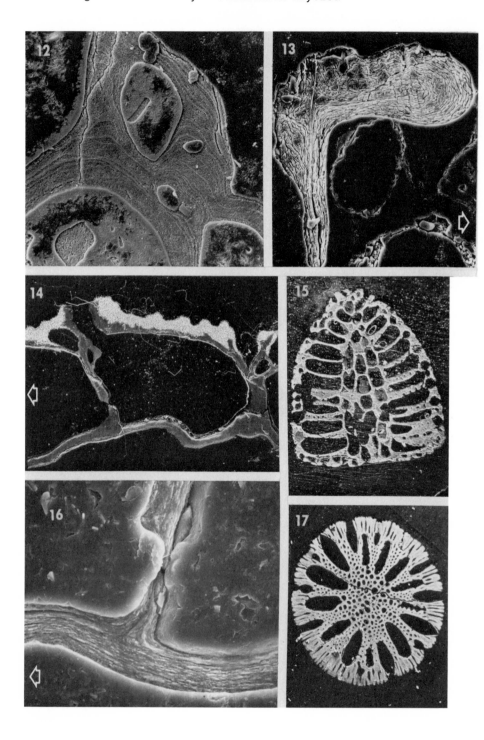

Figures 1 and 9

Metrarabdotos tenue (Busk). Modern, Caroline Station 68, off northeastern Puerto Rico. USNM 184149. 1. Longitudinal section of a basal-transverse wall junction. Arrow in this and later figures indicates the distal direction, ×475. 9. Longitudinal section. Extreme thickening of frontal superficial layer has occluded some zooecia, ×30.

Figures 2, 5, and 17

Myriapora subgracile (d'Orbigny). Modern, Albatross Station D2494, off Nova Scotia. USNM 184150. 2. Transverse section of longitudinal tubules between zooecia. Note noncuticulate walls and zooecial linings, ×475. 5. Longitudinal section of branch tip. Note interruption of growth, zooecial tubules, and lack of distinct zooecial boundaries, ×30. 17. Transverse section of branch. Note indistinct nature of zooecial limits, as well as the numerous pores between individuals in linear series, ×20.

Figures 3 and 4

Mamillopora cupula (Smitt). Modern, Gulf of Panama. USNM 184151. 3. Longitudinal section along the approximate center of a lateral wall. Cones of aragonite needles fan out distally (to the left), ×1190. 4. Transverse section of a lateral wall looking distally along the axes of aragonite fiber cones, ×945.

Figure 6

Porella compressa (Busk). Modern, Albatross Station D2494, off Nova Scotia. USNM 184152. Transverse section of basal-lateral wall junction. Note double cuticles and zooecial linings, ×800.

Figures 7 and 13

Acanthodesia grandicella (Canu and Bassler). Modern, Albatross Station D5134, Philippines. USNM 184153. 7. Longitudinal section of basal-transverse wall junction. Several plastic-filled boring holes interrupt some of the layers, ×480. 13. Longitudinal section of transverse wall and cryptocyst. Note continuity of layers from the transverse wall around lower and upper surfaces of cryptocyst with rapid unlayered initial growth distally and generally thicker deposits on the upper surface, ×250.

Figures 8, 10, and 11

Triphyllozoon sp. Modern, Albatross Station D5179, Philippines. USNM 184154. 11. Transverse section of branch. Note extreme basal thickening, ×65. 10 (detail of Fig. 11). Note sutured boundary between regions of extrazooidal deposits on basal wall and the discontinuity, followed by a lateral offset of those lateral boundaries. Easily distinguishable skeletal units include primary skeletal meterial (P), zooecial lining (L), and basal wall thickening deposits (K), ×180. 8 (detail of Fig. 10). Note the absence of an intercalary cuticle between adjacent lateral series of zooecia, as well as the fine accretionary structure in the relatively massive zooecial lining deposits, ×850.

Figure 12

Flabellopora elegans d'Orbigny. Modern, Albatross Station D5134, Philippines. USNM 184155. Transverse section through two autozooecia (upper and lower left) and two avicularian chambers (center and lower right). Frontal surface runs diagonally down through upper right. Note infrequent growth lines in otherwise homogeneous skeleton with no interzooecial boundaries, ×715.

Figure 14

Metrarabdotos unguiculatum (Canu and Bassler). Modern, Albatross Station D2405, Gulf of Mexico. USNM 184156. Longitudinal section, treated with Feigl solution. Lighter, grainier areas are aragonite, smooth, darker areas are calcite, ×70.

Figure 15

Conescharellina grandiporosa (Canu and Bassler). Modern, Albatross Station D5179, Philippines. USNM 184157. Note axial kenozooecia, vertical elongation of zooecia, and indistinct zooecial boundaries, ×20.

Figure 16

Schizoporella floridana (Osburn). Modern, Florida. USNM 184158. Longitudinal section of basal-transverse wall junction. Note the thick basal platform. The clarity of the layering is somewhat obscured by organic skeletal matrix remnants left from the skeletal material removed during etching, ×475.

p. 5–7) in numerous other taxa. At the moment, there appears to be little relationship taxonomically between cheilostomes possessing that growth mode. Rather it seems to be of adaptive significance in development of certain colonial morphologies which lack the substrate support found in encrusting forms or which are alternative solutions to the stress problems of an erect habit (Cheetham, 1971). It may be anticipated that other genera of such groups as the mamilloporids and conescharellinids (e.g., *Trochosodon, Kionidella*) and, in fact, most of the forms placed in the Suborder Hexapogona by Canu and Bassler (1927) will be found to have non-cuticulate walls and growth according to Borg's cyclostome model.

METHODS

Specimens were prepared and embedded according to techniques presented earlier (Sandberg, 1971). Embedded specimens were cut into 500- to 600-micron-thick slices on an annular blade, ultrathin kerf diamond saw. Slices were mounted on SEM specimen stubs with Lakeside 70 resin and polished on 15-micron (metal bond) and 6-micron (resin bond) diamond laps (water lubricated). Final polishing was on a felt lap with 0.05-micron alumina in water. Oil-based diamond pastes were avoided since they are difficult to wash off completely. Residual oils impair acceptance of a metal vapor coating, and charging defects may result in the SEM images.

All figures in this paper are scanning electron micrographs of polished, etched, Au–Pd–coated slices of embedded cheilostome bryozoans.

ACKNOWLEDGMENTS

This study was supported by a grant (GB 27346) from the National Science Foundation. Specimens from the Albatross and Caroline stations were made available for study by Alan Cheetham of the U.S. National Museum. I am indebted to Elly Brouwers, Susan Wunder, and Lisa Emery for assistance in specimen preparation and photographic processing.

REFERENCES

Banta, W. C. 1968. The body wall of cheilostome Bryozoa, I, The ectocyst of *Watersipora nigra* (Canu and Bassler). Jour. Morph., *125*(4): 497–508, 1 pl., 6 figs.

————. 1969. The body wall of cheilostome Bryozoa, II, Interzooidal communication organs. Jour. Morph., *129*(2): 149–169, 1 pl., 22 figs.

————. 1971. The body wall of cheilostome Bryozoa, IV, The frontal wall of *Schizoporella unicornis* (Johnston). Jour. Morph., *135*(2): 165–184, 6 pls.

Blake, D. B., and K. M. Towe. 1971. Acanthopore ultrastructure in the Paleozoic bryozoan *Idioclema insigne* Girty. Jour. Paleont., *45*(5): 913–917, 2 figs.

Boardman, R. S. 1971. Mode of growth and functional morphology of autozooids in some Recent and Paleozoic tubular Bryozoa. Smithsonian Contr. Paleobiol., (8): 1–51, 11 pls., 6 figs.

————, and A. H. Cheetham. 1969. Skeletal growth, intracolony variation, and evolution in Bryozoa: a review. Jour. Paleont., *43*(2): 205–233, 8 figs.

Borg, Folke. 1926. Studies on recent cyclostomatous Bryozoa. Zool. Bidrag, Uppsala, *10*: 1–507, 14 pls., 109 figs.

Canu, Ferdinand, and R. S. Bassler. 1927. Classification of the cheilostomatous Bryozoa. U.S. Nat. Mus., Proc., *69*(14): 1–42, pl. 1.

Cheetham, A. H. 1971. Functional morphology and biofacies distribution of cheilostome Bryozoa in the Danian Stage (Paleocene) of southern Scandinavia. Smithsonian Contr. Paleobiol., (6): 1–87, 17 pls., 29 figs.

————, J. B. Rucker, and R. E. Carver. 1969. Wall structure and mineralogy of the cheilostome bryozoan *Metrarabdotos*. Jour. Paleont., *43*(1): 129–135, pl. 26, 1 fig.

Friedman, G. M. 1959. Identification of carbonate minerals by staining methods. Jour. Sed. Petrol., *29*(1): 87–97, 2 figs.

Håkansson, E. Mode of growth of the Cupuladriidae (Bryozoa, Cheilostomata) *in* G. P. Larwood, ed., Proceedings of the Second International Conference on Bryozoa, I.B.A., Durham, England, 1971. In press.

Harmer, S. F. 1926. The Polyzoa of the Siboga Expedition, pt. II, Cheilostomata Anasca. Siboga-Expeditie, *28b*: 181–501, 34 pls., 23 figs.

————. 1934. The Polyzoa of the Siboga Expedition, pt. III, Cheilostomata Ascophora I. Family Reteporidae. Siboga-Expeditie, *28c*: 503–640, 7 pls. 25 figs.

Ryland, J. S. 1970. Bryozoans. Hutchinson Univ. Lib., London. 175 p., 21 figs.

Sandberg, P. A. 1971. Scanning electron microscopy of cheilostome bryozoan skeletons; techniques and preliminary observations. Micropaleontology, *17*(2): 129–151, 4 pls., 1 fig.

Schneidermann, Nahum, and P. A. Sandberg, 1971. Calcite-aragonite differentiation by selective staining and scanning electron microscopy. Gulf Coast Assoc. Geol. Soc., Trans., 21st Ann. Meet., p. 349–352, 6 figs.

Tavener-Smith, Ronald. 1969a. Wall structure and acanthopores in the bryozoan *Leioclema asperum*. Lethaia, *2*(2): 89–97, 7 figs.

————. 1969b. Skeletal structure and growth in the Fenestellidae (Bryozoa). Palaeontology, *12*(2): 281–309, pls. 52–56, 9 figs.

————, and Alwyn Williams. 1970. Structure of the compensation sac in two ascophoran bryozoans. Roy. Soc. London, Proc., ser. B., *175*: 235–254, 35 figs.

Mode of Colony Growth, Autozooids, and Polymorphism in the Bryozoan Order Cystoporata

John Utgaard   Southern Illinois University

ABSTRACT

The Cystoporata are a neglected, important group of fossil tubular Bryozoa. Many cystoporates have characteristics (lunaria, extensive extrazooidal vesicular skeletal tissue, communication pores, and budding of isolated autozooecia at the basal layer) that are rarely found in other tubular Bryozoa.

Cystoporates are compared to modern double-walled cyclostomes, probably their closest living analogues, in terms of skeletal microstructure, locus of zooecial budding, colony construction and formation, intrazooecial skeletal structures, extrazooidal skeletal structures, fossilized brown deposits, and cyclical growth features. These comparisons have yielded a double-walled working model for cystoporate colony growth and a model of autozooids with a membraneous sac. Interpretative conclusions on autozooidal functional morphology are presented. Intraautozooecial polymorphic zooids, gonozooids, exilazooids, large monticular zooids, and basal zooids are polymorphs known to date in the Cystoporata. A relatively high degree of polymorphism was reached by the Cystoporata as early as the Middle Ordovician.

Several types of colony-wide or semi-colony-wide features of growth and cycles are interpreted as being the results of (1) colony-wide control of deposition or colony-wide response to environmental cycles; (2) colony-wide control related to reproductive functions; (3) degeneration–regeneration cycles, possibly under semi-colony-wide control; (4) similarity of response of individual adjacent zooids to external environmental

Author's address: Department of Geology, Southern Illinois University, Carbondale, Illinois 62901.

changes; or (5) adjacent zooids attaining the same level of ontogenetic development at the same time.

INTRODUCTION

Cystoporates display several features that apparently are unique among bryozoans and other characters, widespread in the Order Cystoporata, are found in few other tubular bryozoans. In many fistuliporoids, new autozooecia are isolated or partly isolated at the basal or median layer by extrazooidal vesicular tissue. The existence of these new autozooids, isolated from their neighbors, is a feature that is rare in the Bryozoa, and suggests colony-wide control of budding rather than direct parent–daughter autozooecial origins. In many fistuliporoids new autozooecia were budded on top of extrazooidal vesicular tissue in the exozone.

In most double-walled tubular bryozoans the zooecia are completely composed of compound interior walls, secreted from the zooidal side by the zooidal epithelium and from the other side by zooidal epithelium of other zooids or by the hypostegal epithelium in extrazooidal areas of the colony. In many fistuliporoids, the lateral and distal sides of an autozooecium are composed of superimposed vertical parts of extrazooidal vesicular tissue, so that part of the autozooid is bounded by a simple interior wall. This is a rare, if not unique, feature in double-walled tubular bryozoans.

The Ceramoporidae (Ordovician to Lower Devonian) evidently had two means of interzooidal communication: (1) communication between zooids via coelomic fluid in the hypostegal coelom, as in other double-walled tubular bryozoans; and (2) communication between zooids via communication pores in the skeletal zooecial walls. The latter type of communication is rare in Paleozoic tubular bryozoans.

Most genera in the Cystoporata have a lunarium, which projects above the general zoarial surface and above the rim or peristome of the autozooecial orifice. The lunarium is located on the proximal side of each autozooecium and consists of a microstructurally distinct or thicker deposit that is developed through the exozonal extent of the autozooecium. Lunaria are known in some post-Paleozoic double-walled cyclostomes (some lichenoporids and hornerids), and in *Lichenopora* the membraneous sac occupies the proximal half of the living chamber, next to the lunarium.

The cystoporates in the Suborder Fistuliporoidea are characterized by having large amounts of extrazooidal skeletal material called vesicular tissue and stereom. The vesicular tissue is almost invariably composed of simple interior skeletal deposits secreted only from the upper or outer side. The available evidence suggests that the vesicular tissue housed no viable soft parts and served as a buttress between isolated or partly isolated zooecia. No other group of bryozoans has such extensively developed vesiculose extrazooidal structures.

A few cystoporates have intermonticular autozooecia with expanded subspherical oral ends. Their morphology and development in a colony suggests that they housed polymorphic zooids, possibly involved with the production of eggs and the brooding of embryos.

Polymorphic zooecia were common in most ceramoporoids and funnel cystiphragms and flask-shaped chambers in autozooecia in ceramoporoids and fistuliporoids suggest that intraautozooecial polymorphism was widespread in the Cystoporata.

The Cystoporata are the most neglected major group of Paleozoic Bryozoa. Knowledge of their morphology and paleobiology has lagged behind that of other groups of bryozoans, even though cystoporates have a widespread, abundant fossil record. The double-walled cyclostomes seem to be the closest living analogues of cystoporates. Cystoporates can be compared with Recent cyclostomes in terms of (1) studies of skeletal microstructure and ultrastructure, (2) locus of zooecial budding, (3) colony construction and formation, (4) intrazooecial skeletal structures, (5) extrazooecial skeletal structures, and (6) fossilized brown deposits, possibly representing membrane remnants and fossilized brown bodies. The results of these comparisons have yielded working models for cystoporate colony construction and growth, autozooecial functional morphology, polymorphism, and colony control of several morphological, and possibly functional, features and cycles.

Astrova established the Order Cystoporata in 1964 and included the Suborders Ceramoporoidea and Fistuliporoidea. In this paper the Cystoporata includes the Family Ceramoporidae in the Suborder Ceramoporoidea (informally called ceramoporoids). The Families Constellariidae, Anolotichiidae, Fistuliporidae, Hexagonellidae, and Goniocladiidae are members of the Fistuliporoidea. In addition, the small Families Botrylloporidae and Actinotrypidae and some of the genera in the Families Sulcoreteporidae and Rhinoporidae, previously included in the Cryptostomata (Bassler, 1953), are here included in the Fistuliporoidea. Members of the Suborder Fistuliporoidea are informally called fistuliporoids.

DOUBLE-WALLED GROWTH MODEL

Introduction

Borg (1926a) described the double-walled nature of the lichenoporid and hornerid cyclostomes and suggested (p. 596) that the trepostomes had the same kind of body wall but thought that it would be impossible to demonstrate this positively in the fossil forms. Again, Borg (1926b, p. 482) stated that it was evident that the Trepostomata are more closely related to the double-walled cyclostomes than to the single-walled cyclostomes. Elias and Condra (1957, p. 37–38) alluded to the "sclerenchyma" in fenestrate cryptostomes and in trepostomes as apparently being deposited in the same manner as in *Hornera* and related cyclostomes, that is, by an ectoderm that stretched externally over the whole zoariam. Thus, they surmised that fenestrate cryptostomes had a double wall and proposed the new Order Fenestrata, to be included with the Order Cyclostomata and the Order Trepostomata, in Borg's Class Stenolaemata. Borg, in a posthumous publication (1965, p. 3), stressed the relationship of the Fistuliporoidae to the Lichenoporidae and the Trepostomata to the Heteroporidae and stated that he had succeeded in showing that they had a covering of soft tissue over the entire colony surface. Tavener-Smith (1968, p. 86, 88–89; 1969a, p. 291) used the double-walled concept described by Borg as the basis for the construction of a double-walled model for fenestellid growth. Boardman (Boardman and Cheetham, 1969, p. 209, 213) suggested

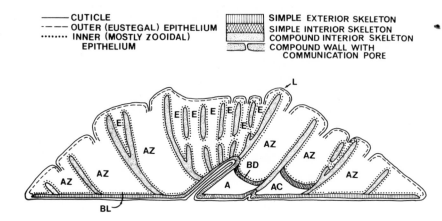

——— CUTICLE
- - - - OUTER (EUSTEGAL) EPITHELIUM
········· INNER (MOSTLY ZOOIDAL)
          EPITHELIUM

SIMPLE EXTERIOR SKELETON
SIMPLE INTERIOR SKELETON
COMPOUND INTERIOR SKELETON
COMPOUND WALL WITH
    COMMUNICATION PORE

Figure 1

Longitudinal section through hypothetical double-walled model for the Ceramoporoidea. The simple skeletal walls include the basal layer (BL) and basal diaphragms (BD) in autozooecia (AZ) and exilazooecia (E) (basal diaphragms in exilazooecia not shown). The basal layer folds back upon itself on the left side of the ancestrula (A). Exilazooecia on the right side of the illustration are shown with a terminal-vestibular membrane, as if they had an extrusable polypide. Exilazooecia on the left side of the illustration are shown with an imperforate terminal membrane. Communication pores are shown in exilazooecial walls. Autozooecial walls do have communication pores, but they are not shown in this illustration because the wall on the proximal side of autozooecia bears the imperforate lunarial deposit. Autozooecia in this hypothetical section are all cut through the lunarial deposit. The position of the lunarium at the surface is indicated by the letter L. An abandoned chamber (AC) is shown below a basal diaphragm (BD). (Modified from Borg, 1926b, p. 308, text fig. 55; Boardman, 1971, p. 7, text fig. 2.)

that the double-walled concept of Borg could be extended to most fossil tubular bryozoans (notably the Trepostomata, Cryptostomata, and Cystoporata) and later presented (Boardman, 1971, p. 6–7) a more detailed account of the double-walled concept as applied to trepostomes. In addition, Boardman (in press) discovered the fossilized remnant of the outer cuticle on the upper surface of a colony of a trepostome from the Ordovician: the most direct proof yet of the double-walled nature of a Paleozoic bryozoan.

The Double-Walled Concept as a Model for Cystoporate Growth

The evidence seems very strong that cystoporates were double-walled bryozoans with an outer cuticle that surrounded the entire colony. On all but the basal incrusting surface, the cuticle probably was underlain by (1) an epithelium (the eustegal epithelium) that secreted the cuticle, (2) a layer of peritoneum, (3) a thin hypostegal coelom between zooids, (4) an inner layer of peritoneum, and (5) an inner epithelium (the hypostegal epithelium) that was responsible for secreting and was adjacent to the calcareous skeleton between zooids. The inner epithelium continued into the zooidal epithelia that lined and secreted the zooecia (Fig. 1–3).

The locus of budding of new autozooecia strongly suggests a double-walled construction

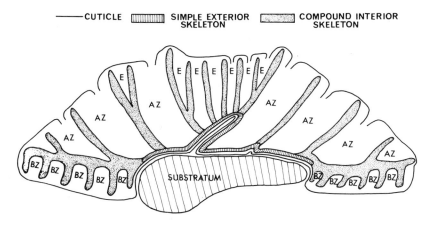

Figure 2

Longitudinal section through hypothetical double-walled *Ceramopora* with a celluliferous base. Terminal pores are as shown in Fig. 1. Epithelia are omitted. The celluliferous base contains relatively short, narrow, diaphragm-less polymorphs (BZ) that probably were some sort of zooid, without a polypide, that opened into the hypostegal coelom on the free basal margins of the colony beyond the substrate. The walls of the basal polymorphs are compound. Only the basal layer in the ancestrula and adjacent to the incrusted substratum is a simple wall.

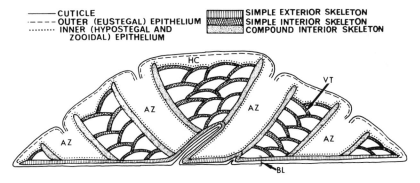

Figure 3

Longitudinal section through hypothetical double-walled fistuliporoid. The basal layer (BL) is simple exterior wall: all other walls are interior, that is, without an outer bounding membrane. The autozooecia (AZ) have compound interior walls and the vesicular tissue (VT) is extrazooidal tissue with simple interior walls secreted by the inner (hypostegal) epithelium immediately below the hypostegal coelom (HC).

in the Cystoporata. In single-walled cyclostomes new autozooecia originate in the common bud, which is the colony-wide growing margin or the growing tips of a colony (Borg, 1926b). New septa or compound walls originate here but grow to reach the terminal membrane (outer membrane) of the common bud. The terminal membrane of part of the common bud becomes the terminal membrane of the zooid. Outward growth from that point on involves isolated zooids, each with its own terminal membrane, and zooids are single-walled with zooecia simple-walled distally (deposited from only one side). Once an area in the colony ceases to be part of the common bud, no new zooids form there. In double-walled cyclostomes septa or compound walls do not reach the terminal membrane of the common bud and, according to Borg (1933, p. 381), ". . . the whole frontal surface of the zoarium must be considered as representing a much widened and highly complicated common bud. Accordingly, new zoids can be formed not only at the budding edge or edges, as in other stenolaemates, but on the whole free surface of the zoarium." The existence of new autozooecia budded in the exozone in virtually any part of the colony in most cystoporates, on old autozooecial walls or on extrazooidal vesicular tissue (Fig. 8), strongly suggests that these bryozoans were double-walled forms. In many fistuliporoids (Fig. 9), autozooecia are isolated or partially isolated at the basal layer by intervening extrazooidal vesicular tissue. The existence of these new autozooids, isolated from their neighbors, suggests colony-wide budding control by an outer membrane, rather than direct parent–daughter autozooecial origins. In addition, autozooecial walls in most cystoporates are compound walls (secreted from both sides), as are interior walls in modern cyclostomes, secreted under an infolding of an inner epithelium into a hypostegal coelomic cavity. Some fistuliporoids can have the lateral and distal sides of an autozooecium composed of superimposed vertical parts of vesicle walls so that part of the autozooecium is simple-walled (Figs. 4 and 10), a rare if not unique relationship in the stenolaemates.

The relatively uniform level of the outer surface (exclusive of the basal layer) of cystoporate colonies suggests a colony-wide epithelium and colony-wide control of growth. The only projections are the relatively short calcite rods (acanthopores) and mural tubulae, some vesicle walls, autozooecial peristomes, and lunaria (Fig. 11). Projections of similar magnitude are known in modern double-walled lichenoporids and hornerids. Individual, isolated autozooecia do not project significant distances above the general surface of the colony in cystoporates as they do in some single-walled cyclostomes.

The nature of the extrazooidal vesicular tissue and stereom in fistuliporoid cystoporates strongly suggests that they were double-walled forms. First, there is no space of relatively constant size and shape and commonly no space for zooids in the vesicular tissue. In only a few forms are vesicle walls formed as compound walls before vesicle roofs are deposited. In most forms, vesicle walls and roofs are essentially one curved structural unit (Fig. 12) which were deposited as one curved plate. Second, thick vesicle roofs or stereom at the zoarial surface or in abandoned zones in the exozone (Fig. 13) indicate colony-wide control of deposition: from one epithelium on the outside of the vesicles. Third, vesicle walls (with few exceptions) and all vesicle roofs are simple walls. They are deposited from only one side, the outer side, at the zoarial surface (Fig. 4). Although some workers have reported the existence of pores in vesicle roofs and walls, undoubted pores have not been seen by the author. The simple-walled nature of vesicles suggests that they contained no living tissue, at least no secretory epithelium, and were not zooids. They are not, as Borg (1965) suggests, structures similar to alveoli in lichenoporid cyclostomes. Alveoli have compound walls with communication pores. The development of autozooecial walls on the outer side of vesicle roofs in the exozone demands an

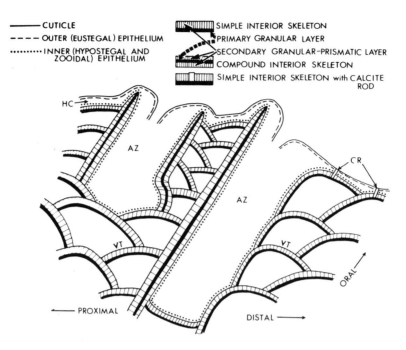

CUTICLE

OUTER (EUSTEGAL) EPITHELIUM

INNER (HYPOSTEGAL AND ZOOIDAL) EPITHELIUM

SIMPLE INTERIOR SKELETON

PRIMARY GRANULAR LAYER

SECONDARY GRANULAR-PRISMATIC LAYER

COMPOUND INTERIOR SKELETON

SIMPLE INTERIOR SKELETON with CALCITE ROD

Figure 4

Longitudinal section through hypothetical fistuliporoid based on observed sections but showing double-walled construction. The autozooecium (AZ) on the left was formed by septa or compound walls forming on old vesicular tissue by folding of inner (hypostegal) epithelium. The autozooid on the right has a compound wall on the lunarial (proximal) side, but the skeleton on the distal side of the autozooid is composed of superimposed vertical simple walls of the extrazooidal vesicular tissue. The autozooid on the left has a peristome and compound walls. The autozooid on the right has no peristome and the inner zooidal epithelium curves up and out to continue as the inner (hypostegal) epithelium below the hypostegal coelom (HC). The walls and roofs of the vesicles can be composed of the inner granular primary layer alone or the primary layer and a secondary granular-prismatic layer. The outer cuticle is shown as being attached to the calcite rods (CR) or "mural tubulae" in the vesicle roofs, as suggested by a similar appearance in a modern *Hornera* (Fig. 15).

epithelium on the outer side. This epithelium would almost have to be nourished by a coelom and that, in turn, protected from the environment by an outer membrane.

What is possibly a partially preserved outer membrane remnant over the zooecial orifice and extending over the vesicular tissue has been observed in a fragment of *Cheilotrypa hispida* (Fig. 14). This brown deposit, preserved beneath an overgrowth, suggests the configuration of a terminal-vestibular membrane which extended over an interzooidal hypostegal coelom. Boardman (in press) discovered a much more extensive membrane in an Ordovician trepostome that is more convincing as a possible outer membrane.

In a recent species of *Hornera* (Fig. 15), sectioned with both the skeleton and soft parts, the outer membrane, beyond the narrow hypostegal coelom, appears to touch the inner (hypostegal) epithelium and skeleton where minute spines project slightly above the general surface of the zoarium. These spines are similar to calcite rods (acanthopores) and mural tubulae (minutopores) in the Cystoporata. This supports the hypothesis of Tavener-Smith (1969b, p. 97) that calcite rods may have functioned in stabilizing or anchoring the outer membrane to the skeleton but perhaps not as a base of attachment of muscles that extended through the hypostegal coelom to the outer epithelium.

Utilizing the double-walled model of growth for the Cystoporata (Figs. 1–4), it is probable that the inner (hypostegal and zooidal) epithelium secreted *all* of the calcareous skeleton and that the outer (eustegal) epithelium secreted only the cuticular cover on the upper surface of the colony (the surface excluding the basal layer), including the terminal-vestibular membranes of the zooids.

AUTOZOOIDS

Recognition of Autozooids in the Cystoporata

Autozooids are the normal individuals in a colony (Borg, 1926b, p. 118) which perform all the usual body functions (Ryland, 1970, p. 29). Autozooids have, at some stage or stages in their ontogeny, a protrusible lophophore (Boardman, 1971, p. 2). If one accepts the evidence that the Cystoporata were bryozoans, they had "normal" individuals in the colony with protrusable lophophores.

Using the definition of *zooid* as a member of the colony, consisting minimally of body wall enclosing a coelomic space (R. S. Boardman, A. H. Cheetham, and P. L. Cook, written communication), and evidence from microstructure, budding, and colony construction, there are several kinds of zooecia in the Cystoporata which could possibly have been autozooecia. In the Fistuliporoidea, the choice is narrowed to one or, usually, two (rarely more) kinds of zooecia. Some fistuliporoids are essentially monomorphic in intermonticular and monticular areas, and these zooecia can be considered to be autozooecia which housed autozooids. Most fistuliporoids have intermonticular autozooecia and slightly larger but essentially similar monticular zooecia. Both types probably housed autozooids, but in this paper the former are considered to be the common kind of autozooecia and the latter are considered to be polymorphs, are termed *large monticular zooids,* and are discussed below. A few fistuliporoids have zooecia with expanded oral ends on the colony surface and probably are autozooecia modified to provide a brooding function and are inferred to be *gonozooids* (see below). In the Ceramoporoidea there are as many as four different kinds of zooecia that are candidates for being called autozooecia. Only two of these types are common to all ceramoporoids: (1) the large intermonticular zooecia and, (2) slightly larger monticular zooecia. The former are here considered to be zooecia of autozooids because they compare in size, number, and position with autozooecia of monomorphic fistuliporoids. As with the fistuliporoids, the larger monticular zooecia probably housed autozooids, but they are considered to be polymorphs and are discussed separately (see below) as *large monticular zooids.* The smaller zooids on the frontal or nonbasal surface of some ceramoporoid colonies could possibly have been another kind of autozooid. They are another kind of polymorph and are here called *exilazooids* (see below). The rare, basal polymorphic zooecia found so far only in some colonies of *Ceramopora* were probably not autozooecia and are termed *basal zooids* (see below).

The zooecia in the Cystoporata that are here considered to be autozooecia compare favorably in terms of relative size, shape, position of origin, extent of living chambers, intrazooecial structures, distribution and abundance in the colony, and similarity to what is considered to be the ancestrula, to zooecia in the Trepostomata (Boardman, 1971) and Cyclostomata (Borg, 1926b, 1933) that are considered to be autozooecia.

Lunaria

Most genera in the Order Cystoporata have structures called *lunaria*. A lunarium is located on the proximal side of each autozooecium and each large monticular zooecium. One zooecium in an Ordovician ceramoporoid has two lunaria (Utgaard, 1968b, p. 1445): in all other observed cystoporates with lunaria there is one lunarium per autozooecium or large monticular zooecium. Lunaria are not present in exilazooecia or in extrazooidal vesicular tissue (cystopores). The lunarium projects above the general zoarial surface (Fig. 11) and above the rim or peristome of the autozooecial orifice. In most genera, lunaria are radially arranged around the monticular centers with the lunarium on the side of the zooecia nearest the monticular center (Utgaard, 1968a, p. 1033).

The lunarium, when viewed in tangential thin sections or acetate peels, generally has a shorter radius of curvature than the remainder of the zooecial orifice (Fig. 16). In some genera the ends of the lunarium project into the zooecial cavity (Fig. 17) and greatly modify the shape of the skeletal living chamber (see below). In other cystoporates the radius of curvature of the lunarium is approximately the same as that of the distal end of the zooecium if the orifice is elongate in the proximal-distal direction (Fig. 18) or is approximately the same as that of the rest of the zooecium if the orifice is circular.

The lunarium is developed early in the ontogenetic development of a zooecium and is a distinct skeletal structure that can be seen in longitudinal and transverse sections as well as in tangential sections. The lunarial deposit in many genera can be seen as a continuous deposit, extending from the outer endozone to the zoarial surface (Fig. 19). In other genera, it is a continuous deposit extending from the inner exozone to the zoarial surface (Utgaard, 1968a, p. 1033). An *Anolotichia,* the structure of the lunarial deposit changes markedly near the zoarial surface (Utgaard, 1968a, p. 1037). In other cystoporates, the structure of the lunarial deposit is relatively uniform throughout the length of its development. It commonly increases in size slightly in the exozone in an oral direction (Utgaard, 1968b, p. 1453). Thus in cystoporates with a lunarium, the lunarium is developed and visible in well-preserved specimens, at the surface of the zoarium, in all zooecia that have reached the stage of development where their oral ends are in the exozone.

The microstructure of the lunarial deposit is not uniform throughout the Order Cystoporata or even within some families. In the Ceramoporidea, the lunarial deposit is distinct from the well-laminated autozooecial and exilazooecial walls. Under the light microscope, the lunarial deposit appears to be dense, light-colored hyaline calcite (Figs. 16 and 19). Indistinct, distantly spaced laminations and cores, or rod-shaped parts of the lunarial deposit, are visible in some forms (Utgaard, 1968b, p. 1445). Preliminary studies of the ultrastructure using the scanning electron microscope reveal that the lunarial deposit is more granular than the well-laminated walls (Fig. 25) and locally shows indistinct signs of lamination. In ceramoporoids with communication pores, the pores have not been observed to pierce the lunarial deposits.

In one constellariid and some fistuliporids that have poorly laminated autozooecial walls, the lunarial deposit appears to be dense, light-colored hyaline calcite, grossly similar to that in the ceramoporoids (Fig. 20). In some fistuliporids that have granular and granular-prismatic wall microstructure, the lunarial deposit is light-colored dense calcite (Fig. 39) and grossly resembles the lunarial deposits in the ceramoporoids, fistuliporids, and constellariids mentioned above.

In most fistuliporoids with granular and granular-prismatic wall microstructure, the microstructure of the lunarial deposit is grossly similar to that of the remainder of the skeleton (Fig. 17). In some genera, the dark, central, primary granular part of the compound autozooecial wall until the time line reaches the zooecial boundary is difficult in trepo-colored granular-prismatic layer commonly is much thicker in the lunarial deposit than in the remainder of the autozooecial wall (Utgaard, 1968a, pl. 131, figs. 1a, 1b, 2b, 3a). In other genera (Fig. 17) the dark-colored, primary granular layer in the lunarial deposit is not continuous into the central, primary, dark granular layer in the remainder of the autozooecial wall. The proximal light-colored granular-prismatic layer is generally the thickest layer in the lunarial deposit. Studies of the ultrastructure of the above forms is being undertaken by the author using the scanning electron microscope. The question as to whether the granular and granular-prismatic microstructure represents a primary microstructure or is due to recrystallization has not been adequately answered.

Borg (1965) and Utgaard (1968a) reported lunaria in post-Paleozoic hornerids and lichenoporids. Boardman (1971) described the position of the membraneous sac in a lichenoporid with respect to the proximal (lunarial) side. The possible functional significance of the lunarium is discussed below in terms of modification of the living chamber and position of the polypide and in terms of its possible effect on the configuration of the terminal-vestibular membrane and the flow of water from the tentacular bell.

Autozooecial Living Chambers

Introduction. In modern cyclostomes the *living chamber* is that part of a zooid that is lined by zooidal epithelium and houses the functional organs of the zooid, if any were present. Skeletal remains of living chambers in the Cystoporata, as in the Trepostomata (Boardman, 1971, p. 5), can be studied in unworn specimens or beneath protective overgrowths. Estimating the minimum extent of a living chamber by trying to trace time lines of skeletal deposition from a basal diaphragm into and along the autozooecial wall until the time line reaches the zooecial boundary is difficult in trepo-stomes (Boardman, 1971, p. 5) and is virtually impossible in the Cystoporata. In the Fistuliporoidea the wall laminae are commonly obscure or walls have a granular or granular-prismatic microstructure. In the Ceramoporoidea, this method could be successful but wall laminae were apparently deposited in bundles in the form of partial cylinders and not as complete cylinders lining the zooecium, as was usually the situation in the trepostomes.

A minimum estimate of the length of the living chamber can be gained from studying relatively unabraded specimens, but the best estimate comes from studying living chambers preserved beneath overgrowths (Boardman, 1971, p. 5).

Autozooecial Living Chambers in the Cystoporata. Nearly all living chambers in cystoporates are (1) modified cone-cylinder shapes with the smaller, modified cone-shaped aboral end on the basal lamina or mesotheca and the larger, cylindrical portion in the exozone, or (2) cylindrical with a nearly flat basal diaphragm. The length of living chambers in the Cystoporata ranges from approximately two autozooecial diameters to about seven autozooecial diameters. One form, *Botryllopora,* with very narrow living chambers for a cystoporate, has living chambers up to nine times as long as wide. Most living chambers in the cystoporates are approximately three to five times as long as wide.

The cross-sectional shape of living chambers of the modified cone-cylinder type can be hemispherical, mushroom, or subcircular in the aboral modified cone end. The cross-sectional shape in the cylindrical end or in cylindrical living chambers can be circular to elliptical and these basic shapes can be modified depending upon the radius of curvature of the lunarium and whether or not the ends of the lunarium project into the autozooecial cavity. In many cystoporates the lunarium encloses a partial circular to elliptical cylinder shape which is on the proximal side of a larger circular to elliptical cylinder shape enclosed by the remainder of the autozooecial walls (Figs. 17 and 66).

In a very few forms of the Constellariidae the basal structures of the living chambers can be curved or cystoidal diaphragms or a combination of cystoidal diaphragms and flat diaphragms. In these forms the living chamber generally has a bisected funnel shape and a smaller cross-sectional area in its lower or aboral portion and is cylindrical in its longer, aboral portion. The deepest part of the living chamber is on the proximal side of the autozooecium, next to the lunarium, if one is present (Fig. 20).

In many of the bifoliate and a few incrusting species of fistuliporoids, a superior (proximal) hemiseptum partially divides the living chamber into a modified cone-shaped aboral portion and a cylindrical outer portion (Fig. 21). Species of *Strotopora, Cliotrypa,* and *Fistuliphragma* have alternating hemiphragms, which are triangular and platelike or laterally curved spines, which protrude into the living chamber (Figs. 22, 24, and 30).

Basal Diaphragms and Abandoned Chambers. As an autozooecium grows and the living chamber reaches a certain length (which is not constant in a colony), a new basal diaphragm is formed. The lengthening of the oral end of the autozooecia and, especially, the formation of a new basal structure are probably related to a degeneration–regeneration cycle in an autozooid (see below.) As in the trepostomes (Boardman, 1971, p. 18), the spacing of basal diaphragms in the cystoporates is such that *abandoned chambers,* between successive basal diaphragms, are usually much shorter than the living chamber in the same autozooecium. Abandoned chamber length in the cystoporates generally ranges from less than one to slightly more than three autozooecial diameters and generally is less than two autozooecial diameters.

Formation of a new basal diaphragm probably involved proliferation of a new epithelium and peritoneum from the lateral walls of the zooid across the zooecium at a more oral level than the preceding basal diaphragm. Brown bodies, some other cellular material, and coelomic fluid probably were left behind in the abandoned chamber. If the basal (aboral) zooidal epithelium were drawn orally to a new position, intact, brown bodies or fossilized brown deposits would not be found in abandoned chambers, but they are. In the cystoporates, basal structures are simple-walled in construction, being deposited by an epithelium on only the oral side. Except for some of the ceramoporoids, where an abandoned chamber might be connected to the living chamber of an adjacent autozooecium or exilazooecium through communication pores in autozooecial walls, living tissue would probably not have been present in abandoned chambers. Brown deposits encapsuled with membrane, diffuse brown deposits and rare membraneous linings found in abandoned chambers are probably the remnants of brown bodies, cellular material, coelomic fluid, and membrane left behind in the abandoned chamber when a new basal zooidal epithelium was proliferated and a new basal skeletal diaphragm was formed.

In the cystoporates, as in the trepostomes (Boardman, 1971, p. 5), the new living chamber consisted of the old living chamber minus the abandoned segment and new space where new zooecial walls were secreted orally while the polypide was degenerated.

Terminal and Subterminal Diaphragms. The existence of subterminal diaphragms, with reverse curvature and which were deposited from the inner (aboral) side, was reported (Utgaard, 1968b, pp. 1445, 1446) in some ceramoporoids. They are common in autozooecia and some exilazooecia, especially in species of *Ceramoporella* (Fig. 23). The geometry of the junction of these diaphragms with zooecial walls indicates that they were deposited, at least partly, from the aboral side. It is possible that some of them are compound, being deposited by the zooidal epithelium on the inside and by the inner (hypostegal) epithelium on the outside (see Fig. 5 for further explanation), but most appear to have been deposited from the inside only. Borg (1933) reported terminal and subterminal diaphragms in autozooecia and kenozooecia in heteroporoid cyclostomes. Considerable variation exists, within and among genera of heteroporoids, in the abundance of terminal and subterminal diaphragms. In the heteroporoids, these diaphragms are compound and can bear pseudopores. Nye (1968, p. 112) reported pore-bearing terminal and subterminal diaphragms and imperforate intermediate diaphragms in some post-Paleozoic cyclostomes. Both the terminal and intermediate diaphragms have aborally flexed laminae as the diaphragm joins the zooecial walls, indicating at least partial deposition from the aboral side. Pores have not been observed in the terminal and subterminal diaphragms in the ceramoporoids.

Terminal and subterminal diaphragms have been observed, to date, in 11 genera of the Fistuliporoidea. The terminal and subterminal diaphragms in the fistuliporoids are of simple construction (Fig. 26), but, unlike those in the ceramoporoids, they were deposited by an epithelium on the oral (outer) side of the diaphragm (the inner or hypostegal epithelium below the hypostegal coelom in Fig. 5). Pores have not been observed in these diaphragms in fistuliporoids.

Ulrich (1883, p. 85) reported a "pellicle" in a species of *Dekayia* (Order Trepostomata) which he thought ". . . was developed at the close of the existence of the zooids of each layer of cells, so as to form the floor of the succeeding layers, and ultimately the diaphragms which cross the tubes." Further, Ulrich (1890, p. 314–316)) reported a variety of "opercular structures" in Paleozoic fenestrate cryptostomes, trepostomes,

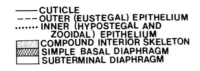

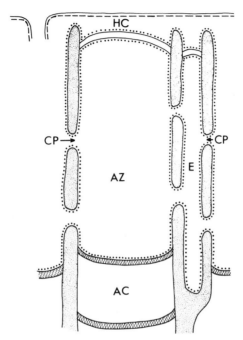

Figure 5

Idealized diagram of living chambers in an autozooecium (AZ) and an exilazooecium (E) in the ceramoporoid *Ceramoporella,* which are closed by a subterminal diaphragm. The subterminal diaphragms could be simple or compound but clearly are (at least mostly) deposited from their aboral side. The inner (hypostegal) epithelium below the hypostegal coelom (HC) could have had a part in secreting the subterminal diaphragms from the oral side. The abandoned chamber (AC), between successive basal diaphragms, contained no viable tissue while the chambers below the subterminal diaphragms contained at least a living epithelium, peritoneum, and coelomic fluid by virtue of their communication with adjacent zooids via the communication pores (CP). Some abandoned chambers could have had a similar epithelium, peritoneum, and coelomic fluid if they communicated through communication pores with an adjacent living chamber before its aboral end became an abandoned chamber.

and cystoporates. Most of the "opercular structures" Ulrich reported as being perforate, but some, such as those in *Fistulipora foordi,* were solid. He again (1890, p. 315) suggested that the opercula in trepostomes ". . . at one time served the purpose of a cover to the zooecium, next formed the floor of the succeeding cell; in other words, became a 'diaphragm'." He suggested (p. 316) that the perforation in the operculum became closed when it became a basal diaphragm. Borg (1933, p. 288, 316, 317) found extensive development of terminal diaphragms in two species of *Heteropora* and reported

(p. 317) that Busk had called these areas a calcareous pellicle or epitheca. The similarity in position suggests that the structures Ulrich (1883, p. 85) referred to as a pellicle and at least some of the opercular structures (Ulrich, 1890, p. 314–316) are terminal or subterminal diaphragms. More detailed studies are needed to determine if any of these diaphragms are ever perforated.

Borg (1933, p. 320) thought that the formation of terminal diaphragms was a result of the degeneration–regeneration cycle. It seems likely that the polypide would degenerate or decompose in a ceramoporoid autozooid when the terminal or subterminal diaphragm was formed, especially if the subterminal diaphragm was formed in the region of the vestibule. It is possible that the aboral part of a polypide could lie dormant, nourished through communication pores. The entire zooid would probably be abandoned to decompose in a fistuliporoid below a terminal or subterminal diaphragm, if the diaphragm lacked a pore. The existence of possible polypide remnants (see below) in large "abandoned" chambers (capped living chambers) below terminal diaphragms in fistuliporoids supports the latter inference: that they underwent decomposition without complete degeneration to form a brown body.

Living chambers capped with terminal or subterminal diaphragms are much longer than normal abandoned chambers in the same colony and, even the same zooecium, indicating that terminal or subterminal diaphragms do not, as a matter of course, become the next basal diaphragm when a polypide is regenerated (as Ulrich thought). It is possible that terminal diaphragms were produced during degeneration and resorbed during regeneration in the degeneration–regeneration cycle. Borg (1933, p. 298, 302, 303) reported resorption of the subterminal diaphragms in parts of colonies of the cyclostome *Heteropora*. However, the existence of some long-abandoned chambers, including those containing flask-shaped chambers, funnel cystiphragms, and partial funnel cystiphragms (see below), suggest that some of these longer-than-normal abandoned chambers may possibly be capped by terminal diaphragms and not by the usual basal diaphragms. The relative rarity of obvious terminal diaphragms and longer-than-normal abandoned chambers suggests that they were not consistently part of a normal degeneration–regeneration cycle. In addition, membraneous remnants possibly representing membraneous sacs of undegenerated polypides have been found in several living chambers capped by subterminal diaphragms.

Interpretive Functional Morphology of Cystoporate Autozooids

Recent Cyclostomes as a Model for Functional Morphology. Similarities of mode of growth led Borg (1926a, p. 596; 1965, p. 3) and Boardman (Boardman and Cheetham, 1969, p. 209, 213; Boardman, 1971, p. 6–7) to look for a growth model for Paleozoic tubular bryozoans in the post-Paleozoic double-walled cyclostomes. The evidence for a double-walled colony construction for tubular bryozoans and Borg's Order Stenolaemata, which included the cyclostomes, trepostomes, and cystoporates, led Boardman (1971, p. 6) to recent cyclostomes as a logical first approximation for a model for zooid form and function in tubular bryozoans. A generalized autozooid, based on that of a lichenoporoid cyclostome (Fig. 6), is used here as a model for autozooids in the Cystoporata.

Evidence from Paleozoic Cystoporates. In addition to the evidence from budding locations and growth of the skeleton, the existence of HCl-resistant preserved organic matter in living chambers below overgrowths or terminal diaphragms gives some indication of the nature and extent of zooidal tissues. In addition, the nature of preserved organic material in abandoned chambers and flask-shaped chambers suggests certain characteristics of cystoporate autozooids that are similar to features of autozooids of trepostomes and cyclostomes.

The Terminal-Vestibular Membrane. A few cystoporates have remnants (Fig. 27) that may have been organic membranes which are in a position similar to that of the terminal-vestibular membrane. Although these specimens are not well enough preserved to be convincing evidence, the remnants suggest the shape of and are in the position of a terminal-vestibular membrane in a *Lichenopora* (Boardman, 1971, pl. 1, fig. 1). The funnel shape of the terminal-vestibular membrane (Fig. 28) could have been produced by postmortem contraction of the muscles that widen the vestibule. It is possible that with the tentacles retracted and the muscles that extend the vestibule in a relaxed condition,

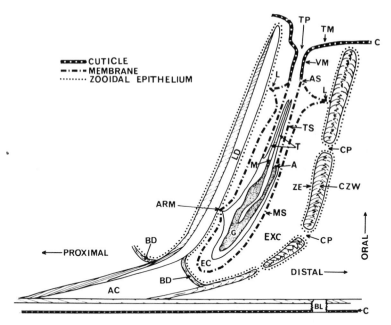

Figure 6

Longitudinal section through hypothetical model of ceramoporoid autozooid based on soft tissue of a recent lichenoporoid cyclostome. A, anus; AC, abandoned chamber; ARM, attachment of retractor muscles; AS, atrial sphincter; BD, basal diaphragm; BL, basal layer; C, cuticle; CP, communication pore; CZW, compound zooecial wall; EC, endosaccal coelom; EXC, exosaccal coelom; G, gut; L, ligament (or attachment organ); LD, lunarial deposit; M, mouth; MS, membraneous sac; T, tentacles; TM, terminal membrane; TP, terminal pore; TS, tentacle sheath; VM, vestibular membrane; ZE, zooidal epithelium. The eustegal epithelium, peritoneum, and muscles that widen the vestibule are omitted from the drawing.

the terminal-vestibular membrane did not assume a funnel shape or such a broad funnel shape.

Borg (1926b, p. 244) reported that when the tentacles are out, the level of the mouth is at or below the zooecial orifice in *Lichenopora*. In *Lichenopora,* as in most cystoporates, the lunarium forms a proximal projection beyond the orifice (Fig. 11). During extrusion of the tentacles, the muscles would tend to pull the terminal-vestibular membrane down and toward the zooecial walls, probably producing a funnel-shaped depression. The mouth at that low level (Borg, 1926b, p. 244) suggests that the lunarium would divert the excurrent water moving through the tentacular bell on the proximal side to distolateral directions.

Possible Polypide Remnants in Cystoporates. Of the several thousand slides of cystoporates examined, 16 have autozooecia containing long, tubular fossilized brown deposits that range from interrupted patches of brown granular material (Figs. 30 and 31) to fairly complete tubular membranes (Figs. 32 and 37). The brown deposits and membrane remnants are found in living chambers below overgrowths or terminal or subterminal diaphragms. It is probably that they represent polypide remnants rather than brown bodies from a degenerated zooid (compare Figs. 32 and 39). Compact fossilized brown deposits (Figs. 38 and 39), more likely remnants of brown bodies, are found in abandoned chambers in several of the colonies containing long, tubular membraneous brown remnants. The latter deposits (Fig. 32) suggest that at least those cystoporates had autozooids with a membraneous sac. The size, shape, and position of the long tubular membraneous remnants in the fossil cystoporates compare favorably with those of the remnant of the membraneous sac in a recent specimen of *Lichenopora* (Fig. 33) and to the organic structure that McKinney (1969) believed to be the remnant of a membraneous sac in the trepostome genus *Tabulipora*. A few of the tubular brown deposits and membraneous remnants occupy most of the width of the living chamber. Most of the tubular brown deposits occupy only a part, some less than one half, of the width of the living chamber (Fig. 42), a situation which is commonly seen in recent cyclostomes. Some of these organic remnants and brown deposits, as seen in longitudinal sections, are closer to the proximal or lunarial side (Fig. 37), at least toward their oral end, a situation found in recent *Lichenopora* (Boardman, 1971, pl. 2, figs. 1, 2; Fig. 33, this paper.)

Several possible polypide remnants have been found in a cystoporate with hemiphragms: *Fistuliphragma spinulifera.* The diffuse brown remnants zigzag around the triangular plate-like to curved spine-like hemiphragms, suggesting that the membraneous sac snaked through the living chamber around the ends of the curved spines (Figs. 30 and 31). R. S. Boardman and J. G. Harmelin are currently working on recent species of the cyclostome *Tubulipora* and have discovered that this form has hemiphragm-like structures (curved spine-like plates) and that the membraneous sac and its contained organs bend around the spines (Fig. 29).

A membraneous remnant in one of the cystoporates (Fig. 34) displays a slight constriction just aboral to a level where the remnant may have been attached to the autozooecial wall. Just oral to this possible attachment level and aboral to the subterminal diaphragm is a major constriction in the membraneous remnant. The most likely comparable shape in zooids of recent cyclostomes is the level of ligament attachment of the membraneous sac (oral to the slight constriction) and the atrial sphincter (oral to the possible attachment level). Another ceramoporoid with ill-defined, large fossilized brown deposits in the living chambers (Fig. 40) displays an apparent attachment to the autozooecial wall just aboral to a rather large constriction, again suggestive of the ligament level and atrial sphincter level, respectively, in recent cyclostomes. A third ceramoporoid (Fig. 42) has a tubular membraneous remnant with a funnel shape just below the subterminal diaphragm. The level of the funnel and subterminal diaphragm in the living chamber is grossly comparable in position and the funnel is grossly comparable in shape to a membraneous sac in a recent *Lichenopora* (Fig. 33), suggesting that the funnel shape is possibly just below the ligament level. The terminal-vestibular membrane in recent *Lichenopora* is more delicate than the membraneous sac and is rarely seen. The funnel shape in the recent *Lichenopora* (Fig. 33) almost certainly represents the membraneous sac just below ligament level. These two subterminal diaphragms (Figs. 34 and 42), which were at least partly secreted by an epithelium on their aboral side, probably did not form close enough to the oral end of the living chamber to preserve a funnel shape related to a terminal-vestibular membrane, but that is a possibility. It is also possible that the funnel shape relates to a level where a vestibular membrane was pulled down toward the autozooecial walls by the muscles that widen the vestibule. It seems likely that these two subterminal diaphragms were formed at least as far orally as the ligament level and possibly at a level somewhere in the vestibule below the terminal membrane. It seems less likely that they formed orally to an abandoned terminal-vestibular membrane.

Some of the more compact fossilized brown deposits in abandoned chambers and in the proximal ends of living chambers in some cystoporates (Utgaard, 1968b) (Figs. 35, 36, 38, and 39) bear a strong resemblance to brown bodies in recent cyclostomes (Fig. 41). These fossilized brown deposits are encapsuled in a membrane and contain masses of dark brown granules. Borg (1933, p. 366) described brown bodies of similar construction, in recent cyclostomes, and presumed that the membrane enclosing each brown body is the remnant of the membraneous sac. However, other bryozoans without membraneous sacs evidently can have membrane-encapsuled brown bodies.

Other Membraneous Remnants. Incomplete remnants to complete organic linings have been observed in abandoned chambers in autozooecia in a few cystoporates. The complete linings line the basal diaphragms and autozooecial walls and some are present on the aboral side of the diaphragm so that they completely line the abandoned chamber. Some have been observed on the basal diaphragm and the aboral, preserved end of incomplete living chambers and a few completely line living chambers and appear to extend over onto vesicle roofs. Some forms display dark brown to black crystalline linings or incomplete pyritic linings that may be altered organic linings.

It is possible that some of the apparently organic linings or that the oral part of the organic linings in abandoned chambers are secondary. They possibly could have been produced by accumulations of organic residues on the autozooecial walls and diaphragms in abandoned chambers. The residues could have been produced from decomposed organic matter left in the abandoned chamber when a new basal diaphragm was formed and/or by diagenetic removal and alteration of an organic matrix in the skeleton.

Some of the linings that are more complete and appear to be membraneous (Fig.

43) are most likely primary. They could have been secreted by the zooidal epithelium lining the living chamber in an autozooecium. Some of them appear to pass along the autozooecial wall between the skeletal wall and succeeding diaphragms (which also have a thick organic component). Here, the organic linings appear to predate the diaphragm in their formation. Others in the same autozooecium or zoarium do not pass between clearly skeletal diaphragms and autozooecial walls and are secondary or postdate diaphragm formation. Boardman (1971, p. 8) reported an organic membraneous layer in the zooidal epithelium in a degenerated autozooid in a recent disporellid cyclostome. In addition, he reported (Boardman, 1971, p. 9, 11) organic zooecial linings in a Devonian trepostome, including linings on the aboral side of membraneous diaphragms.

Questionable membraneous diaphragms were seen in two ceramoporoids (Fig. 34). Membraneous diaphragms have been reported in a few trepostomes (Boardman, 1971, p. 11).

Boardman (1971, p. 21) described and illustrated (pl. 10, fig. 5b) membranes passing from one autozooecium into another without an apparent break in the wall. Similar but smaller threadlike membraneous remnants have been observed to cross autozooecial walls and the skeleton of vesicular tissue in cystoporates. Some of them appear to be associated with a subspherical membraneous swelling (Figs. 44 and 45). The threads cross autozooecial walls and diaphragms with no good indication of a pore. A few instances where membraneous material crosses skeletal structures appear to be artifacts of sectioning but some may not be. The association of threads and subspherical cysts suggests that these membraneous structures may have been produced by an organism other than the bryozoan. Their nature, affinity, and significance needs to be investigated further.

Degeneration–Regeneration Cycles

Introduction. Using post-Paleozoic cyclostomes as a model for autozooecial form and function in the Cystoporata, one might assume that cystoporate colonies may have undergone many degeneration–regeneration cycles. This cycle probably would involve, among other things: (1) degeneration of organs enclosed in the membraneous sac and associated muscles and ligaments, resulting in a brown body that commonly is enclosed by a membrane; (2) closure of the terminal pore in the terminal-vestibular membrane; (3) persistence of the zooidal epithelium and associated peritoneum essentially intact; (4) maintainence of a coelomic space inside the living chamber's body wall (this coelomic space would be continuous with the hypostegal coelom, and with exosaccal coelomic cavities of adjacent zooids in the ceramoporoids where communication pores are present); and (5) formation of a new polypide bud from cells in the terminal membrane and a new terminal pore and vestibular membrane. Oral extension of the autozooecial walls and formation of a new basal diaphragm could also be associated with the latter half of the cycle.

Evidence for Degeneration–Regeneration Cycles in Cystoporates.   Ontogenetic elongation of autozooecia in the oral direction and sealing off of successive abandoned chambers by basal diaphragms is common in the cystoporates. The intermittent nature of the basal diaphragms suggests a possible relationship with a degeneration–regeneration cycle. Other aspects of intrazooidal cyclical growth suggest colony-wide control or semi-colony-wide control and may be related to degeneration–regeneration cycles, semi-colony-wide cycles or environmental fluctuations. Such cycles include (1) formation of spinelike hemiphragms at nearly the same level in adjacent autozooids, (2) formation of funnel cystiphragms and flask-shaped chambers which are intraautozooecial polymorphs (see below), (3) formation of terminal or subterminal diaphragms, (4) formation of mural spines, and (5) development of gonozooecia. Extrazooidal cyclical growth features include (6) development of zones of thick vesicle roofs and stereom. Some of these cycles (such as 3 and 6) probably involve colony-wide or semi-colony-wide response or near-simultaneous responses of individual adjacent zooids to external change. Others (such as 2 and 5) probably represent colony-wide or semi-colony-wide cycles related to reproduction. Others (such as 1 and 4) probably represent independent attainment of a similar stage in ontogenetic development in adjacent zooids. All need further investigation as to their extent and possible origin.

The strongest evidence for degeneration–regeneration cycles in cystoporates is the existence of certain brown deposits. Concentrated brown deposits (Figs. 38 and 39), particularily those with encapsuling membrane (Fig. 35), strongly resemble brown bodies formed from degeneration of a polypide in recent cyclostomes (Fig. 41) and most probably are fossilized brown bodies. The presence of these brown deposits in abandoned chambers suggests that formation of basal diaphragms and abandoned chambers was related to or accompanied a degeneration–regeneration cycle. Boardman (1971, p. 18) reported the occurrence of fossilized brown deposits, possibly the remnants of brown bodies, in every abandoned chamber in one specimen of trepostome, suggesting a one-to-one relationship of degeneration–regeneration with formation of basal diaphragms, at least for that colony. A one-to-one relationship between possible brown bodies and basal diaphragms (and abandoned chambers) is rarely seen in areas of a colony in the cystoporates. Some of these occurrences are zonal in a colony, suggesting possible control, at least by part of the colony, of degeneration–regeneration cycles.

In modern cyclostomes, skeletal basal diaphragms are generally not developed, so extrapolation on their possible significance in cystoporates is not possible. Brown-body formation may correlate with the formation of terminal diaphragms in some modern cyclostomes, but these terminal diaphragms are commonly resorbed (Borg, 1933, p. 367).

Flask-Shaped and Modified Flask-Shaped Chambers

Previous Studies. Ulrich (1890, p. 319) apparently was the first to observe and describe flask-shaped chambers and funnel cystiphragms in Paleozoic tubular bryozoans. He illustrated (1890, p. 318, fig. 7d) a flask-shaped chamber with an outer funnel in *Amplexopora robusta*. He was unsure of their function but suggested that they were in some way connected with reproductive functions. Cumings and Galloway (1915) illustrated flask-shaped chambers and funnel cystiphragms (which they termed "infundibular diaphragms") in several trepostomes, some of which were associated with outer funnels. They (1915, p. 353–356) interpreted the funnel cystiphragms as being deposited by a zooidal epithelium which was withdrawing or retreating and which enveloped a developing brown body. They rejected (1915, p. 356) an interpretation of the flask-shaped chambers representing a zooecium within a zooecium, as might result from total regeneration of a zooecium.

Boardman (1971) described and illustrated a variety of funnel cystiphragms, outer funnels, and flask-shaped chambers in trepostomes and concluded (p. 25, 26) that they contained something other than feeding organs and thus represented intrazooidal polymorphic forms. He discussed the possible origins of such chambers and functions of such zooids and suggested that they were possibly a male reproductive zooid. He also (p. 21) described and illustrated cystoidal diaphragms and brown deposits in some living chambers in an undescribed Ordovician cystoporate. Later, Boardman (in press) included the speculations that the flask-shaped chambers and outer funnels might have been produced by progressive calcification of membranes during ontogeny or that possibly the funnel cystiphragms were secreted by the zooidal epithelium of a polymorphic zooid and that the associated outer funnels were calcified terminal-vestibular membranes. No examples were reported from modern cyclostomes, however.

Funnel Cystiphragms, Flask-shaped Chambers, and Modified Flask-shaped Chambers in the Cystoporata.

A few flask-shaped chambers with a centered funnel cystiphragm and a relatively large vaselike portion aboral to the funnel cystiphragm have been observed in the cystoporates (Fig. 46). This one has an outer funnel with a small neck and is comparable in gross form to the one figured by Ulrich in 1890 (p. 318) and to many of those found in trepostomes by Cumings and Galloway (1915) and Boardman (1971).

Several modified flask-shaped chambers are characterized by having off-centered and oblique funnel cystiphragms (Figs. 47–49). One or both sides of the oblique funnel cystiphragms can abut against the basal diaphragm that marks the aboral limit of the modified flask-shaped chamber. One of them (Fig. 48) has a relatively large vaselike expansion aboral to the greatest constriction of the funnel cystiphragm, but most (Figs. 47 and 49) have relatively small expansions of the living chamber aboral to the neck of the funnel cystiphragms. Two of them (Figs. 47 and 48) have outer funnels associated with the funnel cystiphragms and one has an ontogenetic series of multiple funnel cystiphragms (Fig. 49).

Most modified flask-shaped chambers in cystoporates have a partial funnel cystiphragm, generally attached to the distal and lateral walls of the autozooecium. Some (Figs. 50 and 55) are attached well above the basal diaphragm so that the flask-shaped chamber has considerable volume below the partial funnel cystiphragm. Most (Figs. 51, 53, 54, 56, and 61) are attached so that their aboral ends are on or very near the basal diaphragm. Some (Figs. 54 and 64) have partial funnel cystiphragms in ontogenetic series.

A few modified flask-shaped chambers in the cystoporates have funnel cystiphragms with a closed bottom (Figs. 57, 58, and 59). The aboral extremeties of the funnel cystiphragm may touch the autozooecial wall if the funnel cystiphragm is oblique. In some (Figs. 57 and 59), the aboral extremeties and bottom are near the center of the autozooecium and well away from the autozooecial walls.

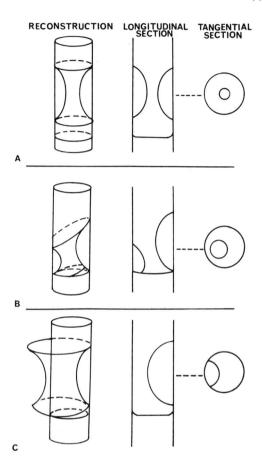

RECONSTRUCTION   LONGITUDINAL   TANGENTIAL
                   SECTION   SECTION

A

B

C

Figure 7

Variations in shape and size of flask-shaped chambers, seen in longitudinal and tangential thin sections, that could be produced by changing the inclination and position of the axis of a funnel cystiphragm. (a) Axis of funnel cystiphragm centered on autozooecial axis and not inclined; (b) axis of funnel cystiphragm inclined to axis of autozooecium; (c) axis of funnel cystiphragm parallel to but offset from axis of autozooecium.

Despite great variation in the size and shape of the chambers here called flask-shaped chambers or modified flask-shaped chambers (Figs. 46–59, 61–66), they are all interpreted to be of similar construction, origin, and function. Variations are interpreted to be the result of (1) variations in the inclination of the axis of the funnel cystiphragm with respect to the axis of the autozooecium; (2) variations in the proximity (for example, centered versus off-centered) of the axis of the funnel cystiphragm to the axis of the autozooecium; (3) position of the funnel cystiphragm with respect to the basal diaphragm aboral to it; (4) the presence or absence of a bottom that is part of the funnel cystiphragm rather than a basal diaphragm; (5) the presence of a single funnel cystiphragm or an ontogenetic series of funnel cystiphragms; and (6) the presence or absence of an outer funnel. Hypothetical variations that would result from 1–3 above are illustrated in Fig. 7. Variation that could have resulted from factors 1, 2, 3, 5, or 6 have been observed in one colony and, in some cases, in one autozooecium (Fig. 62).

Interpretive Functional Morphology of Funnel Cystiphragms, Flask-shaped Chambers, and Modified Flask-shaped Chambers in the Cystoporata.

Flask-shaped chambers beneath overgrowths are comparable in length to normal autozooecial living chambers in the same colony. Abandoned flask-shaped chambers are generally more comparable in length to normal living chambers and flask-shaped living chambers than to normal abandoned chambers. Figure 62 illustrates a flask-shaped living chamber that is slightly longer than an abandoned flask-shaped chamber in the same autozooecium and both of these are considerably longer than normal abandoned chambers in the same autozooecium or in other autozooecia in the same colony. The capping structure in the abandoned flask-shaped chamber is a basal diaphragm of the abandoned chamber just oral to the abandoned flask-shaped chamber. Most likely this basal diaphragm served as the basal structure of a normal living chamber. It is possible that the diaphragm formed originally as a subterminal diaphragm to the flask-shaped chamber, but decisive evidence is lacking. A few flask-shaped chambers evidently are converted ontogenetically to normal living chambers and, subsequently, to abandoned chambers by formation of a partial basal diaphragm closing the funnel. The evidence indicates that flask-shaped chambers are much less abundant than normal living chambers and abandoned chambers and can form, alternatively, with normal living chambers in one autozooecium. To date, flask-shaped chambers have been seen only in autozooecia and in large monticular zooecia.

An important question in trying to determine the origin and functional morphology of flask-shaped chambers is: Are they the product of normal, "usual" autozooids? The evidence is quite strong that the flask-shaped chambers were not the product of normal autozooids, either as a result of an unusual mode of preservation, a common mode of preservation, or as part of their normal functional cycle. Borg (1926b, pls. 3, 8, 9) illustrates young developing polypides in recent cyclostomes that have a flask shape. It is possible that the membraneous sac might be calcified, or that the zooidal epithelium might retreat and secrete the funnel cystiphragms of a flask-shaped chamber at an early, aborted state. This seems an unlikely origin for flask-shaped chambers in cystoporates because the funnel cystiphragm is usually near the aboral end of living and abandoned flask-shaped chambers. Flask-shaped young polypide buds in cyclostomes, and presumably in cystoporates, are developed at the terminal membrane and are located near the oral end of a living chamber.

A second possibility that Cumings and Galloway (1915) suggested is that funnel cystiphragms were formed by zooidal epithelium which was retreating and enveloping a developing brown body. A partly degenerated (?) polypide in a recent heteroporoid cyclostome (Fig. 60) has a flask shape to the membraneous sac. The existence of brown deposits, which are possible fossilized brown bodies, in normal abandoned chambers and in flask-shaped chambers (Figs. 35, 36, 38, 39, 56, 63) suggests that flask-shaped chambers are not associated with the normal degeneration–regeneration cycle of autozooids. Further, the existence of funnel cystiphragms in ontogenetic series (Figs.

49, 52, 54, and 64) and a possible polypide remnant in a flask-shaped chamber (Figs. 64–66) found in the same colony as possible polypide remnants in normal living chambers (Fig. 32) suggests that flask-shaped chambers and funnel cystiphragms were associated with a viable zooid. If these are, in fact, polypide remnants, there is a polymorphic difference in size, shape, and position between those in normal living chambers and those in flask-shaped chambers. The existence of zones of funnel cystiphragms and flask-shaped chambers in a colony suggests that their origin is related to colony-wide or local cycles that involve colony control or response. The evidence suggests that degeneration–regeneration cycles were much more frequent than cycles involving funnel-cystiphragm formation.

A third possibility is that flask-shaped chambers and funnel cystiphragms were produced by progressive calcification of membranes or secretion by zooidal epithelium of a polymorphic zooid and that outer funnels are calcified terminal vestibular membranes. The outer funnel in cystoporates is simple-walled and deposited from the outside (Fig. 47), a feature that suggests that outer funnels are not calcified terminal-vestibular membranes. Many outer funnels and funnel cystiphragms do show progressively less calcification in an aboral direction (Figs. 46, 47, and 49) and this might suggest progressive calcification of membrane. The existence of possible brown bodies and, possibly, polypide remnants in flask-shaped chambers and formation of funnel cystiphragms in ontogenetic series suggests that funnel cystiphragms were secreted by zooidal epithelium and are not calcified membraneous sacs.

Thus the evidence so far seen suggests that funnel cystiphragms and flask-shaped chambers are not formed by normal autozooids but can alternate with normal autozooids in one autozooecium. Thus, they appear to be intraautozooecial polymorphs, as Boardman suggested in 1971 (p. 25, 26). The funnel cystiphragms probably were secreted by retracted or newly proliferated zooidal epithelium. The outer funnels may be partially formed funnel cystiphragms developed in ontogenetic series, although some of those in the trepostomes possibly were not. The flask-shaped chamber appears to have contained a modified polypide that could grow or degenerate and regenerate as more funnel cystiphragms were formed. Apparently it could degenerate and be replaced ontogenetically by a normal autozooid. The function of these presumably modified polymorphic autozooids remains uncertain. Boardman's suggestion (1971, p. 25, 26) that they might represent male reproductive zooids is reasonable. In the cystoporates that have zooids modified into what are here interpreted as gonozooids, presumably to house the developing embryos formed from the eggs developed in those zooids, flask-shaped chambers have not been observed. The flask-shaped chambers do appear to be related to a cyclical need or response of the colony and reproduction seems to be as reasonable a function as any.

ZOOECIAL POLYMORPHS

Exilazooids

Introduction. Exilazooecia are found in most genera of the Ceramoporidae but are not developed in fistuliporoid cystoporates. The term *exilazooecium* is a modification of the term "exilapore" (Dunaeva and Morozova, 1967, p. 87) originally used for mesopores in some Trepostomata that are relatively narrow and long and that lack or have few, distantly spaced basal diaphragms. Using the definition of zooid used in this paper there is little doubt that the exilazooecia in ceramoporoids are zooecia of some smaller kind of zooid. Since the soft parts and function may never be known, the descriptive term "exilazooecium" is preferred to the terms based on function, tissues, or organs, such as kenozooecium, nanozooecium, and heterozooecium. Exilazooecia in ceramoporoids differ from mesozooecia and alveoli. The descriptive term "cystopore" does not fit these structures as well as the term "exilazooecia."

Exilazooecia in Ceramoporoids. Exilazooecia in the Ceramoporoidea arise by the formation of the compound wall proximally to a semirecumbent autozooecial wall in the outer endozone or inner exozone (Fig. 19). Exilazooecia have never been observed to originate by formation of a compound wall on the basal lamina or mesotheca. Exilazooecia extend from their locus of origin to the zoarial surface. They have a relatively long, narrow tubular cavity with a rounded cross section. In some genera they are slightly more subangular and larger in cross section in the inner exozone and become progressively more circular and smaller orally. Exilazooecia can be absent to abundant between autozooecia in intermonticular areas in the ceramoporoids and generally a cluster of exilazooecia forms the center of a monticule. They generally lack intrazooecial skeletal structures, but basal diaphragms are present in some long exilazooecia and subterminal diaphragms can be present, generally at the same level as subterminal diaphragms in adjacent autozooecia (Fig. 23).

Exilazooecial wall microstructure is similar to that of autozooecial walls in the Ceramoporidae; that is, with broadly curved laminae and generally a broadly serrated boundary zone. The cortex is thinner and a zooecial lining is thin or absent in the exilazooecia. Exilazooecia do not have lunaria and lunarial deposits are lacking (Utgaard, 1968b, p. 1449). Calcite rods or acanthopores can be present in exilazooecial walls (Utgaard, 1968b, pl. 181, fig. 3).

In genera of the Ceramoporidae that have communication pores, these structures are present in exilazooecial walls as well as in autozooecial walls. Pores can connect the cavities of adjacent autozooecia, adjacent exilazooecia, or an autozooecium and an exilazooecium that are contiguous. Commonly there are fewer communication pores between exilazooecia than between autozooecia. In one Middle Ordovician species of *Acanthoceramoporella* (Fig. 71), the communication pores are enlarged to huge gaps in the walls, and flat and curved basal diaphragms can extend across several autozooecia and exilazooecia.

Interpretive Functional Morphology of Exilazooecia in Ceramoporoids. The presence of compound exilazooecial walls, zooecial linings, basal diaphragms, terminal diaphragms that were deposited from their aboral side, and communication pores in exilazooecia indicate the presence of secretory zooidal epithelium. The colony construction and locus of budding indicates that ceramoporoids were double-walled bryozoans and that coelomic fluid in the exilazooids could communicate, through the hypostegal coeloem, with the (presumably exosaccal) coelomic fluid in adjacent autozooids and exilazooids as well as through communication pores. When subterminal diaphragms without pores were formed, coelomic communication could still take place through communication pores in the zooecial walls.

Encapsuled brown structures, reminiscent of brown bodies formed from degeneration of a polypide, have been reported in exilazooecia in *Ceramophylla vaupeli* (Utgaard, 1968b, p. 1446, pl. 183, figs. 4a, 4c; Fig. 36, this paper). These brown structures strongly suggest the presence of live tissue, in addition to zooidal epithelium, in the exilazooecia. Kenozooids in modern heteroporoid cyclostomes (Borg, 1933, p. 362, 368) have a zooidal epithelium, peritoneum, coelomic fluid, and various cells in the coelomic fluid, but they do not degenerate to form an encapsuled brown body because they do not have a polypide. Nanozooids in recent cyclostomes do have a reduced extrusible polypide and a membraneous sac and do degenerate to form brown bodies (Borg, 1926b, p. 234, 236). Thus some exilazooecia may have contained a modified polypide or other organs.

Gonozooids

Intermonticular autozooecia with expanded subspherical oral ends have been known for some time (Ulrich, 1890, p. 383) in a few cystoporates. To date they have been found only in the genera *Strotopora* and *Cliotrypa*. Spaces in vesicular tissue interpreted to be gonocysts have also been reported (Shulga-Nesterenko, 1933, p. 49).

An autozooecium of normal living chamber diameter opens into an enlarged oral chamber. The enlarged oral ends of gonozooecia commonly are formed in several zooecia at the same level in a colony. Colonies that died with gonozooecial expansions at the surface display low, rounded hemispherical blisters, each with a small subcircular terminal to marginal pore. Many are broken, presumably after death of the colony, and appear as large, hemispherical depressions with elevated rims. In section, the oral end of the gonozooecium is seen to expand and cover adjacent extrazooidal vesicular tissue (Fig. 70) and, in some cases, adjacent autozooecia. Some adjacent autozooecia adjusted their direction of growth and grew up around the expanded part of the gonozooecium (Fig. 72). Calcified structures within the expanded portion of gonozooecia include centripital shelflike structures in some forms (Fig. 67), and small, slightly to strongly curved plates in the oral-lateral margins of the enlarged space (Figs. 73 and 74), at least partly dividing the enlarged end into a small and large chamber. The shelflike structures and outer wall of the expanded space have compound walls, walls deposited by epithelium from both sides. The inner part of the wall would have been deposited by the gonozooidal epithelium, and the shelflike structures probably were deposited as a compound wall under an infolding of the gonozooidal epithelium into the coelomic space in the large cavity. The outer part of the outer wall commonly is thicker than the inner part (Fig. 68) and locally contains noncalcified areas (Fig. 69). Presumably the outer part of the outer wall would have been deposited by the hypostegal epithelium, under a hypostegal coelom which was connected laterally to the exosaccal coelomic cavities of adjacent feeding autozooids.

One large gonozooecial cavity contains an irregular calcified cyst (related in origin to the curved plates?) and stringers of diffuse fossilized brown material (Fig. 68). Further details of gonozooecial structure must await more sectioning and the study of serial peels.

Older colonies can contain a zone of abandoned gonozooecial chambers and a younger zone of gonozooecia. Although the case certainly is not proved, the zonal, cyclical arrangement, the large chambers connected to normal-sized autozooecia aborally, the pores opening through the surface of the large blisters, and the smaller number of gonozooecia compared to normal autozooecia strongly suggest that these zooecial polymorphs were involved with the production of eggs and the brooding of embryos. They certainly housed polymorphic zooids, and their similarities with gonozooecia in some recent cyclostomes prompts the use of the term ''gonozooecia'' for these structures.

Large Monticular Zooids

The zooecia that immediately surround the monticular centers in most cystoporates are slightly larger than the common intermonticular zooecia, which are interpreted to have housed the normal feeding autozooids. They have comparable wall structure and thickness, living chamber length, abandoned chamber length, and intrazooecial skeletal structures, including funnel cystiphragms and flask-shaped chambers, as do the intermonticular autozooecia. The only differences they show with intermonticular autozooecia are (1) their slightly larger diameter, and (2) the fairly widespread occurrence of the lunarium and proximal side of the living chamber being radially arranged around the monticular center. When they show radial arrangement, the lunarium is on the side of the large monticular autozooecia nearest the monticular center. Intermonticular autozooecia also display radial arrangement of lunaria in many cystoporates. It seems likely that the larger monticular zooecia housed slightly larger feeding autozooids. The functional significance of the larger monticular autozooids and the monticules is not yet clear but it seems likely that they were not related to a reproductive or brooding function. Forms that do have zooecia interpreted to be gonozooecia also have monticules.

Basal Zooids

Basal polymorphic zooecia are known only in some colonies of *Ceramopora imbricata* (Utgaard, 1969, p. 289). They develop in the free margins of encrusting colonies, beyond the incrusting part of the colony, which has a simple-walled basal layer. The basal polymorphs have relatively short, narrow cavities which are subcircular in outline and most closely resemble exilazooecia in shape and in size. They have compound walls and most likely opened into a basal, centripital expansion of the hypostegal coelom that continued from the frontal surface of the colony around the growing margin to the peripheral part of the base of the colony (Fig. 2). There is no evidence, to date, of the possible existence of a modified polypide or a membraneous sac in these basal polymorphs, but their mode of growth suggests that they were lined by a secretory zooidal epithelium and contained coelomic fluid which was in communication with a hypostegal coelom.

SUMMARY

The evidence suggests that by Middle Ordovician time, cystoporates were already highly developed colonial animals. The nature of skeletal growth and colony formation indicates that they had a double-walled colony construction similar to that of some recent cyclostomes and two basic methods of interzooidal communication: (1) through a hypostegal coelom in all cystoporates, and (2) an additional means of communication through gaps and communication pores in the zooidal walls of ceramoporoids.

A relatively high degree of polymorphism developed by Middle Ordovician time in the ceramoporoids. In addition to what are most likely the normal feeding autozooids, they possessed larger monticular autozooids, smaller exilazooids (possibly containing a polypide), and intraautozooecial polymorphs (possibly with a reproductive function.) One Silurian species also developed basal polymorphs. The fistuliporoids generally have normal feeding autozooids and some have larger monticular autozooids. A few fistuliporoids developed zooecia with a normal autozooecial aboral portion and an expanded, subspherical oral portion and probably housed gonozooids and brooded embryos. Many fistuliporoids also have intraautozooecial polymorphic zooids which may have had a reproductive function. Fistuliporoids are also characterized by large amounts of extrazooidal vesicular tissue that probably buttressed the colony between autozooids.

Lunaria, found in nearly all the Cystoporata, were a persistent feature in each autozooecium and large monticular zooecium through their exozonal development. In many forms the lunarium partially delimited a smaller cylindrical chamber within the larger skeletal living chamber. There is some evidence to suggest that, in some forms, normal autozooidal polypides and polypides associated with funnel cystiphragms and modified flask-shaped chambers occupied the smaller cylindrical part of the skeletal living chamber next to the lunarium on the proximal side. The lunarium possibly redirected part of the centrifugal excurrent flow from the tentacular bell.

The existence of various types of brown deposits suggests the existence of autozooids with a membraneous sac and which periodically underwent degeneration–regeneration cycles.

Several types of colony-wide or semi-colony-wide zones of cyclical growth are present in cystoporates. Zones of thick-walled vesicles and stereom probably were the result of colony control of deposition, possibly as a reaction to external environmental fluctuations. Zones of gonozooecia and funnel cystiphragms are possibly the result of colony-wide or near-colony-wide control related to reproductive functions. Zones of fossilized brown deposits, possibly representing brown bodies and zones of basal diaphragms, possibly are related to degeneration–regeneration cycles. Whether such cycles were under semi-colony-wide control or were the result of independent action or response of adjacent zooids is not yet known. Formation of zones of intrazooecial structures such as hemiphragms and mural spines may have been the result of adjacent zooids reaching the same stage of ontogenetic development at nearly the same time and not of semi-colony-wide control of growth. Formation of zones of terminal or subterminal diaphragms may have been the result of colony-wide or semi-colony-wide response to environmental change or of independent reaction of adjacent zooids to environmental change. All of these and other, less obvious cyclical features need further investigation.

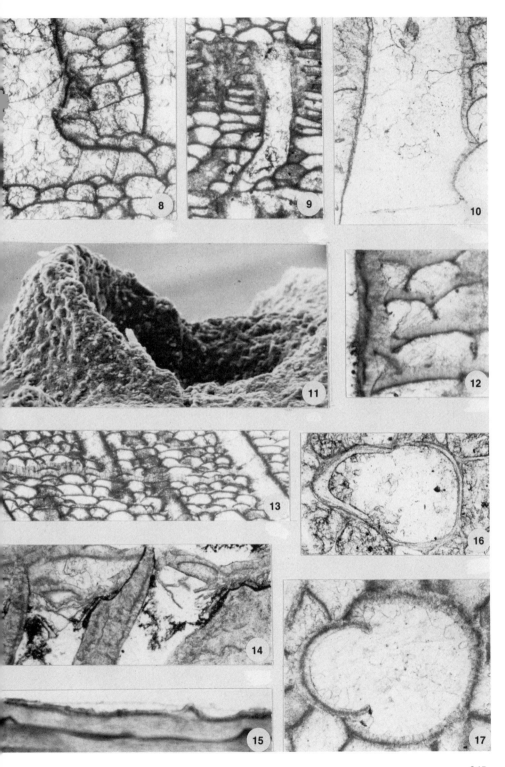

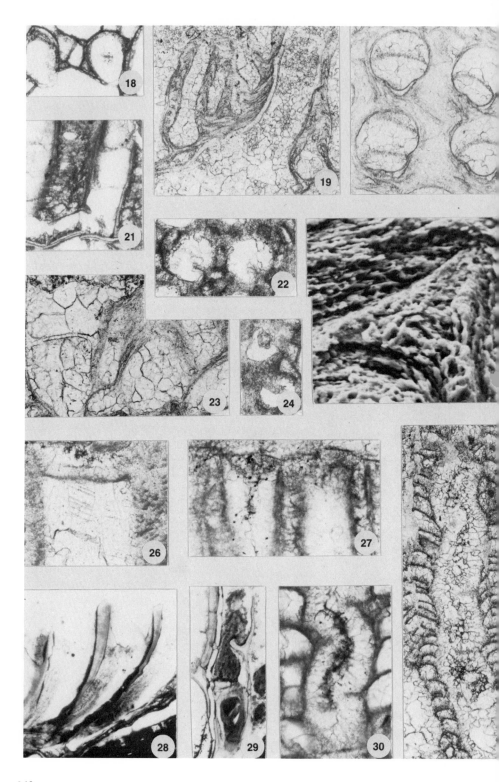

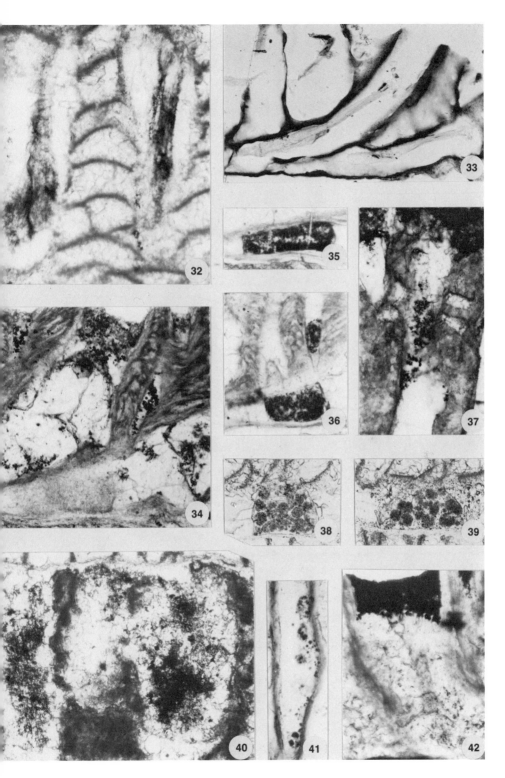

347

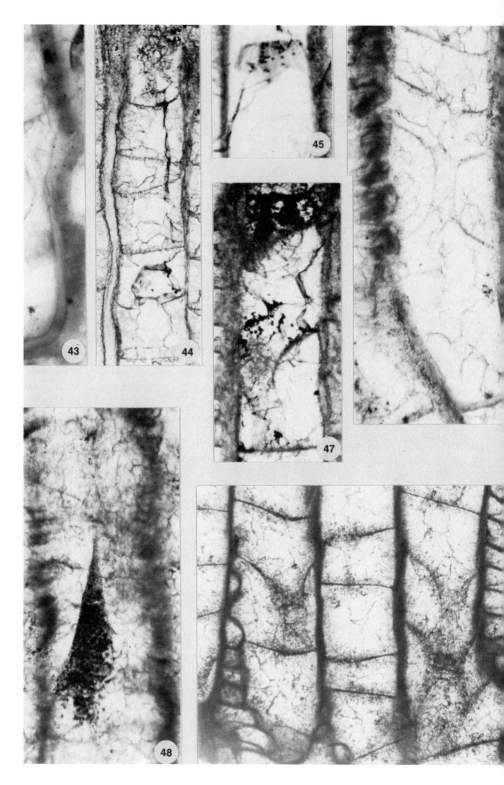

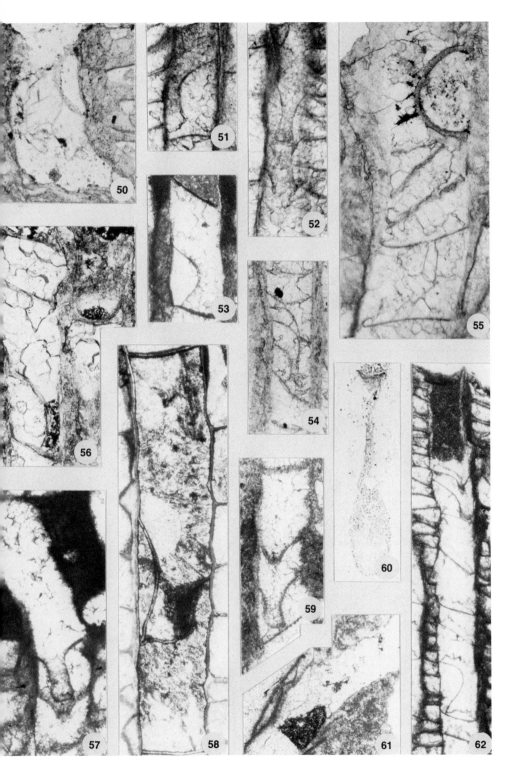

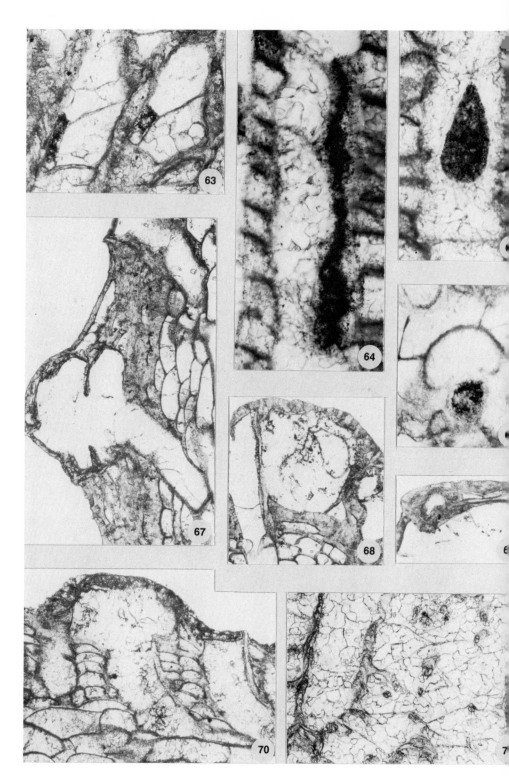

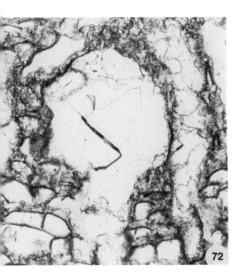

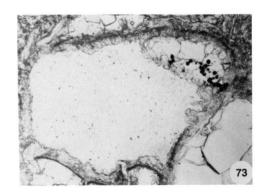

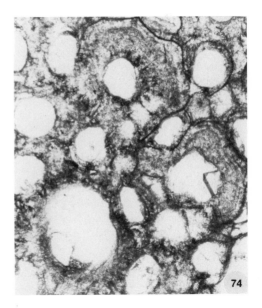

Figure 8

Longitudinal section of *Fistulipora waageniana* Girty, showing autozooecium with recumbent aboral section budded off on top of extrazooidal vesicular tissue. Pennsylvanian, Wu-shan Limestone, near Taning-hién, China, USNM 61922, ×50.

Figure 9

Longitudinal section of fistuliporoid showing autozooecia isolated by vesicular tissue at basal layer. Middle Devonian, "Hamilton," Petosky, Michigan, USNM 54839, ×30.

Figure 10

Longitudinal section of *Meekopora clausa* Ulrich, showing distal side (right) of autozooecial living chamber bounded by superimposed vertical parts of vesicle walls. Mississippian, Glen Dean Limestone, Sloans Valley, Kentucky, SIU 3000, ×100.

Figure 11

External view of lunarium on proximal side of autozooecium of *Cheilotrypa hispida* Ulrich. Note small protusions (''mural tubulae'') on proximal side of lunarium. Mississippian, Glen Dean Limestone, Sloans Valley, Kentucky , SIU 3001, scanning electron microphotograph, 20 kV, wd. 5.5 mm, 100–μ aperture, ×260.

Figure 12

Longitudinal section of *Fistulipora waageniana* Girty, showing simple-walled vesicle roof-wall structural units and compound autozooecial wall on left. Pennsylvanian, Wu-shan Limestone, near Ta-ning-hién, China, USNM 61922, ×100.

Figure 13

Oblique longitudinal section of *Meekopora eximina* Ulrich, showing zone of thick vesicle roofs in exozone. Mississippian, ''Chester Group,'' Monroe Co., Illinois, USNM 159766, ×30.

Figure 14

Oblique longitudinal section of *Cheilotrypa hispida* Ulrich, showing organic deposits beneath overgrowth. Inner layer may possibly represent the outer membrane of the incrusted colony and outer deposit (left) may possibly represent the basal cuticle of the incrusting overgrowth. Mississippian, Glen Dean Limestone, Sloans Valley, Kentucky, USNM 159755, ×100.

Figure 15

Longitudinal section of *Hornera* sp., showing outer colonial cuticle adjacent to calcite rods (mural tubulae) on frontal side of colony. Thin hypostegal coelom is visible under cuticle between calcite rods. Recent, Antarctic, Blacken Coll. 2.6, USNM 159768, ×300.

Figure 16

Tangential section of *Ceramoporella distincta* Ulrich, showing light-colored dense lunarial deposit, and lunarium with shorter radius of curvature modifying cross-sectional shape of skeletal living chamber. Upper Ordovician, Eden Formation, McMicken Member, Cincinnati, Ohio, USNM 159711, ×100.

Figure 17

Tangential section of *Dybowskiella* sp., showing granular-prismatic microstructure of autozooecial wall and lunarium, ends of lunarium indenting autozooecial cavity, and lunarium partially delimiting smaller cylinder in skeletal living chamber. Permian, Upper Productus Limestone, Jabbi, Pakistan, USNM 159769, ×100.

Figure 18

Tangential section of *Strotopora foveolata* Ulrich, showing ovate cross section of skeletal living chamber and lunarial deposit of light-colored, dense calcite. Mississippian, Keokuk Group, Bentonsport, Iowa, USNM 159770, ×50.

Figure 19

Longitudinal section of *Ceramoporella flabellata* (Ulrich), showing light-colored, dense lunarial deposit extending from endozone through exozone and exilazooecia arising by compound wall formation on the proximal side of a semirecumbent autozooecial wall. Upper Ordovician, Maysville Group, Corryville Member, Cincinnati, Ohio, USNM 159713, ×50.

Figure 20

Tangential section of and undescribed constellariid, showing light-colored, dense lunarial deposit and smaller living chamber on proximal side of autozooecia bounded by lunarium and curved diaphragm. Middle Ordovician, Marcem Limestone, Marcem Quarry, Gate City quadrangle, Virginia, USNM 159771, ×100.

Figure 21

Longitudinal section of *Dichotrypa foliata* Ulrich, showing hemiseptum partially separating aboral modified-cone end of living chamber from oral cylindrical end. Middle Devonian, Buffalo, Iowa, USNM 159772, ×50.

Figure 22

Tangential section of *Fistuliphragma spinulifera* (Rominger), showing curved, spine-like hemiphragms. Middle Devonian, Traverse Group, Norway Point, Thunder Bay River, Michigan, USNM 159773, ×50.

Figure 23

Oblique transverse section of *Ceramoporella distincta* Ulrich, showing subterminal diaphragms with aborally recurved ends in autozooecium and exilazooecia. Upper Ordovician, Eden Formation, McMicken Member, Cincinnati, Ohio, USNM 159712, ×100.

Figure 24

Tangential section of *Fistuliphragma spinulifera* (Rominger), showing plate-like to triangular hemiphragms. Middle Devonian, Traverse Group, Norway Point, Thunder Bay River, Michigan, USNM 159740, ×50.

Figure 25

Transverse etched surface of *Ceramoporella flabellata* (Ulrich), showing well-laminated autozooecial wall and granular lunarial deposit with proximal rib at zooecial bend region. Upper Ordovician, Maysville Group, Corryville Member, Jefferson Lake, Indiana, SIU 3002, scanning electron microphotograph, 20 kV, wd. 7 mm, 100-μ aperture, ×650.

Figure 26

Longitudinal section of *Fistulocladia typicalis* Bassler, showing terminal diaphragm of simple construction (deposited from the outside) in an autozooecium. Permian, Left Bank, Noel Boewen, Timor, USNM 159775, ×100.

Figure 27

Longitudinal section of *Constellaria* sp., showing incomplete brown deposits that possibly represent a terminal-vestibular membrane in autozooecia immediately below an overgrowth. Middle Ordovician, Lexington Limestone, Fayette, Kentucky, U.S.G.S. Loc. 4859 CO 7, ×100.

Figure 28

Longitudinal section of *Lichenopora* sp., showing collapsed terminal-vestibular membrane oral to polypide on proximal side of autozooecium. Recent, Galapagos Islands, British Museum (Natural History), St. George Collection, 42-A, −21, ×100.

Figure 29

Longitudinal section of *Tubulipora* sp., showing polypide in membraneous sac curving around spine-like hemiphragms in living chamber. Recent, Marseilles, France, USNM 159767, ×100.

Figure 30

Transverse section of *Fistuliphragma spinulifera* (Rominger), showing brown deposit, possibly representing a remnant of a polypide, bending around spine-like hemiphragms in living chamber preserved beneath overgrowth. Devonian, Alpena, Michigan, USNM 54823, ×100.

Figure 31

Transverse section of *Fistuliphragma spinulifera* (Rominger), showing diffuse brown deposit, possibly representing the remnants of a polypide, curving around hemiphragm-like spines in living chamber beneath overgrowth. Devonian, Traverse Group, Norway Point, Thunder Bay River, Michigan, USNM 159773, ×50.

Figure 32

Oblique longitudinal section of *Fistuliporella* sp., showing tubular to sac-like brown fossilized deposits, possibly representing fossilized membraneous sacs, in autozooecial living chambers beneath overgrowth. Middle Devonian, Calceola beds, Gees, Germany, USNM 113984-1, ×100.

Figure 33

Longitudinal section of *Lichenopora* sp., showing remnants of membraneous sacs along proximal sides of autozooecia. Recent, Galapagos Islands, British Museum (Natural History), St. George Collection, 42-A, −30, ×100.

Figure 34

Longitudinal section of *Ceramophylla vaupeli* (Ulrich), showing fossilized brown deposit in living chamber below subterminal diaphragm. Deposit may possible represent oral portion of membraneous sac with the oral constriction possibly analogous to an atrial sphincter and, just aboral to that, a possible attachment level. In autozooecium to right, transverse brown deposit may represent membraneous diaphragm. Upper Ordovician, "Utica Group," Brown Street, Cincinnati, Ohio, USNM 159720, ×100.

Figure 35

Longitudinal section of *Ceramophylla vaupeli* (Ulrich), showing encapsuled fossilized brown deposit, possibly representing a membrane-encapsuled brown body, in the aboral end of an autozooecium. Upper Ordovician, "Utica Group," Brown Street, Cincinnati, Ohio, USNM 159718, ×100.

Figure 36

Longitudinal section of *Ceramophylla vaupeli* (Ulrich), showing membrane encapsuled brown deposits, possibly representing fossilized brown bodies in the aboral ends of an autozooecium and an exilazooecium. Upper Ordovician, "Utica Group," Brown Street, Cincinnati, Ohio, USNM 159718, ×100.

Figure 37

Longitudinal section of a fistuliporoid, showing fossilized brown deposit, possibly representing part of a membraneous sac in the living chamber of an autozooecium. Middle Devonian, "Hamilton Group," Thunder Bay, Michigan, USNM 159774, ×100.

Figure 38

Longitudinal section of *Fistuliporella* sp., showing tightly bounded brown deposit containing subspherical, small brown masses. This deposit possibly represents a fossilized brown body that may have been encapsuled by a membrane. Middle Devonian, Calceola beds, Gees, Germany, USNM 113984-6, ×100.

Figure 39

Longitudinal section of *Fistuliporella* sp., showing possible fossilized brown body above light-colored, dense lunarial deposit. Middle Devonian, Calceola beds, Gees, Germany, USNM 113984-6, ×100.

Figure 40

Oblique longitudinal section of *Favositella interpuncta* (Ulrich and Bassler), showing diffuse fossilized brown deposits, possibly representing polypide remnants, in living chambers beneath overgrowth. Thin brown deposit touching autozooecial wall (right) may possibly represent attachment area and constriction at oral end of brown deposit (right) may represent sphincter area. Devonian, "Helderberg," Cash Valley, Maryland, USNM 53667, ×50.

Figure 41

Longitudinal section of *Hornera* sp., showing membrane-excapsuled brown bodies near aboral portion of autozooecium. Recent, Arctic, Blacken Collection 2.6, USNM 159776, ×100.

Figure 42

Oblique longitudinal section of *Ceramoporella* sp., showing tubular, fossilized brown deposit, possibly representing a membraneous sac, below a subterminal diaphragm closing a living chamber. Upper Ordovician, Maysville, Pleasant Ridge, Ohio, USNM 159777, ×100.

Figure 43

Longitudinal section of fistuliporoid showing organic autozooecial lining passing between zooecial wall (right) and skeletal diaphragm (upper left). Middle Devonian, "Hamilton," Petosky, Michigan, USNM 54839, ×300.

Figure 44

Longitudinal section of *Constellaria* sp., showing organic thread crossing basal diaphragms and associated larger organic structure in abandoned chamber. The thread and subspherical deposit may be the remains of an organism other than the bryozoan. Middle Ordovician, Cynthiana Formation, Cynthiana, Kentucky, USNM 159778, ×100.

Figure 45

Longitudinal section of *Constellaria florida* Ulrich, showing organic thread crossing autozooecial wall, and associated larger organic remnant. These possibly represents a fossilized organism other then the bryozoan. Middle Ordovician, "Lower Lorraine," Covington, Kentucky, USNM 159779, ×100.

Figure 46

Oblique longitudinal section through outer funnel, funnel cystiphragm and abandoned flask-shaped chamber in an autozooecium of *Crepipora venusta* Ulrich. The abandoned flask-shaped chamber has a basal diaphragm and is capped by a basal diaphragm. Orally and aborally from this abandoned flask-shaped chamber are normal abandoned chambers. The flask-shaped chamber is approximately four times as long as normal abandoned chambers in this colony. Upper Ordovician, Eden Shale, Covington, Kentucky, USNM 159780, ×100.

Figure 47

Longitudinal section of fistuliporoid, through abandoned flask-shaped chamber having off-centered outer funnel and off-centered and oblique funnel cystiphragm. Note the narrow expansion at the aboral end above the basal diaphragm and the progressively thinner calcification in the aboral direction of the funnel cystiphragm. Middle Devonian, "Hamilton Group," Buffalo, Iowa, USNM 159781, ×100.

Figure 48

Longitudinal section of *Anolotichia deckeri* Loeblich, through abandoned flask-shaped chamber having indistinct, thinly calcified outer funnel, oblique funnel cystiphragm and fossilized brown deposit in flask-shaped chamber. Middle Ordovician, Bromide Formation, Carter Co., Oklahoma, USNM 114560, ×100.

Figure 49

Longitudinal section of *Fistulipora carbonaria* Ulrich, through abandoned flask-shaped chamber with oblique and off-centered funnel cystiphragm abutting basal diaphragm (left). Chamber is capped orally by basal diaphragm below normal abandoned chambers. Length of normal abandoned chambers suggested by spacing of basal diaphragms in adjacent autozooecia. Note progressively weaker calcification of funnel cystiphragm in aboral direction. Autozooecium at right has abandoned flask-shaped chamber with multiple, off-centered, oblique funnel cystiphragms in ontogenetic series. Pennsylvanian, Kansas City, Missouri, USNM 159782, ×50.

Figure 50

Oblique longitudinal section of *Actinotrypa pecularis* (Rominger), through modified flask-shaped chamber with partial funnel cystiphragm attached well orally to base of chamber. Mississippian, Keokuk Formation, Keokuk, Iowa, USNM 159783, ×100.

Figure 51

Longitudinal section of *Constellaria* sp., showing modified flask-shaped chamber with a single partial funnel cystiphragm. Middle Ordovician, Catheys Formation, Tennessee, USNM 159784, ×100.

Figure 52

Longitudinal section of *Constellaria* sp., showing multiple funnel cystiphragms in ontogenetic series. Same colony as Fig. 51. Middle Ordovician, Catheys Formation, Tennessee, USNM 159784, ×100.

Figure 53

Longitudinal section of *Fistulipora crassa* Rominger, through abandoned modified flask-shaped chamber with partial funnel cystiphragm. Middle Devonian, Hamilton Group, Thedford, Ontario, USNM 37756, ×50.

Figure 54

Longitudinal section of *Constellari fisheri* Ulrich, showing abandoned modified flask-shaped chamber with ontogenetic series of partial funnel cystiphragms and one partial diaphragm. Middle Ordovician, Cynthiana Formation, Winchester, Kentucky, USNM 44068, ×100.

Figure 55

Longitudinal section of *Canutrypa francqana* Bassler, showing large distal cyst probably comparable to partial funnel cystiphragm. Note secondary lining below distal cyst and ontogenetic series of diaphragms and partial diaphragms. Upper Devonian, Ferques, France, USNM 116417, ×100.

Figure 56

Longitudinal section of undescribed ceramoporoid with partial funnel cystiphragm. Brown deposits, possibly representing fossilized brown bodies are in the aboral ends of the abandoned modified flask-shaped chamber (left) and an abandoned living chamber (upper right). Capping structure of abandoned flask-shaped chamber is a basal diaphragm. Middle Ordovician, Tulip Creek Formation, W. Branch, Sycamore Creek, Oklahoma, USNM 159785, ×50.

Figure 57

Oblique longitudinal section through flask-shaped chamber in a fistuliporoid. Note inner obliquely cut funnel (part of a closed-bottomed funnel cystiphragm?) and outer (oral) closed-bottomed funnel cystiphragm. Middle Devonian, Centerfield Limestone, 3 miles north of East Bethany, New York, USNM 159786, ×50.

Figure 58

Longitudinal section of *Fistulipora vacuolata* Crockford, showing closed-bottomed funnel cystiphragm at base of abandoned flask-shaped chamber. Note secondary crystalline lining on walls of autozooecia, funnel cystiphragm, and walls and roofs of vesicular tissue. Permian, Nooncanbah series, Mt. Marmion, Western Australia, USNM 112456, ×50.

Figure 59

Longitudinal section of fistuliporoid through abandoned flask-shaped chamber. Note normal basal diaphragms at base and top of abandoned chamber, outer funnel cut obliquely, and, aborally, oblique closed-bottomed funnel cystiphragm. Base of abandoned flask-shaped chamber is on bottom of funnel cystiphragm rather than on basal diaphragm. Middle Devonian, Centerfield Limestone, 3 miles north of East Bethany, New York, USNM 159787, ×50.

Figure 60

Longitudinal section of a decalcified *Neofungella claviformis* (Waters), showing flask-shaped membraneous sac, below attachment organ, of polypide which possibly has started to degenerate. Recent, one of Folke Borg's slides, No. 2.19, ×100.

Figure 61

Longitudinal section of a fistuliporoid through modified flask-shaped chamber with obscure partial funnel cystiphragm above brown mass, possibly representing a fossilized brown body, resting on basal diaphragm. Permian, Capinoto, Bolivia, USNM 159788, ×50.

Figure 62

Longitudinal section of fistuliporoid, showing off-centered and oblique funnel cystiphragm and outer funnel in modified flask-shaped living chamber (top) and abandoned flask-shaped chamber with partial funnel cystiphragm (below) in same autozooecium. Note normal basal diaphragms and abandoned chambers between flask-shaped chambers. Middle Devonian, Hamilton Group, Arkona, Ontario, USNM 159789, ×30.

Figure 63

Longitudinal section of undescribed ceramoporoid, through modified flask-shaped chambers with compact brown masses (fossilized brown bodies?) in their aboral ends. Middle Ordovician, Tulip Creek Formation, Carter Co., Oklahoma, USNM 159790, ×50.

Figure 64

Longitudinal section of *Fistuliporella* sp., showing ontogenetic series of partial funnel cystiphragms on distal side of autozooecium and tubular brown deposit, possibly representing the fossilized remnants of a polypide, in the abandoned modified flask-shaped chamber. Middle Devonian, Calceola beds, Gees, Germany, USNM 113984-1, ×100.

Figure 65

Oblique transverse section of *Fistuliporella* sp., showing dark brown deposit in aboral end of an abandoned modified flask-shaped chamber. Note the very narrow neck in the funnel cystiphragm cut at this level and the complete, curved bottom of the funnel cystiphragm. Middle Devonian, Calceola beds, Gees, Germany, USNM 113984-2, ×100.

Figure 66

Tangential section of *Fistuliporella* sp., showing dark brown deposit in modified flask-shaped chamber on proximal side of zooecium adjacent to lunarium. Middle Devonian, Calceola beds, Gees, Germany, USNM 113984-2, ×100.

Figure 67

Longitudinal section of *Cliotrypa ramosa* Ulrich and Bassler, showing shelf-like projections into gonozooecial cavity and blister-like expansion of gonozooecium at zoarial surface. Mississippian, New Providence Formation, Curiosity Hollow, 5 miles east of Martinsville, Indiana, USNM 159791, ×50.

Figure 68

Transverse section of *Cliotrypa ramosa* Ulrich and Bassler, showing calcified cyst in outer part of gonozooecial expansion, diffuse stringers of brown deposits, and thin portion of outer gonozooecial wall (where plane of section approaches pore?). Mississippian, New Providence Formation, Curiosity Hollow, 5 miles east of Martinsville, Indiana, USNM 159792, ×50.

Figure 69

Longitudinal section of *Cliotrypa ramosa* Ulrich and Bassler, showing noncalcified areas in outer gonozooecial wall. Mississippian, New Providence Formation, Curiosity Hollow, 5 miles east of Martinsville, Indiana, USNM 159791, ×100.

Figure 70

Longitudinal section of *Cliotrypa ramosa* Ulrich and Bassler, showing expansion from normal autozooecial diameter into large subspherical chamber at surface of colony. Expanded oral end of this gonozooecium covers adjacent extrazooidal vesicular tissue. Note hyaline lunarial deposit on proximal (right) side of autozooecium and gonozooecium. Mississippian, New Providence Formation, Curiosity Hollow, 5 miles east of Martinsville, Indiana, USNM 159793, ×50.

Figure 71

Longitudinal section of *Acanthoceramoporella* sp., showing gaps in autozooecial and exilazooecial walls. Note basal diaphragms crossing several zooecia. Middle Ordovician, Arline Formation, Friendsville, Tennessee, USNM 159794, ×50.

Figure 72

Oblique longitudinal section of *Strotopora foveolata* Ulrich, showing curved skeletal plate (upper left) dividing enlarged end of gonozooecium. Autozooecia on left and right grew around gonozooecial expansion, presumably at least in part, when gonozooecial outer wall was being deposited. Note thin area in gonozooecial wall, above curved plate, where autozooecium curves around gonozooecium. Mississippian, Koekuk Group, Warsaw, Illinois, USNM 159795, ×50.

Figure 73

Transverse section of *Strotopora foveolata* Ulrich, showing expanded part of gonozooecium resting on vesicular tissue (lower left) and curved plate (upper right) separating off small chamber in oral-lateral part of gonozooecium. Mississippian, Keokuk Group, Warsaw, Illinois, USNM 159795, ×100.

Figure 74

Tangential section of *Strotopora foveolata* Ulrich, showing gonozooecia intersected at different levels by plane of section. Mississippian, Keokuk Group, Bentonsport, Iowa, USNM 159770, ×50.

ACKNOWLEDGMENTS

This paper is an outgrowth of studies on cystoporates aimed at revision of Part G of the Treatise on Invertebrate Paleontology. Much of the study and writing were done while the author was on sabbatical leave from Southern Illinois University and was a Smithsonian Fellow at the U.S. National Museum of Natural History. Thanks are extended to the Office of Academic Studies and the Department of Paleobiology of the Smithsonian Institution for the use of space and equipment and access to the collections. Thanks are extended to the Office of Research and Projects at Southern Illinois University and to Curt Teichert, Editor, Supplements and Revised Editions of the Treatise on Invertebrate Paleontology, for financial assistance (from N. S. F. Grant GB-31331x). Richard S. Boardman of the U.S. National Museum of Natural History read the manuscript and made several helpful suggestions. Thanks are also extended to Dr. Boardman for permission to examine and photograph thin sections of recent cyclostomes (containing both the skeleton and soft parts) currently being studied by Dr. Boardman. Donald Dean of the U.S. National Museum of Natural History made the cyclostome thin sections. Thanks are extended to Judy Murphy, Director of the Center for Electron Microscopy, Southern Illinois University at Carbondale, who helped the author with his preliminary investigations of ultrastructure using the scanning electron microscope. Fredda Burton drafted the illustrations. Olgerts Karklins of the U.S. Geological Survey permitted the author to examine and photograph thin sections of cystoporates in the U.S. Geological Survey collections from Kentucky.

REFERENCES

Astrova, G. G. 1964. O novum otryade Paleozoyskikh Mshanok. Paleont. Zhurnal, *1964*(2): 22–31.

Bassler, R. S. 1953. Bryozoa, *in* R. C. Moore, ed., Treatise on Invertebrate Paleontology, Univ. Kansas Press, and Geol. Soc. America. pt. G., p. G1–G253, 175 text-figs.

Boardman, R. S. 1971. Mode of growth and functional morphology of autozooids in some Recent and Paleozoic tubular Bryozoa. Smithsonian Contr. Paleobiol., *8*: 51 p., 11 pls., 6 figs.

———. Body walls and attachment organs in some cyclostomes and Paleozoic trepostomes. Proceedings of the Second International Conference on Bryozoa, I.B.A., Durham, England, 1971. In press.

———, and A. H. Cheetham. 1969. Skeletal growth, intracolony variation, and evolution in Bryozoa: a review. Jour. Paleont. *43*(2): 205–233, pls. 27–30, 8 text-figs.

Borg, Folke. 1926a. On the body-wall in Bryozoa. Micros. Sci. Quart. Jour., *70*: 583–598, 6 text-figs.

———. 1926b. Studies on Recent cyclostomatous Bryozoa. Zool. Bidrag. Uppsala, *10*: 181–507, pls. 1–14, 109 text-figs.

———. 1933. A Revision of the Recent Heteroporidae (Bryozoa). Zool. Bidrag. Uppsala, *14:* 253–394, pls. 1–14, 29 text-figs.

———. 1965. A Comparative and phyletic study on fossil and Recent Bryozoa of the Suborders Cyclostomata and Trepostomata. Arkiv. für Zoologi, ser. 2, *17*(1): 1–91, pls. 1–14. (Posthumously ed. by Lars Silén, and Nils Spjeldnaes.

Cummings, E. R., and J. J. Galloway. 1915. Studies of the morphology and histology of the trepostomata or monticuliporoids. Geol Soc. America, Bull., *23:* 357–370, pls. 10–15.

Dunaeva, N. N., and I. P. Morozova. 1967. Osobennosti razvitiya i sistematicheskoye polozheniye nekotorykh pozdne-Paleozoyskikh Trepostomat. Paleont. Zhurnal. *1967*(4): 86–94, pl. 5, 2 text-figs.

Elias, M. K., and G. E. Condra. 1957. *Fenestella* from the Permian of West Texas. Geol. Soc. America, Mem., *70*: 158 p., 23 pls., 17 figs., 10 tables.

McKinney, F. K. 1969. Organic structures in a late Mississippian trepostomatous ectoproct (bryozoan), Jour. Paleont., *43*(2): 285–288, pl. 50, 1 text-fig.

Nye, O. B. 1968. Aspects of microstructure in post-Paleozoic Cyclostomata. Atti Soc. Ital. Sci. Nat. Milano, *108*: 111–114.

Ryland, J. S. 1970. Bryozoans. Hutchinson Univ. Lib., London. 175 p., 21 figs.

Shulga-Nesterenko, M. J. 1933. Bryozoa from the coal-bearing and subjacent series of Pechora Land. Trans. United Geol. Prospct. Service USSR, Fac., *259*: 64 p., 9 pls.

Tavener-Smith, R. 1968. Skeletal structure and growth in the Fenestellidae (Bryozoa) (preliminary report). Atti Soc. Ital. Sci. Nat. Milano, *108*: 85–92.

———. 1969a. Skeletal structure and growth in the Fenestellidae (Bryozoa). Paleontology, *12*, pt. 2: 281–309, pls. 52–56, 9 text-figs.

———. 1969b. Wall structure and acanthopores in the bryozoan *Leioclema asperum*. Lethaia, *2*(2): 89–97, 7 figs.

Ulrich, E. O. 1883. American Paleozoic Bryozoa. Jour. Cincinnati Soc. Nat. Hist., *6*: 82–92, pl. 1.

———. 1890. Paleozoic Bryozoa. Illinois Geol. Survey, *8*: 283–688, pls. 29–78, 17 text-figs.

Utgaard, John. 1968a. A Revision of North American genera of ceramoporoid bryozoans (Ectopocta), pt. I, Anolotichiidae. Jour. Paleont., *42*(4): 1033–1041, pls. 129–132.

———. 1968b. A Revision of North American genera of ceramoporoid bryozoans (Ectoprocta), pt. 2, *Crepipora, Ceramoporella, Acanthoceramoporella,* and *Ceramophylla.* Jour. Paleont., *42*(6): 1444–1455, pls. 181–184.

———. 1969. A Revision of North American genera of ceramoporoid bryozoans (Ectoprocta), pt. III, the ceramoporoid genera *Ceramopora, Papillalunaria, Favositella,* and *Haplotrypa.* Jour. Paleont., *43*(2): 289–297, pls. 51–54.

Coloniality and Polymorphism in Bryozoa of the Families Rhabdomesidae and Hyphasmoporidae (Order Rhabdomesonata)

Daniel B. Blake   University of Illinois at Urbana-Champaign

ABSTRACT

Bryozoa of the families Rhabdomesidae and Hyphasmoporidae in general appear to have been highly integrated colonies of limited zooidal differentiation. Many species are very regular in development of such features as branch shape and zooecial arrangement, suggesting strong genetic control of overall colony form. Two types of larger structures other than autozooecia are present. The spine-like acanthopores are considered to have been supports for surficial tissues. Mesopore-like structures may have evolved as space-filling structures that contributed to zoarial strength, or they may have helped support surficial tissues. The double-walled model of Borg (1926), proposed for certain extant cyclostomes, also provides a reasonable model for the coordination of growth and the formation of skeletal material in isolated areas of rhabdomesontid zooaria.

INTRODUCTION

Rhabdomesontids are the typically small regularly cylindrical Bryozoa common in many Paleozoic sediments. Although their abundance and complexity suggests they might be useful in a variety of paleobiologic and biostratigraphic studies, rhabdomesontids

Author's address: Department of Geology, University of Illinois, Urbana, Illinois 61801.

generally are poorly known. Most species have been described from very small suites; in addition, descriptions and illustrations of many taxa, including type species, are not adequate for a clear understanding of the taxon involved. Probably in part because of inadequate study, and in part because of the lack of a truly sharp demarcation between more or less typical trepostome development and more or less typical rhabdomesontid development, some difference of opinion in the literature exists as to the scope and definition of the group. Bassler (1953) recognized the Rhabdomesidae and the Arthrostylidae, both now assigned to the Rhabdomesonata, as families in the Order Cryptostomata. In his diagnosis of the Rhabdomesidae, he stressed such features as a ramose growth mode, the presence of a distinct boundary between the well-defined endozone and the exozone, the shape and orientation of the zooecia, the development of hemisepta, and the size and abundance of acanthopores.

In recent years, a number of Russian workers have considered the family to be distinctive enough to be raised to higher categorical levels. Astrova and Morozova (1956) recognized the Suborder Rhabdomesoidea, which included the Family Arthrostylidae, the very slender articulated predominantly lower Paleozoic family (Figs. 8 and 9), and the Family Rhabdomesidae. Shishova (1966) restored the Family Hyphasmoporidae of Vine (1885) and assigned to the family those rhabdomesid genera which possess small tubular cavities or pits between apertures; these depressions, termed "metapores" by Shishova, may partially or completely fill the interspaces. The taxonomic content of the Arthrostylidae was not changed. Shishova (1968) then elevated the suborder to ordinal rank but did not alter the three contained families. In her diagnoses, Shishova employed most of the characters suggested by Bassler, but also stressed the importance of budding taking place around a central cavity, or an axial bundle of zooecia, or some form of median axis. Shishova considered the Order Rhabdomesonata to be closer to the Trepostomata than to the Cryptostomata, differing from the former in the manner of budding about some form of axis, the lack of maculae, and the nature of the polymorphism.

REGULARITY OF GROWTH AND COLONIALITY IN THE RHABDOMESONATA

Zoarial growth patterns and skeletal development within many species of rhabdomesontid Bryozoa are very regular. Most aspects of colony development do not appear to have been controlled significantly by microenvironmental conditions, rather a relatively high degree of presumably genetic control appears to have been exercised. Overall coordination and control of growth is expressed in a variety of features of skeletal development:

1. The budding pattern is regular. Zooecia of specimens of most genera assigned to the group arise at or close to a reasonably well defined central axis, tube, or bundle of zooecia (Figs. 13 and 25). Successive zooecia commonly arise in a spiral pattern about the axis, although they apparently may also arise in cycles (Figs. 18 and 19); budding sequence therefore may not be important. Cyclical patterns of zooecial addition may be discerned by means of serial peels of transverse sections and by means of tracing ranks of apertures about the surface of the zoarium.

2. Zooecial growth patterns following budding are relatively consistent within a species. The endozonal portions of the zooecia lie at approximately constant orientations relative

to the zoarial axis; zooecia of many species are of relatively constant length. The arrangement seen in the illustration of a specimen of *Saffordotaxis incrassata* (Fig. 25) is thus fairly typical of the species. The change from endozonal to exozonal growth apparently was controlled to occur at approximately constant endozonal diameters. Because zoarial growth is continuous, the change is a continuing event progressing distally through the time of colony growth. Exozonal laminae are not always deposited parallel to endozonal laminae (Fig. 16), thus an apparently significant change in growth pattern occurred.

3. A distinct boundary between adjacent zooecia commonly is lacking, although a boundary may be incompletely developed in certain species. In some species, rows of acanthopores may suggest a zooecial boundary (Fig. 24), although no discontinuity in laminae appears to be present. As viewed in thin sections, the exozonal walls are generally consistent in shape within a species. For example, the exozonal walls of *S. incrassata* in longitudinal section are typically elongate, cylindrical, and rounded (Figs. 24 and 25); those of *Acanthoclema* (Fig. 13) are rectangular. These outlines are controlled at least in part by the fact that laminae secretion does not necessarily produce continuous layers across the zoarial surface; the exozonal wall may be shaped by local differential secretion (Figs. 17 and 29).

4. Fairly large interautozooidal areas in which skeletal material continued to be added at the surface at least through much of colony life occur (Figs. 6, 7, 27, and 28). Secretion was regular, maintaining symmetry; disproportionate amounts of material were not added locally. Skeletal material addition therefore was controlled over the colony as a whole.

Regularity, within a colony and a species, is also expressed in a number of features of general zoarial morphology:

1. Colonies of many species are regular cylinders (Fig. 6) whose diameters generally are constant, at least through the relatively short intervals of zoaria typically preserved intact. For example, a segment of a colony of *Saffordotaxis incrassata* (USNM 56303) which was studied is 52 mm long and had maintained a zoarial diameter of approximately 1.5 mm throughout the fragment length. As would be expected, a greater range of diameter values occurs among fragments presumably derived from different colonies. For example, among approximately 500 available zoarial fragments of *S. incrassata* from three localities, diameters range from approximately 0.8 mm to 2.2 mm. The apparently relatively constant mature zoarial diameter within a single colony suggests rhabdomesontid zooecia reached an approximate mature size, then growth became very slow or perhaps stopped. Ability for colony and zoarial growth and repair, however, did not cease. A colony of *S. incrassata* which was broken during life (Fig. 3) renewed growth, and formed a new branch of the same general orientation as the lost tip. The distal portions of some autozooids were repaired. Renewed growth of the broken tip of a colony of mature size suggests that normal growth was not simply terminated by the end of a growing season followed by colony death. The colony reached a mature size, was broken, but then was capable of further significant growth which did not affect older portions of the colony. Secondary growth over the surface as a whole appears uncommon and can often be recognized by skeletal discontinuity, either on complete specimens (Fig. 5) or in thin sections (Fig. 11).

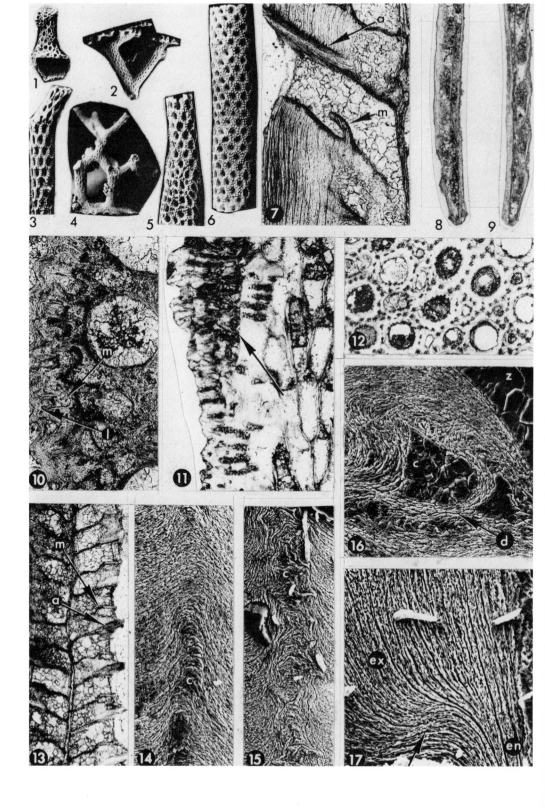

2. The zooecial apertures are of consistent shape and are commonly arranged at the zoarial surface in a rhombic pattern (Fig. 6). This surface expression of regularity presumably is a product of a regular budding pattern, discussed above. Irregularities in form of zooecial arrangement appear to be local and of limited development; they may be microenvironmentally induced.

3. Acanthopores and mesopores are regularly arranged in many species; for example,

---

Figures 1–17

Growth and skeletal features of rhabdomesontid Bryozoa.

1. Base of attachment of an unidentified rhabdomesontid showing the rapid development of the ramose growth mode; Mulde Marl, Silurian, Gotland, Sweden, UI-X4852, ×8.

2. *Acanthoclema* (?) sp. specimen forming a brace-like encrustation on a fenestellid colony; Jefferson Ls. (?), Devonian, Falls of the Ohio, near Louisville, Ky., USNM 178558; ×8.

3. *Saffordotaxis incrassata* (Ulrich) specimen in which the growing tip was lost, and replaced by subsequent growth; New Providence Shale, Mississippian, near Louisville, Ky.; USNM 178563; ×8.

4. Base of attachment and colony fusion in an unidentified rhabdomesontid; the interlocking growth would provide greater strength than one stalk of a single colony; Jeffersonville Limestone (?), Devonian, Falls of the Ohio, near Louisville, Ky.; USNM 178559; ×8.

5, 11. *Saffordotaxis incrassata* (Ulrich) specimen showing: 5, the irregular pattern of secondary growth on the proximal portion of the fragment, ×8; 11, the discontinuity (arrow) between initial and secondary growth in thin section; the discontinuity becomes much less distinct proximally; New Providence Shale, near Louisville, Ky.; UI-X4853; ×30.

6. *Saffordotaxis incrassata* (Ulrich) fragment showing regular mode of growth reflected in zooecial and acanthopore arrangement; New Providence Shale, near Louisville, Ky.; ×8.

7. *Acanthoclema scutulatum* (Hall) showing acanthopore (a) and intraautozooecial (?) mesopore-like structure whose wall (m) does not extend to the zoarial surface; Wanakah Shale, Devonian, New York State; USNM 178562; ×70.

8, 9. *Arthrostylus tenuis* (James) showing typical mode of development of members of the Family Arthrostylidae; Utica Group, Ordovician, Newport, Ky.; ×40.

10. *Clausotrypa conferta* Bassler, transverse section showing shallow metapores (m) closed by thick laminae sequences (1); Permian, Timor; GPI Bonn 21; ×70.

12. *Megacanthopora fallacis* Moore, tangential section showing interautozooecial position of mesopores; Weyland Shale, Pennsylvanian, Texas; KU 58444; ×30.

13. *Acanthoclema scutulatum* (Hall), longitudinal section showing acanthopores (a) and regular arrangement of zooecia and development of the exozonal wall; Wanakah Shale, Devonian, New York State; USNM 178561; ×30.

14. *Goldfussitrypa esthonia* (Bassler), longitudinal section of acanthopore, core (c) passes out of the plane of section distally; Jewe beds, Ordovician, Esthonia; USNM 178557; ×425.

15. *Saffordotaxis incrassata*; longitudinal section of acanthopore, core formed primarily of laminae; New Providence Shale, Louisville, Ky.; ×275.

16. *Goldfussitrypa esthonia* (Bassler), transverse section showing chambers (c) and discontinuity (d) between endozone and exozone; z, zooecial cavity; USNM 178557; ×425.

17. *Saffordotaxis incrassata* (Ulrich), longitudinal section of an area of exozonal (ex) and endozonal (en) wall distal to a zooecial cavity (z) showing discontinuous laminae horizons (arrow); the shape of the exozonal wall is therefore determined by differential secretion; New Providence Shale, Louisville, Ky.; ×405.

Figures 7–12 are light micrographs of thin sections; 13–17 are scanning electron micrographs.

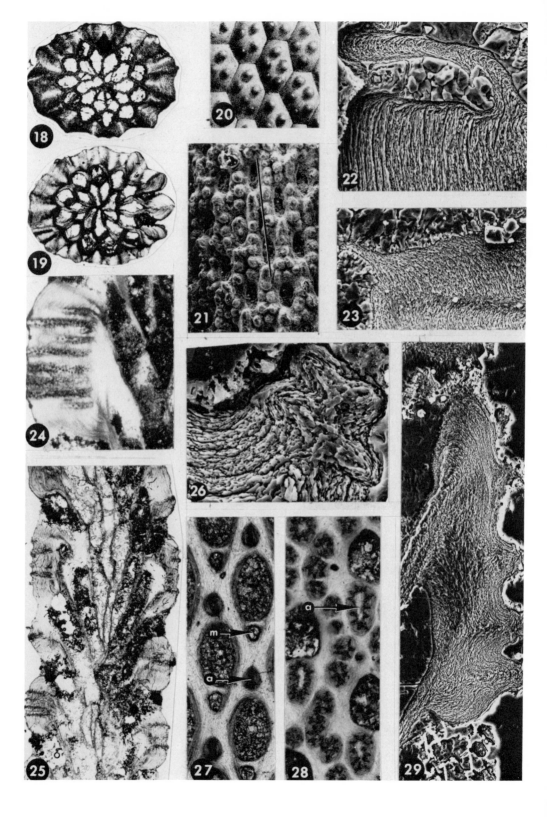

they may be concentrated or occur only proximal to the zooecia (Fig. 27). Enlarged acanthopores may occur only in a single position relative to the zooecia (Fig. 28).

4. Regularity of colony growth is attained early in astogeny. Although bases of attachment are not known for many taxa, rhabdomesontids assumed the ramose growth mode (Fig. 1) after formation of relatively small attachments. Colonies apparently also could support themselves by anastomosing growth patterns near their bases of attachment (Fig. 4), while maintaining the regular growth mode. In the illustrated specimen, three stems arise from an encrusting base. Because no boundaries between colonies are apparent in the base, the specimen appears to represent a single colony; however, preservation is not complete and detailed enough to establish with certainty whether the three stems represent three discrete colonies (that is, derived from three ancestrulas) or whether all originated as separate stalks from a single base of attachment. Branches of different stems grew against and braced one another; the soft tissues of the branches of the separate stems apparently did not merge because distinct discontinuities are present between branches. The crowded growth mode did not affect colony diameter, zooecial shape, and branching angles, all of which remained constant. Certain advantages may have been derived from an anastomosing growth pattern compared to a single stalk; presumably more support would be provided which would allow more extensive branching and larger colonies. A colony apparently also could provide itself with a brace (Fig. 2); during growth, the zoarium illustrated encountered a fenestellid, then formed an anchor but

---

Figures 18–29

Growth and skeletal features of rhabdomesontid Bryozoa.

18, 19. *Saffordotaxis incrassata* (Ulrich) showing cyclic (18) and spiral (19) budding patterns in transverse views; New Providence Sh., Mississippian, near Louisville, Ky.; UI-X-4854, X-4855, ×25.

20. *Steginopora ornata* d'Orbigny, showing protuberances on basal surface; note similarity to acanthopores in Fig. 21; taken from d'Orbigny, 1852, pl. 721; magnification uncertain.

21, 26, 29. Undescribed rhabdomesontid; 21, zoarial surface showing acanthopore development, Fig. 29 lies along a plane comparable to the line drawn on the surface, ×55; 26, acanthopore showing prominent lateral spines, approx. ×1000; 29, longitudinal section showing arrangement of acanthopores and zooecial walls, approx. ×220; Warsaw Shale, Mississippian, near Valmeyer, Ill.

22. *Acanthoclema* sp., longitudinal section showing development of mesopore wall; Wanakah Shale, Devonian, New York State; approx. ×300.

23. *Orthopora* sp., longitudinal section showing development of hemiseptum wall; Keyser Limestone, Silurian–Devonian, Keyser, W. Va.; ×300.

24, 25. *Saffordotaxis incrassata* (Ulrich) showing distribution of acanthopores and arrangement and shape of zooecia; New Providence Shale, Mississipian, near Louisville, Ky., UI-X-4856, 4857; 24, ×40; 25, ×25.

27. *Acanthoclema scutulatum* (Hall), tangential view showing arrangement of zooecia, acanthopores (a), and mesopore-like structures (m); Ludlowville Formation, Devonian, near East Bethany, New York State; USNM 168344, ×100.

28. Undescribed rhabdomesontid, tangential view showing enlarged acanthopores (a) proximal to the zooecia; note the prominent lateral spines on the acanthopores; Warsaw Formation, Mississippian, near Valmeyer, Ill.; USNM 168355, ×100.

Figures 18, 19, 24, 25, 27, and **28** are light micrographs of thin sections; **21–23, 26,** and **29** are scanning electron micrographs.

not an extensive encrustation. Again, the regular zoarial shape was maintained and the colony was provided with support at the expense of the growing tip. Although zooecia were formed in the overgrowth, these zooecia appear too shallow to have supported polypides.

ACANTHOPORES, MESOPORES, AND THEIR FUNCTIONS

Acanthopores and mesopore-like depressions are the larger differentiated structures in addition to the autozooecia present within rhabdomesontid zoaria; both have been interpreted as representing polymorphic zooids. Areas of differentiated zooecia, such as monticules, appear to be lacking in these Bryozoa. Acanthopores (Figs. 7, 13–15, and 24–29) occur in several morphologies in rhabdomesontids, but all are formed in a pattern of outwardly directed cone-in-cone or sheath-like laminae enveloping an axial structure, or core. The core is built of more or less complexly intergrown laminae and calcite crystals. Inferences on acanthopore function frequently have been based in part on the assumption that the acanthopore core was unoccupied by skeletal material during colony life. Many acanthopores, however, could not have contained a continuous axial opening because laminae are present intermittently across the core axis (Figs. 14 and 15). Others contain a continuous central axial space crossed by few if any laminae. In fossils, this space generally is filled by a single calcite crystal (or very fine grained crystal aggregate). For a number of reasons (Armstrong, 1970; Brood, 1970), the core appears to have been solid during colony life; even if cores skeletally open during colony life existed, however, the very narrow diameter of the core in rhabdomesontid acanthopores (mostly 0.003 to 0.015 mm) would seem to preclude the extensive development of differentiated organs.

Acanthopore function remains uncertain. Their prominence has long suggested some protective function to various authors, whereas more recently Tavener-Smith (1969) suggested that acanthopores may have served to help stabilize the soft tissues over the surface of the colony. Harmer (1902a, 1902b), in discussing the morphology of the modern cheilostome *Euthyrisella* (called *Euthyris* by Harmer), as well as that of various Cretaceous cheilostomes, described calcareous papillae on the surface of *Euthyrisella obtecta* which serve, along with the margins of the orifices of the zooecia, to support an epitheca above and below the surface of the zoarium. This epitheca encloses spaces above the surface of the zoarium. Figure 30, modified from Harmer (1902b), illustrates the arrangement of the structures. Harmer notes the presence of similar papillae on certain Cretaceous specimens (Fig. 20) and suggests that these species may also have had such an epitheca. He suggests that the function of the suspended epitheca may have been "to protect the calcareous walls from the attacks of boring organisms (e.g., the Infusorian *Folliculina*) which infest many calcareous Polyzoa." The papillae on some of the Cretaceous species are very similar in appearance to certain acanthopores in rhabdomesontids (compare Figs. 20 and 21). As discussed below, a soft tissue layer had to cover the surface of the rhabdomesontid zoarium; it would seem possible, in spite of the presumably relatively distinct history of cheilostomes and rhabdomesontids, that the acanthopores performed combined support and protective functions similar to those suggested for the papillae on the surface of *Euthyrisella*. In the reconstructions

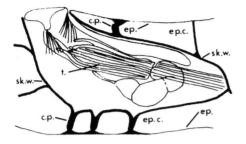

Figure 30

Euthyrisella obtecta (Hincks). Longitudinal section showing positional relationships among epitheca (ep.), epithecal cavity (ep. c.), skeletal wall (sk. w.) and the calcareous papillae (c.p.); t., tentacles. (After Harmer, 1902a.)

(Fig. 31) the soft surficial tissues are therefore shown suspended from the tips of the acanthopores.

This function for acanthopores would explain the prominent lateral spines along the axes of many acanthopores (Fig. 26) as well as the fact that in certain acanthopores, the zoarial laminae and the acanthopore axial laminae may be complexly intergrown (Fig. 33). The lateral spines would have functioned to provide greater stability to the soft tissues; the intergrown laminae would be a result of intermittent growth of a structure not developed as a distinct polymorph but rather as a part of the general zoarium, formed by essentially the same tissues which secreted the remainder of the interspace skeleton. Acanthopores which branch have been observed in some zoaria; this development might be explained by the need for more support of the surficial tissue in an expanding colony.

Although most readily applied to acanthopores of the general morphology of those of *Saffordotaxis incrassata* and the undescribed species of Figs. 29 and 31A, this interpretation of function can also be applied to other acanthopore morphologies. The low, rounded acanthopores present in *Goldfussitrypa* (Figs. 32 and 37) can be reconstructed to provide support for a relatively low surficial tissue layer and a relatively thin hypostegal coelom (Fig. 31B). The core of the *Goldfussitrypa* acanthopore consists of alternate intervals of laminae and enlarged single crystals (or crystalline aggregates) (Fig. 14). If the larger crystal marks the site and time of tissue attachment, then the intermittent development of layers might mark intermittent times of attachment.

Long prominent acanthopores, such as those of *Acanthoclema* (Fig. 35), present more of a problem in interpretation. If the surficial tissue was to provide an enlarged hypostegal coelom for protection and communication, then this tissue could not have adhered closely to the steep acanthopore sides. Harmer (1902b) describes the epitheca in *Euthyrisella* as being stretched flat between supports; this arrangement is used as the analogy for the reconstructions of Fig. 31C and D. The *Euthyrisella* model leads to seemingly highly exposed surficial tissue (Fig. 31D) and therefore this reconstruction is more suspect, although the tissues would be only about 0.2–0.3 mm above the general zoarial surface (based on the prominence of the acanthopore shown in Fig. 35). The elongate single crystal core (or crystalline aggregate) again, as in *Goldfussitrypa*, may mark the site

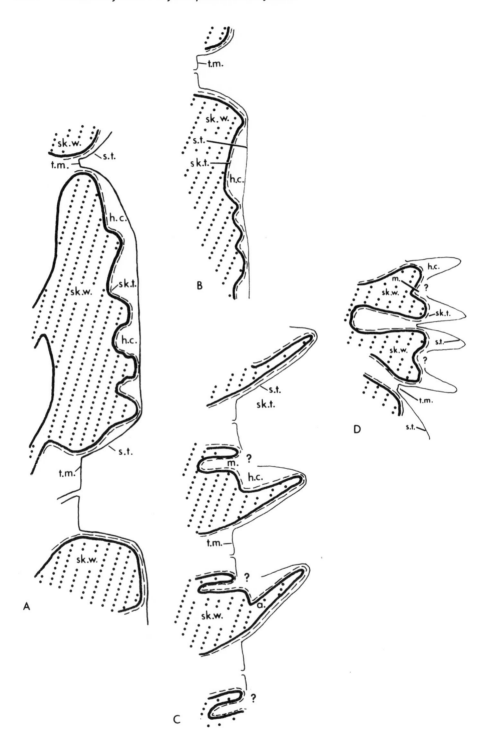

of tissue attachment; variation in core diameter, either within an acanthopore (Blake and Towe, 1971) or among acanthopores of a single zoarium, might mark periodic variation in attachment size.

Function of the mesopore-like structure in *Acanthoclema* is strictly conjectural; this structure may or may not have had an opening to the environment through the surficial tissue, and therefore a question mark is placed in the surficial tissue above the structure (Fig. 31C, D).

Explanation for the differences in acanthopore morphologies as support mechanisms requires more study. The acanthopores of *Saffordotaxis incrassata* morphology are more numerous and complexly spined; such a morphology suggests that these acanthopores would provide greater support than could be provided by the elongate smooth acanthopores of *Acanthoclema*. Perhaps rhabdomesontids with acanthopores of the *S. incrassata* prominence and morphology lived in higher energy environments than those with the *Acanthoclema* morphology and prominence; such a hypothesis could be partially tested with environmental studies. Acanthopore morphology variability presumably also could be a response to such probably untestable factors as predator and parasite behavior.

The suggested pattern of surficial tissue development would have served to partially isolate neighboring polypides and therefore possibly reduce interference during feeding.

Figure 31

Inferred relationships among certain hard and soft parts in rhabdomesontid Bryozoa. 31A, undescribed species, reconstruction from Fig. 29, along a line of section which is approximately comparable to the line drawn on the zoarial surface in Fig. 21. Surficial tissues (s.t.; cuticle, epidermis, and peritoneum of Ryland, 1970) are inferred to have been supported by the tips of the acanthopores above the general zoarial surface. The tips of two acanthopores (those immediately distal to the proximal zooecium) lie in the plane of section; the surficial tissues are approximately tangent to the acanthopore tips. The tips of the next two acanthopores lie beyond the plane of section; therefore the contacts between surficial tissue and the acanthopores are not visible. The surficial tissue extends toward the distal zooecial aperture just beyond the tissue contact with the distal acanthopore. Proximal to the proximal aperture, the surficial tissue and the supraskeletal tissue layer (sk. t., epidermis and peritoneum) are reconstructed passing over an enlarged acanthopore (examples are visible in Fig. 21, but not in the section of Fig. 29). The plane of section passes through the center of the proximal zooecial cavity; the terminal membrane (t.m.) is placed at the narrowest zooecial diameter. Only the edge of the distal zooecial aperture lies in the plane of section; a small portion of the edge of the terminal membrane is shown. h.c., hypostegal coelom. 31B, *Goldfussitrypa esthonia* (Bassler), reconstruction from Fig. 37. 31C, *Acanthoclema scutulatum* (Hall) reconstruction from Figs. 13 and 35 and an unillustrated specimen; the surficial tissues are shown passing over the tip of the acanthopore, then extending toward the zooecial aperture. Because the function of the mesopore-like structure is uncertain, a question mark is placed above the structure in the position of the surficial tissue. 31D, *Acanthoclema alternatum* (Hall) reconstructed from the transverse section of Fig. 36; the surficial tissues are reconstructed well above the general zoarial surface. The difference in elevation between the acanthopore tips and the apertures leads to a highly irregular surface; the surfaces between points of support is reconstructed as having been flat. This model follows the development in *Euthyrisella obtecta* (Hincks), in which the epitheca is stretched flat between supports, well above the zoarial surface. In Fig. 31D, the acanthopore tips do not lie in the plane of section and therefore are not illustrated.

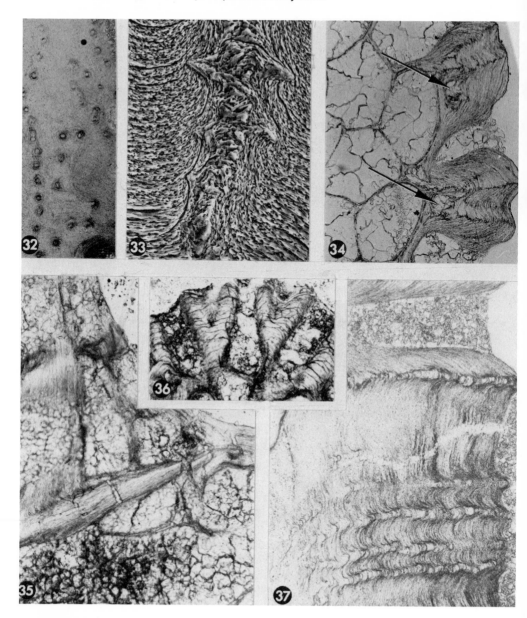

Figures 32–37

Growth and skeletal features of rhabdomesontid Bryozoa.

32, 34, 37. *Goldfussitrypa esthonia* (Bassler); 32, tangential section showing acanthopore development and distribution, ×50; 34, transverse section showing development of chambers (arrow) at the endozonal–exozonal boundary, ×100; 37, longitudinal section showing acanthopore development, ×100; Jewe beds, Ordovician, Esthonia,; USNM 168363.

33. *Saffordotaxis incrassata* (Ulrich), longitudinal section of acanthopore showing nature of laminae development; New Providence Shale, Mississippian, near Louisville, Ky.; approx. ×550.

Mesopore-like structures, chambers smaller than autozooecia into which the zoarial laminae are directed, are a second group of larger differentiated structures present in rhabdomesontids. As noted above, the Family Hyphasmoporidae is distinguished by the presence of "metapores," depressions which more or less completely fill the interspaces. They typically extend through the exozone, as in *Ogbinopora* Shishova (1966, pl. 1, fig. 3). In *Clausotrypa conferta* Bassler, however, metapores were intermittently closed with thick sequences of laminae (Fig. 10).

Mesopore-like structures are apparently rare in the Family Rhabdomesidae. Those of *Acanthoclema* (Fig. 13) have an intraautozooecial position and possibly origin. Laminae orientation of the mesopore wall and of the zoarial wall near the base of the mesopore is very similar to the orientation of the laminae in proximal hemisepta of *Orthopora* (Figs. 22 and 23). In addition, in some *Acanthoclema* autozooecia, the mesopore wall does not extend to the zoarial surface (Fig. 7). These structures in *Acanthoclema* therefore are positioned within the zooecium and may have been phylogenetically of intraautozooecial origin, created by deflection of the direction of hemiseptum growth.

Mesopores of a second general morphology occur in *Megacanthopora* and *Idioclema*, two genera which have been assigned to the Rhabdomesidae (Bassler, 1953) but not included in any of the rhabdomesontid families by Shishova (1968). Mesopores in these genera appear to be interautozooecial in position (Fig. 12) and presumably could have originated phylogenetically from autozooids. In addition, however, they may have originated both phylogenetically and ontogenetically as open pockets left in the skeleton by intermittent secretion of skeletal material or by changes in orientation of the secreting surface during colony growth. Such an origin is suggested by closed chambers which are present within the skeletal wall of *Goldfussitrypa* (Figs. 16 and 34) between the endozonal and exozonal laminae. Such openings, if not subsequently closed by laminae, would yield mesopore-like open structures.

Evidence bearing on the function of mesopore-like structures is difficult to obtain. The distribution of metapores in the Hyphasmoporidae suggests they may have originated as strengthening devices (Shishova, 1968). Boardman (in Boardman and Cheetham, 1969, pp. 213–214) suggested mesopores in general may have originated as structures which fill space yet conserve material, rather than as zooecia for heterozooids. The irregular ridges which typically separate neighboring metapores in the Hyphasmoporidae also may have functioned as a support system for a suspended epitheca, much as the acanthopores are suggested to have done. Acanthopores are not present in all members of the metapore-bearing Hyphasmoporidae; the metapores may have obviated the need for acanthopores by providing the same support and isolating functions.

The presence of other skeletal structures in rhabdomesontid Bryozoa has been noted (for example, "capillaries"; Shishova, 1968) but such structures appear too small to have supported heterozooids. There is no skeletal structure in Bryozoa of the Order Rhabdomesonata which necessitates the existence of highly differentiated heterozooids.

---

35. *Acanthoclema scutulatum* (Hall), longitudinal section showing prominent nature of acanthopore under overgrowth; Wanakah Shale, Devonian, New York State, USNM 168340, ×150.

36. *Acanthoclema alternatum* (Hall), transverse view used as basis of reconstruction of Fig. 31D; NYSM 579; ×100.

Figure 33 is a scanning electron micrograph; all others are light micrographs.

POSITIONS OF CERTAIN SOFT TISSUES

As described above, the tissues which secreted the skeleton apparently were present over the surface of the zoarium; the colony would have required some means of supplying nutrients to these tissues. Borg (1926) described a complex "double-walled" pattern of soft tissue development in certain modern cyclostomes. In these Bryozoa, the zoarial surface is covered by two tissue layers, an inner skeleton secreting layer and an outer layer at the colony surface; the two layers are separated by a coelomic space which thus extends under the surface of the colony. This double-walled arrangement would appear to provide a reasonable mechanism for maintenance of continued coordinated growth through colony life in rhabdomesontids and is the basis for the reconstruction of the position of soft tissues over the zoarial surface shown in Fig. 31.

Position of the polypide within the zooecium is difficult to ascertain. Studies of zoaria with the scanning electron microscope have not yet revealed differentiated structures suggestive of attachment areas. As has been noted by Ryland (1970), the orifice in Bryozoa which have been assigned to the Cryptostomata, including the rhabdomesontids, has been inferred by some authors to have been well within the zoarium at the proximal end of the vestibule. Hemisepta, present only in some genera, have been considered to mark the position of the orifice. As Ryland pointed out, there is no apparent evidence for such an orifice position, which would require the tentacles to move a considerable distance during extrusion. Further, the endozonal interval of many rhabdomesontid zooecia is very short, ½ mm or less (Fig. 13). Borg (1926, p. 190) states that polypides in cyclostomes average 1 mm in length; the space available in the endozone of many rhabdomesontids appears limited, although Hyman (1959) reports modern zooecia in a size range under ½ mm. Apparently no living cyclostome has been described in which the orifice is positioned deep within the zoarium. The hemisepta, rather than functioning as a part of the orifice, may have provided attachment surfaces or marked the base of the mature zooid. Although the surficial tissue is reconstructed at the tips of the acanthopores (Fig. 31), the orifice is not placed at this level for several reasons: (1) There appears to be no described example of a cyclostome with such a suspended orifice; (2) if the terminal membrane were located at the level of the distal surfaces of the acanthopores, support problems for the vestibule and tentacle extrusion mechanisms would appear to arise because the skeleton would not confine the zooid and therefore could not function as a rigid chamber to increase pressure for tentacle extension; and (3) by analogy with the cheilostome *Euthyrisella,* the tissues are drawn approaching the level of the skeletal aperture; such a position presumably would help support and provide rigidity for the soft tissues of the rhabdomesontids, as Harmer suggested for *Euthyrisella.* In the reconstructions, the terminal membranes are therefore somewhat arbitrarily placed at the level of the narrowest diameter of the exozone, which would require the smallest possible and therefore presumably least vulnerable terminal membrane.

SUMMARY

Bryozoa of the Order Rhabdomesonata display considerable intraspecific regularity of overall form and zooecial arrangement, suggesting a high degree of genetic control over growth; rhabdomesontid colonies developed as entities rather than as many individual zooecia growing independently of one another. Regularity is expressed in such features as budding pattern, endozonal–exozonal relationships, and colony shape. The extensive interspaces between adjacent zooecia requires the existence of soft tissues over the zoarium surface to account for skeletal secretion. The double-walled construction described in certain cyclostomes by Borg (1926) provides a suitable model to explain this secretion. Evidence derived from fossils bearing on terminal membrane position is lacking; the terminal membrane is reconstructed near the zooecial aperture primarily by analogy with living cyclostomes. Larger secondary structures are of two types in rhabdomesontids, acanthopores, and mesopore-like structures. A possible functional explanation for both structures can be based upon morphology of the extant cheilostome *Euthyrisella obtecta*. Specimens of this species possess a surficial layer of soft tissue over the zoarial surface; this tissue is supported by the zooecial orifices and calcareous papillae. Acanthopores may have functioned as supports for an analogous surficial layer; the rims of mesopore-like structures also may have served such a function, as well as providing a space-filling function.

ACKNOWLEDGMENTS

The scanning electron microscope used in the study, a Cambridge Steroscan Mk IIa, is located in the Central Electron Microscope Laboratory of the University of Illinois. It was acquired with support of the National Science Foundation (GA-1239), the U.S. Public Health Service (FR-07030), and the University of Illinois Research Board. Figured specimens are from the collections of the University of Kansas Museum of Invertebrate Paleontology (KU); the Geology-Palaeontology Institute of the University of Bonn, Germany (GPI Bonn); the New York State Museum (NYSM); the U.S. National Museum (USNM); and the Department of Geology of the University of Illinois at Urbana-Champaign. The writer is grateful to R. S. Boardman and A. H. Cheetham of the U.S. National Museum for many helpful comments.

REFERENCES

Armstrong, J. 1970. Zoarial microstructures of two Permian species of the bryozoan genus *Stenopora*. Palaeontology, *13*: 581–587.

Astrova, G. G., and I. P. Morozova. 1956. Systematics of bryozoans of the Order Cryptostomata. Doklady Adak. Nauk SSSR, n. s., *110*(4): 661–664 (in Russian).

Bassler, R. S. 1953. Bryozoa, p. G1-G253, *in* R. C. Moore, ed., Treatise on Invertebrate Paleontology. Geol. Soc. America, and Kansas Univ. Press.

Blake, D. B., and K. M. Towe. 1971. Acanthopore ultrastructure in the Paleozoic bryozoan *Idioclema insigne* Girty. Jour. Paleont., *45*(5): 913–917.

Boardman, R. S., and A. H. Cheetham. 1969. Skeletal growth, intracolony variation, and evolution in Bryozoa: a review. Jour. Paleont., *43*(2): 205–233.

Borg, F. 1926. Studies on Recent cyclostomatous Bryozoa. Zool. Bidrag. Från. Uppsala, *10*: 181–507.

Brood, K. 1970. On two species of *Saffordotaxis* (Bryozoa) from the Silurian of Gotland, Sweden. Stockholm Contrib. Geol., *21*: 57–68.

Harmer, S. F. 1902a. On the structure and classification of the cheilostomatous Polyzoa. Cambridge Phil. Soc. Proc., *11*: 11–17.

————. 1902b. On the morphology of the cheilostomata. Quart. Jour. Micros. Sci., n.s., *46*: 263–450.

Hyman, L. H. 1959. The Invertebrates, v. 5, The Smaller Coelomate Groups. McGraw-Hill, New York. 783 p.

d'Orbigny, A. 1852. Paléontologie française. v. 5, Terrains crétacés. Bryozoaires. 1192 p.

Ryland, J. S. 1970. Bryozoans. Hutchinson & Co., London. 176 p.

Shishova, N. A. 1966. Systematic position and scope of the Family Hyphasmoporidae. Internat. Geol. Rev., *8*(1): 64–70 (transl. from Paleont. Zhur.).

————. 1968. New order of Paleozoic bryozoans. Paleont. Jour., 2: 117–121 (trans. from Paleon. Zhur.).

Tavener-Smith, R. 1969. Wall structure and acanthopores in the bryozoan *Leioclema asperum*. Lethaia, 2: 89–97.

Variation in the Bryozoan *Fistulipora decora* (Moore and Dudley) from the Beil Limestone of Kansas

Jack D. Farmer   University of Kansas

Albert J. Rowell   University of Kansas

ABSTRACT

The study is based on five selected characters, each character being measured on 25 randomly selected zooecia in comparable ontogenetic and astogenetic stages from randomly chosen colonies at each of four localities. Data are typically normally distributed for four of the five characters, but the colony variances are not homogeneous when data for a given character are considered for all localities. This failure of the assumptions inherent in a nested analysis of variance would lead to grossly misleading conclusions if the technique were applied to the present data.

Nonparametric tests show that there are significant differences between localities for four of the five characters. These differences are rather uniformly distributed among localities and characters. They are believed to be caused by small differences in the average genetic composition of the populations from different localities, but the effects of differences in "gross" environment may have had some influence.

Highly significant differences exist within all four localities, as shown by analysis of variance. Differences are not confined to one or two colonies; this, combined with relative uniformity of "gross" environment implied by field evidence, suggests that the differences are caused by a high degree of genetic diversity between colonies at one locality. Comparison of the intercolony and intracolony components of variance

Authors' present addresses: J.D.F., Department of Geology, University of California, Davis, California 95616; A.J.R., Department of Geology, University of Kansas, Lawrence, Kansas 66044.

suggests that the latter typically accounts for a larger proportion of the total variance than the former. This is believed to reflect the strong influence of microenvironmental factors on the phenetic expression of zooecia within a single colony.

INTRODUCTION

In the past decade there has been renewed interest in the study of Bryozoa. Numerous authors have contributed to our understanding of the paleobiology of these animals, and, although much still remains to be learned, various aspects of growth have been explored and our conceptual understanding of the causes of variation have been improved (e.g., Boardman and Cheetham, 1969; Boardman, Cheetham, and Cook, 1970). Simultaneously, there has been an increased usage of biometrical techniques in bryozoan investigations, both as descriptive tools and in testing for significant differences between colonies. This field has recently been reviewed by Anstey and Perry (1970).

The principal objective of the present study involves joint consideration of both of the above aspects: to assess how variation is distributed in a number of characters of a particular species of bryozoan from one horizon over a restricted region, to determine how much of the total variation is contributed by variation within a colony, how much by variation between colonies at one locality and to ascertain whether significant differences exist between localities. As expressed, the problem clearly requires a statistical approach and an ordered sampling plan, but the general conclusions have an equally obvious application in any conventional systematic study of comparable colonial organisms. Related secondary objectives of the investigation are an assessment of character correlation and the degree of redundancy in the data, together with an evaluation of the limitations imposed on this type of investigation by the methods utilized.

MATERIAL AND METHODS

The choice of a species for study was pragmatic; we wished to keep the model as simple as possible and confine our investigation to zooecia in the same ontogenetic and astogenetic stages. Individual zooecia of *Fistulipora decora* (originally *Cyclotrypa decora* Moore and Dudley, 1944, p. 275) are well suited for this purpose because they rapidly attain a stable adult form in the lenticular to subhemispherical zoaria, thus diminishing the possibility of inadvertently including measurements of ontogenetically immature individuals (Fig. 1).

F. decora is a relatively abundant faunal element of the Beil Limestone Member of the Lecompton Limestone (Virgilian) of northeastern Kansas. In addition, the stratigraphy of the Beil Limestone is known through the work of Brown (1958). Although one could not claim exact correlation from one locality to another, the units sampled are certainly of closely comparable age, since the total thickness of the member is usually less than 3 m.

At the outset, consideration must be given to the sampling plan for any study of this type, for although differing statistical techniques may have somewhat different underlying assumptions, all are predicated on the assumption of random sampling. As is well known in geological situations (Krumbein and Graybill, 1965), very rarely is the target

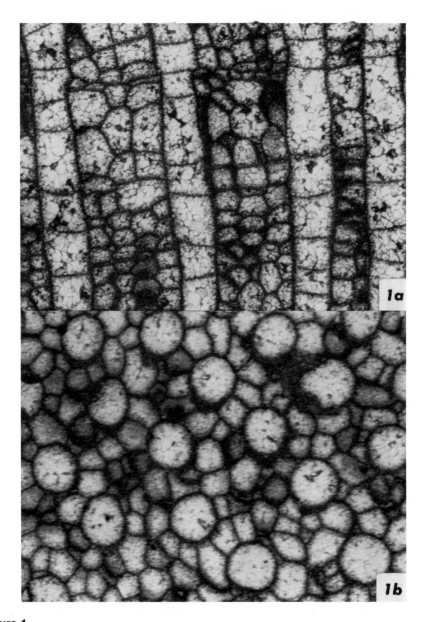

Figure 1

a. Longitudinal acetate peel of *Fistulipora decora*. Beil Limestone Member, Lecompton Formation. Stull locality. KU. 72000. ×35. b. Transverse acetate peel of *F. decora*. Beil Limestone Member, Lecompton Formation. Stull locality. KU. 72686. ×35.

population (in this case, all specimens of *F. decora*) available for sampling. The choice of collecting localities is not random but is determined by the available exposures. Consequently, the statistical inferences apply only to the available population, defined as all well-preserved specimens of *F. decora* exposed on selected bedding planes at selected

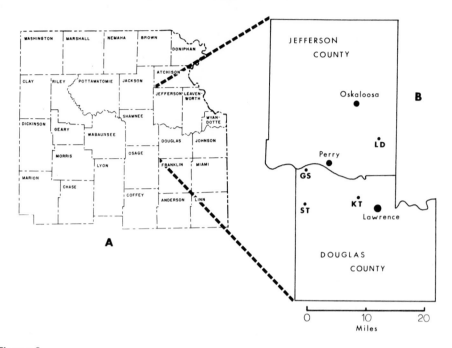

Figure 2

Collection localities in eastern Kansas: LD, Lake Dabinawa; ST, Stull; KT, Kansas Turnpike; GS, Grover Station. These abbreviations are also used in Tables 3 and 5 and Fig. 8.

localities. However, since there is no *a priori* reason for believing that the available population of *F. decora* should be significantly different from the target population, it becomes possible to extend the conclusions derived from the available population to the target population by substantive geological argument (Krumbein and Graybill, 1965). It must be recognized, however, that the statistical conclusions apply rigorously only to the available population, and their extention, although reasonable, has no statistical basis.

Collections from the Beil Limestone Member were made at four localities in northeastern Kansas (Fig. 2), the localities forming a rough parallelogram with sides approximately 15 km in length. The four localities are referred to in subsequent discussion as Stull Road, Grover Station, Lake Dabinawa, and Kansas Turnpike. At each locality, specimens were collected *in situ* from a single bedding surface in order to avoid the inadvertent mixing of material foreign to the chosen horizon. Specimens embedded in matrix and with the growth surfaces of the lenticular zoaria oriented upward were assumed to be *in situ*. The assumption is at least reasonable, because most of the associated brachiopod fauna is unworn and consists of complete bivalved shells.

As much material as was feasible, within the bounds of reasonable expenditure of time and money, was collected from each locality. Subsequently, it was found that many of the specimens were unusable due to poor preservation, mainly a consequence of dolomitization or secondary recrystallization of calcite. Of the available specimens, five colonies were chosen randomly from each locality, using a random number table.

One of the four localities sampled (Kansas Turnpike) failed to provide the desired number of usable specimens, and in this particular instance only four colonies were analyzed.

Measurements

The acetate peel technique outlined by Boardman and Utgaard (1964) was employed. It was desirable to evaluate the significance of distortion introduced during the process of removing an acetate replica from a specimen. Measurements of an arbitrarily chosen colony dimension were made directly from a specimen and compared to measurements of the same dimension taken from an acetate peel. Statistical analysis of the data using a simple t-test revealed no significant differences between the two sample means at the $\alpha = 0.001$ level of significance.

Measurements were made from the acetate peels by projecting character images at a known scale ($\times 49.3$) through a standard petrographic microscope onto a sheet of white tracing paper. Characters were measured directly from the projected image using a pair of Helios calipers (J and S Precision Scientific Measuring Instrument Company, Brooklyn, N.Y.), graduated to 1/20 mm.

Twenty-five measurements for each character for each colony were taken along randomly chosen traverses utilizing a calibrated mechanical stage. Traverse coordinates were chosen from a random number table, recorded, and each value set on the appropriate scale of the calibrated stage. The traverse was carried out and as many measurements as possible were made. If, after completing a traverse, less than the desired number of measurements had been obtained, a new set of traverse coordinates was chosen in an identical manner, and the process repeated until the required number was recorded. Traverses were consistently carried out in the same direction in order to avoid the possible introduction of bias by making arbitrary choices during the data-gathering process.

Choice of Characters

Owing to the relatively simple structural morphology of fistuliporoid bryozoans, only a modest number of phenetic characters are available for study. This investigation is based upon five characters illustrated diagrammatically in Fig. 3 (cf. Fig. 1a and b).

In tangential sections, zooecial diameter (ZD) in millimeters was determined as the minimum distance between zooecial walls. The interzooecial distance (IZD) is the distance between nearest-neighbor zooecia in millimeters, as measured in tangential section. Related to this character is the number of vesicles between nearest-neighbor zooecia (VCT), also measured in tangential section. In longitudinal section, two characters were obtained: the number of diaphragms in a distance of 1 mm (DC/MM), and the number of complete vesicles enclosed in a circle of radius 0.25 mm (VC/0.25).

RESULTS

Twenty-five measurements were obtained for each of five characters from a total of 19 colonies representing four localities. The raw data are given in Appendix 2 of Farmer (1971), and copies may be obtained as computer tabulation from the authors.

The nature of the principal question posed—how is the variation distributed—suggests

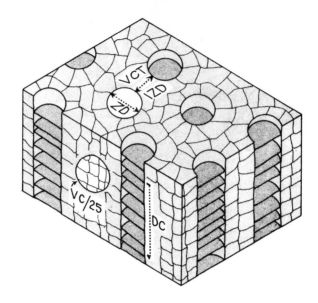

Figure 3

Diagrammatic representation of a fistuliporoid bryozoan showing the five characters utilized in the study. For tangential sections they include: zooecial diameter (ZD), interzooecial distance for nearest-neighbor zooecia (IZD), and counts of the number of vesicles between nearest-neighbor zooecia (VCT). For longitudinal sections they include diaphragm counts per millimeter (DC/MM) and the number of complete vesicles in circle of radius 0.25 mm (VC/0.25). These abbreviations are also used in Tables 1, 2, 5, and 8.

that the data should be analyzed as a nested analysis of variance. However, parametric analysis of variance (anova) makes certain assumptions about the data that must be met before its use can be regarded as appropriate or the tests exact. These assumptions are discussed in detail by Sokal and Rohlf (1969); they include the need for the error term to be a normally distributed, independent variable, and the variance of the samples to be equal (homoscedastic).

The consequences of failure to satisfy the assumptions inherent in analysis of variance may be serious (Bradley, 1968), particularly if the variances of the samples are heteroscedastic and only a few degrees of freedom are involved. The data should always be tested to see if use of the method is justified.

The Kolmogorov–Smirnov test for goodness of fit showed that for all 19 colonies the data are normally distributed ($P = 0.05$) for zooecial diameter (ZD), minimum interzooecial distance (IZD), the number of vesicles per unit area (VC/0.25), and the number of diaphragms per millimeter (DC/MM). Only one character, the number of vesicles between nearest-neighbor zooecia (VCT), deviates consistently from normality. Values of d_{max} for this character are all significant at $P < 0.01$. This is not surprising due to the small number of classes involved (counts ranged from 0 to 2) and the relatively low frequencies in classes 0 and 2 for most colonies. For the data as a whole the required equality of variances does not exist; the variances are markedly heteroscedastic for the four characters studied (VCT was not tested as it had previously failed the test for normality). None of the transformations employed succeeded in solving this problem of inhomogeneity of variances. Consequently, the nested analysis of variance could not be employed.

Between Locality Variance

The significance of variation between localities was tested using the Kruskal–Wallis nonparametric test (Table 1). For purposes of comparison only, a nested anova was carried out for the four normally distributed characters (Table 2). Comparison of the two tables reveals the serious limitations imposed on the parametric analysis of variance of these data by deviations from homoscedascity. At the highest level in the nested anova (Table 2), between-locality effects are all seemingly nonsignificant (at $P = 0.05$) for all characters. Yet the Krusal–Wallis test, which is less powerful than an analysis of variance when all the assumptions of the latter are met, consistently shows highly significant differences between localities ($P < 0.005$). This example reemphasizes the importance of testing the assumptions of anova; failure to meet them may give rise to spurious F values and subsequent gross misinterpretations.

The results in Table 1 give no indication how the significant differences between localities are distributed. This may be overcome by performing a Krusal–Wallis test for all possible pairs of localities for each character (Table 3). These tests reveal that the differences arise not by one locality differing consistently from the remainder but by a rather uniform distribution of the differences. The Stull and Kansas Turnpike localities both differ significantly from other localities in eight of a maximum of fifteen character/locality combinations (each locality is compared with three others, for five characters). Grover Station and Lake Dabinawa differ similarly six out of a maximum of fifteen combinations. With the exception of the character "number of vesicles between nearest-neighbor zooecia," all the localities commonly differ significantly from each other. Clearly they do not belong to one statistical population. However, Dice diagrams of the normally distributed characters (Figs. 4–7) show no clearly marked discontinuities between localities. A better overall impression based on simultaneous consideration of all five characters can be obtained using principal-components analysis and projecting the colony means into the reduced character space defined by the first three principal components (Rohlf, 1968). This technique has previously been used in paleontological work by Kaesler (1970) and Rowell (1970), both of whom provide more detailed accounts of the method. Reducing the dimensionality of the data inevitably introduces some distortion; this is often modest and its extent is always known. Moreover, the distortion is not uniformly distributed: the small phenetic distances are more heavily distorted; the larger ones, giving the overall view of phenetic relationships, suffer least. The amount of distortion introduced in the present model is very small (Table 4).

This model (Fig. 8) reveals no marked tendency for colonies to cluster together by locality. The lack of any clearly defined discontinuities suggests that although the colonies from the four localities are not part of a single statistical population, nonetheless they are conspecific. Several hypotheses may be advanced to explain the significant differences that exist between the localities. The samples could be from four relatively localized populations, the morphological differences being reflections of differences in average

Table 1

Results for the Nonparametric Kruskal–Wallis Test, Between Localities

Character[a]	Kruskal–Wallis
ZD	$P < 0.005$
IZD	$P < 0.005$
VCT	Not significant
VC/0.25	$P < 0.005$
DC/MM	$P < 0.005$

[a]Characters abbreviated as in Fig. 3.

Table 2

Results of Nested Analysis of Variance for Four Characters

Source of Variation	F ratio for each character[a]			
	ZD	IZD	DC/MM	VC/0.25
Among localities	0.6276[b]	0.8588[b]	0.4312[b]	1.3164[b]
Within localities	43.2857[c]	11.0138[c]	16.3007[c]	15.0016[c]

[a]Characters abbreviated as in Fig. 3.
[b]Not significant.
[c]$P < 0.001$.

genetic composition built up by the effects of isolation. The short distances between localities and the presence of the species in most outcrops of the Beil Limestone member suggest, however, that prolonged isolation of populations is unlikely. Alternatively, small-scale geographic variation in environmental factors may be responsible for the differences, either producing differential selection, or acting directly on the phenotype without causing notable changes in the genetic composition of the populations. Another explanation may be that the differences are primarily genetically controlled and are a reflection of slight differences in geologic age. It is not, and probably never will be, possible to choose between the latter three potential mechanisms; indeed, all three may have operated.

Within-Locality Analysis of Variance

Although the assumptions for a nested analysis of variance are not met, the conditions for a single-way analysis of variance are satisfied at some localities for all characters except "number of vesicles between nearest-neighbor zooecia." This analysis was performed for each character/locality combination where appropriate (Table 5), and both the Kruskal–Wallis anova analog and modified Snedecor test for the equality of means (Sokal and Rohlf, 1969, p. 376) were employed for all character/locality combinations. The three methods consistently reveal highly significant differences between colonies at each locality for the four characters examined.

In the 11 cases for which it was possible to run a single-way analysis of variance,

Table 3

Kruskal–Wallis Test Between All Possible Pairs of Localities[a]

	ST	GS	LD	KT
Character: Zooecial Diameter (ZD)				
ST	ns[b]			
GS	c	ns		
LD	ns	c	ns	
KT	c	ns	c	ns
Character: Interzooecial Distance (IZD)				
ST	ns			
GS	d	ns		
LD	c	ns	ns	
KT	c	ns	ns	ns
Character: No. Diaphragms/mm (DC/MM)				
ST	ns			
GS	ns	ns		
LD	ns	ns	ns	
KT	c	e	d	ns
Character: No. Vesicles/Unit Area (VC/0.25)				
ST	ns			
GS	ns	ns		
LD	c	c	ns	
KT	c	c	ns	ns
Character: No. Vesicles Between Nearest-Neighbor Zooecia (VCT)				
ST	ns			
GS	ns	ns		
LD	ns	ns	ns	
KT	ns	ns	ns	ns

[a]Localities abbreviated as in Fig. 2.
[b]Not significant.
[c]$0.005 > P$.
[d]$0.01 > P > 0.005$.
[e]$0.05 > P > 0.01$.

one may analyze the data still further. Recall that this technique enables the variance to be partitioned into two parts, that due to variation within colonies and a component caused by variation between colonies at one locality. In all cases but one, the estimate of the within-colony variance is greater than the estimate of the between-colony variance (Table 6). One can claim only that the unbiased estimates of these variances are so related, their real values are unknown. Because 10 of the 11 pairs of variances have this relationship, one may feel some confidence in making the generalization that in

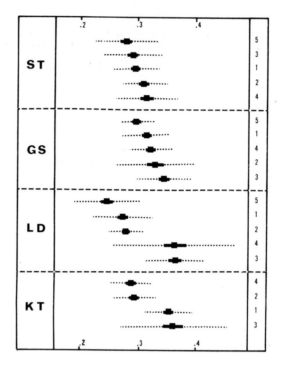

Figure 4

Dice diagrams for zooecial diameter for each colony, dimensions in millimeters. Black square is location of mean, black bar is 95 percent confidence limit of mean, and broken bar is 95 percent confidence limit of character for colony.

Table 4

Number of characters	% Variance explained by first three principal components	Correlation between distances of all possible pairs of OTUs in n space and three-principal-component space
5	93.84	0.993

F. decora the within-colony variance is typically greater than the between-colony variance for any one locality. However, for our data, even statements of this nature should be handled with caution. This is readily seen by calculating 95 percent confidence limits for the variance components of the most extreme case in Table 6, the interzooecial distances at Grover Station (Table 7). Confidence limits for variance components are skewed and the upper 95 percent limit for a between-colony variance overlaps the lower limit of the within-colony variance. Clearly, if the parametric value of s_A lay close to the upper 95 percent confidence limit of s_A, then the intercolony variance component

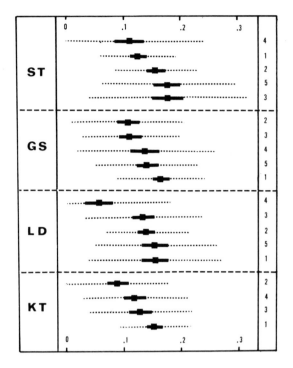

Figure 5
Dice diagrams for interzooecial distance for each colony. Dimensions in millimeters. Black square is location of mean, black bar is 95 percent confidence limit of mean, and broken bar is 95 percent confidence limit of character for colony.

would exceed the intracolony component, the converse of the relationship suggested by the unbiased estimates of these statistics.

Comparable methodological limitations apply in utilizing the coefficient of variation to examine the relative magnitude of within- and between-colony variation, a technique that has been used in some earlier studies of colonial animals. This approach, although having the advantage of simplicity, suffers from two difficulties. If the original measurements are markedly nonnormal, the coefficient of variation is not a very meaningful statistic, being based on the standard deviation and mean, both inapproporate statistics for nonnormal distributions. The second difficulty is comparable to that experienced in examining the relative size of the partitioned variance components. It is desirable to ascertain whether observed differences between the estimated intra- and intercolony coefficients of variation are indeed significant; they may be more apparent than real. It is necessary that confidence limits for the estimates of coefficients of variation be calculated to give some control to subsequent paleobiological speculation. The appropriate statistics are discussed by Sokal and Rohlf (1969, p. 137).

The biological explanation of the within- and between-colony variances may be examined further. The within-colony variance is an expression of microenvironmentally induced variations (Boardman, Cheetham, and Cook, 1970). As is well known, genetically

Table 5

Comparison of Results for Parametric and Nonparametric Tests Between Colonies, Within Localities[a]

Character[a]	Locality[b]	Anova	Kruskal–Wallis	Snedecor
ZD	ST	c	c	c
	GS	na[d]	c	c
	LD	na	c	c
	KT	na	c	c
IZD	ST	na	c	c
	GS	c	c	c
	LD	c	c	c
	KT	c	c	c
DC/MM	ST	na[c]	c	c
	GS	na	c	c
	LD	c	c	c
	KT	c	c	c
VC/0.25	ST	c	c	c
	GS	c	c	c
	LD	na[c]	c	c
	KT	c	c	c

[a]Characters abbreviated as in Fig. 3.
[b]Localities abbreviated as in Fig. 2.
[c]$P < 0.001$; [c], brackets enclose results for $\log_{10}Y$ transformed data.
[d]Test not applicable.

the individual zooids of a colony were identical (barring mutations), having originated by vegetative budding from a single larvae. In detail, however, the individuals were not exposed to identical microenvironmental situations. The animals responded phenetically to these differences, but the response to a given microenvironmental situation will have been governed by the genetic features of the colony, which differ from colony to colony, Thus, the within-colony variance may be thought of statistically as consisting of two components—a "nongenetic (microenvironmental)" effect and a "nongenetic–genetic" interaction term. Unfortunately, with fossil material, it is not possible to isolate these components and we can only group them under the accepted term of "extragenetic" effects, as defined by Boardman, Cheetham, and Cook (1970).

Field evidence suggests that the environment at any one collecting locality was relatively uniform. Consequently, the between-colony variance at any one locality is probably best regarded as a measure of genetic diversity between individual colonies of the type found in any population.

Correlations Between Characters

In preceding discussions, characters have been treated as though they were independent variables. However, it can be argued on geometrical grounds that some characters (for

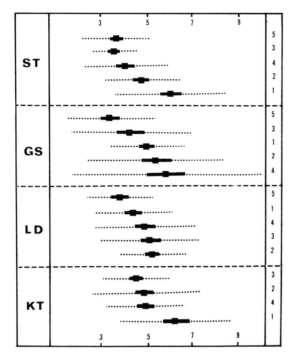

Figure 6

Dice diagrams for the number of diaphragms in a 1-mm distance. Black square is location of mean, black bar is 95 percent confidence limit of mean, and broken bar is 95 percent confidence limit of character for colony.

example, IZD and VCT) must be correlated to some degree. Particularly for studies limited to only a few characters, it is desirable to reduce the amount of redundancy (in the form of highly correlated characters) to a minimum in order to obtain a maximum amount of meaningful information. A matrix of Pearson product-moment correlation coefficients (r) was calculated for all possible pairs of character means and variances (Table 8). The variance of VCT (number of vesicles between nearest-neighbor zooecia) was not used because of the pronounced deviation of the data from normality.

Two characters stand out in displaying a high degree of independence from the other variables. Correlation coefficients for DC/MM (diaphragms per millimeter) and VC/0.25 (vesicles per unit area) are not significantly correlated with any of the other four principal characters. Both are count data, rather easily obtained and, for this study, normally distributed.

It is interesting to examine the probable cause of the negative correlations of ZD (zooecial diameter) with IZD (interzooecial distance) and VCT (vesicles between nearest neighbors). Biologically, these correlations are not entirely unexpected. As zooecial diameter increases, crowding occurs with a decrease in the interzooecial distance, also reflected by a decrease in number of vesicles between zooecia. The high positive correlation

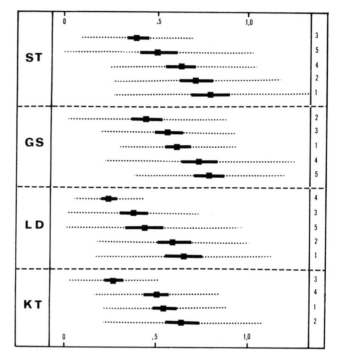

Figure 7

Dice diagrams for the number of complete vesicles in a circle of radius 0.25 mm, as measured in longitudinal section. Black square is location of mean, black bar is 95 percent confidence limit of mean, and broken bar is 95 percent confidence limit of character for colony.

between IZD and VCT ($r = .811$) seemingly reflects some uniformity in the size of vesicles, although this is not apparent through cursory observation. The relatively high negative correlation (-0.77) of VC/0.25 and the variance of ZD is also notable. This can seemingly be interpreted as an increase in the average vesicle size as the variance of zooecial diameters increases. Perhaps related to this is the positive correlation between VCT and the variance of VC/0.25, interpreted as an increase in the number of vesicles between zooecia with an increased variation in the size of the vesicles. A clear-cut biological explanation for these correlations is not apparent, but they may possibly reflect the influence of monticular areas on those characters.

Although most characters are correlated to some extent with one another, the fact that no r value is 1.0 indicates that varying degrees of independence exist; thus, varying amounts of information are obtainable from all characters. However, in evaluating the usefulness of a particular character it is important to consider not only the degree of independence but also the nature of the data obtained. The data obtained for VCT, as discussed earlier, could not be handled well statistically because of the lack of normality of the data and the limited number of size classes. This, coupled with the fact that

Table 6

Partitioning of Variance Components by Single-Classification Anova[a]

		Zooecial Diameter (ZD)			
		ST[b]	GS	LD	KT
s_A^2	Between colonies	24.96	na[c]	na	na
s^2	Within colonies	75.04	na	na	na

		Interzooecial Distance (IZD)			
		ST	GS	LD	KT
s_A^2	Between	na	17.47	41.46	25.82
s^2	Within	na	82.53	58.53	74.18

		Diaphragm Counts per Millimeter (DC/MM)			
		ST	GS	LD	KT
s_A^2	Between	$[50.43]^d$	na	28.36	35.30
s^2	Within	$[49.57]^d$	na	71.64	64.70

		Vesicle Counts per Area (VC/0.25)			
		ST	GS	LD	KT
s_A^2	Between	31.82	29.54	42.41	45.78
s^2	Within	68.18	70.46	$[57.59]^d$	54.22

[a]Values are expressed as a percentage of the sum of the variance components.
[b]Localities abbreviated as in fig. 2.
[c]Anova not applicable due to failure of assumptions.
[d]Values given in brackets are for $\log_{10}Y$ transformed data.

Table 7

Anova Table—Interzooecial Distance (IZD), Grover Station[a]

	d.f.	Mean Square	Expected Mean Square
Between colonies	4	0.3388	$s^2 + 5s_A^2$
Within colonies	120	0.0538	s^2

$$s^2 = 0.0538 \qquad s_A^2 = 0.0114$$

95% confidence interval for $s^2 = 0.041 - 0.072$
95% confidence interval for $s_A^2 = 0.003 - 0.110$

[a]Five colonies each with 25 measurements.

it has a moderately high correlation with ZD (-0.633) and IZD (0.811), makes it a relatively undesirable character. It is clear that the potential usefulness of IZD is much greater because it is a continuous variable; moreover, it can be more effectively handled statistically.

Table 8

Matrix of Correlation Coefficients Between Character Means and Variances

	ZD[a]	VAR/ZD	IZD	VAR/IZD	VCT	VC/0.25	VAR/VC/0.25	DC/MM	VAR/DC/MM
ZD	**1.00**[b]								
VAR/ZD	**0.51**	**1.00**							
IZD	**-0.55**	**0.46**	**1.00**						
VAR/IZD	-0.12	0.33	0.02	**1.00**					
VCT	**-0.63**	**-0.59**	**0.81**	0.05	**1.00**				
VC/.25	-0.42	**-0.77**	0.30	-0.36	0.42	**1.00**			
VAR/VC/.25	**-0.64**	**-0.50**	0.36	-0.00	**0.55**	**0.59**	**1.00**		
DC/MM	0.40	-0.07	-0.27	**-0.47**	-0.13	0.14	0.01	**1.00**	
VAR/DC/MM	0.33	-0.06	-0.29	0.03	-0.10	0.25	0.27	**0.55**	**1.00**

[a]Characters abbreviated as in Fig. 3.
[b]Values in boldface significantly different from 0 at $P = 0.05$.

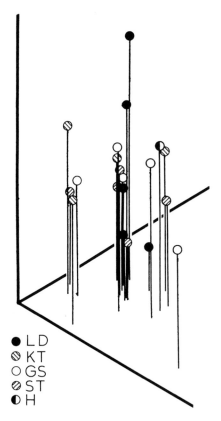

Figure 8
Projection of colony mean values into first three principal component space based on five characters.
Holotype of *F. decora* shown by half-hollow circle labeled H. Other colonies labeled by locality,
abbreviated as in Fig. 2.

CONCLUSIONS

In this study (as in most previous studies of Paleozoic Bryozoa) data are found to
be normally distributed for the majority of characters utilized, thus fulfilling one fundamental assumption of analysis of variance. However, when data from all 19 colonies were
considered together, the variances of the selected characters were not homogeneous.
It is not yet known how widespread deviation from homoscedasticity is among Bryozoa;
Anstey and Perry (1969) found that it occurred in two out of seven characters, but
their study was based on a smaller number of colonies. The need to test for the assumptions
of anova is emphasized; with the present data, any interpretation based on a nested
analysis of variance would be grossly misleading.

The available population of *Fistulipora decora* is characterized by extensive and significant variation between individuals within a colony, between colonies at one locality,

and between localities. Phenetically the group is quite flexible, responding readily to differences in environment and genetic makeup. The extent of variation between colonies at each locality implies that the population exhibited a high degree of genetic diversity. The available data also suggest, but do not conclusively prove, that microenvironmental effects typically account for more than half of the observable variation at a locality.

The extent of morphological variation, and more particularly its distribution, poses problems for the systematist. These are not insurmountable problems; they can be overcome by utilization of a logical sampling plan.

ACKNOWLEDGMENTS

One of us (J.D.F.) wishes to acknowledge the financial support afforded by a Chevron Fellowship in Geology. Computations were performed at the University of Kansas Computation Center on a Honeywell 635. We are indebted to R. S. Boardman and R. J. Cuffey, who critically read earlier drafts of the manuscript; but this is not to imply that they necessarily agree with all our conclusions.

REFERENCES

Anstey, R. L., and T. G. Perry. 1969. Redescription of cotypes of *Peronopora vera* Ulrich, a Cincinnatian (Late Ordovician) ectoproct species. Jour. Paleont., *43*: 245–251.

———, and T. G. Perry. 1970. Biometric procedures in taxonomic studies of Paleozoic bryozoans. Jour. Paleont., *44*: 383–398.

Boardman, R. S., and A. H. Cheetham. 1969. Skeletal growth, intracolony variation, and evolution in Bryozoa: a review. Jour. Paeont., *43*: 205–233.

———, A. H. Cheetham, and P. L. Cook. 1970. Intracolony variation and the genus concept in Bryozoa, *in* N. Am. Paleont. Convention Proc., pt. C, p. 294–320.

———, and J. Utgaard. 1964. Modifications of study methods for Paleozoic Bryozoa. Jour. Paleont., *38*: 768–770.

Bradley, J. V. 1968. Distribution-Free Statistical Tests. Prentice-Hall Inc., New Jersey. 338 p.

Brown, W. G. 1958. Stratigraphy of the Beil Limestone, Virgilian of eastern Kansas. M.S. thesis, Univ. Kansas, 189 p.

Farmer, J. D. 1971. A study of variation in the bryozoan *Fistulipora decora* (Moore and Dudley) from the Beil Limestone member of the Lecompton Limestone of Kansas. M.S. thesis, Univ. Kansas. 120 p.

Kaesler, R. L. 1970. Numerical taxonomy in paleontology: classification, ordination and reconstruction of phylogenies, *in* N. Am. Paleont. Convention, Proc., pt. B, p. 84–100.

Krumbein, W. C., and F. A. Graybill. 1965. An Introduction to Statistical Models in Geology. McGraw-Hill Book Co., New York. 475 p.

Moore, R. S., and R. M. Dudley. 1944. Cheilotrypid bryozoans from Pennsylvanian and Permian rocks of the Midcontinent Region. State Geol. Survey Kansas, Bull., *52*, pt. 6: 229–408.

Rohlf, F. J. 1968. Stereograms in numerical taxonomy. Syst. Zool., *17*: 246–255.

Rowell, A. J. 1970. The contribution of numerical taxonomy to the genus concept, *in*, N. Am. Paleont. Convention, Proc., pt. C, p. 264–293.

Sokal, R. R., and F. J. Rohlf. 1969. Biometry: The Principles and Practice of Statistics in Biological Research. W. H. Freeman & Co., San Francisco. 776 p.

Colony Development in Species of *Plumatella* and *Fredericella* (Ectoprocta: Phylactolaemata)

Timothy S. Wood

Wright State University

ABSTRACT

Growth and development are described for colonies of *Plumatella repens* Linnaeus, *Plumatella emarginata* Oka, and *Fredericella sultana* Blumenbach grown under both field and laboratory conditions. Field colonies were established in open petri dishes inverted and suspended in a pond near Boulder, Colorado. Data for over 3000 individual zooids include dates of asexual budding, survivorship, and directional orientation with respect to the parent.

In all species an increase in colony age was accompanied by decreasing zooid longevity and a progressively smaller fraction of the colony engaged in asexual reproduction. In older colonies the zooids formed buds at an increasingly earlier age, including both buds for new zooids and for sclerotized "resting" structures (statoblasts).

Colony and zooid longevity were greatest for *Fredericella sultana*. Although this species produced a maximum of only two statoblasts per zooid, free branches of the colonies broke away spontaneously and were effective in establishing new colonies elsewhere. Statoblasts of *Plumatella casmiana* were formed by only 25 percent of the zooids, but reproductive potential was enhanced by an unusual statoblast type (leptoblast), which germinated within a few hours of its release from the colony.

Author's address: Department of Biological Sciences, Wright State University, Dayton, Ohio 45431.

Zooids of *P. repens* not only produced more statoblasts than the other species, but did so in the shortest length of time. All these statoblasts, however, required a diapause of at least 3 weeks.

Differences in colony morphology among the *Plumatella* species were due mainly to a temporal difference in the budding sequences. The erect and largely free aspect of *Fredericella* colonies is attributed to considerable, slow elongation of the maturing zooid, and to relatively weak and transitory adhesive qualities of young individuals.

In the field, the production and germination of four different types of statoblasts were noted for *P. casmiana*. Some appeared throughout the growing season, others were produced during a much shorter time. Statoblast formation generally occurred throughout single colonies within a period of approximately 10 days, although in *F. sultana* it often extended over many weeks.

Sexual activity in field colonies was severely limited. Neither larvae nor larval colonies were ever observed, although sperm were present in May colonies of *F. sultana,* and in small portions of *P. casmiana* colonies collected in June 1970.

Colony growth was clearly geometric in *P. repens* but very erratic in the other two species. A stable colony size was rarely maintained by any species, and very active budding was usually soon followed by mass polypide death throughout the colony. Survivorship curves indicate a constant mortality among polypides of all ages.

INTRODUCTION

Among several groups of invertebrates there has evolved independently a level of organization at which individuals and their offspring are organically continuous and share a common pool of metabolites. Such colonies, both permanent and temporary, occur among Protozoa, Porifera, Cnidaria, annelids, ascidians, and ectoprocts. In the latter they are characteristic of the entire phylum, with the possible exception of the curious *Monobryozoon* spp. of European seas. This level of organization requires mechanisms in development and differentiation which do not occur in solitary animals. In the Cnidaria these have been studied by such investigators as Berrill (1953), Braverman (1962, 1963), and Loomis (1959). Patterns of colony formation in the marine Ectoprocta have been studied by Borg (1926), Cheetham (1968), Soule (1954), and others. However, in the freshwater Ectoprocta colony growth and development have received little attention.

Of the approximately 4000 living species of Ectoprocta only 45–50 occur in fresh water, and most of these are in the Class Phylactolaemata. This class is generally characterized by a horseshoe-shaped lophophore, unrestricted flow of coelomic fluid among the polypides, and colony form ranging from branching tubes to a compact gelatinous mass.

An ectoproct polypide is composed for the most part of a ciliated lophophore and its attached U-shaped digestive tract. A cord of tissue, the funiculus, hangs loosely between the blind end of the caecum and the outer body wall (Figs. 10–16). The body wall secretes a nonliving external ectocyst, and together these make up the zooecium. A zooid of the colony is thus formed by a polypide and its associated zooecium.

A new polypide bud originates in the body wall at a point just ventral to the maternal polypide, and as it grows the surrounding zooecium enlarges proportionately. In general, no budding will occur without the presence of an actively feeding maternal polypide. In species of *Plumatella* and *Fredericella* the death and disintegration of a polypide lead to the pulling away of the body wall from the ectocyst, and the living tissue usually withdraws from the ectocyst.

A second type of bud, the statoblast, is formed only in the Class Phylactolaemata. It is composed of a mass of undifferentiated tissue and yolky material enclosed in a chitinous bivalve capsule. Statoblasts serve to maintain the population through suboptimal conditions and often function as agents of dispersal, carried by wind, water, or animals. They develop within the funiculus of a polypide and are released to the collective coelom (metacoel) of the colony. From here they may exit either through the vestibular pore of any polypide or, more commonly, from a disintegrated portion of the zooecium. There are several morphological types of statoblasts, and these will be described as they are encountered.

Until recently, most studies of freshwater Ectoprocta have dealt largely with systematics, zoogeography, morphogenesis, and nature of the statoblasts. Information regarding total colony growth and development is fragmentary, possibly because of the difficulties in maintaining colonies in a way that permits long-term observation. Bushnell (1966) gave extensive treatment to colony growth and polypide longevity in natural populations of *Plumatella repens,* and Rogick (1935, 1945) obtained limited data for laboratory-reared colonies of *Lophopodella carteri* and *Hyalinella punctata*. The formation of colonies of *L. carteri* from embryo and statoblast has been graphically illustrated by Oda (1960), although substantial data are missing.

There are likewise few long-term records of colony development at any single locality during a growing season. Kraepelin (1887) suggested one statoblast generation each year for *Fredericella sultana;* Brown (1933) had evidence of two statoblast generations of *Plumatella repens,* and Bushnell (1966) observed three for this species. Several types of statoblasts have been described for *Plumatella casmiana* by Rogick (1941, 1943), Toriumi (1955b), Wiebach (1963), and Viganò (1968), and in many cases their relative abundance during the year has been recorded.

The present investigation was made as a first step toward an understanding of developmental mechanisms in the freshwater Ectoprocta. It involves primarily an intensive study of three species inhabiting a single pond near Boulder, Colorado. An effort was made to describe and record the pattern in which total energy resources of a colony are used for sexual and asexual activities. During 1970 careful records were kept of 155 naturally occurring colonies of *Plumatella repens* Linnaeus, *Plumatella casmiana* Oka, and *Fredericella sultana* Blumenbach, including colony longevity, budding rates, and polypide survivorship. Data for individual zooids also included dates of budding and statoblast formation, directional orientation, numbers of buds, and types and numbers of statoblasts. All three species were routinely collected from the pond during 1969 and 1970, and *P. repens* and *F. sultana* were collected during 1968 and 1969 from a montane lake (elevation 10,500 feet) for comparative purposes. Finally, all three species were grown and studied in the laboratory from 1968 to 1971.

It seemed appropriate that these particular species be used in the investigation. *F. sultana* is considered morphologically the most primitive phylactolaemate (Lacourt, 1968) because of its simple thin zooecial tubes and widely spaced polypides. There is a trend toward shorter zooecia and correspondingly more compact colonies in *P. repens* and *P. casmiana,* although both species maintain the identity of the individual zooecia. More advanced species, such as *Pectinatella magnifica,* tend toward closely packed individuals sharing a single coelomic cavity in a gelatinous mass, and their patterns of budding and colony formation are generally more elaborate (Brien, 1954). In the present study it is supposed that growth and development data taken from relatively simple colony systems will be initially more useful than those obtained from more advanced ones.

HABITAT DESCRIPTION

The Sawhill Ponds are a series of 19 shallow excavations created by a local gravel industry during 1950–1967. They have become filled by groundwater, surface runoff, and by seepage from nearby Boulder Creek. Located in Section 23, T1N, R70W of Boulder County, Colorado, the ponds are of various sizes, and many are stocked regularly for sport fishing. While nearly all the ponds contain small populations of *P. repens,* only one has in addition flourishing colonies of *P. casmiana* and *F. sultana* and a recently discovered population of *Plumatella emarginata.* This pond, designated as Pond 9 by the Colorado Game and Fish Commission (unpublished report, 1966) covers 1/6 hectare and has a maximum depth of approximately 3 m. Occasionally the pond is flooded by Boulder Creek during heavy spring rains, creating turbid conditions which prevail throughout the summer months and resulting in generally unstable plant and animal communities. *Plumatella emarginata,* for example, has been collected from the pond only since the 1969 flood, when it was apparently carried in from Boulder Creek. The species is most common in lotic waters, and it may not survive many more seasons in the pond.

In past years there have been several species of rooted aquatic plants in Pond 9, although these have not reappeared since the 1969 flood. Often there are thick algal blooms of *Aphanizomenon* and *Rhizoclonium* during July and August. Ice reaches a thickness of approximately 15 cm during brief periods in the winter, and the maximum recorded summer temperature is 29°C. The water is moderately hard and slightly alkaline, with a mean of 56.4 ppm bound CO_2.

At the narrow eastern end of the pond is an assembly of logs which are submerged, floating, or projecting into the water from the shore. Water depth in this area ranges from 0.8 to 1.2 m during the years when no flooding has occurred. The bottom is a fine organic ooze. It is in this area that great numbers of ectoproct colonies are found, and most of the present field study was conducted here.

DESCRIPTION OF ECTOPROCT SPECIES STUDIED

Plumatella repens L.

Plumatella repens is probably the most cosmopolitan ectoproct species in the freshwater family Plumatellidae. It has been collected on every continent, and is usually abundant wherever it occurs. A fairly euryokous species, it ranges from brackish water (Lacourt, 1968) and very eutrophic conditions (Bushnell, 1966) to relatively unproductive glacier-fed montane lakes.

As in other species of *Plumatella* two types of statoblasts are produced: a buoyant floatoblast and an attached sessoblast. The floatoblast has a periphery of gas-filled cells (Figs. 20–22), which apparently enhance its effectiveness in passive dispersal from the water surface. The sessoblast is larger and heavier than the floatoblast and is cemented to the substrate through the zooecium. Long after the colony has disintegrated its sessoblasts may remain, appearing as black specks on the substrate.

Isolated colonies of *P. repens* are usually open and repent, forming thin lines of successive zooids across the underside of a flat submerged substrate (Fig. 1-B). Toriumi (1955a) attributes the zooid density of colonies to budding frequency, zooecial elongation, and physical limits of the substrate, as well as to the density of statoblasts germinating simultaneously. These and related factors contribute to the remarkable plasticity in colony form exhibited by the species, which includes dense fungoid masses. Bushnell (1965b) reports the origin of dense colony growths not only from many contiguous sessoblasts, but also from the simultaneous spring germination of many floatoblasts held within the ectocyst of the old parent colony.

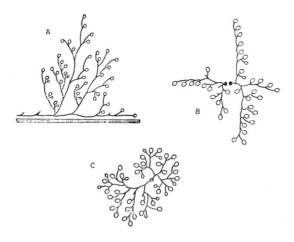

Figure 1

Schematic diagrams of colony form in the three species of freshwater Ectoprocta used in the present study. A, *Fredericella sultana;* B, *Plumatella repens;* C, *Plumatella casmiana.*

Dense growths of *P. repens* are rare in Sawhill Pond 9. Substrates are usually temporary, and most colonies develop from floatoblasts, not from masses of closely packed sessoblasts. During the summer months zooecia tend to disintegrate soon after the death of polypides, releasing trapped floatoblasts. There is generally uniform distribution of floatoblasts on submerged surfaces in the pond, and even where many are clustered together, as often occurs in the early spring, mortality of the young colonies is so high that dense growths rarely occur.

Allman (1856), Toriumi (1955a), and Bushnell (1965b) have all remarked on the relatively small amount of foreign particles adherent to the ectocyst of *P. repens*. In the Sawhill specimens such incrustation varies considerably, appearing heaviest when water is turbid and budding frequency low.

During 1970, 46 colonies of *P. repens* were studied at Sawhill Pond. Thirty-eight of these developed three or more zooids and had a combined total of 1402 zooids, 667 floatoblasts, and 80 sessoblasts. Eighteen colonies survived beyond 30 days with 20 or more zooids, and these were used in compiling data for zooid budding and statoblast production.

Plumatella casmiana Oka

First reported from a lake near Tokyo, Japan (Oka, 1907), *P. casmiana* was subsequently found in Java, the U.S.S.R., and in North America in the vicinity of Lake Erie. More recently this species has been discovered to be abundant in certain shallow Colorado plains lakes (Bushnell, 1968), and it is undoubtedly more common than previously believed. Of the 117 colonies studied at Sawhill Pond 9 during 1970, only 28 produced statoblasts, and 30 produced twenty or more zooids.

Isolated colonies of *P. casmiana* adhere firmly to the substrate and typically form compact mats, often with a nearly circular outline (Fig. 1-C), and with a zooid density far greater than isolated colonies of *P. repens*. During periods of rapid budding, zooecia may become contiguous and vertical, giving a honeycomb appearance to the colony (Bushnell and Wood, 1971). Zooecia are usually yellow to light brown, with a strong hyaline keel, and in all respects the Sawhill specimens were in close agreement with descriptions by Oka (1907), Rogick (1941, 1943), Toriumi (1955b), Wiebach (1963), and Viganò (1968).

Two types of statoblasts were attributed to the first described specimens of *P. casmiana:* a sessoblast and a distinctive long thin-walled floatoblast with a uniformly narrow annulus (Fig. 22). This floatoblast was later termed a "leptoblast" by Wiebach (1963) after its habit of germinating soon after release from the coelom of the parent colony.

A third statoblast type, the "capsuled" floatoblast (Fig. 21), discovered by Rogick and later called a "pyknoblast" by Wiebach, was attributed to *P. casmiana;* Viganò (1958) describes several other capsuled floatoblasts of various shapes for this species. Such wide variation in statoblast types is unique to this species. No other group of ectoprocts produces more than two kinds.

Fredericella sultana Blumenbach

F. sultana shows greater tolerance to low temperatures than does any other freshwater ectoproct. At Sawhill it flourishes in small numbers throughout the year, although stato-blasts are not formed during the coldest winter months. At the outlet of Long Lake (Boulder County, Colorado) it grows luxuriantly where the winter temperature may drop below 2°C for over 7 consecutive months. Zschokke (1906) reports this species at a depth of 170 m in Lake Lucerne, and Forel (1884) collected it from the bottoms of several Swiss and Italian alpine lakes. While rarely found in tropical waters, it has a worldwide distribution and is a frequent member of the benthos of many lakes.

Morphologically, the Sawhill colonies of *F. sultana* are in excellent agreement with descriptions by Toriumi (1951), Rogick (1940), and Bushnell (1965a). They are charac-teristically dendritic and antler-like, and while only a small portion of the colony is attached to the substrate, the free branches may adhere at several points. The zooid density in these colonies reflects relative growth of the zooecia, which is greatest during the summer months. Winter growth is slow, and colonies collected in February are relatively compact. Incrustation varies throughout the season, and the author confirms Toriumi's (1951) observation that hyaline ectocysts occur primarily in laboratory-grown colonies.

Statoblasts of *F. sultana* are usually called sessoblasts since they lack a buoyant periphery of float cells and remain with the parent ectocyst until germination. However, this term does not distinguish between freely movable *Fredericella* statoblasts and the firmly cemented sessoblasts of the Plumatellidae. *Fredericella* statoblasts actually more closely resemble floatoblasts than sessoblasts, both in their development on the funiculus and in their predictable appearance within the colony. There is good reason for replacing the term "sessoblast" with Marcus' (1955) "piptoblast" in referring to statoblasts which are neither buoyant nor adherent to a substrate.

Fredericella sultana was never abundant in the Sawhill pond, and complete records are available for only four colonies, representing 354 zooids and 58 piptoblasts.

MATERIALS AND METHODS

In the field it was essential that microscopical examination be thorough and accurate, yet cause minimal disturbance to the colonies. The colonies were grown on the inside of 4-inch-diameter glass petri plate covers suspended in the water open-side down. Supports for the glass covers were fashioned from 2-inch slices of 4-inch-diameter ABS plastic sewer pipe into which the covers fit snugly (Fig. 2). The pipe was heated at certain points along its edge and puckered in to prevent the petri covers from falling through. The entire apparatus was held 25–30 cm from the underside of floating logs.

A disc of tar paper resting on top of each dish blocked light from above and thus prevented a growth of algae, which would have impaired microscopical examination. It is unlikely that the black disc ever absorbed sufficient direct radiant energy to warm the substrate beneath it significantly, although slight warming may have occurred occasionally.

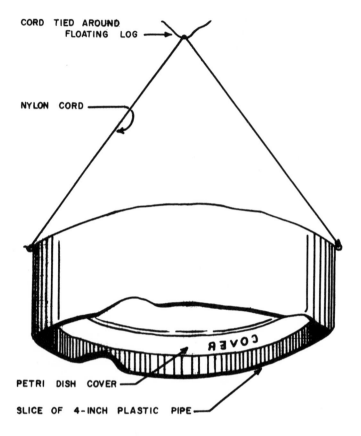

Figure 2

Apparatus used to suspend petri plate covers on which ectoproct colonies were grown. The cover fits inside a slice of 4-inch plastic pipe and is covered by a disc of tar paper to block light from above. The device is suspended from the underside of a floating log.

During this study many floatoblasts in the water became trapped under the inverted petri covers, where they germinated and gave rise to new colonies. Often, however, it was desirable to observe the development of certain floatoblasts dissected from specific colonies. These were usually pipetted under water to the inside of a petri cover, and the cover was then carefully fitted over a petri dish. The entire apparatus was then suspended in a plastic holder under the logs with other dishes. The petri dish bottom served both to prevent the establishment of foreign floatoblasts and to keep small eddies from carrying away the floatoblasts inside.

For sessoblasts and piptoblasts the petri dish and cover were inverted before being placed in the plastic holder, so that the statoblast would germinate against the cover. Once germinated and firmly attached to the substrate, the petri dish was removed and the cover reinverted to the usual position.

Ectoproct colonies were removed from the pond at least twice each week for examination, and the petri covers were kept top-side down so that the colonies they contained were always submerged in water. A Leitz dissecting microscope magnifying six diameters was used for most of the field examinations, and occasional close observations were made with a Bausch and Lomb compound microscope with 100× magnification.

Once out of the pond the dishes were protected from direct sunlight, but no attempt was made to maintain the original water temperature. What little debris collected around the colonies was removed only to permit more accurate observation. The ectocyst usually remained sufficiently transparent to allow observation of statoblasts and occasional sperm inside the zoecium.

The presence of a statoblast was recorded only when it became mature. A new bud was noted after it had become an actively feeding member of the colony, judged by its gut content. Polypides were considered dead if the lophophore retractor muscles failed to contract when poked with a small probe.

Dates of budding and of polypide death were estimated for days between observations on the basis of the number of buds or statoblasts produced in the interim, or by extrapolating from an established rhythm of budding or mortality rate. Where such estimates could not be justified, polypides were arbitrarily assumed to have become functional 1 day before observation and to have died 2 days after observation. The longevity estimated for such polypides is probably conservative.

Laboratory culture of ectoprocts was performed with a modified inverted-petri-dish technique (Wood, 1971). Growing colonies under somewhat controlled conditions at various temperatures was attempted, but invariably only those colonies at one temperature grew well and produced statoblasts. In all species, tolerance to temperature extremes in the laboratory was markedly less than to the same temperatures in the natural environment.

COLONY GROWTH AND LONGEVITY

Plumatella repens

Colonies of *Plumatella repens* at Sawhill Pond 9 lived for as long as 59 days, and among colonies surviving more than 10 days the mean longevity was 35.1. There was no strong correlation between a colony's longevity and the season during which it lived. However, in the second summer generation a number of colonies became fragmented as they disintegrated in August. Later in the month many of the small surviving pieces underwent rapid budding and formed new colonies of relatively short-lived polypides which formed no statoblasts. Such secondary growth in *P. repens* was observed by Bushnell (1966) in his Michigan colonies, and it has been noted by the author in laboratory-reared specimens.

Colonies of *P. repens* having no more than 15 zooids show very symmetrical growth curves. These colonies slowly attained a maximum size and then gradually died. In only a few of these did budding rate equal mortality for more than 1 week, briefly maintaining a constant polypide number. All the small colonies originated from overwintered floatoblasts and grew in a temperature range of 10–22°. Zooids were slow in forming new buds, and the polypide longevity was several days greater than the mean for those occurring later in the year. Most of the miniature colonies produced relatively large numbers of floatoblasts and sessoblasts, and as they grew older the ectocyst became increasingly opaque, as it usually does in larger colonies.

The exact causes of the stunted growth are not known. Low temperatures inhibit budding in *P. repens,* but while most of the ancestrulae first appeared at 14°, the budding rate did not increase appreciably with a later rise in temperature. It is possible that the seston in the pond was not of the type or abundance that supports optimum budding rate. Crowding of the colonies was not a likely factor in suppressing growth, although a certain amount of debris tended to accumulate on the glass substrate during this part of the year.

Growth curves for second-generation colonies are more heterogeneous than for the first. Consequently absent was any tendency to maintain an equilibrium of budding and polypide mortality rates. Budding rate usually accelerated as the colony matured, and a final brief period of very rapid budding was often followed by sweeping mortality in which polypides of all ages died within a few days. Although ectocyst disintegration occurred with this mortality, there was no evidence of gross predation on the colonies.

The geometric growth attributed to *P. repens* by Bushnell (1966) was evident in many of the Sawhill colonies. It has also been described in colonies of *Podocoryne carnea* (Braverman, 1963) and *Cordylophora* (Fulton, 1960).

Individual colonies of *P. repens* have been reported with several hundred living zooids, although in the present study the greatest number was only 151. Collections made in the pond throughout the year showed no significant differences between colonies on petri covers and those growing on other substrates.

Colony disintegration was definitely accelerated by the activity of many tendipedid larvae, which tended to reside in empty zooecial tubes. The vigorous movement of these larvae apparently ruptured the peritoneum of adjacent living polypides, and it was not uncommon for several dozen polypides to die following the habitation of an empty part of their ectocyst by a tendipedid larva. Marcus (1925, 1926) has cited tendipedids as predators of ectoprocts, but in frequent observations of tendipedid–ectoproct associations at the Sawhill pond the larvae seemed invariably to be seeking a substrate, not nourishment. Bushnell (1966) found Trichoptera larvae to be heavy predators of *P. repens,* but these did not occur at Sawhill. Gross predation, in fact, was never indicated during this study.

Since both polypide and statoblast formation presumably require substantial amounts of energy, one would expect an inverse reciprocal relationship between the rates of these activities. However, floatoblast production in 69 percent of the colonies occurred during periods of most rapid polypide budding. The appearance of sessoblasts did not appear to correspond with either polypide or floatoblast formation, and there is evidence from both laboratory and field data that sessoblasts tend to occur in response to unfavorable conditions.

Plumatella casmiana

Longevity of *P. casmiana* colonies at Sawhill ranged from 2 to over 64 days. Growth was erratic in this species, frequently with two or more periods of rapid budding followed by relatively low levels of activity. In only 48 percent of the colonies did periods of greatest floatoblast production tend to occur at the time of active budding. Growth curves are irregular for all colonies recorded, often showing no clean geometric tendencies as in *P. repens*. Like *P. repens,* colony disintegration was frequently aided by the activity of tendipedid larvae, with the difference that the larvae tended more often to build their tubes against the exterior of the ectocyst than penetrate the interior of the zooecium. This may well have been due to the smaller zooecial diameter, frequent branching, and the resulting compactness of the colonies. Destruction of zooecia by tendipedids was consequently less here than in *P. repens*.

Zooid budding and polypide mortality rates in *P. casmiana* frequently varied together, so that when a colony entered a phase of active budding it showed a concomitant increase in mortality rate. Such increase would normally be expected to follow changes in budding rate by several days, reflecting a natural polypide longevity. Since, however, both budding and mortality rates often changed together, a closer relationship is indicated. Either the two phenomena are regulated by common factors, or there is a causal relationship between them.

Fredericella sultana

Growth and longevity data for *Fredericella sultana* are necessarily incomplete. The colony fragments easily, and living branches are passively carried away to establish new colonies elsewhere. The author found no satisfactory way to retain these pieces and still have growth data comparable to those of the two *Plumatella* species. Hence the fragments were lost, and data for colony growth and size are somewhat conservative.

Furthermore, the vertical aspect of *Fredericella* colonies often hinders observation of individual zooids. Upright zooecia become interlaced, adhering at several points, and it is difficult to examine each zooid without damaging the colonies. Consequently, only relatively small colonies were studied, and close observations were discontinued when accuracy of the data could no longer be assured.

Longevity of the four colonies investigated was recorded as 89, 53, 73, and over 47 days, all significantly greater than for either of the *Plumatella* species, not even considering the daughter colonies established by living fragments. Records for the last colony were discontinued after 47 days, when the colony became quite large and complex. It remained under observation, however, throughout the winter months, when it continued to grow at temperatures as low as 3°C. Piptoblasts were not formed below 8°.

Unlike the situation in *Plumatella* species, statoblast development in *F. sultana* does not appear to be closely related to other colony activities, such as budding. Also characteristic of *F. sultana* was the relatively long period of piptoblast production, ranging from 25 to well over 35 days. This stands in contrast to the *Plumatella* species, where floatoblasts are formed throughout the colony only during a period averaging less than 10 days. The significance of these interspecific differences is not clear but is possibly related to the considerably greater polypide longevity in *F. sultana*.

In the Sawhill Pond budding and mortality rates in *F. sultana* fluctuated within single colonies but did not vary directly with changes in temperature. Budding was usually least frequent where detritus around the ectocysts was fairly dense. These fluctuations, combined with occasional loss of colony fragments, resulted in a very uneven pattern of growth.

COLONY MORPHOLOGY

Each of the three species in Sawhill Pond had a characteristic morphology that made the colonies easily distinguishable. *F. sultana* is anchored very lightly, and most of the colony is erect and bushy. Because of its freedom from the substrate, this species is occasionally found growing on the bottom sediments of lakes (Bushnell, 1966). *P. repens* adheres closely to the substrate and is usually very spread out. *P. casmiana,* also firmly adherent, forms dense and compact mats.

Colony shape and density are a function of budding and growth patterns. *F. sultana* is the only species of the three in which the zooecium elongates considerably after the bud has matured. It is also the one least adherent to the substrate. Together these factors may account for the free and upright aspect of the colonies. The mean age at which a zooid of this species forms its first bud is only one day less than the mean age for the second bud, and this would account for the frequently high polypide density. Furthermore, up to four buds may be formed by a single zooid, and only 28.6 percent of 321 zooids over 2 days old form no buds at all.

Plumatella repens and *P. casmiana* are strikingly similar in the morphology of polypides, sessoblasts, and certain floatoblasts, yet the shapes of isolated colonies are quite different (Fig. 1B, C). Among zooids of *P. repens* at Sawhill, the primary bud is formed at a mean age of 2.7 days, and the secondary bud appears at 10.4 days. Thus, as shown in Fig. 3, a zooid may have its primary bud within 4 days, which will in turn produce

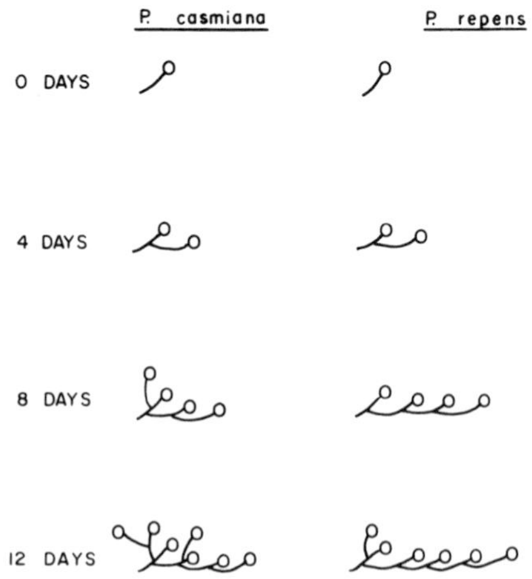

Figure 3

Comparison of budding patterns in *Plumatella repens* and *P. casmiana,* based on mean data from 265 zooids.

a second zooid, which will then produce a third. It is not until the tenth day, however, after the fourth successive bud has appeared, that the original zooid will form its secondary bud. This delay results in a very spread-out colony. The effect is heightened when only 22 percent of the zooids in colonies less than 30 days old produce secondary buds at all, and 2.3 percent have tertiary buds.

By contrast, the primary and secondary buds of *P. casmiana* are given off at a mean zooid age of 3.5 and 7.2 days, respectively. In colonies less than 30 days old 24.4 percent of the zooids bud a second time, and 5.9 percent have tertiary buds. Quaternary buds also have been observed at Sawhill Pond 9. Since zooecial elongation in this species does not always keep pace with the rapid succession of buds, the identity of each zooid is often reduced, and the colonies become very compact.

In many marine ectoprocts the direction in which a new zooid develops relative to the parent is the basis for colony morphology characteristic of the species. A simple pattern of zooid orientation has been described for the freshwater ectoproct *Lophopodella carteri* by Oda (1960). The shapes of *P. repens* and *P. casmiana* colonies suggest that *P. casmiana* may bud in alternate directions, while *P. repens* may bud primarily in one (Fig. 1B, C). Among 500 zooids of both species the ratio of right to left budding was 34:33 in *P. repens* and 33:39 in *P. casmiana*. In the former species, however, there is a definite tendency for a series of zooids to bud more consistently in one direction. In a linear series of zooids this direction is difficult to determine, since a new polypide always appears ventral to the parent. However, branches which later develop lateral to the original series all tend to project in the same direction. This direction varies among different parts of the same colony and follows no apparent pattern. On the other hand, budding direction in *P. casmiana* seems to be entirely random.

Discussing *P. casmiana*, Rogick (1941) commented on the sharp horizontal angle of the first few zooids relative to the ancestrula. This angle is accentuated by the close adherence of all zooecia to the substrate throughout their length. Horizontal angles, either large or small, are not clearly evident in young colonies of *P. repens* for the same reason that budding direction is obscure: zooecial tips are usually erect in a linear series of polypides. In larger colonies of both species, however, the budding angle between established zooids is easily measured because zooecia are firmly affixed to the substrate. In both species it ranges from 50° to 90° with no significant interspecific differences.

POLYPIDE LONGEVITY

Most longevity data for ectoproct polypides have been gathered from laboratory-reared colonies. Rogick (1945) measured the longevity of *Hyalinella punctata* ancestrulae as 3 to 21 days. Ancestrulae of *Lophopodella carteri* (Rogick, 1938) lived considerably longer, from 6 to 47 days, and the range might have been extended had the colony not been accidentally killed. Brooks (1929) kept the ancestrulae of *Pectinella magnifica* alive for 6 weeks, but they eventually died without forming a single bud. Brown (1934) found few polypides of *P. repens* surviving more than 10 days in his laboratory, and Marcus (1941) estimated polypide longevity in *Stollela evelinae* at 3–4 weeks. In natural populations of *P. repens* Bushnell (1966) found the range in polypide longevity to be 4–53 days. Rey (1927) recorded polypide longevity of 3–4 weeks in the marine ectoproct *Flustrella hispida.*

Laboratory Studies

Among the author's laboratory-grown colonies of *P. repens* and *P. casmiana* the upper limit of polypide longevity was approximately 16 days. There is, however, a single record of a polypide of *P. repens* surviving for 25 days at 18°C. All these long-lived individuals were either ancestrulae or the primary buds of ancestrulae, and they were never members of colonies having more than three zooids. The suboptimal conditions in a laboratory culture dish may thus have somehow prolonged polypide life by inhibiting bud formation. Such may also have been the case in the stunted spring colonies of *P. repens* in the Sawhill pond, where a low budding rate was accompanied by above-average polypide longevity.

Fredericella sultana, which has never grown well in the laboratory, had a maximum recorded polypide longevity of 18 days under laboratory conditions. Both here and in the field studies, however, its mean longevity was slightly greater than for polypides of either *Plumatella* species.

Field Studies

In Sawhill Pond 9 the ranges in polypide longevity for the three species are as follows: *P. repens,* 1–44 days; *P. casmiana,* 2–31 days; *F. sultana,* 2–39 days. Because the mean polypide longevity varied considerably with colony age, it was calculated independently for polypides appearing in 10-day intervals throughout the life of the colony (Fig. 4). These and the following data were gathered only from colonies surviving more than 20 days and having at least 20 zooids, thus eliminating many of the stunted spring colonies. Colony age was measured from the date of appearance of the second polypide, since in many cases the exact date of statoblast germination was not known.

Figure 4 is a composite graph for polypide longevities in all three species. In every case the greatest values are among polypides appearing during the first 10 days, and the general trend is toward decreasing longevity as the colony grows older. *P. repens,* the simplest species in terms of numbers of buds per zooid, shows the steadiest decline in longevity. The curve for *P. casmiana* is not as definite, and possibly this reflects the characteristic variation in budding rates in this species. Mean polypide longevity was 6.8 days in *P. casmiana* and 7.4 in *P. repens.* Although mean polypide longevity in *F. sultana* was 7.8 days, this species clearly showed capacity for greater longevity, with 68 percent of the polypides living beyond 7 days and 10 percent surviving more than 14 days. This significantly longer polypide life span, evident in Fig. 5, may be an important part of the success of the species.

Given a hypothetical ectoproct colony in which new polypides are formed at a constant rate until the colony is suddenly destroyed, one would expect the original polypides to have lived longer than those which appeared shortly before the end. This is not analogous to the situation at Sawhill, however, since no polypides are included in the

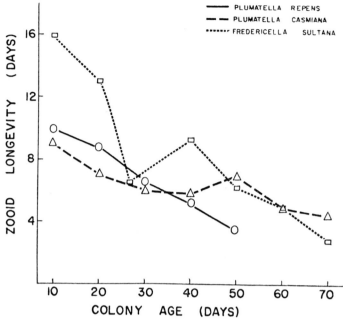

Figure 4

Mean zooid longevity is plotted against colony age. Data are compiled from zooids appearing during 10-day intervals throughout the life of a colony. Colony age is measured from the date of appearance of the second zooid.

longevity data where mortality could be directly attributed to an external factor, e.g., destruction by tendipedid larvae. Nor can the decrease in longevity be a result of changes in the external environment, since the data represent polypides living at different times throughout the season. It is more likely that the decrease in longevity reflects actual physiological changes within the colony, and this hypothesis is supported by data for budding and statoblast formation which follow.

Survivorship

The survivorship curves (Fig. 5) are distinctive for each of the ectoproct species. They are compiled from all polypides observed at Sawhill Pond 9 throughout 1970. Every species begins with an immediate drop in survivors, which, in *P. repens,* accelerates steadily, especially beyond 15 days. In *P. casmiana* the decline is fairly constant until it begins to ease somewhat at 25 days. In *F. sultana* the polypide survivorship curve also is very straight, although not as steep as in either *Plumatella* species because of its greater polypide longevity. The three survivorship curves indicate a constant mortality rate among polypides of all ages, lending further support to the idea of steadily decreasing polypide longevity with increasing colony age.

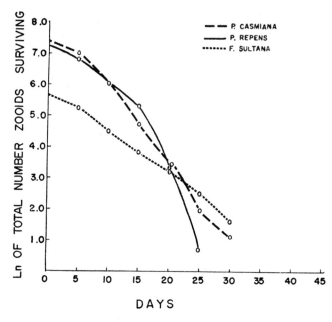

Figure 5

Survivorship curves for 1439 zooids of *Plumatella repens*, 1680 zooids of *P. casmiana*, and 318 zooids of *Fredericella sultana*. The natural logarithm of zooids surviving is plotted against time.

The only other survivorship curve for ectoproct polypides is given by Bushnell (1966) for natural populations of *P. repens* in Michigan, extending over 53 days. It shows an initial drop in the number of survivors which begins to level off slightly before declining more steadily beyond 20 days. Bushnell attributes the initial dip to possible damage to the delicate and vulnerable young zooids. The absence of such an early drop in the curve for Sawhill colonies may be partly due to greater protection afforded young zooids by the clean glass substrate of the petri plate cover.

ZOOID BUDDING

Ectoproct colonies increase in size by budding, and the rate and distribution of this budding activity are reflected in the size and shape of the colony. To observe how this activity is shared among individual zooids, data were gathered on the budding of 3300 zooids of *P. repens, P. casmiana,* and *F. sultana.*

In all three species the percentage of living zooids forming buds is greatest during the first 10 days after appearance of the colony's second zooid (Figs. 6 and 7). As the colonies mature, progressively fewer zooids form buds, and this trend appears to be correlated with the decrease in polypide longevity (Fig. 4).

The overall percentage of zooids producing buds is shown in Table 1. An appreciably higher value for single zooid buds in *P. repens* is consistent with the very rapid growth

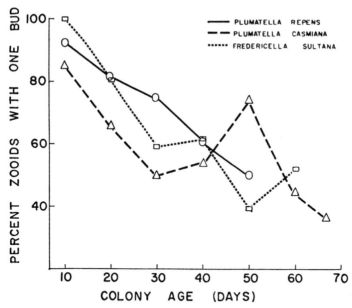

Figure 6

Percentage of zooids in a colony having at least one bud is plotted against colony age. Mean data represent only those colonies with more than 20 zooids and having a longevity greater than 20 days.

observed for this species (Jónasson, 1963; Dendy 1963). *P. casmiana* and *F. sultana* show similar percentages for the first two buds, but a relatively larger number of *F. sultana* zooids form three and four buds than in either of the *Plumatella* species. This is undoubtedly explained by the greater polypide longevity in *F. sultana,* since all the *P. casmiana* zooids forming three and four buds did so at an age well beyond the mean polypide life span.

Table 1

Percentage of All Zooids Involved in Budding Throughout the Life of the Colony

Species	One Bud	Two Buds	Three Buds	Four Buds
Plumatella repens	75.5	22.4	1.9	—
Plumatella casmiana	61.8	24.3	4.9	0.4
Fredericella sultana	62.5	23.6	9.8	1.5

There are very marked interspecific differences in the age at which zooids form their first buds. *P. repens* buds develop at a mean zooid age of approximately 2 days, and this value changes little during the life of the colony. In *P. casmiana* the mean zooid age is 3.0 days, and for *F. sultana* it is 3.7 days.

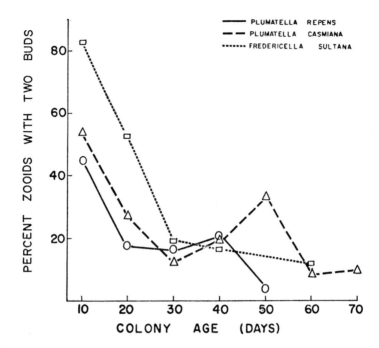

Figure 7

Percentage of zooids in a colony having at least two buds is plotted against colony age. Mean data represent only those colonies with more than 20 zooids and having a longevity greater than 20 days.

In *F. sultana* there seems to be little direct relationship between zooid budding and colony age. However, as the *Plumatella* colonies grow older their zooids tend to form new buds at an increasingly earlier age (Figs. 8 and 9). This trend also appears in Rogick's (1938) growth data for laboratory-reared colonies of *Lophopodella carteri,* and it has been observed by the author in laboratory colonies of *P. repens* and *P. casmiana.* In the Sawhill colonies this effect cannot be directly attributed to changes in the external environment, since the data were compiled through a wide range of environmental conditions spanning nearly 7 months. It is more likely a phenomenon related to progressive changes in longevity mentioned previously, based on physiological changes within the colony. Small fragments of 1–3 polypides separated completely from a mature colony of *P. repens* show patterns of longevity and budding very similar to those of young colonies, suggesting perhaps the presence of endocrine metabolic stimulant factor(s) within the metacoel of older colonies. Because the effects are progressive, it is possible that such factor(s) are metabolites produced by each zooid and eventually broken down within the colony or excreted to the external environment. Excretory mechanisms in freshwater Ectoprocta have been postulated by Marcus (1926, 1934), Gerwerzhagen (1913), and Becker (1937), but the evidence is fragmentary and indirect.

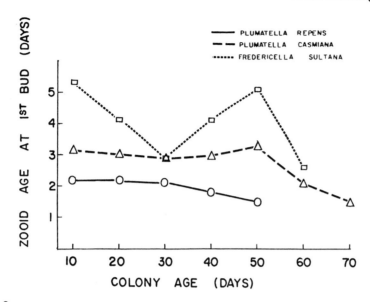

Figure 8

Mean age at which a zooid forms its primary bud is shown for 10-day intervals throughout the life of the colony. Data represent only those colonies with more than 20 zooids and having a longevity greater than 20 days.

Budding rate in ectoprocts is known to be closely associated with both temperature and the availability of suitable food. In the Sawhill pond, budding ceased in both *Plumatella* species at a temperature of 10° in October, which is close to the 9° cited by Bushnell (1966) for *Plumatella repens*. Under carefully controlled laboratory conditions both *P. repens* and *P. casmiana* grow more rapidly at 30° than at 24°, and they grow not at all at 18°, even though all live in water continuously circulated from a common source.

Rogick (1945) noted a sharp increase in the budding rate of her *Hyalinella punctata* colonies after changing the culture medium and adding yeast to the water. Tenney and Woolcott (1966) considered the availability of suspended food a primary factor in the distribution of *P. repens* in streams of North and South Carolina, and they found particularly large colonies in areas having the greatest density of potential food organisms.

Replacement Budding in *Plumatella casmiana*

In the course of maintaining complete records of colony development it was discovered that degenerating polypides of *P. casmiana* may be replaced by new buds occupying the same zooecium. This had never before been observed in a freshwater species. Figures 10–16 show the sequence of the replacement budding. The polypide is withdrawn into the zooecium, and the lophophore rapidly disintegrates. The entire gut shortens and pulls away from the tentacle sheath. All structures lose their identities, and only a small dark mass remains conspicuous in the coelom, often with the funiculus intact.

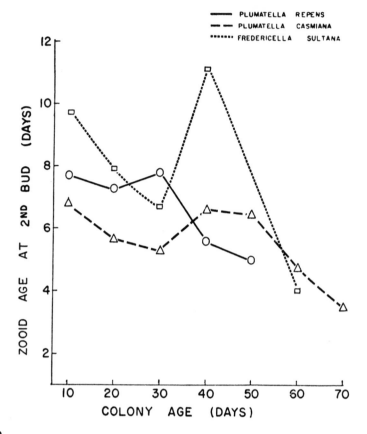

Figure 9

Mean age at which a zooid forms its secondary bud is shown for 10-day intervals throughout the life of the colony. Data represent only those colonies with more than 20 zooids and having a longevity greater than 20 days.

This pheonenon was observed repeatedly in both laboratory and field colonies, usually those in which normal budding processes had ceased and the colonies were in early stages of degeneration. According to the current interpretation of zooid morphogenesis (Brien, 1936, 1954), it would be unlikely for the replacing bud to extend through the same foramen of the ectocyst formerly occupied by the mother polypide. While this certainly appears to be the case here (fig. 16), it has yet to be confirmed from prepared sections. An interesting feature of the process is the apparent migration of the point of insertion of the new lophophore retractor muscles on the zooecial wall (Figs. 15 and 16).

Replacement budding in *P. casmiana* is strongly reminiscent of a similar process in marine species (Römer, 1906; Gerwerzhagen, 1913; Rey, 1927). In both cases the degenerated polypide is transformed into a small dark mass ("brown body"), and a new polypide appears in its place. In many marine species the brown body is ingested

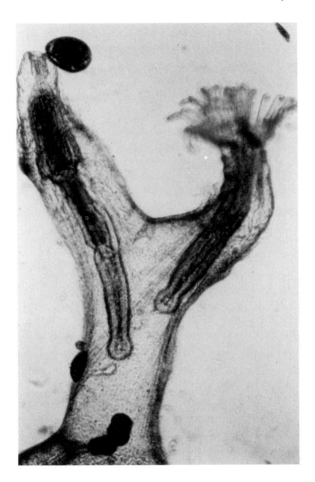

Figure 10

Both polypides are shown intact although they have stopped feeding and the guts are empty. The polypide at the top of the photograph will be replaced by a new bud. The structure within the zooecial tube at the far left is the rounded-up remains of a former polypide. An ostracod is beside the orifice of the upper polypide. Photographed Nov. 15, 1969, at 9:30 A.M., ×42.

by phagocytic cells; in others it is taken into the digestive tract and expelled through the anus. The entire process apparently serves simply to maintain portions of the colony beyond natural polypide longevity. Although it was never a common event in the Sawhill colonies of *P. casmiana,* it may occur more frequently under different conditions.

SEXUAL REPRODUCTION

Sexual activity among the ectoprocts in Sawhill Pond 9 was extremely limited. *F. sultana* colonies passed through a phase of sperm production during several weeks in

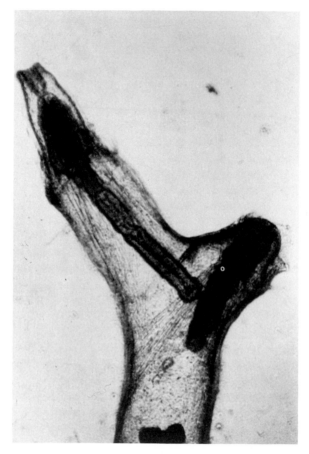

Figure 11

The lower polypide has begun to disintegrate; the upper one remains intact. Photographed Nov. 16, 1969, at 1:10 P.M., ×42.

May of 1969 and 1970, but no ova were observed. Sperm were also found in a small portion of a *P. casmiana* colony collected June 18, 1970. This colony had developed from an intermediate floatoblast. Although colonies of all species were frequently collected and studied in the pond for 2 years, neither larvae nor larval colonies were ever observed.

There are many scattered reports of sexuality in freshwater ectoprocts, although nothing is known of the conditions favoring sexual activity. In her laboratory Rogick (1945) noted the development of sperm masses in certain 10- to 24-day-old colonies of *Hyalinella punctata*. In the present investigation, 323 isolated floatoblasts were treated in various ways, including 135 chilled and dried soon after collection and stored at 7–13°C. Eleven percent of the 36 stored the longest time (846 days) gave rise to colonies in which sperm developed. Under no other conditions did sexual activity appear. Curiously, this

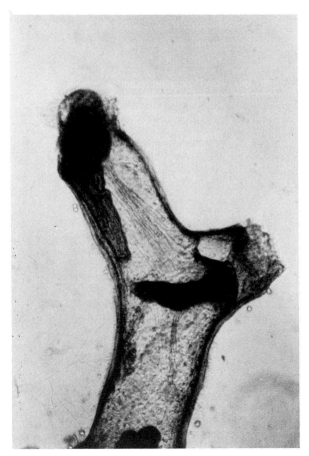

Figure 12

The lophophore of the upper polypide has disintegrated, and the lophophore retractor muscles are slack. Photographed Nov. 17, 1969, at 8:45 P.M., ×42.

would indicate that the stimulus for sexuality is received by the statoblast. A carefully controlled laboratory investigation would clarify this question and possibly indicate other factors involved in the initiation of sexuality.

STATOBLASTS

Floatoblasts and Piptoblasts

The production of floatoblasts and piptoblasts follows much the same pattern as zooid budding. As colonies of all species become older there is a progressive decrease in the percentage of individuals forming floatoblasts (Figs. 17 and 18). While zooids of

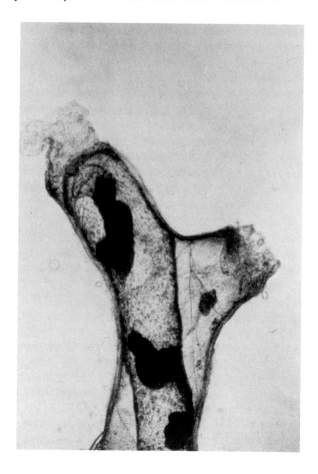

Figure 13

The digestive tracts of both polypides have pulled away from the tentacle sheaths, and much of the peritoneum has separated from the ectocyst. The embryonic bud to replace the upper polypide is visible. Photographed Nov. 18, 1969, at 8 P.M., ×42.

F. sultana form statoblasts more routinely than either of the other species, *P. repens* produces the larger number of floatoblasts per zooid. *P. casmiana* generally forms fewer statoblasts than either of the other species, although its leptoblasts are released long before colony disintegration through the vestibular pore of living zooids.

Not unexpectedly, as a colony becomes older each polypide forms its first statoblast at an increasingly earlier age (Fig. 19). This tendency is similar to that in zooid longevity and the age at bud formation, and presumably these phenomena are all related. In *Plumatella* colonies the first floatoblast matures close to the time of mean zooid longevity. However, in *F. sultana* the mean zooid longevity is actually less than the age at which the first piptoblast usually matures, and statoblast production in this species is correspondingly low.

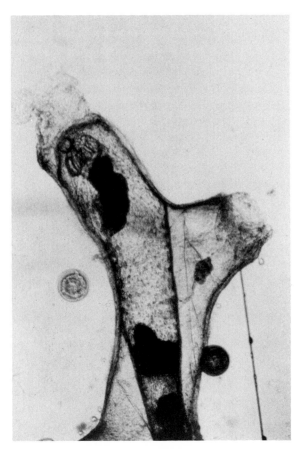

Figure 14

Tentacles of the developing lophophore on the new bud may be distinguished. Photographed Nov. 19, 1969 at 4:55 P.M., ×42.

Less than 2 percent of the zooids in any species produced a floatoblast or piptoblast without having first formed at least one new zooid. The initial priority given to zooid budding caused production of floatoblasts and piptoblasts to lag behind increases in zooid number by 2–10 days. In the *Plumatella* species this gap was greatest in new and rapidly growing colonies. Among colonies of *P. repens* the total number of statoblasts often exceeded the number of zooids, but no more than eight statoblasts were formed by any single zooid.

In rapidly growing colonies of *P. repens* it was common for new buds and statoblast Anlagen to form even before the young parent zooid had begun to feed. Although this has never before been reported, it was frequently observed in both laboratory and field colonies. It illustrates the essential unity of the colony in terms of nutrition, for the energy required by rapidly growing embryonic tissues can come only from materials

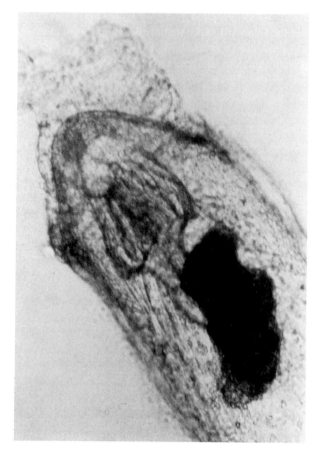

Figure 15

The new polypide is shown with all parts developed. Photographed Nov. 20, 1969, at 9:10 A.M.,
×130.

circulated within the metacoel. It is not clear why this phenomenon should be so common in *P. repens* yet be rare in *P. casmiana* and absent in *F. sultana*.

Rogick (1943) reported that *P. casmiana* in Lake Erie had an abundance of leptoblasts during late June, becoming less frequent in July and August. The pyknoblasts (''capsuled floatoblasts'') appeared in June and became most abundant in August. Rogick (1943) concluded from these observations that leptoblasts and pyknoblasts represented early and late summer forms.

Toriumi (1955b) examined specimens of *P. casmiana* collected in Japan, Korea, Formosa, and Manchuria. Leptoblasts were most abundant in colonies collected in July and August, with pyknoblasts very frequent in September and October.

Although both Rogick and Toriumi believed the type of floatoblast produced varied with the season, it has not yet been possible to identify the precise evnironmental factors

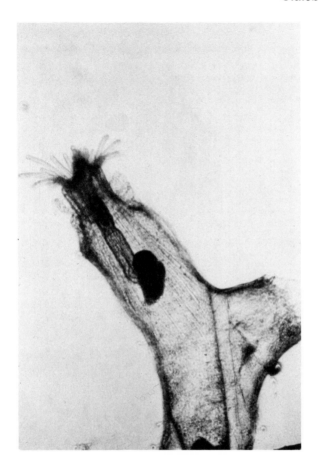

Figure 16

The new polypide is shown with food in the gut. In this particular case the polypide was the only surviving member of the colony. It fed actively for 12 days, then died without having produced a single new zooid or statoblast. Photographed Nov. 21, 1971, at 2:30 P.M., ×42.

involved, nor to distinguish between extrinsic and intrinsic mechanisms in the production of different statoblast types. Colonies of *P. casmiana* raised in the laboratory from intermediate floatoblasts or leptoblasts at 24–30° with an abundance of food organisms form only leptoblasts. Colonies not grown under these conditions form no floatoblasts at all. The development of proper rearing procedures will undoubtedly help identify the mechanism of differential floatoblast production in this species.

At Sawhill Pond 9 *P. casmiana* formed four kinds of statoblasts: sessoblasts, intermediate floatoblasts (Fig. 20), pyknoblasts (Fig. 21), and leptoblasts (Fig. 22). Among the floatoblasts during 1969 and 1970 only the intermediate type was available to germinate in the spring, and the colonies they formed produced both leptoblasts and more intermediate floatoblasts. When a single colony of *P. casmiana* produced both types it did so at

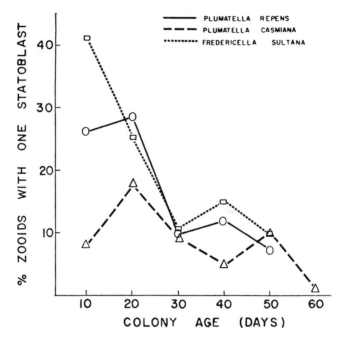

Figure 17

The percentage of zooids in a colony having at least one floatoblast or piptoblast is plotted against colony age. Mean data represent only those colonies with more than 20 zooids and having a longevity of greater than 20 days.

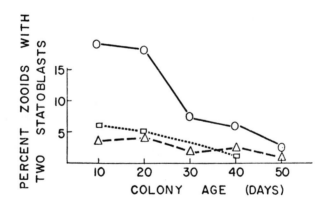

Figure 18

The percentage of zooids in a colony having at least two floatoblasts or piptoblasts is plotted against colony age. Data represent only those colonies with more than 20 zooids and having a longevity of greater than 20 days.

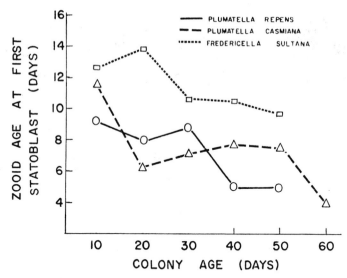

Figure 19

The mean age at which a zooid forms its first floatoblast or piptoblast is shown for 10-day intervals throughout the life of the colony. Data represent only those colonies with more than 20 zooids and having a longevity greater than 20 days.

different times, never simultaneously, although it was not usual to find both types within the same colony. Leptoblasts were usually formed prior to the appearance of intermediate floatoblasts.

Leptoblast colonies produced both more leptoblasts and pyknoblasts, although these were never found together in the same colony at the Sawhill pond. Rogick (1943) pictures a colony from Lake Erie containing both types. During June and July pyknoblasts were fairly common, and they germinated within 2–3 weeks. Few records were made of pyknoblasts giving rise to statoblast-bearing colonies, but those available show the production of only intermediate floatoblasts.

In summary, colonies from intermediate floatoblasts produced both leptoblasts and intermediate floatoblasts; leptoblast colonies formed both pyknoblasts and more leptoblasts; and pyknoblast colonies produced only intermediate floatoblasts. Only the intermediate floatoblasts overwintered to germinate the following spring. No morphological differences were found among colonies arising from different types of statoblasts.

Figure 23 illustrates the relative number of various types of floatoblasts germinating in petri dishes during 1970 at the Sawhill pond. These data are consistent with the relative abundance of floatoblast types found in colonies of *P. casmiana* collected in the pond during 1969 and 1970.

Species Differences in the Production of Floatoblasts and Piptoblasts

Table 2 compares the reproductive potential of three ectoproct species in Sawhill Pond 9 with respect to the production of floatoblasts and piptoblasts. *P. repens* shows

Figure 20

Intermediate floatoblasts of *Plumatella casmiana* collected from Sawhill Pond 9, Oct. 23, 1969.
×87.

Table 2

Comparison of the Reproductive Potential of Three Species of Ectoprocta in Sawhill Pond 9 with Respect to the Production of Floatoblasts and Piptoblasts

Ectoproct Species	Max. No. Statoblasts per Zooid	Mean Minimum Zooid Age for Statoblast Production	% Zooids Over the Minimum Age Having Statoblasts
Plumatella repens	8	5.6	42.7
Plumatella casmiana	3	7.4	25.6
Fredericella sultana	2	8.0	43.7

the greatest maximum number of floatoblasts per zooid (8), and this is well below a previously reported maximum of 25 for the species (Bushnell, 1966). Not only is *P. repens* more prolific than the other two species in floatoblast formation, but it produces floatoblasts at an earlier zooid age. Brown (1933) estimated 800,000 floatoblasts per square meter of plant zone on the basis of 4.5 floatoblasts per zooid.

From the data presented in Table 2, *P. casmiana* would seem to have a lower reproductive potential than *P. repens*. Its polypides require more time to form a floatoblast, they produce fewer of them, and a smaller portion of the colony is engaged in floatoblast formation. However, *P. casmiana* has a great reproductive advantage in its ability to produce leptoblasts which germinate immediately following release, with no intervening diapause. Viganò (1968) observed germination of leptoblasts 1½ to 7 hours after release from the parent colony. Appearing in the Sawhill pond from May through September, leptoblasts effectively enhance the reproductive potential of *P. casmiana* and make it the dominant ectoproct species during most of the year. Since *P. repens* forms only statoblasts requiring a diapause, there is a period between the two summer generations during which colonies of this species are rare.

The consistently small number of piptoblasts found in single zooecia of *F. sultana* belies its great success in many parts of the world. The zooids have a natural longevity greater than zooids of either *Plumatella* species, and *Fredericella* tolerates the greatest diversity of aquatic habitats of all freshwater ectoprocts. The low production of piptoblasts is compensated by the ability of the colony to fragment easily. The fragments, including both living zooids and often several piptoblasts, are passively transported to other sustrates, where they attach lightly by the adhesive tips of young zooecia. Except for their lack of buoyancy, these colony fragments perform a function similar to the leptoblasts of *P. casmiana,* and thus maintain an impressive reproductive potential for the species.

Sessoblasts

The sessoblasts of *P. repens* and *P. casmiana* remain enigmatic. In both species their appearance is sporadic, but most frequent during the spring and fall months. In *P. casmiana* at the Sawhill pond sessoblasts were extremely rare until late August, and by September in most colonies they formed continuous lines throughout the zooecia,

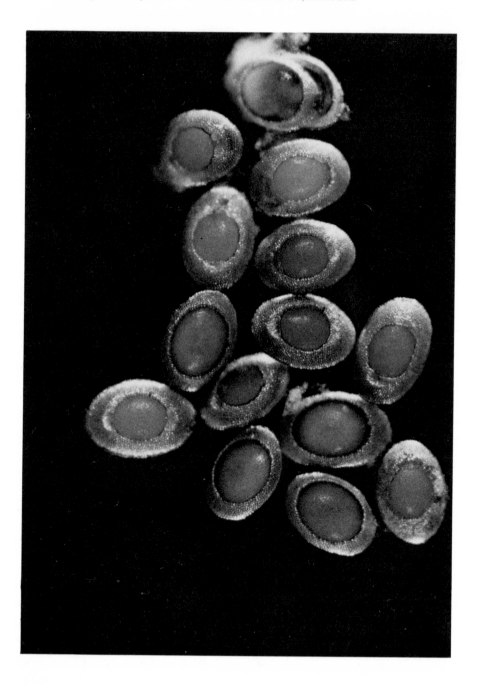

Figure 21

Pyknoblasts of *Plumatella casmiana* collected from Sawhill Pond 9, Oct. 6, 1969. ×87.

Figure 22
Leptoblasts of *Plumatella casmiana* collected from Sawhill Pond 9, June 23, 1970. ×87.

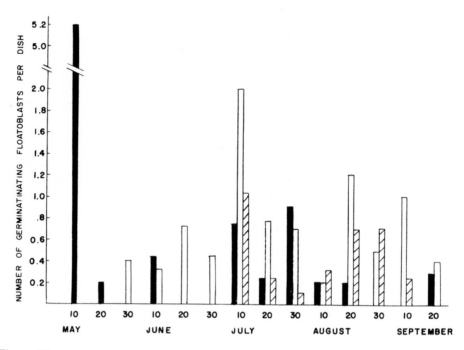

Figure 23

Numbers of floatoblasts of *Plumatella casmiana* germinating per dish during 1970 in Sawhill Pond 9. Solid bars represent intermediate floatoblasts, open bars represent leptoblasts, and hatched bars represent pyknoblasts.

with as many as three to every zooid. In both species sessoblasts occurred with approximately equal frequency in zooids of all ages, and their appearance was usually followed shortly by the death of the zooids. In the laboratory the development of sessoblasts has so far been unpredictable, occurring in some individuals and not in others, but never in actively growing colonies. Laboratory studies confirm the phenomenon of zooid death soon after sessoblast formation, and while a causal mechanism is not necessarily implied, certainly these events cannot be unrelated. In several cases a sessoblast completed development after the zooid had been recorded as dead.

This evidence suggests that sessoblasts form in response to suboptimal environmental conditions, or to general physiological deterioration of individual zooids. Unlike plumatellid floatoblasts, sessoblasts function primarily in holding a favorable substrate during unfavorable conditions for future colony generations. Phylactolaemate ectoprocts lacking sessoblasts always produce floatoblasts with small hooks or spines which perform the same function. It is therefore quite reasonable that sessoblast formation be responsive to environmental conditions, so that they be produced only when likely to be needed.

Sessoblast morphogenesis has never been described, although accounts of floatoblast origins are many (Braem, 1890; Oka, 1891; Kraepelin, 1892; von Buddenbrock, 1910; Brien, 1954). The unpredictable appearance of sessoblasts in the laboratory has so far been an obstacle to studying early developmental stages. The tacit assumption has usually been that sessoblasts form along the funiculus in a similar fashion to piptoblasts and floatoblasts. From observations of sessoblasts halfway through development, however, it is apparent that no direct connection exists between the funiculus and the sessoblasts. It is not yet clear whether sessoblasts develop initially in the funiculus or whether they form under the peritoneum lining the metacoel.

SUMMARY

Colony growth and development were studied intensively in three species of Ectoprocta coinhabiting the small Sawhill Pond 9 near Boulder, Colorado. During 1970 *Plumatella repens, Plumatella casmiana,* and *Fredericella sultana* were grown on inverted glass petri dishes suspended in the water, and were examined twice each week throughout the growing season. Regular collections were made of the three species during 1969 and 1970 at the Sawhill pond, and all were maintained under laboratory conditions for detailed study.

In the Sawhill pond niches were well defined for the three ectoproct species. *Fredericella sultana,* for example, differed from the *Plumatella* species in several significant respects:

1. *F. sultana* showed the greatest colony longevity, extending in one case through at least 7 months. Much of this longevity is explained by a tolerance of low temperatures, but it was also partly due to a remarkably long zooid life span.

2. The percentage of zooids producing three and four buds was significantly greater than in the *Plumatella* species, possibly because high polypide longevity allowed sufficient time for late buds to mature. However, a parallel situation was not observed in piptoblast development. A second successive piptoblast appeared only at a mean zooid age of 16.3 days, and no zooid was seen to form three piptoblasts.

3. *F. sultana* was the only species in the pond showing noticeable sexual activity. Sperm was circulating within colonies collected during May of 1969 and 1970, although neither ova nor larvae were ever observed.

4. The use of living colony fragments as agents of dispersal is unique to *F. sultana* and was commonplace in the Sawhill colonies. Since this species can thrive at low temperatures that retard piptoblast development, fragmentation would seem to make it well adapted to perenially cold waters.

5. The erect branching form of *F. sultana* colonies, which makes possible dispersal by fragmentation, is apparently the result of two factors: (a) the considerable, slow elongation of young zooecia which would tend to discourage firm attachment to the substrate; and (b) the relatively weak and short-lived adhesive properties of young zooecia.

6. Piptoblasts were produced at a constant rate throughout the life of the colony. In contrast to the situation in the *Plumatella* species, in which most statoblasts are formed during a relatively brief time, piptoblasts seem to occur only when the individual zooid reaches a certain physiological condition.

P. repens and *P. casmiana* have many structural similarities, and there is probably little difference in the food organisms taken in by their lophophores. Definition of their niches ultimately involves differences in their reproductive activities.

1. The difference in colony form of *P. repens* and *P. casmiana* (repent versus compact) reflects primarily a temporal difference between the formation of successive buds by

a single zooid. In *P. casmiana* the secondary bud appears at a mean zooid age of 7.3 days, compared to 10.4 days in *P. repens*. A compact colony may afford some protection from damage by tendipedid larvae.

2. While colony growth in *P. repens* tended to follow a simple geometric pattern in the Sawhill pond, in *P. casmiana* it was generally very irregular. Budding rates fluctuated in *P. casmiana* from week to week throughout the colony, and to a lesser extent within various regions of a single colony.

3. All statoblasts of *P. repens* have a diapause of at least 3 weeks. After the spring colonies have died, there is an interval during which few colonies of this species are found until the spring crop of statoblasts germinates. *P. casmiana* releases leptoblasts through a vestibular pore of living polypides, which germinate almost immediately. Because of these leptoblasts, colonies of *P. casmiana* persist from May through September at Sawhill Pond 9. Theoretically this species may have up to seven leptoblast generations per year in the pond.

4. Although several types of statoblasts were formed by *P. casmiana,* the overall statoblast productivity was low, involving only 10.5 percent of all zooids. Very few zooids formed more than two floatoblasts. By contrast, *P. repens* was prolific in statoblast production, with 18.8 percent of the zooids having as many as eight each.

Despite the difference among these three species, all show similar trends in colony development. In general, an increase in colony age was accompanied by (1) a reduction in zooid longevity; (2) a decrease in the percentage of zooids involved in statoblast and bud production; and (3) the production of statoblasts and buds at an increasingly earlier zooid age. The pattern of these progressive changes suggests control by one or more substances circulated and accumulated within the metacoel of the colony. The asexual behavior of young isolated colony fragments is consistent with this hypothesis. Confirmation must come from physiological studies of the developing colony.

In all species the mortality and budding rates tended to change together, suggesting a common mechanism of control. It is not clear why this should be so, and further study is indicated.

ACKNOWLEDGMENTS

This work represents a major portion of the author's dissertation research at the University of Colorado. I am particularly grateful to John H. Bushnell for his valuable comments and criticism through all stages of the investigation. Robert W. Pennak was very helpful in his critical review of the original manuscript. Transportation expenses involved in the study were met by a grant from the Kathy Lichty Memorial Fund, Department of Biology, University of Colorado.

REFERENCES

Allman, G. J. 1856. A monograph of the fresh-water Polyzoa including all the known species, both British and foreign. Ray Society, London, *28*: 1–119.

Becker, G. 1937. Untersuchunger über den Darm und die Verdauung von Kamptozoen, Bryozoen und Phoroniden. Zeitsch. Morph. Okol Tiere, *33*: 72–127.

Berrill, N. J. 1953. Growth and form in gymnoblastic hydroids, VII, Growth and reproduction in *Syncorne* and *Coryne*. Jour. Morph. *92*: 273–302.

Borg, F. 1926. Studies on recent cyclostomatous Bryozoa. Zool. Bidrag, Uppsala, *10*: 181–507.

Braem, F. 1890. Untersuchungen über die Bryozoen des süssen Wassers. Bibl. Zool., *2*: 1–134.

Braverman, M. h. 1962. Studies in hydroid differentiation I. *Podocoryne carnea*: culture methods and carbon dioxide induced sexuality. Exp. Cell Res., *27*: 301–306.

———. 1963. Studies on hydroid differentiation II. Colony growth and the initiation of sexuality. Jour. Embryol. & Exp. Morph., *11*: 239–253.

Brien, P. 1936. Contribution à l'étude de la reproduction asexuée des Phylactolémates. Mém. Mus. nat. Belg., *3*: 569–625.

———. 1954. Etude sur les Phylactolémates. Evolution de la zoécie-—Bourgeonnement d'accroissement—Bourgeonnement statoblastique—Embryogenése—l'Ontogenèse multiple. Ann. Soc. Zool. France, *79*: 203–239.

Brooks, C. 1929. Notes on the statoblasts and polypides of *Pectinatella magnifica*. Acad. Nat. Sci. Phila., Proc., *81*: 427–441.

Brown, C. J. D. 1933. A limnological study of certain fresh-water Polyzoa with special reference to their statoblasts. Am. Micros. Soc., Trans., *52*: 271–313.

———. 1934. Internal budding: with suggestions for a laboratory study of the fresh-water Polyzoa. Am. Micros. Soc., Trans., *53*: 425–430.

Buddenbrock, W. von. 1910. Beiträge zur Entwicklung der Statoblasten der Bryozoen. Zeitsch. Wissensch. Zool., *96*: 447–524.

Bushnell, J. H. 1965a. On the taxonomy and distribution of freshwater Ectoprocta in Michigan, pt. I. Am. Micros. Soc., Trans., *84*: 231–244.

———. 1965b. On the taxonomy and distribution of freshwater Ectoprocta in Michigan, pt. II. Am. Micros. Soc., Trans., *84*: 339–358.

———. 1966. Environmental relations of Michigan Ectoprocta and dynamics of natural populations of *Plumatella repens*. Ecol. Monogr., *36*: 95–123.

———. 1968. Aspects of architecture, ecology and zoogeography of freshwater Ectoprocta (Bryozoa). Atti Soc. Ital. Sci. Nat. Mus. Civ. St. Nat. Milano, *108*: 129–151.

———. and T. S. Wood. 1971. Honeycomb colonies of *Plumatella casmiana* Oka. (Ectoprocta: Phylactolaemata). Am. Micros. Soc., Trans., *90*: 229–231.

Cheetham, A. H. 1968. Evolution of zooecial asymmetry and origin of poricellariid cheilostomes (Bryozoa). Atti Soc. Ital. Sci. Nat. Mus. Civ. St. Nat. Milano, *108*: 185–194.

Dendy, J. S. 1963. Observations on bryozoan ecology in farm ponds. Limn. and Oceanog., *8*: 478–482.

Forel, F. 1884. La faune profonde des lacs suisses. Neue Denkschr. Allg. Schweiz. Gesellsch. Gasamt. Naturwiss., *29*: 1–234.

Fulton, C. 1960. Culture of a colonial hydroid under controlled conditions. Science, *132*: 473–474.

Gerwerzhagen, A. 1913. Das Nervensystem von *Cristatella mucedo* Cuv. Beiträge zur Kenntnis der Bryozoen I. Zeitsch. Wissensch. Zool., *107*: 309–345.

Jónasson, P. M. 1963. The growth of *Plumatella repens* and *P. fungosa* (Bryozoa Ectoprocta) in relation to external factors in Danish eutrophic lakes. Oikos, *14*: 121–137.

Kraepelin, K. 1887. Die deutschen Süsswasserbryozoen I. Anatomisch systematischer Teil. Abhandl. Gebiete Naturwiss. Verein Hamburg, *10*: 1–168.

———. 1892. Die deutschen Süsswasserbryozoen. II. Entwicklungsgeschichtlicher Teil. Abhandl. Gabiete Naturwiss. Verein Hamburg, *12*: 7–67.

Lacourt, A. W. 1968. A monograph of the freshwater Bryozoa—Phylactolaemata. Zool. Verhandl., *93*: 1–159.

Loomis, W. F. 1959. Feedback control of growth and differentiation by carbon dioxide tension and related metabolic variables, *in* D. Rudnick, ed., Cell, Organism and Milieu. The Ronald Press Co. p. 253–293.

Marcus, Ernst. 1925. Tentaculata. Kranzfühler: Bryozoa. Moostiere. Biologie der Tiere Deutschlands. Lieferung 14, Teil, *47*: 46 p.

———. 1926. Beobachtungen und Versuche an lebenden Süsswasserbryozoen. Zool. Jahrb. Syst., *52*: 270–350.

———. 1934. Uber *Lophopus crystallinus* (Pall.). Zool. Jahrb. Abt. Anat., *58*: 501–606.

———. 1941. Sôbre Bryozoa do Brasil. Bol. Fac. Fil. Cienc. Letr. Sao Paulo Zool., *5*: 3–208.

Marcus, Eveline. 1955. Polyzoa. Percy Sladen Trust Expedition to Lake Titicaca. Linn. Soc. London, Trans., ser. 3, *1*: 355–357.

Oda, S. 1960. Relation between sexual and asexual reproduction in freshwater Bryozoa. Bull. Mar. Biol. Stat. Asamuchi, Tôhoku Univ., *10*: 111–116.

Oka. A. 1891. Observations on fresh-water Polyzoa. Jour. Coll. Sci. Imp. Univ. Tokyo, *4*: 89–150.

———. 1907. Zur Kenntnis der Süsswasser-Bryozoenfauna von Japan. Annot. Zool. Jap., *6*: 117–123.

Rey, P. 1927. Observation sur le orps brun des Bryozoaires Ectoproctes. Bull. Soc. Zool. France, *52*: 367–379.

Rogick, M. D. 1935. Studies on fresh-watr Bryozoa, III, Development of *Lophopodella carteri* var. *typica.* Ohio Jour. Sci., *35*: 457–467.

———. 1938. Studies on fresh-water Bryozoa, VII, On the viability of dried statoblasts of *Lophopodella carteri* var. *typica.* Am. Micros. Soc., Trans., *57*: 178–199.

———. 1940. Studies on fresh-water Bryozoa, IX, Additions to New York Bryozoa. Am. Micros. Soc., Trans., *59*: 187–204.

———. 1941. Studies on fresh-water Bryozoa, X, The occurrence of *Plumatella casmiana* in North America. Am. Micros. Soc., Trans., *60*: 211–220.

———. 1943. Studies on fresh-water Bryozoa, XIII, Additional *Plumatella casmiana* data. Am. Micros. Soc., Trans., *62*: 265–270.

———. 1945. Studies on fresh-water Bryozoa, XV, *Hyalinella punctata* growth data. Ohio Jour. Sci., *45*: 55–79.

Römer, O. 1906. Knospung Degeneration, und Regeneration von einigen marinen ectoprocten Bryozoen. Zeitsch. Wissensch. Zool., *84*: 446–478.

Soule, J. 1954. Post-larval development in relation to the classification of the Ctenostomata. Bull. South. Calif. Acad. Sci., *56*: 13–34.

Tenney, W. R., and W. S. Woolcott. 1966. The occurrence and ecology of freshwater bryozoans in the headwaters of the Tennessee, Savannah and Saluda river systems. Am. Micros. Soc., Trans., *85*: 241–245.

Toriumi, M. 1951. Taxonomical study on fresh-water Bryozoa. I. *Fredericella sultana* (Blumenbach). Sci. Rep. Tôhoku Univ., *19*: 167–177.

———. 1955a. Taxonomical study on fresh-water Bryozoa, IX, *Plumatella repens* (L.). Sci. Rep. Tôhoku Univ., *21*: 51–66.

———. 1955b. Taxonomical study on fresh-water Bryozoa, X, *Plumatella casmiana* Oka. Sci. Rep. Tôhoku Univ., *21*: 67–77.

Viganò, A. 1968. Note su *Plumatella casmiana* Oka (Bryozoa). Riv. Idrobiol., *7*: 421–468.

Wiebach, F. 1963. Studien über *Plumatella casmiana* Oka (Bryozoa). Vie et Milieu, *14*: 579–598.

Wood T. 1971. Laboratory culture of fresh-water Ectoprocta. Am. Micros. Soc., Trans., *90*: 90–94.

Zschokke, F. 1906. Ubersicht über die Tiefenfauna des Vierwaldstattersees. Arch. Hydrobiol., *2*: 1–8.

Effects of Colony Size on Feeding by *Lophopodella carteri* (Hyatt)

John W. Bishop

University of Richmond

Leonard M. Bahr

University of Georgia

ABSTRACT

Currents and feeding by the bryozoan *L. carteri* were examined to ascertain the influence of aggregate feeding in a compact colony. Regions of currents of contiguous zooids overlapped. Clearance rate (microliter/zooid/minute) decreased as colony size (number of zooids) increased. Interactions of currents are discussed with reference to the decreased clearance rates.

INTRODUCTION

Suspension feeders commonly procure food from currents of water that they create. When animals aggregate, their currents may interfere with one another and inhibit feeding or they may become more extensive and enhance feeding (Bidder, 1923; Mackie, 1963). Interactions of currents could be related both to the shape and size of an aggregation within a particular environment. In the present study, currents and feeding of the freshwater bryozoan *Lophopodella carteri* (Hyatt) were examined to ascertain the influence of aggregate feeding in a compact colony.

Authors' addresses: J. W. B., Department of Biology, University of Richmond, Richmond, Virginia 23173; L. M. B., Department of Biology, University of Georgia, Athens, Georgia 30601.

METHODS

L. carteri was collected from the Virginia State Fish Cultural Station in Stevensville, Virginia, and maintained without food in glass containers of spring water at 23–26°C for 1 to 2 days before each experiment.

To measure currents, isolated zooids of *L. carteri* were placed in gridded petri dishes with a diameter of 5 cm. The dishes contained suspensions of *Euglena gracilis* or *Saccharomyces cerevisiae*. The zooids and suspensions were observed under a dissection microscope and photographed with a motion-picture camera. The movements of suspensions were used to estimate the directions and extents of the currents.

To measure feeding, colonies of *L. carteri* of known size were placed into vials with a diameter of 22 mm. Each vial contained a 10-ml suspension of ^{14}C-tagged *E. gracilis*. Three control vials containing *E. gracilis* and no *L. carteri* were set up for each experiment. After 15 minutes, a drop of Lugol's solution was added to each control and experimental vial. A 5-ml aliquot was removed and filtered through a Millipore filter. The filter was analyzed for radioactivity in a Geiger-Müller gas flow counter. A 0.1-ml aliquot was removed from each control vial, placed in a Palmer chamber, and used to estimate the concentration of *E. gracilis* cells.

The numbers of cells ingested by *L. carteri* were estimated from differences in radioactivity between the experimental and control samples. Clearance rates, microliters of water cleared of food per zooid per minute, were calculated by the method described in Gauld (1951), in which food was assumed to decrease exponentially with time.

RESULTS AND DISCUSSION

Currents approached the lophophore of a zooid frontally (Fig. 1). The afferent currents extended 5 mm from the oral side and the efferent currents were limited to 4 mm beyond the aboral side of the lophophore. Some water, after passing between tentacles, formed eddy currents of 2 mm diameter on all sides of the tentacular crown. Colonies of *L. carteri* were roughly circular and consisted of 2–66 zooids. The average distance between neighboring lophophores was approximately 0.2 mm. The diameters of the colonies were roughly proportional to the numbers of zooids within the colonies (Table 1).

At high concentrations of *E. gracilis*, 40–159 cells/microliter, clearance rates were unrelated to colony size. Visual observations suggested that food was procured faster than it was ingested at these concentrations. The currents persisted and food was swallowed, but excess food built up around the mouth and spilled over between the tentacles about every minute. Any influences of currents on clearance rates, therefore, may have been masked by limited ingestion rates.

At low concentrations of *E. gracilis*, 4–16 cells/microliter, clearance rates decreased as colony size increased (Table 2). The decrease was most dramatic among colonies of less than 10 zooids. Because of ample space within the vials, confinement of *L. carteri* was not considered an important influence on clearance within this size range. The distances between the periphery of a colony of 10 zooids and the sides of the

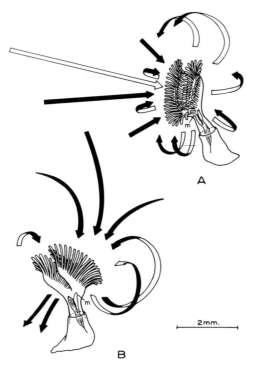

Figure 1

An isolated zooid of *L. carteri* showing ciliary currents: A, laterofrontal view; B, lateroabfront
view. Dark surfaces of arrows are convex; light surfaces are concave; m indicates mouth.

Table 1

Average Diameter For Colonies of *L. carteri* Consisting of Various Numbers of Zooids

Zooids (number/colony)	Diameter (mm)
11–15 (1)[a]	7
16–20 (1)	8
21–25 (2)	10
26–30 (3)	10
31–35 (5)	12
36–40 (3)	14
41–45 (1)	13
46–50 (4)	14
51–55 (5)	16
56–60 (2)	15

[a]The number of observations is in parentheses.

Table 2

Average Clearance Rates for Colonies of *L. carteri* Within Various Size Classes

Colony Size (number of zooids)	Clearance Rate (microliter/zooid/minute)
1–2 (1)[a]	102
3–4 (4)	84[b]
5–6 (4)	56
7–8 (4)	46
9–10 (2)	36
11–15 (5)	45
16–20 (7)	42
21–25 (2)	38
26–30 (3)	40
31–35 (5)	28
41–45 (5)	21
65–66 (3)	20

[a]The number of observations within each size class is in parentheses. The concentrations of *E. gracilis* were 4–16 cells/microliter.

[b]A clearance rate of 543/microliters zooid/minute for a colony comprised of three zooids appears spurious and is not included in the calculations.

vial and surface of the water were 7 and 20 mm, respectively. The currents of a zooid extended at most to a distance of 5 mm.

The reasons for the decreased rates are not clear. The decrease could result from the behavior of the zooids. Large colonies consist of older zooids, which may be less active feeders than young zooids. Second, a zooid in a large colony would be subjected to more tactile stimulation from its neighbors. It can be demonstrated that a zooid stops feeding when touched, and this response could account for a lower clearance rate.

Mutual interference of currents of contiguous zooids could decrease the rates. We found that an isolated zooid created currents within a dome-shaped region about its lophophore. Clearance rate, therefore, depended on the surface area of this region and the velocity with which water was drawn toward the lophophore. The diameter of the region was approximately 5 mm, so that within a colony, the currents of contiguous zooids overlapped. Such overlapping might decrease the surface area available for each zooid, thereby limiting clearance rate.

Currents of a colony also may form a dome-shaped region about the colony. The surface area/unit volume would be inversely related to the volume of this region. Likewise, the surface area/zooid would be inversely related to colony size, assuming that colony size were proportional to the volume of the dome-shaped region. If clearance rate was limited by the surface area/zooid, clearance rate would be expected to be inversely related to colony size. Equation (1) shows such a relationship based on our data:

$$Y = 25 + 181X^{-1} \tag{1}$$

Here Y is clearance rate, X is colony size, and the numerical values are constants. Although much of the variability in the data still was unexplained, the relationship between clearance rate and colony size expressed in this fashion was statistically significant ($r = 0.64$, $p < 0.05$).

Interactions of currents do not always decrease feeding and under certain conditions may enhance feeding. Mackie (1963) reports that zooids of the bryozoan *Plumatella fungosa* can orient together. As a result, their currents may interact and extend over greater distances than when the zooids act independently. Such a phenomenon would tend to increase the area available per zooid and thus increase clearance rate. Our data suggest that as a colony of *L. carteri* approaches 10 zooids, processes enhancing clearance increase with respect to those that decrease clearance. In larger colonies, the two processes appear to counterbalance each other.

ACKNOWLEDGMENTS

We would like to thank William S. Woolcott and Wilton R. tenney of the University of Richmond for their advice on the research and manuscript.

REFERENCES

Bidder, G. P. 1923. The relation of the form of a sponge to its currents. Quart. Jour. Micros. Sci., *67*: 293–323.

Gauld, D. T. 1951. The grazing rate of planktonic copepods. Jour. Mar. Biol. Ass. U. K., *29*: 695–706.

Mackie, G. O. 1963. Siphonophores, bud colonies and superorganisms, *in* E. C. Dougherty, ed., The Lower Metazoa Comparative Biology and Phylogeny. U. Calif. Press, Berkeley.

Introduction to Graptolites

Graptolites comprise an extinct group of wholly colonial organisms that are considered to be most closely related to the living Hemichordata. They flourished during early Paleozoic time. Numerous detailed studies have elucidated their morphology and permitted functional interpretations that are more advanced than those for any other extinct Paleozoic group of animals. With respect to coloniality, graptolites exhibit certain phenomena that are unknown or only weakly developed in other colonial animals.

Urbanek discusses in detail the organization of graptolite colonies with emphasis on their morphological polarity. This is interpreted in terms of physiological gradients in the distribution of morphogenetic substances and of penetrance and expressivity of the genetic factors involved. The whole is placed in context by a broad discussion of graptolite evolution and of coloniality in general.

Jaanusson interprets morphological discontinuities in graptolite evolution in terms of genetic polymorphism. **Erdtmann** develops a picture of a functioning graptolite colony through analogy with living pterobranchs. **Berry** and **Takagi** present a generalized summary of graptolite coloniality through analysis of skeletal fine structure.

Organization and Evolution
of Graptolite Colonies

Adam Urbanek   University of Warsaw and Smithsonian Institution

ABSTRACT

A graptolite colony is a collective unit integrated by physical continuity of zooids
which are subject to common morphogenetic control and share a common genotype.
Heteropolar organization of the colony is considered an important feature of Graptoloidea. Regular astogenetic succession of thecae in graptoloids is interpreted as
an effect of morphophysiological gradients, based on spreading of a morphogenetic
substance produced by the siculozooid. Observations on multiramous and regenerative
bipolar rhabdosomes give evidence in favor of this hypothesis. New phylogenetic
characters occur polarly, and the spreading of new thecal types in successive links
of a lineage is interpreted as change in penetrance and expressivity. The peculiar
pattern of colonies in some Idiotubidae is explained as an expression of a complexity
gradient, comparable with that in Plumulariidae (Hydrozoa). Colonies are collective
biological systems, although in certain aspects they might be considered as single
individuals. Therefore, a dualistic approach to the problem of coloniality is suggested.
Description of each instance of colony evolution in terms of its individuality increase
and individuality transfer is considered misleading. Comparison with highly integrated
pseudocolonial populations seems equally heuristic, and in some aspects more adequate.
While graptoloid zooids were possibly hermaphroditic, patterns of their evolution
suggest rather normal crossbreeding. Hypothetical mechanisms preventing inbreeding
and ensuring outcrossing which could operate in graptoloid colonies are therefore
discussed.

Author's address: Katedra Paleontologii U.W., Uniwersytet Warszawski, Warszawa, Poland.

BIOLOGICAL SIGNIFICANCE OF COLONIAL ORGANISMS

There is a renewed and growing interest in the investigation of colonial organisms including their fossil representatives. The biology of colonial organisms ("Kormenbiologie" of German zoologists) in indeed a separate branch of the life sciences and a rather promising field of investigation. Until recently, little attention was paid to this circle of problems in spite of its great importance. A number of recent and rather fashionable problems of theoretical biology such as integration, individuality, wholeness, and levels of organization are closely related to this field of investigation.

It is therefore truly surprising that so little is known, not only of the genetics and physiology, but even the morphogenesis and development of Recent colonial animals. The understanding of evolutionary patterns has been limited to rather general and outmoded speculations of the application of biogenetic law to the colonies. With the exception of an excellent study of Beklemishev (1950, 1964), no attempt has been made more recently to synthesize the data to create a general theoretical concept of colony organization and evolution.

Paleontology may contribute to the improvement of our knowledge of colonial organisms through a better understanding of fossil colonial animals. Some extinct groups formed particular patterns of colonies unknown among Recent organisms. Graptolites can serve as an example of such a fossil group, and knowledge of them may help to complete our understanding of the organization and evolution of colonies.

Investigations made on three-dimensionally preserved material etched by chemical treatment, combined with studies of specimens flattened and preserved on rock surfaces, have supplied important data concerning the development, structure, and evolution of graptolite colonies. In the light of accumulated data, a graptolite colony permits one to trace the interrelations of such phenomena as *ontogeny* and *metamorphosis* of the primary zooid and the founder of the colony (the siculozooid, which originates by sexual reproduction), and the budding and growth of a number of daughter individuals produced asexually and connected throughout the colony by soft tissues. The development of the primary oozooid and the sequential origin and arrangement of the asexually produced blastozooids contribute to an event of higher rank—the development of the colony as a whole (astogeny). This permits an analysis of different morphological patterns of interaction between development of individual zooids and the entire colony.

The significance of fossil material is increased by the fact that it is often possible to trace reliable phyletic lineages showing evolutionary modifications of astogeny. Comparison of astogeny in closely related taxa, producing a series with a regular morphological and stratigraphical sequence, enables one to recognize main trends and modes of phylogenetic change in astogeny. These changes constitute the basic events of evolution in colonial organisms, for which the term *phylastogeny* has been suggested (Urbanek, 1963).

These considerations indicate that a graptolite colony is an almost unique object which permits one to analyze changes going on at different levels of integration. These relationships are (1) siculozooid (oozooid) and its ontogeny; (2) blastozooids, their budding (blastogeny) and growth; (3) development of the entire colony (astogeny); (4) phylogenetic modifications of astogeny (phylastogeny); (5) changes on the species-population level (changes of composition of populations, speciation); and (6) changes in particular trends and lineages.

These properties of graptolite material throw a new light on mechanisms involved in the evolution of colonial animals and complete to some extent our picture of evolutionary mechanisms based primarily on solitary organisms.

The starting point for this review of problems concerning the organization of graptolite colonies are the specialized Silurian graptoloids of the order Monograptina. Whereas non-graptoloid primitive and benthonic orders of Graptolitina, especially the Dendroidea, supplied decisive data which may help to solve the problem of the origin and kinship of the whole class with the other invertebrates (Kozlowski, 1949, 1966), the specialized and late representatives of planktonic Graptoloidea were especially significant in the interpretation of the organization of the colony and for an understanding of the morphogenetic relationships involved in colony formation. Monograptids probably attained the highest level of colony integration among graptolites, and a number of phenomena, important for recognizing their general organization, are emphasized. The simple methodological principles employed here may be useful in determining the proper mode of approach to similar problems in other groups. For example Schenck (1965) has arrived at important results concerning coloniality in Recent hydroids through studies on their most specialized group—the Plumulariidea.

Any realistic approach to problems of coloniality in graptolites must take into account their collective nature, so astogenetic and phylogenetic changes can be recognized only through changes in particular zooids. The total number of zooids in a graptolite colony varies over a large range, being greatest in Dendroidea and smallest in certain Graptoloidea which display progressive tendencies toward reduction in the number of zooids. As estimated by Bulman (1955), this number is 20,000–30,000 in large *Dictyonema flabelliforme* and only 2000–3000 in multiramous anisograptids (as *Clonograptus*) and in large multibrachiate dichograptids. In the remaining graptoloids the total number of zooids ranges from three (with only one normally developed blastozooid!) in *Corynites* (Kozlowski, 1953, 1956) to approximately 1500 as estimated by Jaeger (1959) in the four-branch growth stage in *Linograptus posthumus* (observed number of one branch—360 zooids). Bulman (1955) has rightly stressed that as a result of the loss of bithecae and the reduction of the number of stipes, the total number of zooids in graptoloid colonies has been strikingly reduced as compared with that in dendroids. Averages for most leptograptids, dicellograptids, and diplograptids are 100–200, and perhaps 50–100 in typical monograptids, being only 20–40 in Early Devonian graptoloids. On the other hand, Jaeger (1959) emphasized the fact that there is no evidence for an overall tendency toward progressive reduction of the colony in monograptids. This is indicated by the appearance of late Silurian linograptids showing a strong increase in the total number of thecae in the rhabdosome. However, even in "normal," one-stiped monograptids, some giant rhabdosomes have been found, which strongly exceed the average size observed in the majority of representatives of a given species, and include some 350–1000 zooids (Regnéll, 1949). In most monograptids the colony is composed of 20–350 zooids, mutually interconnected by the stolon. Environmental factors seem to be responsible to a great extent for the size of rhabdosomes, as is indicated by dwarfed faunulae under extreme conditions (Jaeger, 1959; Urbanek, 1970).

POLAR ORGANIZATION OF GRAPTOLOID COLONIES

The skeleton of a monograptid colony, the rhabdosome, is under normal conditions the only part of the colony capable of preservation in the fossil state. It consists of an organic substance, probably a scleroprotein, whose amino acid composition has been preliminarily established in a few cases (Foucart and Jeuniaux, 1965) and for which the term "graptin" was suggested (Kozlowski, personal communication). Its ultrastructure has recently been examined (Berry and Takagi, 1970, 1971), but is still inadequately

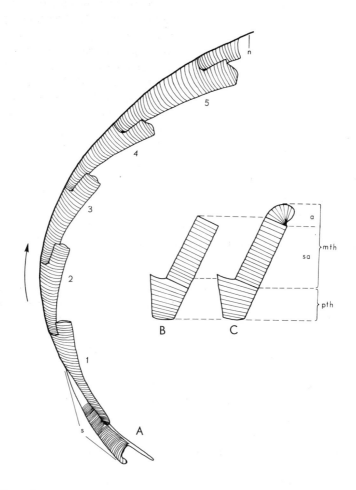

Figure 1

Diagrams illustrating structure and terminology of an uniaxiate monograptid rhabdosome.
A. Proximal part of rhabdosome, showing sicula (s); a number of thecae (1–5); and nema or virgula (n), a thread-like prolongation of apex of sicula which may serve as axis for float structures.
B, C. Constitution of a theca, composed of apertural portion (C) or devoid of it (B), an apertural part of metatheca, metatheca (mth), protheca (pth), subapertural part of metatheca (sa).

understood.* Happily enough, secretion of this skeleton has been rather intimately related to the soft tissues of particular zooids, and all skeletal parts are usually informative for understanding the structure of particular zooids, their mode of budding and interconnections, and for comprehension of the development of the entire rhabdosome.

From the anatomical point of view, the rhabdosome (Fig. 1A) is an assemblage of a number of interconnected thecae, each of them being a tube or lodging for an individual zooid. The initial theca of the colony is termed the *sicula*. The initial zooid is termed the *siculozoid* and is the only individual formed by sexual reproduction and therefore is the only oozooid of the colony. A sicula differs distinctly from the other thecae of the rhabdosome. All remaining zooids were formed by successive budding and are therefore all blastozooids. Their zooidal tubes are simply called thecae. Thecae are often straight tubes with a simple aperture, but the majority of monograptids developed more complicated thecal types for which a number of technical terms have been applied (see Bulman, 1955, 1970). Independent of the particular thecal elaboration, each theca may be divided into two parts—the *basal protheca* (Bulman, 1951), through which openings are connected with internal cavities of adjacent thecae, and an upper part, the *metatheca* (Urbanek, 1953), which is occupied by the zooid proper. In monograptids the terminal part of the metatheca is commonly elaborated as a hook, or provided with apertural lobes, lappets, or spines projecting over the aperture proper. The presence of terminal structures makes desirable the subdivision of the metatheca into two segments, subapertural and apertural (Fig. 1B, C; Urbanek, 1960).

In a vast majority of monograptids the rhabdosome produced only one stipe (the procladium of Urbanek, 1963); such colonies are called *uniaxiate* or *uniramous*. In certain monograptids, however, more stipes were formed by the addition of secondary branches (metacladia of Urbanek, 1963) to the primary stipe. These are compound *multiaxiate* or *multiramous* rhabdosomes, which may differ in the number and position of secondary branches as well as in mode of origin.

Uniaxiate, one-stiped rhabdosomes are morphologically simplest and thus a most convenient starting point for any consideration of colonial morphogenesis in monograptids. The most striking and important fact, which may be observed here is a gradual, progressive distal increase in the size of thecae. Successive thecae become gradually but distinctly larger as the growth of the rhabdosome progresses. As the changes of thecal size are rarely isometric, they are often correlated with certain, generally minor, allometric changes in length/width ratio, or changes in proportion of thecal segments. Size and proportion of thecae are therefore different at the proximal and distal ends of the rhabdosome, although in the course of astogeny contrasting thecal types are attained in each case by gradual modifications in the successive thecae. So gradual are these size and proportion changes that application of the term "size gradient" seems to be appropriate.

Size and proportion changes are rarely the only thecal characters involved in astogenetic changes. A number of structural characters such as inclination angle of thecae to the axis of rhabdosome, their overlap, bending of thecae, incurvation of apertural parts,

---

*A new light has been thrown on ultrastructure and composition of graptolite periderm by a recent study by Towe and Urbanek (1972).

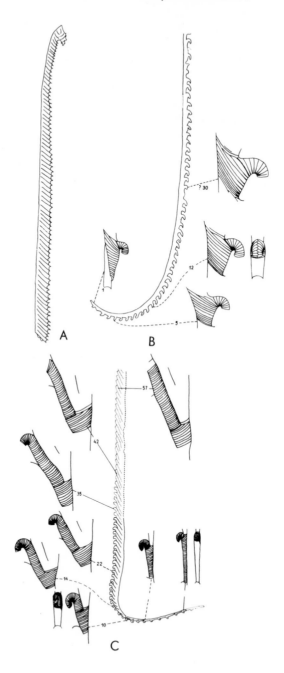

Figure 2

Thecal changes along rhabdosome in A, *Didymograptus pakrianus* Jaanusson, only one branch presented; B, *Monograptus clingani* (Carruthers); C, "*Monograptus (= Pernerograptus)* argenteus (Nicholson). A, B belong to uniform, and C to proximally progressive, biform types. (From Bulman, 1951, and Jaanusson, 1960.)

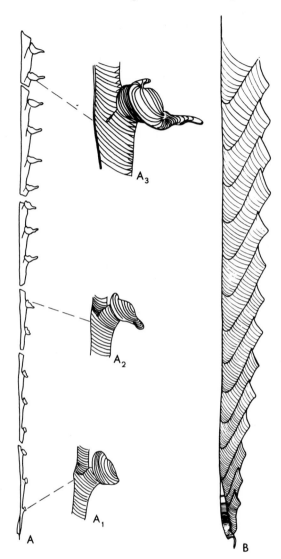

Figure 3

Thecal changes along rhabdosome in A, *Cucullograptus aversus rostratus* Urbanek; A_1–A_3, details of structure of apertural apparatus in proximal, medial, and distal thecae; B, *Pristiograptus dubius* (Suess), fusellar structure shown diagrammatically. A is a distally progressive biform type and B, a uniform type. (After Urbanek, 1958, 1966.)

isolation of thecae, their spacing, and presence or absence of apertural additions in addition to their size and shape are commonly subject to astogenetic variation. Figures 2 and 3 show some examples of structural characters of thecae involved in intracolonial morphological variation. In some forms structural differences are so extreme between proximal and distal thecae of the same rhabdosome that such thecae hardly seem conspecific (Fig. 3A). Monograptids in which thecae represent two extreme morphological types were called "biform" by Elles and Wood (1901–1918). Nevertheless, in each case investigated the extreme thecal types on opposite ends of the rhabdosome have been connected by a great number of intermediates, which serves to confirm the previous conclusion that astogenetic changes in monograptids were expressed by gradual modification of successive thecae. When the size and proportion changes are accompanied by structural ones, the term "structural gradient" appears justified.

On the whole, structural differences due to astogenetic variation of thecae, in a comparative-anatomical aspect, may be based on different modifications of the same thecal segment present in proximal and distal thecae (Figs. 4 and 5C, D), or based on presence or absence of respective thecal segments (Fig. 5A, B) in particular thecae of the rhabdosome (Urbanek, 1960). These facts are of importance for a discrimination between alternative interpretations of supposed causes of morphological gradients in graptolite colonies (see below).

Generally in monograptids astogenetic variation most frequently involved (1) a size gradient, which is fundamental and present in each case examined; (2) a structural gradient which is frequently present and secondarily superimposed on the size gradient and is expressed by different structural characters peculiar to given taxa; and (3) changes in shape of the entire rhabdosome, often provided with different curvature, coiling, etc., in proximal and distal parts, or in the position and spacing of lateral branches (thecal cladia) in compound rhabdosomes.

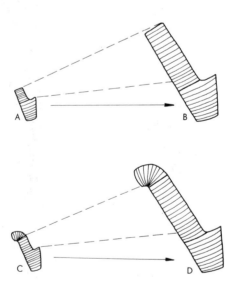

Figure 4

Morphological analysis of changes involved in astogeny of uniform monograptids. A, B. Thecae without the apertural portion of metatheca throughout the rhabdosome. C, D. Thecae with the apertural portion throughout the rhabdosome.

Broken lines delimit the subapertural portion of metatheca, arrows indicate distal direction; note increase in size distalward and change in proportion of thecal segments between proximal and distal thecae.

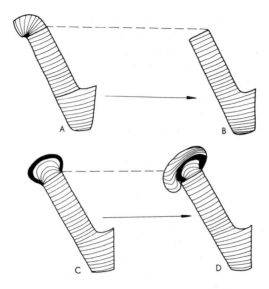

Figure 5

Morphological analysis of structural changes involved in astogeny of biform monograptids.
A, B. Proximally progressive type based on presence in proximal (A) and absence in distal thecae (B) of the apertural portion of metatheca.
C, D. Distally progressive type based on modifications (symmetry or asymmetry of apertural lobes) of apertural portion present in proximal and distal thecae. Broken lines indicate apertural level to distinguish apertural and subapertural portions of metatheca; arrows indicate distal direction; size differences between proximal and distal thecae were neglected.

   Taking into consideration all the evidence available on astogenetic variation in monograptids, certain general patterns of uniaxiate colonies may be distinguised. (1) Colonies which show only a size gradient, the structural pattern of thecae being the same throughout the rhabdosome; such rhabdosomes can be termed *uniform* (Figs. 2A, B and 3B). (2) Colonies which display a structural gradient superimposed on the size-gradient system (Figs. 2C and 3A); they are called *biform*, especially when differences in thecae on opposite ends of the colony are pronounced. Two types of biform colonies may be distinguised: (1) those with proximal thecae displaying certain characters absent in distal ones, or showing more pronounced expression of certain characters, as compared with distal thecae (Fig. 2C); and (2) those with distal thecae showing some characters absent in proximal ones, or displaying more pronounced expression of given characters as compared with proximal thecae (Fig. 3A).
   Discrimination of these two types of biform colonies has great significance for further consideration of the mode of evolution of graptoloids (see below). They may be called the *proximal progressive* and *distal progressive* types respectively. The recognition of proximal progressive types was made earlier (Elles, 1922), but the clear understanding of the other type of morphological organization of the colonies has been elaborated only in the last decade (Urbanek, 1960, 1966, 1970).

Essential for an understanding of the morphological organization of uniaxiate monograptid colonies is the fact that due to a regular, directive, and gradual modification of size and shape of thecae in the course of astogeny the proximal and distal thecae differ conspicuously. The presence of contrasting thecal types on both ends of the rhabdosome is indicative of heteropolar organization of monograptid colonies. These contrasting, and often extreme thecal types appear through small structural intergradations of adjacent thecae to display a definite succession, i.e., direction of changes. The lack of sharp discontinuities and the regular spatial pattern of changes involved permits the conclusion that we are dealing here with a regular modification of the morphogenetic properties of zooids. This depends on their position in the colony and in the case of uniaxiate colonies is determined by the order of budding.

So conspicuous are these morphological facts that several attempts have been made in the past to explain this polar organization of the graptolite colony (especially Elles, 1922; Westoll, 1950; Bulman, 1951, 1958). Some of them, such as Elles (1922) and especially Bulman (1951), made important progress in interpreting thecal variation and suggesting stimulating ideas. Nevertheless, entirely new light has been shed on the problem of the biological nature of graptolite colonies following the application of Child's (1915, 1941) concept of physiological gradients.

Mutatis mutandis, application of this concept has contributed much to our understanding of several problems of development, morphogenesis, and regeneration in Recent organisms. It is not surprising therefore that the fundamental points of the classical physiological gradient theory* proved also to be very helpful in interpreting the organization of graptolite colonies (Urbanek, 1960).

The main idea of the interpretation suggested lies in the supposition that regular morphological gradients in graptolite colonies have been expressions of underlying physiological gradients, i.e., a graded change of physiological properties and conditions of growth of successive zooids. As in many Recent solitary and colonial organisms, in the monograptids and other graptoloids there were gradual and directional changes of physiological properties from the regions of highest to the regions of lowest intensity. Evidence available for Recent colonial organisms indicates that several factors are responsible for the existence of physiological gradients. For graptolite colonies there is the production and spreading of appropriate substances along the rhabdosome, which may be suggested as most probable. Spreading (transport) of such substances might be effected through the stolon, it being the permanent tissue connection between all zooids of the colony, or through extrathecal membranes, spread among them. While in Recent organisms similar agents are usually responsible only for changes in certain physiological

---

*The physiological gradient theory is taken here as a very broad generalization concerning the spatial organization of biological systems. It is independent of certain special points of Child's classical concept which are not necessarily involved in our considerations (e.g., Child's postulate that physiological gradients are generally and solely founded on differences in metabolic activity and are primarily expressed only by axial respiratory gradient; see also Urbanek, 1960). Although the causes of morphophysiological gradients are still only imperfectly understood, recent investigations indicate that other approaches to this problem are also possible, e.g., those presuming double gradient systems, especially those based on the notion of induction and inducer (organizer of Spemann). For a contemporary evaluation of the heuristic role and significance of the physiological gradient theory, see also Tokin, 1970.

properties (respiratory intensity, regeneration and growth abilities, etc.), Urbanek (1960) has suggested their morphogenetic influence on particular zooids in graptoloids.

In the light of Urbanek's considerations, physiological gradients in graptolite colonies appear due to the production and spreading along the rhabdosome and the gradual change in the amount of certain substances, which at the same time show an appropriate morphogenetic influence on the growing zooids. Size and shape of successive zooids, and therefore size and shape of their thecae, have been determined by the amount of active substances available at the moment of their budding and growth. A morphogenetic role suggested for this substance is therefore comparable with induction in embryonic development. A close correlation between a gradient of certain physiological substances and the morphological gradient displayed by thecae in the rhabdosome, through suggestion of a morphogenetic role for the former, is the main thesis of the morphophysiological theory of organization of graptolite colonies as formulated by Urbanek (1960).

Elementary discussion shows that morphogenetically active substances may be produced in graptolite colonies in two ways: (1) *diffused*, the appropriate substance being produced in an additive way by all zooids of the colony, and accumulated subsequently in tissues of the colony, or (2) in a *localized* way (Fig. 6). The latter mode of production of morphogenetic agents, according to heteropolar organization of the graptoloid colony, may exist in two regions (poles) of the rhabdosome: (a) the proximal end (sicular region), corresponding to a *basipetal* type of organization as recognized for many athecate hydroids (Fig. 6B), and (b) the distal end (growing tip), corresponding to an *acropetal* type of organization typical of plants (Fig. 6B).

A diffused mode of production of morphogenetic agents (substances) means that the latter were certain metabolites formed by all zooids, each of which contributed to a common pool which gradually accumulated and diffused into the tissue of all individuals (Fig. 6C). A morphogenetic action of such metabolites would be inhibiting or stimulating for certain characters. Urbanek (1966) argued against this supposition and emphasized the regular morphological pattern of graptoloid colonies showing limited intercolonial variation in structural gradients and branching patterns (cladogeny), which are relatively stable specific characters. Determination of such characters, independent of variable metabolic activity and trophic conditions of the colony, through centers of common control appears far more convincing.

Braverman (in Braverman and Schrandt, 1966) suggested that discontinuous (periodic) feeding in *Podocoryne*, cultured in an otherwise stable microenvironment, may result in a somewhat discontinuous growth pattern, the latter being programmed according to a generally invariable geometric rule. Similar nutrient conditions would result in a great lability of astogeny in graptolites, but this is not observable in fossil material (Urbanek, 1966).

Urbanek (1960, 1963, 1966) discriminated between two alternative positions for localized centers of production of morphogenetic substances, favoring the sicular region. Its acropetal place, at the distal end of the colony, was eliminated because of the sympodial type of growth in graptoloid colonies. All available evidence on the mode of budding in Graptolithina indicates that it closely paralleled the sympodial budding of hydroid colonies (Kozlowski, 1949; Bulman, 1955, 1970). There was no permanently growing terminal bud, but each theca in turn was the terminal theca of its branch. The growing

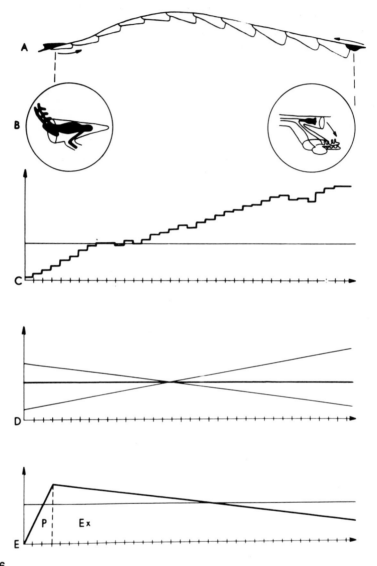

Figure 6

Diagrams illustrating theoretical possibilities of biological interpretation of polar organization of graptoloids.

A. Monograptid rhabdosome showing heteropolar organization, opposite poles of uniaxiate colony in solid black.

B. Alternative positions of inducing centers possible in heteropolar colony: right, a growing tip; left, siculozooid (oozoid); arrows in A indicate corresponding directions of induction.

C. Interpretation presuming diffused mode of production of morphogenetic substance, as a metabolite accumulating in tissues through contribution of each blastozooid (heavy line), at the stable level of reactivity of their tissues (horizontal light line). Irregularities in accumulation of such a morphogenetic substance approach realistically the variation of certain parameters observed

tip in graptolites was not a permanent center but a temporary structure. From these conditions of growth, Urbanek (1960) deduced that, taking successive terminal zooids as inductors, the intensity of their action (i.e., amount of active substance produced) would be very much the same for all zooids of the colony, resulting in an absence of a morphologic gradient (Fig. 6D). The supposition of a changing activity at the growing tip [e.g., decrease of physiological activity through aging of the colony as a whole (Fig. 6D)] is contradicted by the evidence showing a distal increase in the size of thecae and often also a distal increase in complexity of structure as in the distal progressive type of biform monograptids (e.g., Cucullograptinae and Neocucullograptinae among monograptids). Except for certain somewhat problematic irregularities in terminal autothecae on the branches of some dendroids [described and interpreted by Kozlowski (1949) as senile changes], and the regular terminal structures, indicating the stoppage of growth in reduced and highly specialized colonies of Retiolitidae (*Gothograptus, Holoretiolites*), there is no indication of any senescent changes in graptolite rhabdosomes. The occurrence of large-sized rhabdosomes (e.g., Regnéll, 1949) indicates that under suitable conditions the growth of monograptids continued far beyond the statistically most frequent size limits.

A progressive increase of activity from a center placed distally (Fig. 6D) would be hardly understandable physiologically (progressive "rejuvenation"). Figure 6D suggests progressive increase in reactivity of tissues in newly formed zooids (decrease in threshold level) as a possible but unlikely mechanism of rejuvenation. Such a supposition agrees well with size gradients but is contradicted by certain patterns of structural gradients, namely by proximally progressive types of biform colonies (e.g., Saetograptinae, certain Monograptinae). Some additional, ad hoc hypothesis would be necessary to make this supposition fit with the real organizational picture.

Far more convincing is the concept of the basal, sicular region (siculozooid itself), as a localized center of production of morphogenetic substances. Urbanek (1960) offered the following reasons to support this suggestion. (1) The siculozooid is the only zooid in the colony formed by sexual process (oozooid); this enables more precise genetic control over the amounts of morphogenetic substances produced. (2) Siculae are morphologically distinct from the remaining thecae and in the vast majority of graptolites are of notably constant shape; furthermore, with few exceptions, they were scarcely affected by evolutionary changes. This contrast between variability of thecae and constancy

---

in astogeny. However, changing trophic conditions may result in much greater irregularity than suggested by the graph (see text).
D. Explanation of growing tip as inducing center. The horizontal heavy line indicates a steady power of induction, as suggested by sympodial type of growth in graptolites. Morphological gradient may result from decrease of reactivity of tissues ("aging") or increase of such reactivity ("rejuvenation"), indicated by oblique light lines.
E. Explanation of alternative hypothesis presuming localized production of morphogenetic substance by siculozooid. Horizontal light line, threshold level of reactivity of tissues; heavy line, amount of morphogenetic substance present; P, period of accumulation and production of this substance; Ex, period of gradual exhaustion of it, resulting in decrease of the amount of morphogenetic substance below the threshold level. In C–E: ordinates, amount of activity of morphogenetic agent; abscissae, successive thecae.

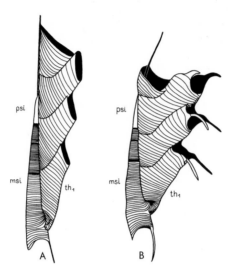

Figure 7

Principle of "noninvolvement" as illustrated by *Saetograptus chimaera* Barrande (b), descending from pristiograptids of *Pristiograptus dubius* (Suess) group (A). While thecae show profound changes in structure, siculae remain unaffected. (Combined after Bulman, 1955, Walker, 1953, and Urbanek, 1953, 1958.)

of siculae suggests an analogy to the reacting tissues and organizer in the phenomena of embryonic induction in solitary organisms. In graptolite colonies the biological role of the former factor would be performed by the blastozooid tissues, reacting on morphogenetic substances produced by the siculozooid, which represents the role of an inductor (Fig. 6E).

The problem of contrasting the evolutionary behavior of the siculae and the blastozooid thecae has been analyzed in some detail by Urbanek (1966, 1970). In the majority of described lineages the sicula is not involved in evolutionary changes such as displayed by thecae, being much the same in successive links of a given lineage. At most its changes are strongly limited. While thecae are subject to changes due to a proximal introduction, the sicula is simply omitted in a sequence of changes, th_1 being the first theca affected by them ("noninvolvement," Fig. 7). In the case of distal introduction a sicula may be unaffected or involved in the last changes (Archiretiolitinae) or affected with change to a lesser extent as compared with the remaining thecae (*Neocucullograptus*) (Fig. 8). Less common are progressive changes in the sicula, which differ in their morphological nature from those in the thecae (*Monograptus hercynicus* lineage, Linograptinae, a number of monograptids, Fig. 9). Most instructive, however, is the evolution of the sicula which follows a certain "exclusion principle" as compared with thecae (Fig. 10). Here belong the examples of strong enlargement and differentiation of the apertural apparatus in siculae, with simultaneous simplification of thecae and extreme decrease in the number of zooids per colony (Corynoididae, *Nanograptus*). Kozlowski (1953, 1956) suggested that colony growth has been replaced here by growth of the

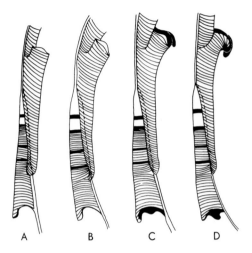

Figure 8

In evolution of *Neocucullograptinae* sicula has been affected by the same changes as thecae (appearance of microfusellar additions at aperture), but to a lesser extent.

A. *Neolobograptus auriculatus* Urbanek. Thecae show apertural elevations not present on sicula ("noninvolvement"); microfusellar additions occur sporadically.

B, C. *Neocucullograptus inexspectatus* Boucek. Microfusellar additions, lacking in early growth stages (B), appear later both on sicula and thecae (C, solid black strips along apertures). While microfusellar additions on thecae are strongly developed and asymmetric, those on sicula are narrow and symmetrical.

D. *Neocucullograptus kozlowskii* Urbanek. Represents more advanced stage in development of microfusellar additions on sicula and thecae, those on sicula remaining symmetrical; those on thecae are strongly asymmetrical and provided with spine-like processes. (After Urbanek, 1966.)

siculozooid. Urbanek (1960) attempted an explanation, suggesting that the siculozooid (organizer) would at the same time become the reacting center, which to a considerable extent used morphogenetic substances produced by itself. This resulted in dwarfism of the colony. Later studies showed, however (Urbanek, 1963), that the morphogenetic substance produced by the siculozooid was not indispensable for the budding of thecae. More likely hypertrophy of the sicula resulted in production of too great an amount of morphogenetic substance, which acted as an inhibitor of budding. Similar paradoxical effects were observed in the action of a high concentration of plant-growth hormones, and in *Hydra* a decrease in control by the dominating region situated in the hypostome is necessary for formation of some kinds of colonial assemblages (Lesh and Laurie, 1971).

"Noninvolvement" and "exclusion" are modes of evolution of siculae which testify to a considerable degree for a particular function assumed by the siculozooid in the colony. Figure 30 attempts to clarify the role played by the siculozooid in light of the concept suggested here.

In some ways a dominating position occupied by the siculozooid in the graptoloid

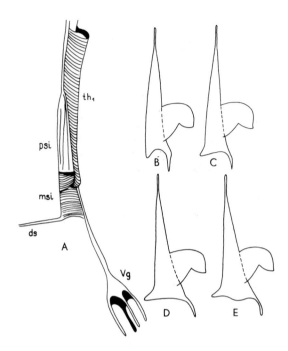

Figure 9

Morphological specializations confined to sicula, and not reflected in other thecae of rhabdosome.
A. *Linograptus posthumus* Rein. Richter. Strong abbreviation of metasicula, early formation of
initial bud, presence of virgellarium (vg) on tip of virgella being probably a part of buoyancy
apparatus, dorsal spine (ds) prolonged into pseudovirgula of the first sicular cladium, prosicula
(psi), metasicula (msi), first theca (th_1).
B–E. Diagram to show progressive changes in shape of sicula in *Monograptus uniformis* Přibyl–*M.*
hercynicus Perner lineage: B, shape of sicula typical for *M. uniformis* and *M. hercynicus,* two
related species which do not differ in shape of sicula; C, advanced morphotype showing certain
expansion at the aperture of prosicula (a trend toward *M. hercynicus*), but found together with
M. uniformis; D, *M. hercynicus* with trumpet-like expansion at the aperture of sicula, lower
zonal form with straight margin of aperture; E, *M. hercynicus,* an upper zonal form showing
in addition to expanded aperture certain elaboration of its margin. Note that the distinct changes
in shape of sicula are not accompanied by corresponding changes in shape of thecae. (After
Urbanek, 1963, and Willefert, 1962. B–E strongly simplified to emphasize the principle of changes.)

colonies is evidence for their "monarchical" organization (in the terminology used by
Beklemishev, 1950, 1964). Particular zooids are not controlled by the common colonial
tissues, tissues to which certain functions (e.g., budding) may be entirely transferred.
Control is transferred to a single zooid which with respect to certain functions plays
a dominating role. Such organization, through a further increase of physiological and
morphogenetic dominance of this particular zooid, may lead finally to a reduction of
coloniality by a gradual decrease in the number of remaining zooids and their morphological
degradation (Beklemishev, 1950, 1964). The Corynoididae seem to represent such an
extreme case among graptoloids and Kozlowski (1953) has suggested that theoretically

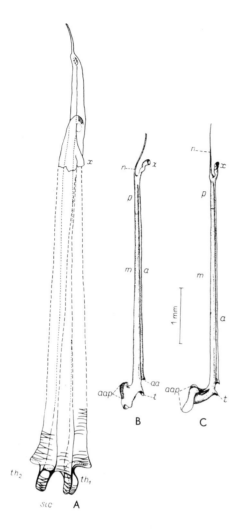

Figure 10

Corynoididae as an example of the "exclusion principle" in evolution of sicula and thecae.
A. *Corynoides calicularis* Nicholson with rhabdosome consisting of sicula (sic), two adnate and normally developed thecae (th$_1$, th$_2$), and a vestigial third theca (x).
B, C. *Corynites divnovensis* Kozlowski and *C. wyszogrodensis* Kozlowski with rhabdosomes composed of sicula (p, m) and only one normal theca (a) adnate to sicula and provided with vestigial and partly coiled second theca (x) of unknown nature. Note complex apertural apparatus of sicula (aap) and simplified aperture of theca (aa) in B and C as compared with A, nema (n), tongue-like outgrowth of sicular aperture (t). (After Bulman, 1955, and Kozlowski, 1953, 1956.)

the next step in the reduction of coloniality in *Corynites* could result in the complete arrest of seriation, or in other words a reduction of the colony to a single individual, the siculozooid. Potentially at least this seems quite possible, as certain premises for secondary transformation of the colony into a solitary organism are hidden in the organization of graptoloid colonies and expressed by a general trend to decrease rather than increase number of zooids per colony.

Urbanek (1960) has suggested that morphogenetic substances produced by tissues of the siculozooid (Fig. 6E, P) were subsequently transported along the stolon, to diffuse and permeate the entire colony, at the same time showing a regular decrease in their amounts as the colony grew. This gradual decrease has been caused by the use and exhaustion of these substances during the processes of budding and growth of subsequent zooids, rather than merely by diffusion and concentration gradients (Fig. 6E, Ex). The gradual exhaustion of the remaining amount of morphogenetic substances would, on the other hand, be responsible for a regular decrease in the power of induction and appearance of morphological gradients. Urbanek (1960) has stressed that this line of reasoning resembles closely the conclusions of a number of other authors (Barth, 1940; Spiegelman, 1945; Waddington, 1956; Tardent, 1956, 1963, 1965; Tardent and Tardent, 1956; Tardent and Eymann, 1959; Tweedell, 1958), on the factors governing the processes of development and regeneration in Recent solitary and colonial organisms (see also Berrill, 1961, for a review of the contemporary state of problems of polarity in Hydrozoa).

Essentially similar ideas about the origin of spatial organization in biological systems have been suggested by Barth (1938) and Spiegelman (1945), who attempted certain mathematical calculations to estimate the results of competition among zooids for the supplies of a morphogenetic substance, thus approaching our postulate of progressive exhaustion of limited resources of such substances. More recently, Smith (1968) has discussed a model of diffusion of a substance through a tube which is applicable, according to him, to the solution of certain problems of spatial organization in morphogenesis. Taking a tissue strip instead of a tube and assuming certain inductive influences of the substance diffusing according to the gradient of concentration, without attaining a stationary state, and furthermore assuming some threshold effect in the action of this substance, we approach closely the task conditions of the graptolite colonies. As the parameters of such processes are entirely unknown for graptolites, speculations beyond those suggested above for the general model (see Fig. 6E) seem unwarranted.

In graptoloid colonies, the morphogenetic role of similar substances has been, according to concepts advocated by Urbanek (1960, 1963), rather manifold. On the one hand, they were responsible for growth inhibition or limitation. Their gradual decrease in the course of astogeny enables the successive zooids to attain progressively larger size. This effect of morphogenetic substances has been undoubtedly the primary one on which certain additional morphogenetic functions were later founded. Inhibitory control (limitation in size) of zooids should be distinguished from bud inhibition, as there is no evidence for the latter, and most probably budding progressed quite easily close to the siculozooid.

Additional functions of the morphogenetic substances which resulted in the appearance of structural gradients are linked with an inductive influence on the presence or absence and degree of expression of certain structural characters. In proximally progressive colonies, morphogenetic substances played the role of stimulating factors with respect to certain characters, their expression being most complete in the sicular region and decreasing distalward. In the case of distally progressive colonies, the opposite was true, and morphogenetic substances behaved as inhibitors of certain characters. The most complete expression of these characters was at the distal end, where the amount of substances in question was the smallest (see Fig. 6E).

In the light of Urbanek's (1960) considerations, the role of the growing tip, although not merely passive, has been limited to be a certain "point of attraction" for morphogenetic agents, similar to the role exerted by buds or growth fields of particular incipient organs ("Anlagen") in embryonic development. Moreover, the tips of growing branches (buds) were centers of highest intensity in cell division and differentiation.

Application of a morphophysiological gradient theory seems more useful in the interpretation of astogenetic changes than the use of concepts borrowed from ontogenies in solitary organisms. According to de Beer (1958) and Jaeger (1959), proximal introduction of a phylogenetic novelty may be explained as an instance of "colonial paedomorphosis" or "proterogeny." Urbanek (1960) has noted that this change, "paedomorphic" with respect to the whole colony, is manifested by changes in a number of zooids, each of them being individually affected according to the "gerontomorphic" mode. For example, in the spreading of hooked thecae distalward, the thecae acquire mature or gerontic characters of the proximal thecae of their ancestors. This paradoxical conclusion seems to indicate that concepts which neglect the collective nature of a colony are misleading (see also Bulman, 1951). Westoll (1950) has suggested a different, and very stimulating

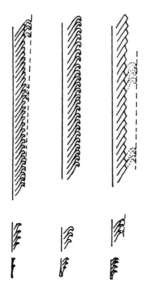

Figure 11

Diagram illustrating the spread and loss of hooked thecae in "progressive" and "retrogressive" series in monograptids, according to the "potential hypermorph" concept of Westoll. Hooked shape of aperture is believed to be linked with a defined growth stage of zooid (marked by broken line), which in some species may be attained by all zooids of a colony (as in *Monograptus priodon*; center figures), in others only by part of them *(M. argenteus,* left; *Colonograptus colonus,* right). In "progressive" series *(argenteus–priodon)* increased number of zooids will attain corresponding growth stage (spread of hooked thecae); in "retrogressive" series *(priodon–colonus)* such stages will be attained by a decreasing number of zooids (disappearance of hooked thecae). (From Westoll, 1950.)

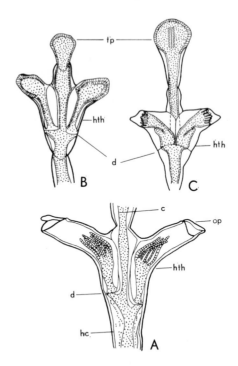

Figure 12

Gradual passage from alternate (A) to opposite (B, C) budding along the branches in Sertulariidae. Illustrated by *Diphasia rosacea* (A and B) and *Sertularia gracilis* (C).

A, B. Alternate and intermediate to opposite budding of hydranths, seen dorsally.

C. Opposite budding seen ventrally; coenosarc (c), diaphragm (d), hydrocaulus (hc), hydrotheca (hth), with hydranths or their buds inside, operculum or lid (op), terminal growing tip (tp). (After Kühn, 1909.)

hypothesis, based on de Beer's (1958) concept of hypermorphosis (a phylogenetic modification of ontogeny due to prolongation of growth through addition of a new stage to the adult stage of an ancestor). According to Westoll, the appearance of certain thecal characters is connected with a definite stage of growth attained by the zooid. In uniform types, such a stage is attained by all zooids of the colony; in biform types only a certain number of zooids reach a corresponding growth stage, the remaining being simplified, owing to a relatively restricted growth ability. In biform colonies a given thecal character might be potentially realized by prolongation of growth, e.g., the addition of a new stage to ontogeny. Abbreviation and prolongation of ontogeny are the main factors responsible for evolution of thecal characters in graptoloids, which according to Westoll (1950) may be "stated in terms of de Beer's paedomorphosis and retardation." Westoll attempted to explain in this way the evolution of "retrogressive series from hook" (Fig. 11). It seems doubtful, however, that such a series has been represented by real phylogenies. But independently of this, Westoll's concept has a certain general significance as an elegant working hypothesis. According to Bulman (1951) the "potential

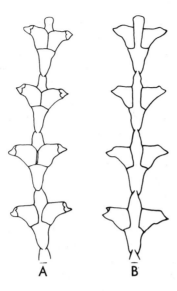

Figure 13

Branches of *Sertularia gracilis* (third to sixth internode above basal stolons).
A, seen from ventral side; B, seen from dorsal side. Note gradual approach of internal walls˙
of hydrothecae in A. (From Kühn, 1909.)

hypermorph'' concept cannot be applied to rhabdosomes where all thecae are composed of the same thecal segments (pro- and metatheca), because this excludes the existence of stadial growth differences among the members of the colony. Urbanek (1960) put more emphasis on the fact that in certain biform graptolites (*Cucullograptus*) the thecae throughout the rhabdosome are composed not only of the same segments, but also of the corresponding parts of them, and in spite of distinct morphological differences in thecae, any growth-stage difference among them is very unlikely (see Figs. 3A and 5C, D). However, when astogenetic thecal variation is expressed by the presence or absence of certain parts of metatheca (as in examples cited on Figs. 2C and 5A, B), certain growth restriction seems to be an adequate explanation. But such an interpretation is only a special case of the gradient theory, which explains in a more general way the variation of thecae, including their spatial arrangement, morphological differentiation, and growth potential. A gradient of growth capacity is even included in the "potential hypermorph" concept. Westoll did not put emphasis on this point, which in the opinion of the present author was the general biological foundation for his observed phenomena.

Although the presence of regular morphological gradients is a feature highly characteristic of graptoloids, it is not unique to them. In Recent Hydrozoa gradients are seldom manifested morphologically, being expressed mainly through intrinsic properties such as in regeneration ability. Judging from the papers of Driesch (1890), Kühn (1909), and Schenck (1965), however, certain hydrozoans display regular changes of some simple morphological characters along the hydrocaulia. In sertulariids, such gradients are expressed in the mode of budding. In *Diphasia (D. pinaster, D. rosacea),* biserially arranged hydrothecae change gradually from alternating positions at the base into opposing positions (vis-à-vis) at the terminal part of a hydrocaulus (Fig. 12). Similar conditions were recognized in *Sertularia secunda,* which seems to indicate that they appeared independently in certain lineages and are an expression of a more general evolutionary trend.

Driesch (1890) has pointed out that in *Sertularella* alternating budding is maintained throughout the hydrocaulus. In *Diphasia* and *Sertularia* it is characteristic of an early stage of cormogeny only. According to Driesch, *Sertularella* represents an ancestral stage of phylogeny, which is recapitulated in the development of the colony of the two latter genera.

Kühn (1909) has described in certain species of *Sertularia, Diphasia,* and *Dynamena* a gradual approach of hydrothecal walls on the ventral side of the hydrocaulus. At the base of a hydrocaulus these walls are widely spaced on the ventral and dorsal side, then more closely spaced on the ventral side, and finally touch in the distal part of a hydrocaulus (Fig. 13). In the majority of species, however, hydrothecal walls are separated throughout the hydrocaulia. These phenomena show a great resemblance to morphological gradients in graptolites and enable one to distinguish among hydrozoans both biform and uniform types with respect to particular features. These most probably have the same significance and origin as in colonies of the Graptolithina. Schenck (1965) has recently described morphological gradients along vertical branches (cormoids) in Plumulariidae, expressed in the spacing of hydrothecae, the number of nemathozooids associated with a hydranth, and branching capacities. The basal parts of particular cormoids may be constructed more simply than the terminal ones.

REGENERATIVE MORPHOSES AS AN EXPERIMENTAL VERIFICATION OF THE MORPHOPHYSIOLOGICAL THEORY OF ORGANIZATION IN GRAPTOLOID COLONIES

The main theses of the morphophysiological theory of organization in graptoloid colonies are somewhat speculative and for this reason have been met with certain reservations. Further evidence in support of the theory has been supplied from observations concerning the fate of fractured rhabdosomes in which the sicula and a number of proximal thecae were broken off. The distal fragments of such rhabdosomes were capable of regeneration, resulting in the formation of a secondary, regenerative branch—the *pseudocladium* (Urbanek, 1963). This structure consists of a series of thecae growing in the opposite direction when compared with those on the primary branch—the *procladium*. This type of bipolarly (bilaterally) growing colony which shows at the point of divergence certain traces of fracture is not uncommon, especially in slender monograptids, susceptible to easy disruption by various factors in their environment. Such bipolar regenerative morphoses (results of heteromorphosis) have been recognized in *Linograptus posthumus,* and the early stages in the formation of the regenerative branch have been described for *Lobograptus scanicus parascanicus* (Urbanek, 1963). Some additional evidence is presented in Figures 14, 15, and 18. These observations were more recently supplemented by the brilliant findings of Palmer (1971), which indicate the presence of bipolar regenerative morphoses in *Neodiversograptus nilssoni.*

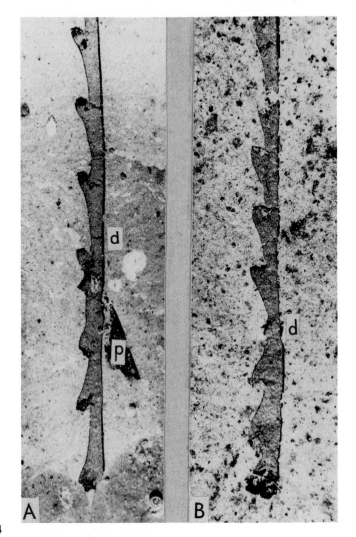

Figure 14

Bipolar regenerative rhabdosomes in *Linograptus posthumus* (Reinh. Richter) without sicula in region of divergence of two thecal series; ca. ×15.

A. Showing no contrast in size of thecae on primary and regenerative branch.

B. Similar rhabdosome, showing contrast at the point mentioned. d, Point of divergence; p, fragment of periderm of another specimen. Upper Silurian, deep boring Chelm, *Pristiograptus bugensis* zone, Poland. (From Urbanek, 1963; specimens obtained by courtesy of L. Teller, Warsaw, Poland.)

Earlier similar regenerative morphoses have been observed in other monograptids (Bouček and Přibyl, 1953; Jaeger, 1959, 1960), and their regenerative nature was suggested by Bouček and Přibyl (1953). The ability to form ''bilateral rhabdosomes'' was erroneously interpreted by the latter authors as a peculiar character of taxonomic significance sufficient

to warrant erection of a separate family to accept such forms. Jaeger (1959, 1960) denied the regenerative nature of "bilateral rhabdosomes," considering them as either separate taxa (1959) or as certain stages of the complex life cycle (metagenesis) in monograptids (1960).

Urbanek (1963) has distinguished these rhabdosomes from bipolar rhabdosomes formed during normal cladogeny (formation of sicular cladia), the existence of which has been wrongly questioned by Bouček and Přibyl (1953); has explained their general biological origin, which does not allow them to be taxonomically ranked; and has recognized their significance for an understanding of the organization of graptoloid colonies (Fig. 16). As indicated by Urbanek (1963), bipolar regenerative morphoses have also been recognized in certain recent organisms (e.g., in *Tubularia* and *Pennaria* among hydrozoans—"janus-like regeneration" of Tardent, 1965).

Assuming that graptoloids were capable of forming similar morphoses, certain morphological effects may be predicted on the basis of the morphophysiological theory of organization of their colonies. The breaking off and detachment of the sicular region is comparable to amputation or "explantation" of this center from the colony. Observation of the fate of the distal part deprived of the sicula simulates a biological experiment. The growth of the regenerative branch, as a result of "amputation," proceeds from the very beginning in complete isolation from any inductive influences of a sicular center and should be apolar, i.e., formed by repetition of completely isomorphic thecae having the same size and shape throughout the pseudocladium.

In light of these considerations, observation on the results of regenerative processes should supply decisive evidence in support of the suggested concept of organization of graptoloid colonies ("*experimentum crucis*"). In the opinion of Van Valen (1969, p. 208), such observation "gives important information on the control of morphogenesis and can be regarded as the first instance of an embryological experiment on a fossil organism".

Both Urbanek (1963) and Palmer (1971) noted a close correspondence between the effects predicted by the concept (Fig. 17A–C) and those observed on fossil material (Figs. 14, 15, and 18). In each case examined, at least one branch was completely apolar and consisted of a series of thecae of stable size and shape. This confirms the predicted effects of growth under conditions of complete isolation from the inducing influences of the siculozooid.

Urbanek (1963) emphasized the occurrence of two different patterns of bipolar, regenerative morphoses. In the first type the rhabdosome consists of two branches growing in opposite directions from one point of divergence (fracture), each branch being formed

---

Figure 15

Bipolar–homopolar rhabdosome without distinct sicula at the point of divergence (d), commonly described as *Diversograptus bohemicus* Bouček. Lack of distinct sicula at the point of divergence, rather close spacing of opposite thecae at this point; apolar seriation of thecae are indicative that the specimen in question is rather a regenerative morphosis. There is no contrast between neighboring thecae at the point of divergence; ca. ×5. Lower Silurian, zone of *Spirograptus spiralis,* Cerecel, Bulgaria, outcrop at the road; specimen obtained by courtesy of Christo Spassov, Sofia, Bulgaria.

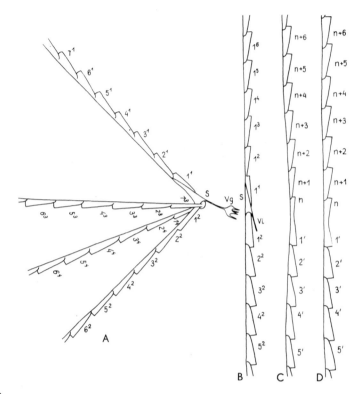

Figure 16

Comparison of compound rhabdosomes of different biological nature.

A, B. Multiramous (*Linograptus*) and bipolar (*Neodiversograptus*) rhabdosomes formed as the result of normal cladogeny. Particular morphological pattern is peculiar for a given taxon. Note the presence of sicula at the point of divergence of branches;

C, D. Bipolar morphoses formed due to regeneration, showing contrast in size of initial thecae on primary and regenerative branch (C) or without such a contrast (D). Note lack of sicula at the point of divergence of branches, supposed point of breakage at the base of theca n, n+6 thecae on preserved part of the primary stipe, 1′–5′ successive thecae of regenerative branch (pseudocladium), 1¹–7¹ thecae on main stipe (procladium), 1² . . . , 1³ . . . , 1⁴, . . . thecae on sicular cladia (metacladia); s, sicula; Vg, virgellarium (see also Fig. 14). (After Urbanek, 1963, modified.)

by a series of isomorphic thecae, i.e., by repetition of thecae of the same size and shape (bipolar–homopolar conditions). In this case there is no contrast at the point of divergence (Figs. 14A and 15). The second type, however, shows a clear contrast at this point (bipolar–heteropolar conditions). One branch is composed of thecae showing a more or less distinct morphological succession, indicating the presence of a gradient. The other branch, in contrast, is formed by uniform, isomorphic thecae, similar from the beginning to the thecae at the distal part of the former branch (Figs. 14B and 18). According to the views presented above, the first type may be predicted on the assumption

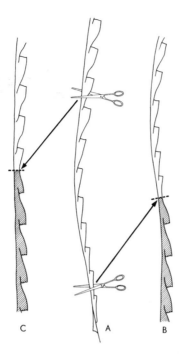

C   A   B

Figure 17

Expected and observed effects of regeneration of the monograptid colony.
A. Normal rhabdosome showing astogenetic succession of thecae, and relation of thecae in bipolarly growing regenerative morphoses (B, C): presuming breakage at proximal part of the primary rhabdosome (lower shears); resulting bipolar regenerative rhabdosome (B) should show distinct contrast between first thecae of the primary stipe and regenerative branch (pseudocladium). Presuming breakage at distal part of A (upper shears); resulting regenerative rhabdosome (C) should show no contrast between adjacent thecae of both branches.
Pseudocladium (regenerative branch) striped, point of breakage indicated by broken lines. (After Urbanek, 1963, from Bulman, 1970, modified.)

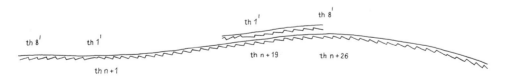

Figure 18

Bipolar rhabdosome in *Neodiversograptus nilssoni* (Barrande) without a sicula at the point of divergence of the two branches; interpreted as a regenerative morphosis, presuming breakage at theca n of the primary stipe. Comparison made between the initial thecae (th1′–th8′) of the regenerated branch (pseudocladium) and the primary branch shows the best fit at the level of thecae n+19–n+26, which seems to indicate their concomitant growth. (From Palmer, 1971, modified.)

that the point of detachment was situated at the distal end of the primary rhabdosome (procladium). The preserved part of the procladium thus consists of distal thecae, formed under relative physiological isolation from the inductive influences of the sicular center and resulting from exhaustion of morphogenetic substances by the preceding zooids and their great distance from the sicula. They are almost uniform in size and shape. The formation of the regenerative branch (pseudocladium), on the other hand, occurs from the very beginning under conditions of complete isolation from the inductive influences of the sicular center, which has been detached ("amputated") from the colony. The expected and observed results are two series of uniform thecae growing in opposite directions from the point of common divergence (see Figs. 17C, 14A, and 15). In the case where detachment has been placed at the proximal part of the primary rhabdosome, regenerative morphosis consists of (1) the preserved primary branch, which grows under conditions of induction and shows a distinct morphological gradient (succession of thecae), and (2) a newly formed regenerative branch. The latter develops under entirely different conditions, being again in complete isolation from the influences of a siculozooid. It is composed of uniform thecae similar to the theca at the distal end of the primary rhabdosome, and shows apolar repetition of isomorphic thecae. This results in a distinct contrast between the thecae of these two branches at the point of divergence (see Figs. 17B, 14B, and to a lesser extent Fig. 18).

The above considerations are slightly simplified, and certain observations indicate that in some biopolar–heteropolar regenerative morphoses, regenerative branches show accelerated and abreviated astogenetic succession, attaining true distal thecae only after two or three thecae were formed. Contrast at the point of divergence of two thecal series is slightly decreased in this case. Explanation of this phenomenon presumes presence of certain remnants of morphogenetic substance at the proximal end of the fractured rhabdosome; this seems especially probable in relatively youthful colonies (Urbanek, 1963). Such rhabdosomes represented still unbalanced morphogenetic systems, with higher potential at the proximal than at the distal end. Residual amounts of the morphogenetic substance present at the proximal end were probably responsible for accelerated and abbreviated astogenetic change of the first thecae of the regenerative branch, and rapidly decreased with budding. This resulted in quick equalization of morphogenetic potentials on both growing tips of a bipolar colony. No information is available at this time on growth rates of primary and regenerative branches, but some delay in formation of the first theca of a pseudocladium might be expected due to necessary healing and repair of tissues at the regenerating end prior to seriation. Traces of such processes were observed by Urbanek (1963).

In each case the absence of a sicula from the rhabdosome results in the apolar growth of a thecal series. The regenerative branch is composed of uniform thecae corresponding to the distal type of thecae on the primary branch. On the other hand, in the presence of a siculozooid in the colony, successive thecae invariably display a distinct morphological gradient. This supplies strong evidence in support of the previously suggested concept of an inductive role for the siculozooid in the colony. It seems that in bipolar regenerative morphoses, budding and growth proceed simultaneously on the primary branch (procladium) and on the regenerative branch (pseudocladium). This contiguous growth on the tips of both branches results in the simultaneous formation of thecae showing the same size and shape, i.e., being morphologically equivalent. Although the thecae of the presumed primary stipe on Fig. 14B do not represent the true proximal part of the rhabdosome, they nevertheless show a distinct contrast to the adjacent thecae of the regenerative branch. This may be explained by assuming that the growth of the latter was simultaneous with the nonpreserved distal part of the primary branch, represented by the thecae of the extreme distal type. This resulted in the formation of thecae of extreme distal appearance in the proximal part of the regenerative branch. Conclusions of Urbanek (1963) have later been confirmed on the large regenerative rhabdosomes of *Neodiversograptus nilssoni* (Fig. 18) by Palmer (1971). The morphological equivalence of simultaneously budding thecae in the same rhabdosome is a regularity of great significance useful also in the explanation of growth in normal multiramous graptoloid colonies, and as expounded by Urbanek (1963) is, in fact, a general morphogenetic rule governing the development of the colonies.

As was mentioned above, bipolarly growing monograptid rhabdosomes do not constitute a single biological category, but may correspond to (1) rhabdosomes showing generation of sicular cladia, or (2) morphoses produced as a result of regeneration. Discrimination between these two categories is relatively easy for Upper Silurian bipolar monograptids (*Neodiversograptus*), where presence or absence of a sicula may be readily recognized at the point of divergence. This is because the virgella remains free, indicating the position of sicular aperture, which is also discernible as a distinct ridge on the surface of the rhabdosome (Urbanek, 1963; Palmer, 1971). Such discrimination is more difficult for Lower Silurian forms, such as *Diversograptus,* which shows obliteration of sicular margin and incorporation of virgella into the ventral wall of theca 1^2 of sicular cladium (R. Thorsteinsson, personal communication). Presence of an oblique line interpreted as a particular type of septation (Bouček and Přibyl, 1953), or as a trace of a fracture (Urbanek, 1963), is probably indicative of an oblique line of junction between theca 1^1 and the sicula itself, thus proving that the form in question shows presence of a true sicular cladium.

The Lower Silurian bipolar monograptids seem therefore to be a highly heterogeneous assemblage, which may consist of (1) bipolar forms showing normal generation of sicular cladium (*Diversograptus* and related forms), (2) bipolar regenerative morphoses produced as a result of fracturing of diversograptid rhabdosomes, and (3) bipolar rhabdosomes formed as a result of regeneration of fractured one-stiped monograptids. Morphological patterns of rhabdosomes formed in cases 2 and 3 have no taxonomic significance.

EXTENSION OF THE MORPHOPHYSIOLOGICAL CONCEPT TO COMPOUND RHABDOSOMES

Compound rhabdosomes, made of a primary branch (procladium) and a number of secondary branches (metacladia), represent another kind of natural experiment which can supply evidence to substantiate the morphophysiological concept of graptoloid organization (Urbanek, 1960). In uniaxiate, one-stiped monograptids the time (order) of budding of particular zooids determines their position on the rhabdosome, which in turn coincides with their distance from the siculozooid. The spatial order of these colonies fully corre-

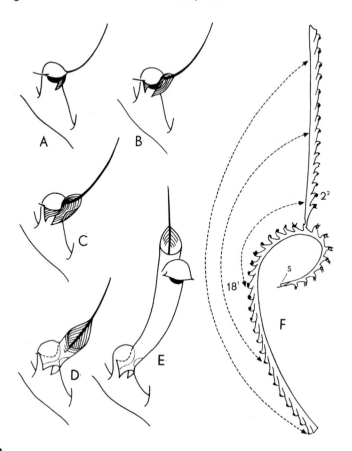

Figure 19

Cladial generation in *Cyrtograptus rigidus* var. A–E. Successive stages of budding of lateral cladium from aperture of a mother theca on the main stipe.

F. Mature rhabdosome; isochronous thecae showing the same size and shape are connected by broken lines.

Note the position of theca 18^1 on main stipe, being equivalent to the first thecae 2^2 on the lateral cladium, which marks the beginning of concomitant growth on main stipe and lateral cladium; s, sicula. (After Thorsteinsson, 1955, from Bulman, modified.)

sponds to their temporal order, or, to use Schenck's (1965) terminology, the "tectonic age" of the thecae agrees with their "topologic age." In such colonies we cannot discriminate between the effects caused by the position of a given theca with respect to the presumed inductive center and the effects caused by the time (order) of its budding, because they coincide exactly.

Different conditions may be observed in multiramous rhabdosomes which consist of a main branch and a number of additional branches, generated with definite delay in respect to the primary branch. Cladogeny in *Cyrtograptus rigidus*, analyzed thoroughly by Thorsteinsson (1955), is a classical example (Fig. 19). Thecae on the procladium

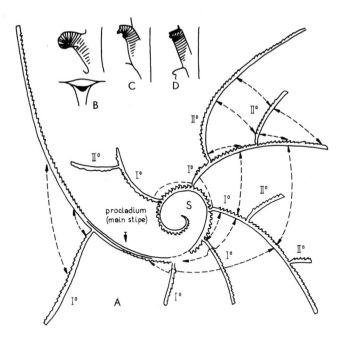

Figure 20

Diagram showing structure of a complex colony in *Cyrtograptus* Carruthers. Approximately isochronic and isomorphic thecae are connected by broken lines. B–D. Shape of proximal, medial, and distal thecae on procladium shown diagrammatically. s, sicula; I°, II°, lateral branches (metacladia) of first and second order; point where twisting of procladium occurs indicated by an arrow. (After Bulman, 1958, modified.)

display a normal astogenetic succession, which may be classified as the proximal progressive type as previously defined. According to Thorsteinsson (1955), the formation of the lateral cladium is delayed when compared with the growth rate of the main branch. Before the first theca of the lateral cladium is completed, some six to eight thecae have appeared on the main stipe.

Thorsteinsson's (1955) precise observations indicate that when this inhibition ceased, further growth on the main stipe and on the lateral cladium is concomitant and occurs at the same rate ("one for one"). The thecae that are simultaneously produced on the main branch and on the lateral cladium are very much the same in size and shape, although the former are distinctly more remote from the sicula than the latter. In compound colonies there is thus a distinct disjunction between their temporal and spatial order, and, in Schenck's (1965) terminology again—the "tectonic age" of the thecae on the lateral cladium does not agree with their "topologic age," the former being "overestimated" in a certain way, which corresponds to an extension of growth inhibition.

Bulman (1958, 1970) has shown that the regularities observed by Thorsteinsson apply also with respect to cyrtograptids having more complex patterns of cladogeny (Fig. 20). These somewhat paradoxical effects represent another "*experimentum crucis*," which may serve to verify the previously suggested hypothesis.

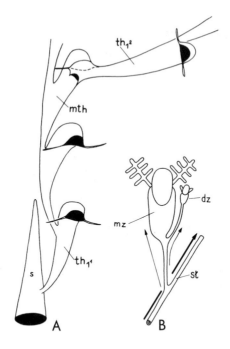

Figure 21

Diagrams to illustrate conditions of growth of lateral (thecal) cladia in *Cyrtograptus*. A, while thecae on main stipe (th_1–mth) are formed due to monograptid budding from prothecal opening of a preceding theca, being interconnected directly by a stolon, first theca of lateral cladium ($th_1{}^2$) is produced by means of cyrtograptid budding, from aperture of its mother theca (mth). This may result in communication of growing daughter zooid (dz) with stolonal system of the main stipe via body of its mother zooid (mz) (B). The latter function like a filter, permitting lesser access (thin arrows) to main stream of nutritive and morphogenetic substances diffusing along the stolon (thick arrow)—a possible cause of delayed growth of $th_1{}^2$. s, sicula; st, stolon.

The mechanisms responsible for these effects seem largely unspecific (Urbanek, 1960), and thus biological analogues can be used based on the physiology of growth in plants. The delayed formation and the retarded growth of the first theca in a lateral cladium of graptoloids may be compared with the growth inhibition of the main stem over the axillary shoots in plants, as described in the classical paper of Child (1915). The mechanisms and physicochemical background of this inhibition have been investigated by a number of authors (Snow, 1925, 1929; Thimann and Skoog, 1933, 1934; Thimann, 1937, 1939; Went and Thimann, 1945), and for the present status of the problem Leopold (1964) may be consulted. Apical dominance and inhibition of growth in axillary shoots have been found to be caused by the distribution of auxins (plant-growth hormones) and by the competition for nutritive (nonhormonal) substances. *Mutatis mutandis,* this point of view has been used to explain the growth inhibition of a newly generated lateral cladium in *Cyrtograptus* (Urbanek, 1960). With the exception that morphogenetic substances have been produced basipetally rather than acropetally as in plants, the mechanisms of inhibition in *Cyrtograptus* are closely comparable. The cause of the inhibition of the bud for the lateral cladium may lie in its "aside position" with respect to the main stream of nutritive and morphogenetic substances distributed by the stolon (Figs. 19A–F and 21). While the thecae of the procladium are directly connected by the stolon, and bud subaperturally, the first theca of the lateral cladium is formed through cyrtograptid budding from a definite aperture of its mother theca on the main branch (Urbanek, 1960, 1963). The growing bud in this latter instance has been supplied with a smaller amount of nutritive and morphogenetically active substance than the thecae of the main stipe which are interconnected by the stolon. These conditions of inhibition cease when the growing lateral cladium attains a level of concentration of substances equal to that in the growing tip of the main branch. Further concomitant growth of both branches may be compared with the simultaneous growth of terminal buds in two-shoot plants occurring at the same rate. The necessary prerequisite of this growth pattern is an equilibrium in morphogenetic potential at the growing tips of the two shoots (equal amount of auxins). Similarly, in multiramous graptoloids, the simultaneously budding and growing thecae must have had at their disposal equal amounts of morphogenetic substances. Any deviation from this state of balance would result in inhibition of growth or a decrease in the growth rate of one of the branches. Equilibrium would be attained when the amount of morphogenetic substance at the tip of the main branch drops, as a result of its use in budding, to the level represented by the lateral bud. However, an equal amount of morphogenetic substance means an equal power of induction, and this in turn determines the equality of size and shape of simultaneously growing thecae in spite of their different distances from the sicula. Disjunction between the spatial and temporal order in *Cyrtograptus* allows one, when discriminating between the position of a bud and the time of budding as the decisive factors determining size and shape of thecae, to favor time. Time, or more likely the amount of active substances available is the decisive factor, although in uniaxiate colonies the spatial, temporal, and quantity parameters coincide completely.

The observations of Bulman (1958) and the earlier data of Bouček (1933) enable one to judge that in the compound, multibrachiate rhabdosomes of some cyrtograptids the lateral cladia of the first order (being subject to inhibition at the side of the main stipe) are in turn the center of dominance with respect to the lateral cladium of the second order, which also may inhibit and control the growth of cladia of still higher order. Such a system of multistage control is responsible for the realization of certain specifically characteristic patterns of cladogeny.

Urbanek (1963) has recognized similar correlative effects in the development of compound colonies of a different type and which are represented by the linograptids. This group is composed of a primary stipe, and a number of secondary branches radiating from the aperture of the sicula (sicular cladia; see Fig. 16). In *Neodiversograptus nilssoni*, only one such cladium may be produced and only after considerable delay with respect to the growth of the main branch. According to Palmer (1971), who investigated some especially rich material, the sicular cladium is initiated most commonly when the procladium reaches the 15^1- to 20^1-thecae stage, although there is great variability in the time of onset (Fig. 22A). The result is a bipolarly growing rhabdosome with a sicula at the point of divergence. Such rhabdosomes, which appear as a result of normal astogeny,

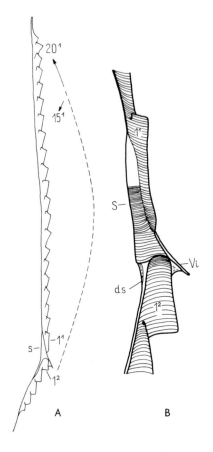

Figure 22

Neodiversograptus nilssoni (Barrande), general appearance of rhabdosome in diversograptid stage of astogeny (A) and structural details at the sicular region (B). Broken lines at A indicate that first theca of sicular cladium (1^2) is most frequently isochronic with thecae 15^1–20^1 of the primary stipe (procladium). ds, dorsal spine of metasicula prolonged to be a pseudovirgula of sicular cladium; s, sicula; vi, virgella, 1^1–20^1 thecae of procladium; 1^2, first theca of sicular cladium. (After Palmer, 1971, and Urbanek, 1963, modified.)

should be distinguished from bipolar morphoses formed during regeneration, and which differ fundamentally in their biological nature from the latter (Urbanek, 1963; see Fig. 16B–D). The only sicular cladium in *Neodiversograptus nilssoni* is formed from the very beginning by thecae resembling those in the distal part of the primary stipe. This results in a sharp contrast with the proximal thecae of the latter (Fig. 22). This effect is caused by the concomitant growth of the distal part of the procladium and the proximal part of the sicular cladium and is indispensably connected with the production of thecae showing the same size and shape (Urbanek, 1963; Palmer, 1971).

The development and growth relations observed in *Cyrtograptus* and *Linograptus,* the development of uniaxiate colonies of numerous monograptids, and the observed course of regeneration of the rhabdosomes in certain monograptids permitted Urbanek (1963) to formulate a general rule governing the morphogenesis of graptoloid colonies. This rule, named the "Thorsteinsson rule," states that in astogeny and in pseudocladogeny (regeneration) any simultaneously budding thecae of the same graptoloid colony will have the same size, shape, and structure, independent of their position on the rhabdosome (Urbanek, 1963).

The Thorsteinsson rule is based on very general regularities in morphogenesis and with certain modifications is tenable also for hydroid colonies. The observations of Child (1941) and the older experiments of Gast and Godlewski (1903) suggest a certain physiological dominance of axial over lateral branches in compound hydroid colonies and indicate that inhibition is based on a system of balance which may be altered by blocking the inhibition with a number of factors (see also Barth, 1938; Spiegelman, 1945; Tardent and Eymann, 1959; Tardent, 1963). It is therefore not surprising that Schenck (1965) found the Thorsteinsson rule applicable to hydroid colonies, suggesting that isochronic zooids are isomorphic and that the lines of simultaneous budding (isochrones) unite zooids of the same size, shape, and degree of complexity (isophenes).

In some graptoloids the generation of cladia may be accelerated, and the additional branches are formed with such a small lag that the structural differences between their proximal thecae and the proximal thecae of the main branch are not detectable. In *Linograptus* (Urbanek, 1963) and multiramous Dichograptidae, the rhabdosomes are composed of a number of cladia radiating from the sicula, or are produced by budding from proximally situated dicalycal thecae, and grow almost simultaneously with the same growth rate. The sequence of thecae on all branches is fully analogous, showing no detectable differences, which again confirms the validity of the rule suggested (Fig. 16A).

APPEARANCE OF POLAR ORGANIZATION IN GRAPTOLOIDEA

There is little doubt that the Graptoloidea emerged from planktonic dendroids of the species group *Dictyonema flabelliforme* (Bulman, 1955, 1970). The foundations of the polar organization of graptoloids may thus be searched for among their immediate ancestors—the dendroids.

A number of facts speak for a high integration within colonies of the Dendroidea. First, there is the character of budding, which follows a strikingly regular pattern determined by the "Wiman rule" (Kozlowski, 1949) and according to which the dendroids differ from the majority of other benthonic graptoloids. Moreover, the distribution of branching divisions in the colony is not haphazard, but rather surprisingly regular (Kozlowski, 1949; Bulman, 1950, 1958). The position of the first bifurcation triad and the frequency of occurrence of later branching generally determine the habitus (pattern) of the whole colony, which may be highly characteristic of a given genus or species (Kozlowski, 1949). In *Dictyonema* and in a number of anisograptids, relatively regular "branching zones" have been recognized which are situated on lines equidistant from the center of the colony (Bulman, 1958). This seems to indicate that the growing tips reacted almost simultaneously, and the frequency of branching has been recognized by Bulman as a specific character. Since the frequency of branching diminishes in the

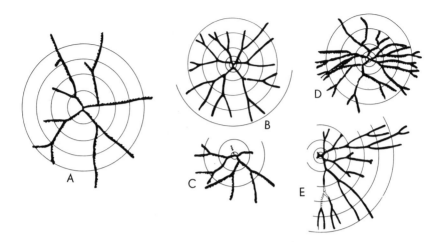

Figure 23

Regular branching pattern in *Anisograptus*.
A–C, *Anisograptus matanensis* Ruedemann; D, E, *A. flexuosus* Bulman. (From Bulman, 1958.)

course of astogeny (centrifugally), we may speak of a certain latent gradient of morphogenetic abilities expressed (Figs. 23 and 24) by regular changes of branching frequency.

Of the features which might indicate a high integration of the colony in Dendroidea, no traces of the occurrence of a size or structural gradient of thecae (auto- or bithecae) have been found. The size and shape of thecae in any category remains stable, being independent of their position in the colony.

In the vast majority of Anisograptidae, which seem to be in many respects truly intermediate between Dendroidea and Graptoloidea ("grapto-dendroids"), there is a resemblance to the Dendroidea in the absence of any traces of a morphological gradient of thecae (Bulman, 1950). One of the most striking features about the later course of evolutionary events has been therefore the appearance of polar organization in true graptoloids, an event which in some way coincides with the origins of this latter group (Bulman, 1958). A regular distal increase in thecal size is the morphologically simplest, and probably also the phylogenetically earliest manifestation of polar organization in primitive graptoloids with simple tubular thecae ("dichograptid thecae"). Only later were structural gradients superimposed on the primary system of size gradient.

The adaptive significance of polar organization in graptoloid colonies is not clear at this time. Different modes of life might be ascribed to younger colonies composed of diminutive zooids. Older colonies consisted also of bigger zooids. Such subdivision of life cycle combined with gradual passage from one to another mode of life may be generally advantageous. Any concrete reasoning in this direction is hindered, however, by our ignorance of graptolite ecology. As the size gradient appeared in such extremely different adaptive types as encrusting Idiotubidae and probably planktonic Graptoloidea, its adaptation to a particular environment does not seem convincing. Polar organization can be considered not to be a single character with individual adaptive significance but as an evolutionary invention which permitted creation of a "morphogenetic field." Par-

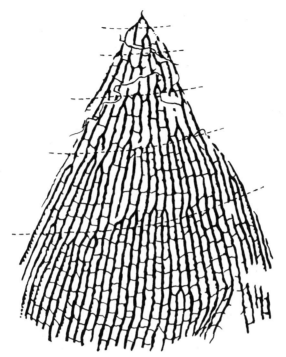

Figure 24

Approximate zones of branching in a rhabdosome of *Dictyonema flabelliforme* (Eichw.); ×1. (From Bulman, 1955.)

ticular mutations were first tested within this field as genes with low morphogenetic activity, affecting only part of the zooids of the colony. Such "piecemeal technology" has been doubtlessly convenient as producing more insurance in introducing phylogenetic novelties, a feature of probably rather general significance.

The appearance of polar organization, a feature highly characteristic of Graptoloidea, was an event having no less significance that the loss of bithecae, which until now was considered to be the crucial change in the earliest evolution of this order. Polar organization, itself of a great biological significance, is a character of great diagnostic value and may serve for easy discrimination between a graptoloid and a nongraptoloid. A distal increase in the width of branches, which is suggestive of a size gradient of thecae, may be easily recognized in fossils, while the presence or absence of bithecae is often hardly discernible.

The available evidence indicates that the mechanism responsible for polar organization has been elaborated independently in a number of lineages belonging to Anisograptidae. Bulman (1955) has found traces of distal increase in *Triograptus,* and considered that a regular increase in the size of thecae "becomes first appreciable with the loss of bithecae" (1955, p. 69).

Some evidence, however, seems to indicate that incipient polarity may appear when bithecae are still present. In *"Didymograptus" primigenius* described by Bulman (1955, pl. VII, fig. 23; pl. VIII, fig. 6) the presence of bithecae is possible (although not recognized surely) and is combined with a distinct distal increase in size of autothecae. In *"Didymograptus"* cf. *primigenius* found by G. Szymański (Geological Institute, Warszawa, Poland; unpublished data), in the Upper Tremadocian of northeastern Poland, bithecae are almost certainly present and are associated with a distinct size gradient of autothecae. In addition, *Triograptus* with indications of a distal increase in the width of branches, as previously noted above, is suspected to have bithecae. Spjeldnaes (1963) drew attention to the fact that elimination of bithecae in *Kiaerograptus kiaeri*, a probable anisograptid which otherwise displays no distal increase in the width of branches, has already been subjected to polarity because of their elimination from the proximal end.* It seems probable therefore that the polar organization of colonies appeared simultaneously with the gradual elimination of bithecae in a number of lineages of Anisograptidae, but prior to their disappearance from the rhabdosome.

Such a course of events is understandable in the light of a probable general biological model for the elimination of bithecae. According to Kozlowski (1949), differentiation into auto- and bithecae in dendroids is an expression of sexual dimorphism, and the transition to a single type of thecae in graptoloids means a transition to hermaphroditism of zooids. In analogy with sex differentiation in Recent animals, such a transitory stage may be imagined, as shown by Kozlowski, as an acquisition of hermaphroditism by former female zooids (= autozooids) combined with the presence of partly reduced complementary male zooids (= bizooids). Elimination of bithecae occupied by male zooids was probably preceded by the appearance of hermaphroditism in previous female individuals. Prior to the elimination of bithecae such a transformation resulted in the labile, unbalanced physiological state so commonly correlated with monoecious conditions. This set up a general biological basis for hormonal interactions which resulted in the appearance of polar organization.

Further speculations on possible sexual conditions in graptoloid colonies are given with the discussion on their reproductive biology.

Polar organization delimits a structural ("grade") boundary between Dendroidea and Graptoloidea, a boundary which almost certainly has been crossed by a number of independent lineages. It is as conventional a criterion as that based on the presence or absence of bithecae, but is far more convenient for practical purposes and is perhaps more fundamental biologically. Spjeldnaes (1963) suggested the use of the widening of the stipe as an additional criterion when the presence or absence of bithecae cannot be ascertained. Since polar organization, however, is a necessary component of any true graptoloid, there is no reason to hesitate in accepting this feature as a key-character indicative for the determination of systematic position. This is not to suggest, however, that we should consider polarity as an "order-making character," which would determine affinity to the Graptoloidea independently of other characters.† Treating the Graptoloidea as

---

*This conclusion is based on data published by Spjeldnaes (1963). Certain distal increase in size of autothecae is indicated, however, by drawings of *Kiaerograptus kiaeri*, accompanying its original description by Monsen (1925), and especially by a new figure published by Bulman (1970, Fig. 19, 3).

†This is especially true if we remember that certain Idiotubidae show a distinct size increase of autothecae within thecal groups, frequently transformed into some sort of branches. Size gradient of autothecae in Tuboidea occurs, however, with an entirely different configuration of characters, determining a quite different pattern of colony organization (type of budding, thecorhiza).

a grade (in the sense introduced by Huxley, and applied to graptolite taxonomy by Bulman, 1963) or a structural stage in morphological evolution, we should as a consequence distinguish a character especially essential for the understanding and delimitation of this grade. There are many premises both theoretical and practical which will point toward polar organization in this respect. Should we think this procedure unsatisfactory, we must keep in mind that the only alternative is to treat the taxon as a clade, which would unite groups representing quite different structural stages on the principle of their common ancestry. In the specific case of the Graptoloidea, it would seem indispensable to place the lower boundary of such a clade within representatives of *Dictyonema* and to separate the *Dictyonema flabelliforme* group as the earliest graptoloids. It follows from the important studies of Bulman (1941, 1955) that "early schism" and later radiation among anisograptids just above the stage represented by *Dictyonema flabelliforme* resulted in the origin of a great number of lineages which independently cross structural boundaries.

Such radical redefinition of the dendroid/graptoloid boundary is theoretically desirable, yet it is not generally accepted, and present usage considers the Graptoloidea as a structural (and organizational) grade.

In viewing the dendroid/graptoloid boundary here, greater emphasis is placed on the appearance of a new character (size gradient of thecae) than on the disappearance of a primitive one (presence of bithecae). At the point of transition between the Dendroidea and Graptoloidea (near the Tremadocian–Arenigian boundary) the suggested criterion may serve to discriminate between truly graptoloid directions of change and nongraptoloid specializations. The former trends should display a proper combination of specializations, including an indispensable elaboration of morphological gradients in autothecae. Some evidence supports the rather early appearance of autothecae before the final loss of bithecae. Pauciramous forms of Tremadocian–Arenigian age may be segregated on this basis into graptoloid and essentially dendroid lineages.

Historically, the much discussed morphological criteria for the reptilian–mammalian boundary support a higher ranking for newly acquired characters than the persistence of some ancestral traits. Although conjectural and partly inadequate in connection with its multiple origin, the dentary-squamosal joint criterion proved to be more useful than the disappearance of such an ancestral character as the articulare complex, which persisted until the Cretaceous among some true mammalian groups. The use of polar organization as an analogous premise for graptoloids is even more satisfactory since this feature also concerns the level of their morphophysiological organization. In the example of the reptilian–mammalian boundary the therapsids are eliminated on the basis of the criterion mentioned, in spite of their essentially prototherian level of organization.

ORGANIZATION OF COLONIES IN TUBOIDEA

Besides dendroids, the Tuboidea are the only group of nongraptoloid Graptolithina in which the development of the colony (Kozlowski, 1949, 1953) is known in some detail. This is especially true for encrusting colonies of tuboidal type represented by the Idiotubidae (Fig. 25). The rhabdosomes consist not only of a basal structure on an encrusting, densely crowded, and irregularly placed mass of thecae, the thecorhiza (Kozlowski, 1949) but also of erect tubular autothecae, either single or gathered into thecal groups.

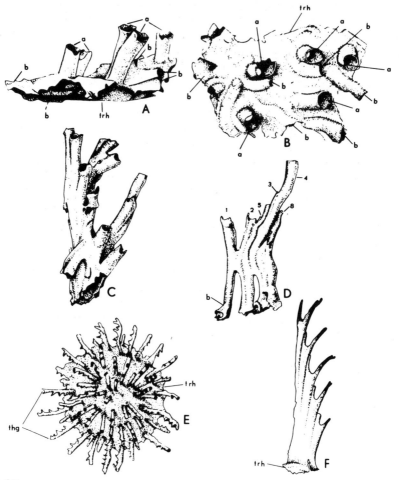

Figure 25

Patterns of tuboid colonies.

A, B. Fragment of colony in *Idiotubus* sp. seen laterally and from above.

C, D. Irregular autothecal groups in *Dendrotubus wimani* Kozlowski, seen frontally (C) and laterally (D).

E. colony in *Discograptus schmidti* Wiman from above, autothecal groups disposed regularly on the surface of thecorhiza.

F. Isolated thecal group of *Discograptus schmidti* Wiman showing regular arrangement of autothecae.

Autothecae (a), bithecae (b), autothecal groups (thg), thecorhiza (trh); 1–8, successive autothecae concentrated into a bundle. (After Kozlowski, 1949; Wiman, 1901; Bulman and Rickards, 1966.)

The thecorhiza is made of basal, adnate portions of autothecae, stolothecae, and bithecae, the latter two being confined to the thecorhiza, with the bithecae partly extending beyond it and opening at the base of the autothecae.

Kozlowski (1949) has grouped colonies of Idiotubidae into a graded series which

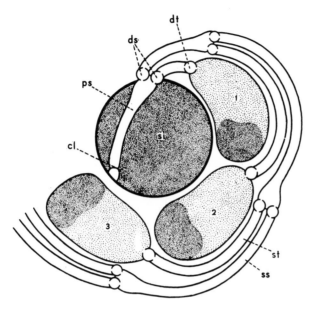

Figure 26

Diagram showing the spiral coiling of stolon in the central part of thecorhiza in *Dendrotubus erraticus* Kozlowski.

Septum inside initial part of prostolon (cl), diaphragms at the base of stolon (ds), diaphragms at the base of thecae (dt), sicula (si), stolothecal stolon (ss), thecal stolon (auto- or bithecal) (st); 1–3, successive thecae.

illustrates successive changes in their spatial organization. These are expressed by the following elementary events: (1) the grouping of primarily single autothecae, scattered irregularly over the surface of the thecorhiza, into thecal groups (*Idiotubus–Dendrotubus* stage (Fig. 25A–D); and (2) an increase in regularity in the distribution of thecal groups within the colony and a progressive increase in complexity of the thecal groups involved (*Discograptus* stage, Fig. 25 E, F).

The first process is responsible for the gathering or concentration of single autothecae (as in *Idiotubus*) into bundles or assemblages of autothecae (as in *Dendrotubus*). Each thecal group consists of various numbers of more or less adpressed autothecae, and within each group their position is irregular, the individual thecae lying in somewhat different planes. Moreover, the distribution of thecal groups over the surface of the colony also remains irregular (*Dendrotubus*) and displays great intracolony variation (Fig. 25C, D). In the next grade of the series, this irregular distribution of thecal groups changes into an ordered arrangement along several radii, showing a highly characteristic radioconcentric pattern (*Discograptus*).

Discograptus schmidti, described by Wiman (1901; Fig. 25E) and analyzed more recently by Bulman and Rickards (1966), shows a highly regular morphological pattern. The thecal groups are arranged along certain lines radiating approximately from the center of the colony. Each thecal group is definitely ordered, with the autotheca lying

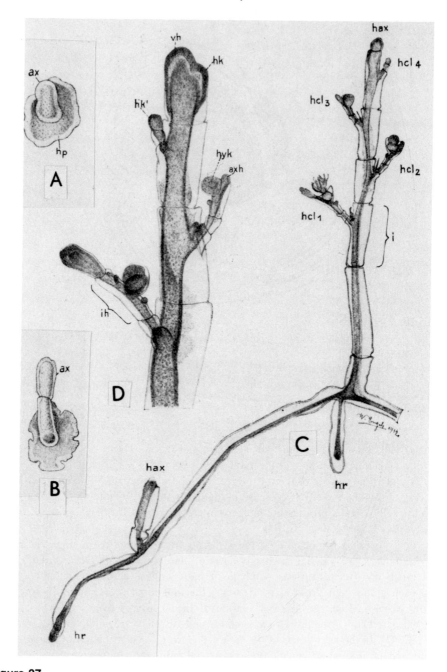

Figure 27

Compound colony in *Plumularia echinulata*.

A, B. Successive growth stages of the first primary monopodium from planula attached to substrate.

C. Fragment of basal stolons (hydrorhiza) with two vertical stipes (cormoids); a number of cormoids in one place with their ventral sides directed centripetally. There is a regular peripheral size increase of the autothecae within each thecal group, the shortest thecae being at the centripetal end (Fig. 25E). Moreover, thecal groups consist of an increasing number of autothecae concentrated into individual bundles peripherally. In the center of the colony there is a zone of single autothecae, and slightly more peripherally the autothecae are assembled into simple groups composed of two thecae only. Still more marginally, a number of autothecae concentrated into a separate group increase successively up to eight and probably more (Kozlowski, 1949; see Fig. 25F). In certain representatives of Idiotubidae, the thecal groups form distinct twigs stretching beyond the thecorhiza, and in *Galeograptus* and *Cyclograptus* they are limited to the margin of the thecorhiza, forming fairly long "sheaves" composed of numerous autothecae which show a distinct size increase peripherally. In certain idiotubids thecal grouping results in formation of erect "branches" stretching distinctly above the thecorhiza. Size gradients of autothecae within such bundles appear very similar to the size gradient in graptoloids. *Parvitubus* is perhaps the most striking example (Skevington, 1963). Its separate branches probably arise from thecorhiza and are comparable in general appearance to a graptoloid cladium.

Both colony development and the stolonal system underlying the thecal grouping have been recognized only in *Dendrotubus erraticus* (Kozlowski, 1963). The stolon which starts from the base of the sicula situated in the middle of the colony is coiled spirally around the sicula, and at regular intervals is divided into two branches (budding by diads; see Fig. 26), leading to a stolotheca and an autotheca or a bitheca. Thecae belonging to different diads are then secondarily accumulated into thecal groups.

In other idiotubids as in *Idiotubus, Discograptus,* and *Galeograptus*, no traces of stolons have been found, and it seems probable that they were not sclerotized (Bulman and Rickards, 1966). This seems to indicate that particular representatives of the graded series presented here, after Kozlowski (1949), are not necessarily closely related. However, the detailed analysis made by Bulman and Rickards (1966) indicates that the budding pattern in *Discograptus* and *Galeograptus*, in spite of the absence of a sclerotized stolon, has been essentially the same, the thecae being produced in diads as in *Dendrotubus*. Certain astogenetic differences recognized in the proximal part of the colony (structure of sicula, budding) therefore have no influence on our thinking about the general trend in the evolution of colony organization in the Idiotubidae.

Taking together both the surface picture and the underlying stolonal system, the morphological pattern of idiotubid colonies may be reinterpreted as centripetal or concentric. One of the single autothecae in the center of the colony in *Dendrotubus* and *Discograptus* is the sicula. The centripetal organization of the colony is therefore only a special case of the basipetal type, previously recognized in graptoloids. To a considerable extent

---

with increasing degree of complexity may occur on the same stolon (see also Fig. 28), showing thus a distinct complexity gradient from proximal to distal end of each stolon.

D. Termination of a cormoid with three incipient lateral cladia. First primary monopodium (ax), axis of a cladium (axh), internode (segment) of primary stipe (i), internode (segment) of cladium (ih), axis of the primary stipe (cormoid) (hax), successive cladia (hcl$_{1-4}$), ultimate and penultimate bud of lateral cladia (hk', hk), attached planula (hp), bud of a hydranth (hyk), growing tip of the primary stipe (cormoid) (vh). (From Kühn, 1909.)

thecal groups are comparable to colonial subunits such as the cormidia in Hydrozoa, even though they are not composed of polymorphic individuals. A peripheral increase in the number of autothecae assembled into a particular thecal group is essentially the same as the gradient of complexity ("Komplexitätsgradient") observed by Schenck (1965) in the cormogenesis of compound colonies in Recent Plumulariidae.

According to Schenck (1965), in colonies of Plumulariidae, cormoids* of different orders of complication may occur on the same stolon (Fig. 27). The simplest cormoids are situated proximally, near the primary monopodium ("Primärmonopodium"), which develops from the attached planula. The most complicated ones morphologically, more advanced and integrated, lie distally. There is a regular increase in the complexity of the cormoids (number of secondary branches, branching pattern), a complexity gradient from the proximal to the distal end of each stolon (Fig. 28). Increase in the complexity of particular cormoids tends to reach a certain upper limit of expression, typical for a given species ("Maximalgestaltung"). Certain plumulariids have colonies which consist of cormoids with a low degree of complexity—all are primary monopodia (according to Schenck, 1965, in *Antennella, Antennelopsis,* and *Corhiza*). These are comparable with *Idiotubus* among the Tuboidea. Cormogenesis in other plumulariids passes only through a simple stage represented by the most proximal cormoids and continues later to produce more distally placed cormoids which are successively more and more complicated. In *Halicornaria* this leads to the appearance after the primary monopodia (no additional branching) of secondary monopodia or rhachis (one lateral cladium from a node) and, later, diplorhachis (two lateral cladia from a node). In more advanced plumulariids a stage of polyrhachis may be attained as the maximal expression of the complexity gradient where a greater number of lateral cladia are produced on a node (Fig. 28F). The more advanced plumulariids are comparable with such advanced tuboids as *Dendrotubus* and *Discograptus,* and to a lesser extent with *Galeograptus* and *Cyclograptus*.

Although expressed by quite different morphological characters, the complexity gradients in colonies of tuboids (increase in number of thecae cumulated into thecal groups, centrifugally-peripherally), and in plumulariids (increase in complexity of branching in cormoids along the same stolon and distalward) correspond to a fundamentally similar pattern of organization. Within each thecal group of tuboids there is a distinct size, or rather height, increase of autothecae, and within a particular plumulariid cormoid there is a morphological gradient expressed in the number of nemathozooids, the spacing of hydranths, or in the branching capacity. The basal parts of separate cormoids may be constructed more simply than those above them.

Another feature, peculiar to colony organization in Tuboidea, is a certain tendency toward intracolonial polymorphism. In *Tubidendrum* the autothecae may occur in two forms: (1) normal, and (2) with constricted aperture, the microthecae of Kozlowski (1949). The latter are uncommon and irregularly spaced. Later Kozlowski found other derivatives of autothecae in *Idiotubus* (unpublished), and Bulman and Rickards (1966) have discovered them in *Retiograptus* and *Discograptus,* suggesting the term "*conothecae*." They, too, are irregularly placed in the rhabdosome. The nature of micro-

---

*Cormoid, as defined by Schenck (1965), is any cormal (subcolonial) unit "which extends inside a vertical branch beyond basal stolon, and which may be composed exclusively of cormidia or stolons, or be composed of both, may be mono- or polysiphonal, branched or unbranched."

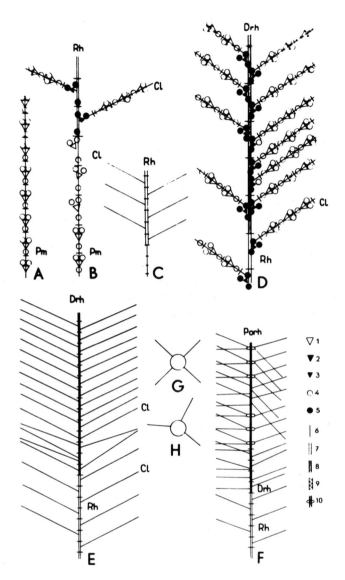

Figure 28

Diagrams showing cormoids, or vertical stipes in compound colonies of plumulariids, of different complexity and branching pattern.

A. Primary monopodium, Pm (unbranched, uniaxiate vertical stipe).

B, C. Rhachis, Rh or secondary monopodium, which consists of unbranched proximal part (basal part or pedunculum) and simply ramified distal part (one lateral cladium, Cl from each segment, C).

D, E. Diplorhachis, Drh, where two lateral cladia are formed in each segment, and proximal (basal) part is made of simpler rhachis.

F, G, H. Polyrhachis, Porh with multiple ramification, 3 or 4 lateral cladia being produced at each node, but which proximally (basally) is constructed simpler, as rhachis or diplorhachis; G and H are sections at different levels of polyrhachis. 1, hydrotheca; 2, rhachis hydrotheca; 3, abortive hydrotheca; 4, nematotheca; 5, rhachis nematotheca; 6, primary monopodium; 7, rhachis; 8, diplorhachis; 9, polyrhachis; 10, ordering of ramification at one plane. (From Schenck, 1965, modified.)

thecae remains obscure, but Kozlowski (personal communication) has arrived at the opinion that conothecae have the role of incubatoria and are functionally comparable to ooecia in bryozoans. Although tendencies toward intracolonial polymorphism are distinct in Tuboidea, they have never reached a scale comparable to that of the Coelenterata.

MODES OF INTRODUCTION OF PHYLOGENETIC NOVELTIES INTO GRAPTOLITE COLONIES

The polar organization of graptoloid colonies affects the mode of appearance of and further changes in phylogenetic novelties,* which are subjected to the principle of morphophysiological gradients (Urbanek, 1960, 1963, 1966, 1970). An analysis of extensive evidence based on lineages and trends established empirically on the basis of regular stratigraphical and morphological sequences of graptoloids led Urbanek (1960, 1966, 1970) to the opinion that phylogenetic novelties were introduced to the graptoloid colonies polarly, i.e., from their proximal end (proximal introduction) or from their distal end (distal introduction). In the vast majority of observed cases, a phylogenetic novelty introduced according to one of the previously mentioned modes affects only a fraction of the zooids in a colony. This results, in the early stages of phylogeny, in the appearance of biform rhabdosomes (as previously defined), which can be either proximally or distally progressive (Figs. 2C and 3A). The former type is produced by proximal introduction, and the latter is the effect of the distal introduction of new phylogenetic characters. In both cases of introduction, subsequent events, described many times from graptoloid lineages, consist of the spreading or progression of a new character (e.g., thecal form) along the rhabdosome toward the opposite pole. Phylogenetically, the new character, which initially affects only a number of zooids concentrated at the proximal or distal end of the rhabdosome (depending on its mode of introduction), progressively involves an increasing number of zooids, and finally may spread over the whole colony (Fig. 29A, B, D, E). Particular links of a given lineage may represent different stages of this process, which when completed, results in transformation of a biform to a uniform rhabdosome, through different stages of "biformism." Such a full cycle of events is by no means an absolute rule for each lineage, and Bulman (1955) rightly emphasized that, frequently, new characters were arrested at certain stages of spreading, for the remainder of a given phyletic line. Urbanek (1960) has stressed and elaborated this point of view, pointing out that some formally distinguished graptoloid taxa (as *Pernerograptus* Přibyl, *Demirastrites* Eisel, or to a lesser extent *Diplograptus* McCoy) are stabilized biform types. The same is true for numerous taxa of specific rank, being "perpetuated" biform types, which until their extinction never attained the uniform stage (e.g., *Cucullograptus aversus rostratus*; see Urbanek, 1966).

The course of evolutionary events as analyzed from the point of view of the number of zooids affected by given characters is as follows: introduction of phylogenetic novelty 1 (biform rhabdosome)—spreading of introduced character 1—eventual spreading of

---

*"Phylogenetic novelty" is here understood as any new evolutionary aquisition, independent of whether it performs entirely new functions or not. Mayr (1963, 1970), using this term, put emphasis on the functional aspect, but de Beer (1958) had used it previously without this restriction.

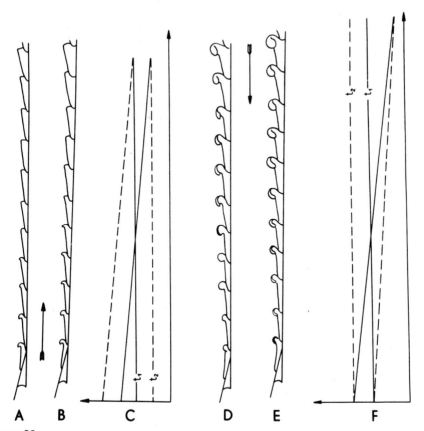

Figure 29

Diagrams illustrating modes of introduction and spread of new thecal types in monograptids (A, B, D, E) with attempted biological interpretation of evolutionary changes involved (C, F).

A, B. Proximal introduction and distal spread progression, indicated by an arrow, of a phylogenetic novelty, interpreted in C as a result of increasing activity of a morphogenetic stimulator produced by siculozooid (change from continuous oblique into broken oblique line) at the stable threshold level of reactivity of tissues (t_1) or through lowering of the threshold level (t_2) (increase of reactivity of tissues) at the stable amount of morphogenetic substance (stimulator) produced.

D, E. Distal introduction and proximal spread (progression) as indicated by an arrow, of a phylogenetic novelty, explained in F as a result of decreasing activity of a morphogenetic inhibitor produced by siculozooid (change from continuous oblique into broken oblique line) at the stable threshold level of reactivity of tissues (t_1) or by rise of the threshold level (t_2) (decrease in reactivity of tissues) at the stable amount of morphogenetic substance (inhibitor) produced. (After Urbanek, 1960, modified.)

character 1 to all zooids of the colony (uniform rhabdosome)—introduction of phylogenetic novelty 2 (secondarily biform rhabdosome)—etc.

The polar introduction of phylogenetic novelties into the graptoloid rhabdosome was in fact recognized earlier by Elles (1922) but without proper understanding of their biological nature. Elles ascribed to each mode of introduction a different phyletic role, much in the style of thinking at the time. A proximal mode of introduction was considered by Elles as progressive evolution (the appearance and development of new characters); a distal one was believed to represent simplification or retrogression (a reduction or disappearance of ancestral characters). While the significance of proximal introduction was supported by a quickly growing body of facts, the presence and function of a distal introduction remained open for a longer time. The "retrogressive lines" postulated by Elles (1922) later proved to be based on a misunderstanding of both morphological data and stratigraphic sequence. Bulman (1951, 1955) provided some evidence to indicate that new features may originate or be dominantly expressed distally, and Thorsteinsson (1955) likewise suggested that changes which effect the distal part of the rhabdosome alone cannot always be classed as retrogressive. Additional evidence for the distal introduction of phylogenetic novelties and their role in progressive evolution was supplied somewhat later (Urbanek, 1960, 1963, 1966, 1970). Cucullograptinae, Linograptinae, and Neocucullograptinae proved to be large groups of Upper Silurian monograptids in which distal introduction is the prevailing mode of appearance of a number of new and progressive characters. Furthermore, Urbanek (1960) pointed out that certain phylogenetic novelties, "retrogressive" in a formally morphological sense, such as reduction, attenuation, and disappearance of periderm in Archiretiolitinae, and simplification of thecae in triangulate monograptids, also appeared to be due to distal introduction. The mode of introduction therefore seems to be largely independent of the morphological nature of the characters introduced. This is especially true with respect to characters introduced distally, which may be both "progressive" and "retrogressive" in the formal morphological sense.

The majority of well-established phylogenetic series in monograptids illustrate the appearance and development of new thecal characters resulting from proximal introduction. This fact may have led to the rather common belief that this is the main method of appearance for all important new features. In spite of the widespread occurrence of proximal introduction, the basic graptoloid character (i.e., size gradient) behaves as a character introduced distally and attains its greatest expression at the distal end of the colony.

Size gradient (i.e., distal increase of size of thecae) is a phylogenetic novelty which is peculiar also in certain other aspects. It is a perpetually stabilized biform character since all graptoloids are biform with respect to thecal size. Moreover, it is a feature which, because it was introduced polarly, probably involved the whole colony simultaneously.

Urbanek (1960, 1966) has also emphasized the fact that other thecal characters, such as asymmetry of apertural lobes in *Lobograptus,* were simultaneously manifested as new characters in all zooids without showing a distinct polar increase. Certain other facts, however, point toward a distal introduction of these novelties (Urbanek, 1966).

Size gradient apart, distal introduction may be considered the main mode of appearance of phylogenetic novelties in Idiotubidae. New characters (greater number of thecae gathered into a thecal group) are primarily expressed peripherally in the rhabdosome, and the greatest size of autothecae is attained within each thecal group at its distal end. It seems, therefore, that in spite of the widely accepted belief, distal introduction has played a great role in colony evolution in Graptolithina.

According to Schenck (1965), the distal introduction of phylogenetic novelties is a prevailing mode of evolution in compound colonies of Plumulariidae. Structural features in the proximal parts of colonies are often more primitive than those situated more distally. In keeping with earlier authors (Driesch, 1890; Kühn, 1909), Schenck points out that development of the colony ("Kormogenese") conforms to "the biogenetic law for cormae" ("biogenetisches Gesetz für Stöcke," "Kormogenetisches Gesetz"). Early stages of colony development, represented by simple, primary monopodia, may be considered on the basis of supposed parallelism between cormogeny and phylogeny as a recapitulation of the ancestral type of colony in the Plumulariidae.

The use of morphological features of zooids in Graptolithina placed in either the proximal or distal part of a colony as potential markers of ancestral characters depends on the mode of evolution governing the fate of such features. The significance of proximal or distal thecae can be completely different with either proximal or distal introduction (recapitulative or prophetic effect; compare Urbanek, 1966). Moreover, there is no one universal mode of introduction for all characters developing simultaneously in the same lineage. Evidence has been assembled (Urbanek, 1966) to show that different trends may operate simultaneously within the same lineage according to opposite modes of introduction and progression. It seems, however, that certain key characters, especially significant taxonomically for a given group, show one prevailing mode of evolution. In Monograptinae and Saetograptinae the corresponding thecal characters evolved according to proximal introduction, while in Cucullograptinae, Neocucullograptinae, and Linograptinae, the prevailing mode for thecal characters was distal introduction. In this rather complex picture of colony evolution in graptoloids, there is no single morphological rule to describe the relations between phylogenetic modifications of astogeny and the course of colony development (astogeny).

Different modes of introduction, and the increase in the number of zooids affected by change, are not the only regularities observed in the evolution of graptoloid colonies. Another important feature correlated with an increase in the number of zooids displaying a given character is an increase in the extent of expression of this character observable within a particular colony. The greater the number of hooked thecae within a rhabdosome, the stronger is the maximal curvature attained by the hooked thecae of this rhabdosome. The larger the number of rastritiform thecae, the greater is the extent of this rastritiform modification shown by strongly affected thecae. The more strongly asymmetric are thecae, the greater is the number of zooids affected by this asymmetry (Bulman, 1951, 1963; Sudbury, 1958; Urbanek, 1960, 1966). These changes in manifestation or degree of greatest observed severity of the effect in general are closely related with changes in the number of zooids affected by the given feature (see Fig. 29A, B, D, E).

Phylogenetic modifications of astogeny (phylastogeny) consist of a set of elementary events: (1) an appearance of a phylogenetic novelty according to one of the possible modes (proximal or distal introduction), and showing subsequently, (2) a gradual progression (spreading) of the new character toward the other pole or, more rarely, its simultaneous manifestation in all zooids of the colony, and (3) displaying a parallel gradual increase in the degree of manifestation of the character introduced. This set of mutually linked phenomena, described above in morphological aspect, may be explained in more general genetic and phenogenetic terms.

GENETICS AND PHENOGENETICS OF GRAPTOLITE COLONIES

As a graptolite colony owes its origin to agamic reproduction by way of budding from a single oozooid (siculozooid), the colony may be referred to as a clone (Urbanek, 1960). Hence a graptolite colony is an assemblage of zooids all displaying the same genotype, that of the oozooid. This simple deduction from previously known facts concerning the development of graptolite colonies bears important consequences for further discussion of the genetic aspects of their organization.

Intracolonial variation in graptolites may only express different grades of the phenotypic manifestation of genetic factors common for all zooids of the colony. It cannot, however, represent genotypic differences among particular zooids of the colony, neglecting the occurrence of somatic mutations in blastozooids. The latter probably have only limited significance and cannot be involved in normal astogeny, as manifested by abrupt changes in thecal succession observed in only a few cases of teratology.

The total observed astogenetic variation in thecae as well as the whole rhabdosome can only be an expression of phenotypic variance, inasmuch as the genetic environment remains stable for a given colony. Thus astogenetic variation and the occurrence of a morphological gradient may be restated in terms of phenogenetics (Urbanek, 1960).

Colonial morphogenesis in graptolites results in formation of rhabdosomes of characteristic patterns, peculiar for particular taxa. Budding and branching, changes in size and shape of thecae, and direction of their growth show sequences highly ordered in space and time. This seems indicative of common genetic control of colonial morphogenesis in graptolites (Bulman, 1951, 1958; Urbanek, 1960). Data available suggest, however, that genetic information contained in the zygote was later sorted in the processes of colonial morphogenesis, the colony acting as complex and differentiated system of graded organization rather than as a uniform entity. As observed intercolonial variation is relatively small, this portion of the phenotypic variance which is caused by factors in the external environment seems to have only secondary significance for phenotypic expression. The great autonomy of these expressions resulting from differences in external environment may suggest that phenotypes react primarily to factors in the internal environment of the colony itself, thus creating different conditions for gene action at a given point in the colony.

Our previous considerations may lead one to the conclusion that these differences in the internal environment of the colony were based on the production, distribution, and change in the amount of certain hormone-like substances, and are the factors which modified in different ways the phenotypic expression of the same genetic material in the tissues of particular zooids. Production of such morphogenetic substances, probably restricted to the tissues of the siculozooid, is a certain aspect of a localized gene effect connected with the cytodifferentiation of cells having the same genotype and essentially the same morphogenetic potential (Waddington, 1940, 1956). Localized production of gene-controlled substances, in the light of current views on the process of differentiation, may be considered as the result of cytoplasmic–nuclear interactions resulting from the segregation of ooplasm during cleavage of the egg cell. Subsequent differentiation of the genetic apparatus of daughter cells connected with the different functional states of their chromosomes results from these cytoplasmic–nuclear interactions. Owing to

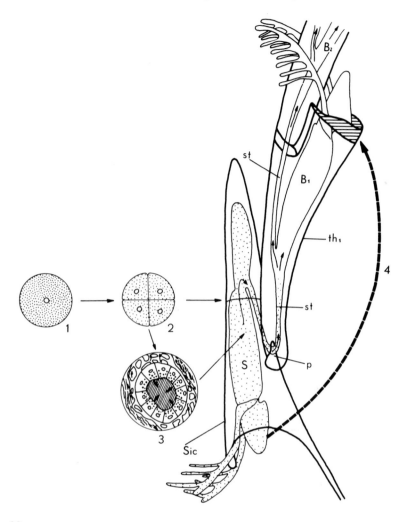

Figure 30

Ideogram to explain the role suggested for siculozooid by morphophysiological theory of organization
of graptoloid colonies through the following course of events: (a) direct transmission of certain
cytodifferentiations through cleavage of an egg cell (1, 2) to the tissues of an oozooid (s, siculozooid);
(b) secretion of a gene controlled morphogenetic substance in tissues of an oozooid (3), through
activation of certain sites in chromosomes of corresponding cells as the result of cytoplasmic–nuclear
interactions; (c) spread of morphogenetic substance produced as indicated by corresponding arrows,
enabled by interconnections of thecae (p, porus) and tissues of zooids (st, stolon) from siculozooid
into tissues of successive blastozooids (B_1, B_2, . . .); (d) specific inductive influence of
morphogenetic substance present onto the responding blastozooid tissues, resulting in appearance
of certain thecal characters (4). In the case suggested, this induction is responsible for appearance
of certain apertural elaboration (obliquely stippled area on the aperture of the first theca, th_1),
as shown by a thick arcuate arrow. Sic, sicula. No attempt was made to restore realistically
soft parts of the zooids.

an activation or repression of particular sites on chromosomes, only some fraction of the available genetic information is actually used by a particular tissue or group of cells (Ebert, 1965). It is possible that similar factors have been responsible for a localized production of morphogenetic substances in the tissues of the siculozooid. This being the only zooid of the colony which developed from a fertilized egg, it occupied a very marked position in the rhabdosome (Fig. 30). Certain substances and differentiations present in the ooplasm would be introduced through cleavage directly into the tissues of the oozooid, later becoming diluted, dissolved, and exhausted during the subsequent cell divisions and tissue differentiation involved in blastogeny. Such agents were most probably responsible for the primary cytodifferentiation of the siculozooid, and subsequently caused a corresponding activation of the chromosomes, which in turn resulted in the production of gene-controlled inductive substances. These would subsequently diffuse into the tissues of blastozooids, permeating the whole colony and acting morphogenetically according to their amount and the reactivity (competence) of the blastozooid tissues. These factors were probably responsible for the special position and function of the siculozooid in the graptolite colony, a position which as a result of localized gene effects attained the role of organizer or inductor. Through transmission of certain gene-controlled inductive substances (''genohormone'') it evoked a corresponding differentiation in the blastozooid tissues, modifying in this way the expression of given phenotypic characters. In order to explain the observed morphological patterns, the suggestion is made that the production of morphogenetic substances in the tissues of the siculozooid were rigidly determined quantitatively and occurred in a single secretionary act of rather short duration (Fig. 6E, P).

Observations on morphoses formed as a result of regeneration of graptoloid rhabdosomes show that the latter were capable of seriation of zooids with siculozooid absent from the colony. Produced phenotypes were, however, uniform under these conditions and were comparable to thecae formed in the course of normal astogeny at the distal end of the rhabdosome. Genetic information contained in blastozooid tissues, in the absence of a siculozooid, resulted always in stereotypic formation of the same phenotype, while in the presence of a siculozooid, different phenotypes were produced in otherwise similar conditions of development. This indicates the significance of the sexual process for realization of the species-specific morphological pattern of colony in graptoloids. It seems that in the course of normal colonial morphogenesis (astogeny) genetic information contained in the zygote was probably subject to segregation into oozooid and blastozooid fractions (probably due to activation or repression of particular sites of the genome in large groups of cells). The blastozooid fraction was capable of formation of the species-specific pattern of the colony only when placed within the ''morphogenetic field'' created by influences of the oozooid (siculozooid). The blastozooid fraction determined the reactivity and competence of blastozooid tissues and their response to inductive action of oozooid. Their interaction resulted in formation of a series of phenotypes, represented by different zooids within the normal colony.

An analysis of certain departures from regular astogenetic changes, expressed in fluctuations of some thecal characters along the rhabdosome or in an accelerated or sudden shift in others, may supply additional information on the mode of action of the morphogenetic agents involved. Based on observations made by Sudbury (1958), Kraatz (1958), and some data obtained by the present author, Fig. 31 shows in contrast to the steady

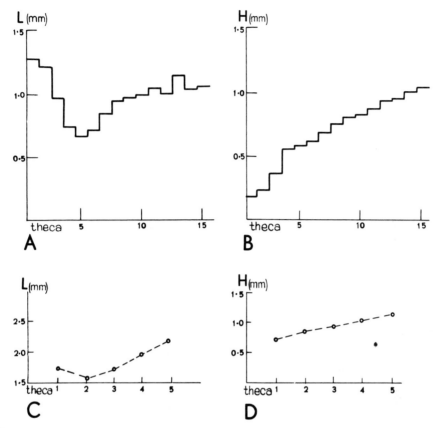

Figure 31

Graphs showing proximal decrease–increase in length (L) of successive thecae of rhabdosome as contrasted with steady increase in their height (H) in *Monograptus communis communis* Lapworth (A, B) and in *"Monograptus kallimorphus"* (C, D). Mean values for several rhabdosomes in each case. (After Sudbury, 1958; Kraatz, 1958.)

and rectilinear progression in the height (width) of thecae (H), a fluctuating behavior of their length (L). A variable number of proximal thecae shows initially a certain decrease in their lengths, which then increase gradually (Fig. 31A) or rather suddenly (Fig. 31C) to continue a rectilinear progression. The lengths of thecae, as a result of their different shapes, are measured differently by the authors cited and are therefore not strictly comparable. But this proximal fluctuation in the length of thecae, treated broadly as a deviation in the astogenetic elongation of thecae, seems to be a common, repeated phenomenon, at least in monograptids. This fluctuation is also a clear deviation from the principle of regular gradients governing the changes of these and other characters, in a majority of instances (Fig. 32).

To explain this departure, the following working hypothesis is suggested: The production of morphogenetic substances was a single secretionary action of the siculozooid, but one which continued over some time span (Fig. 6E, P). This would coincide with the

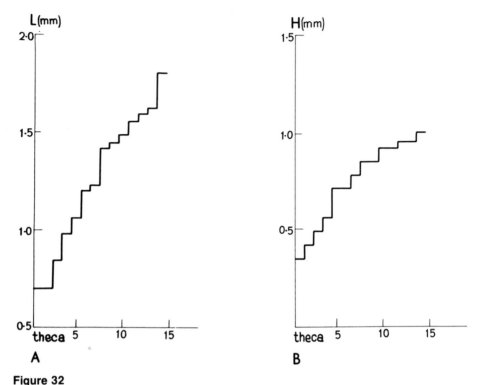

Figure 32

Graphs showing steady although somewhat irregular increase in length (L) and height (H) of thecae in *Pristiograptus dubius* (Suess). Values for single rhabdosome.

early stages of colony formation, causing a relatively large accumulation of morphogenetic substances in the tissues of the most proximal zooids. Too high a concentration of these substances was responsible for a temporarily greater inhibition in elongation of thecae. This was soon stopped when the concentration dropped below a certain threshold due to the use of such substances in budding as well as in the regular decrease in secretionary activity of the siculozooid.

Similar threshold effects may be considered responsible for locally accelerated astogenetic changes, where neighboring thecae show great modifications in their size and shape (Urbanek, 1960). Among numerous examples is the astogeny in *"Monograptus" limatulus* Törnquist, which shows a sudden change from very thin proximal thecae into much wider distal ones, with only one to two thecae being transitional in size and shape (Fig. 33). This abrupt change in thecal characters is accompanied by a strong bending of the rhabdosome, with the greatest curvature situated just at the point of sudden thecal change. Both changes may be explained as a threshold effect in morphology of thecae and direction of their growth.

Still more discontinuous threshold effects were involved in astogenetic changes of the mode of budding in diplograptids (Urbanek, 1960). The sudden change from alternate

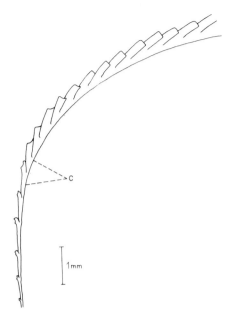

Figure 33

"Monograptus" (= *Pernerograptus?*) *limatulus* Törnquist, a Llandoverian monograptid showing an abrupt astogenetic change at the critical point (c) expressed by rapid widening of rhabdosome, change in shape of apertural part and in proportions of thecae, combined with strong dorsal incurvation of rhabdosome. Investigations of Sudbury (1958) indicate that species in question decends probably from *"M." revolutus,* the latter showing more gradual change from thread-like proximal into wider distal portion of rhabdosome.
Diagrammatical restoration based on examination of Törnquist's (1899) material housed in Department of Paleontology, University of Lund (especially L01472-1473^t).

budding in proximal thecae to two independent thecal series in the distal part of the rhabdosome through the appearance of a dicalical theca may be considered a special case of morphological gradients combined with a threshold effect. Mode of budding probably has been determined by some agent which would display a gradient along the rhabdosome. Where the amount of this required substance drops below a certain level, the tissues react to the character involved by a sudden change in the mode of budding. There are no transitions possible between the alternating and independent budding (the all-or-none effect).

Somewhat different is another departure from the regular size gradient observed in a number of graptoloids (some Dichograptidae, Petalograptinae, some Monograptidae), where at the distal end of the rhabdosome, thecae decrease in size. This phenomenon seems to be linked with corresponding growth relations of thecae on the growing tips of particular stipes. In a number of observed cases (e.g., in *Saetograptus chimaera* Barrande), the rhabdosome ends rather abruptly, the penultimate theca being fully grown and the ultimate theca being incipient (Fig. 34A).

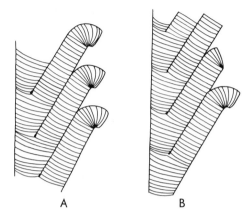

Figure 34

Contrasting patterns of growth relations of thecae at growing tips of rhabdosomes.
A. An abrupt termination, penultimate theca being complete and ultimate incipient.
B. Termination of a growing tip which consists of a series of incompletely developed thecae in different growth stages (growth gradient).

In other instances a series of thecae of different growth stages, therefore showing a gradual decrease in size distally, is always present at the growing tip (as in *Lobograptus scanicus parascanicus, Cucullograptus aversus rostratus, Neocucullograptus inexspectatus*, Fig. 34B). When, for some reason, growth stops, these growth relations seem to be responsible for a size decrease in most distal thecae. This is conspicuous in *Phyllograptus, Tetragraptus,* and *Petalograptus*. This phenomenon, called the growth gradient (Urbanek, 1966), seems to be fully analogous in nature to the proximal gradient of ontogenetic change as defined for bryozoan colonies by Boardman, Cheetham, and Cook (1969).

It is not surely established whether in graptoloids a certain thecal type is reached in astogeny which is entirely stabilized and repeats itself "endlessly" due to further budding, or whether there are steady, although very minor, astogenetic changes to the very end of colony growth. According to Schenck (1965) in colonies of Plumulariidae a certain morphological type of cormoids and hydrothecae is attained, beyond which no further changes are observed ("Maximalgestaltung"). The observations of Regnéll (1949) on the large-size rhabdosome of *Monograptus flemingi* Salter showed no increase in the width of rhabdosome over a long distance (391 mm) of the distal part comprising approximately 350 zooids. According to Jaeger (1959), in large branches of *Linograptus posthumus* the greatest width is attained between the fifth and tenth thecae, and more distally there is only a very restricted increase, if any. Certain observations seem to indicate that restricted increase in width of the rhabdosome, and some minor changes in shape of thecae, continued for a longer time after attainment of the distal type of the thecae. This is why for descriptive purposes the terms "distal thecae" and "true distal thecae" are sometimes used.

Somewhat more obscure are causes of departures from regular morphological gradients, observed in *Glossograptina* (Jaanusson, 1960; Whittington and Rickards, 1969). Their thecae show a normal size gradient. Certain spines, however, situated at different points on a rhabdosome show discontinuous distribution. In the proximal part of a rhabdosome they occur on each second theca. Distally they show less regular and wider spacing. No satisfactory explanation can be offered at this time to explain these phenomena except that being made of microfusellar tissue and formed with considerable delay, they probably were not rigidly controlled by the axial gradient. Microfusellar apertural structures in *Bohemograptus* occurred irregularly, although later in phylogeny they were subordinated by the normal polarity of the rhabdosome (Urbanek, 1970). In glossograptids, existing microfusellar spines seem to exert inhibition over certain areas, thus preventing formation of the same structure in adjacent thecae (see also the extensive discussion of these problems in this volume by Jaanusson).

Phylogenetic modifications of astogeny may be conveniently described, and in some ways explained, by using certain notions and terms of phenogenetics. The appearance of a phylogenetic novelty through a distal introduction seems to indicate that the morphogenetic substances produced by the siculozooid had an inhibiting phenogenetic effect on the character introduced. In contrast, a proximal introduction means that the substance produced had a stimulating influence on the character involved. Simultaneous operation of trends showing opposite directions of progression seems to indicate that the same morphogenetic agent played a different role in relation to various characters, being inhibitor or stimulator respectively. Urbanek (1960, 1966) has suggested that the phenomena involved in the progression (spreading) of new characters may be conveniently described by using the terms "expressivity" and "penetrance" which were introduced by Timoféeff-Ressovsky (1931; see also Timoféeff-Ressovsky and Ivanov, 1966). Both terms are a measure of the effectiveness in the phenotypic manifestation of genetic factors. Penetrance is determined by the percentage of individuals carrying the responsible gene (or genes) which manifests its effects (percentage of phenotypic effect). Expressivity denotes the extent to which an organism carrying a gene and showing it is affected by it. It is a measure of the relative degree (severity) of the phenotypic manifestation.

As a colony is a clone, all zooids have the same genotype, i.e., all carry the same genes and are therefore potentially capable of their phenotypic manifestation. In biform colonies some of these factors are manifested in a part of the zooids only. Hence we may here speak of *incomplete penetrance* of the corresponding genetic factors. In the evolution of numerous graptoloid lineages we may note that these factors are phenotypically manifested in a progressively increasing number of zooids. This phenomneon of progression of a phylogenetic novelty may be explained as *increasing penetrance* of the corresponding factors. Uniform colonies which may be finally produced by these processes represent the state of *complete penetrance* of the corresponding genetic factors.

Major changes in the phenotypic manifestation of a given character attained within a colony (degree of thecal curvature in hooked thecae, degree of asymmetry of apertural apparatus) may be considered as a process of change in expressivity of the corresponding genetic factors.

As has been clearly indicated by our earlier considerations, increase in penetrance and increase in expressivity are distinctly correlated in graptoloid colonies and expressed according to a definite spatial pattern which is subordinate to the polar organization of the colony. In the evolution of graptoloid colonies, changes in penetrance and expressivity (as previously defined) are functions of a gradient in the distribution of morphogenetically active agents within the colony.

In order to explain hypothetically the quantitative changes in the phenotypic manifestation of genetic factors in graptoloid colonies, Urbanek (1960) suggested the use of certain terms and ideas as elaborated in phenogenetics. The inductive mechanism involved in these changes may be explained as an effect of the interrelation between the amount of morphogenetic substances available at a given point on the rhabdosome and the reactivity (competence) of the blastozooid tissues, which show a certain threshold in their response. Figure 29C, F illustrates these considerations, which postulate the increasing activity of a stimulator as the cause of the distal progression of a character introduced proximally, and the diminishing activity of an inhibitor as the cause of the proximal progression of a character introduced distally. An equally acceptable alternative to this reasoning, suggested by Urbanek (1960), is a corresponding change in the reactivity of blastozooid tissues, which on the diagram are conveniently expressed by a lowering, or rise, in the threshold level of phenotypic response. It is evident that these interrelations may change as a result of mutations which could be responsible for a localized production of morphogenetic substances, or for the competence of blastozooid tissues. In the case of a simultaneous manifestation of phylogenetic novelties in all zooids of the colony, the character introduced illustrates complete penetrance from the very beginning. This is not necessarily correlated with its high expressivity. Asymmetry of apertural lobes in *Lobograptus scanicus parascanicus* (Urbanek, 1966) may serve as an example of such a character with complete penetrance but a relatively low degree of expressivity. In the case where a new character starts with incomplete penetrance, however, further increase in the latter is indispensably linked with an increase in the degree of expressivity of the character in question.

Although it is very convenient to describe the changes in the quantitative manifestation of the characters involved in the evolution of graptoloid colonies, the terms "penetrance" and "expressivity" are not sufficient. The appearance of phylogenetic novelties is an expression of a different phenomenon for which the term "specificity" is needed. It determines the localization or "field of operation" of genetic factors, and their particular morphological form of manifestation. The appearance of particular thecal characters (such as apertural processes or spines) is an emergence of a new specificity in the morphogenetic action of the genes, but their subsequent, and usually gradual, changes may be described in such terms as "penetrance" and "expressivity."

It seems obvious that the appearance and progression of phylogenetic novelties in graptoloid colonies have been controlled by nature selection. Selection pressure has probably been responsible for secondary changes in the phenotypic manifestation of genes within the colony, according to the adaptive advantage of resulting phenotypes. Inasmuch as the phylogenetic modifications in graptolite colonies may be largely reduced to changes in penetrance and expressivity, the selection of modifying genes seems to be a powerful means of evolution in this group of colonial organisms (see Urbanek, 1960). The adaptive significance of certain thecal characters has been evidenced in many cases (see discussions in Urbanek, 1958, 1966, 1970). Others are more enigmatic, but the general picture of evolution in graptoloids is most suggestive of control by natural selection. In this respect half-developed biform rhabdosomes may be transient stages toward better adapted uniform types. Stabilized biforms, owing to a balance in their epigenetic systems, probably represent a compromise between the adaptive optimum in the phenotypic manifestation of a gene and some unknown harmful effects. The stabilized low penetrance of hooked thecae in *Monograptus aequabilis notaequabilis*, quoted recently by Jaeger (1969) as a "Gegenbeispiel," an example that argues against the hypothesis that biform types are transient, ill-adapted stages in evolution, may be easily understood following this line of reasoning as a combined effect of linkage between a partly adaptive character and certain harmful effects of its stronger expression; (see the earlier considerations of Bulman, 1955 and Urbanek, 1960).

GRAPTOLITES AND THE WIDER ASPECTS OF COLONY ORGANIZATION

The application of the terms "expressivity" and "penetrance" to colonies has met with some reservations and counterarguments by Van Valen (quoted in Urbanek, 1966; see also Van Valen, 1969). These arguments may be summarized as follows:

1. Colonies (cormae) may be considered as single individuals: (a) because of the interconnection and interdependence of zooids, and especially (b) because all zooids of the colony are genetically identical.

2. If we consider a colony as one individual, neither "expressivity" nor "penetrance" is accurate for the description of its morphological gradient (Van Valen, personal communication, 1960, suggested a hybrid term, "incomplete expression").

The first of these arguments is composed of two objections of a quite different nature. The first (1a) leads us into the troublesome and vague "individual versus colony" dilemma, which Mackie (1963) feels is a barren "Scheinproblem" of biology. But the angry words of Mackie, which on the whole are quite justified, did not free biology from the embarrassing problem of individuality!

One line of reasoning (partly presented in Urbanek, 1960, 1966) puts emphasis on the fact that the mere appearance of a tissue connection does not deprive organisms of their whole individuality, in the same way Siamese twins are not single individuals simply because of tissue interconnection and some physiological interdependence. Thus the present author agrees with Naumov (1960) and Schenck (1965) that the appearance of a slightly integrated cormus does not mean the transfer of the entire individuality of particular zooids to the entire colony. As Schenck (1965, p. 962) rightly says: "Die Zooide eines nicht integrierten (unzentrierten) Primärkormus sind nicht mehr solitär, aber noch immer Autozooide, e.h. jedes ist in sich Träger aller Vitalpotenzen."

In graptolite colonies the degree of integration and common control has been rather low, when compared with the higher level of integration attained by some Recent colonial animals. Even the dependence of the sicula has been only facultative since regenerative morphoses indicate that a graptolite colony may live and develop after a loss of the siculozooid (Urbanek, 1963). It is probable also that isolated single zooids were viable and capable of growth and budding. The inductive influences of the siculozooid are by no means a necessary prerequisite to life and budding in a graptolite colony. Several facts speak for a significant individuality (= autonomy) in the zooids of a graptolite colony: (1) a separate lodging for each zooid, (2) separate tentacular apparatus and oral opening, (3) independent feeding and digestion, (4) early separation of each zooid as a result of blastogeny, and (5) each zooid probably capable of reproduction. The application of certain biological viewpoints as elaborated for solitary organisms seems also to be quite adequate for the description of the properties of particular zooids in such slightly integrated colonies. The degree of interdependence of zooids in such cormae is no greater than in colonial-like ecological populations such as swarms or societies of certain insects. The powerful influence of egzohormones (pheromones) in certain insects on the behavior and morphogenesis (arrest in development of sexual characters) of the individuals in the population is an excellent example of such dependence. Such an influence is exerted by means of air diffusive substances. The close "skin-to-skin" contacts in the echiurid *Bonelia* or the gastropod *Crepidula*, as a means of transmission of sex hormones, are intermediate between air diffusion and the tissue transmission found in true cormae. The individual character of the majority of biological processes is not abolished completely and automatically by mere occurrence of tissue interconnection among zooids, and the degree of integration of simple colonies (cormae) is about the same as in the integrated swarms or pseudocolonies of solitary organisms.

There is, however, a current tendency to describe each instance of increase in the interdependence and integration of zooids in colonies in terms of a decrease in the individuality of the zooids and an increase in the individuality of the colony as a whole, in other words an individuality transfer. This tradition is among the weakest points associated with the problem of coloniality in biology.

One reason for such a conclusion is the vagueness in biology of the term "individual" and especially its derivative "individuality." These terms have often been used with such different meanings and extensions that they could be applied to almost any biological system. The present author is of the opinion that in the majority of cases the use of other terminilogy, such as "autonomy," "integration," or "levels of control" is much safer (see Schenck, 1965). These could be at least partly quantitatively characterized, while there is no objective measure for "individuality." However, these suggested substitutes for the "invalid" terms "individual" and "individuality" are very general and they do not refer to any particular biological system since quite different things may be "integrated." It may, therefore, be convenient to retain these terms but in a redefined form to denote relative autonomy, distinctness, and separateness of biological systems at the organismal level of organization only. As will be shown below, any further extrapolation of these notions to higher levels of organization results in unsafe levels of reasoning. As I understand Mackie (1963), he is of the opinion that the terms "individuals" and "individuality" are generally nonoperative or inapplicable to organisms, partly because the terms themselves imply "something which cannot be divided." I agree that these terms, in their present usage, are vague and too elusive, and it is much safer to speak of distinctness, autonomy, and degree of dependence or integration than of "individuality." Mackie's viewpoint, based on the considerations of von Bertalanffy (1952), is, however, an extreme one. It seems dangerous to judge the properties of objects from their names. The substitution of the term "individuum" by "dividuum" (as once suggested by A. Braun) does not really change the situation. In my opinion, other factors, discussed below, should influence the decision to use, limit, or abandon the term "individual" in biology. What seems really important is that this term may be used with approximately the same uncertainty for zooids of a colony as for solitary organisms.

Still more vague is the meaning of the term "individuality." According to present usage, "individuality" is a property of biological individuals. But it occurs also in other biological systems. One can speak of the "individuality of chromosomes," the "individuality of plastids" or other organelles, the "individuality of a colony," and the "individuality of constituent zooids." Individuality is a very general term and one which means little more than "distinctness" or "integration" of a given system. The presence of an individuality of some degree should not imply that the system in question has the nature of an individual.

Nevertheless, levels or degrees of individuality have been used to distinguish a hierarchy of biological individuals, the latter being viewed as systems which have attained the greatest possible autonomy of some kind, and therefore capable of independent existence (bionts). In addition to the undoubted individuals in the animal kingdom (monoenergid and polyenergid protozoans, solitary metazoans), Beklemishev (1950, 1964) distinguished with some hesitation highly integrated cormae, which at the highest level of colony integration are something more than a colony and perhaps something less than single organisms.

Ivanova-Kazas and Ivanov (1967), Ivanov (1968), and Schellhorn (1969) go further and include highly integrated colonies of Metazoa in a series of biological individuals, referring to them as "individuals-cormae" or the "third level of individuality." According to Ivanov, the fate of multicellular individuals in the integrated cormae of Metazoa repeats the course of events presumed for single unicellular individuals by the colonial (monoenergid) theory of the origin of Metazoa. Although this theory seems to be well founded, as is indicated by a critical analysis of the data (see also Urbanek, 1967), a complete symmetry in the evolution of integration among uni- and multicellular colonies is not indisputable evidence for it. Some evidence indicates that this course of events has not been repeated in the evolution of multicellular colonies where a decrease in the individuality of constituent parts (zooids) never attained a level comparable with that in the cells within multicellular organisms. Taking Siphonophora as the upper level of integration attained by cormae, we find no traces of dissociation or coalescence of zooids, each zooid being developed from a distinct bud and with the budding pattern being the same as in Hydroida (Stepanianc, 1967). Particular zooids in the cormae of the Siphonophora are doubtlessly homologous with individuals of the Hydroida. From the standpoint of morphology and morphogenesis, such cormae are only polymorphic and highly integrated colonies (Stepanianc, 1967). Functionally, however, they approach in some respects a single individual, being a single unit of function (Mackie, 1963; Stepanianc, 1967). Stepanianc emphasized this somewhat ambivalent nature of the Siphonophora, but the inescapable impression is that even at the uppermost level of integration attained, a metazoan colony is a colony which differs quantitatively rather than qualitatively from nonintegrated cormae. In certain important respects cormae always remain collective units, which cannot be described simply as some sorts of individuals.

Cormae always comprise highly specialized zooids, incapable of independent existence and in certain respects therefore are similar to organs and less specialized zooids which might be considered bionts, and therefore comparable biologically to solitary individuals.

The trend prevailing nowadays in the interpretation of colonies is only an extension of the classical haeckelian line, so skillfully elaborated by E. Haeckel (1866, 1906) in his great works on general morphology. The living world is individualized; it is in morphological and physiological aspects a hierarchy of individuals of different order. For example, a colony (cormus) is considered an individual of the sixth order, while a metazoan individual (person) is of the fifth order. An individuality transfer from the unities of a lower order onto higher ones, connected with a lowering of their own individuality rank, is easily understandable in this context. The views of Haeckel were so impressive that the majority of studies on the problem of coloniality developed essentially along haeckelian lines, including the most recent and excellent studies by Beklemishev (1950, 1964). Colonies of graptoloids were considered as individuals of a higher order, or simply as one individual, and this viewpoint has been expressed most clearly by Ruedemann (1902, 1904, 1947). The following criteria were used to substantiate this view: (1) The colony develops as a whole, shows its own ontogenesis as a single individual; (2) particular zooids correspond to organs; and (3) rhabdosomes (colonies) are persons of the second order, being composed of thecae (zooids) or persons of the first order. This concept of graptolite colonies has been silently taken for granted by the majority of graptolitologists of recent generation, with the exception of Bulman (1951), who expressed a somewhat different opinion. The recent suggestions of Van Valen (1969) are much along this line, although they are based on somewhat different premises.

The classification of different biological objects according to the haeckelian concept of individuality orders is, in many instances, of little heuristic value, and is commonly simply inadequate (see also Mackie, 1963; Schenck, 1965). Biologists have been hypnotized to such an extent by certain rather rare and extreme cases of intracolonial specializations, reducing zooids to the role of a kind of colonial organ (although this analogy is not complete), that the triad ''colony–individual–organ'' has grown to a central position in the biology of colonial organisms. But in the vast majority of colonial groups there is no particular difficulty in recognizing these levels of organization as clearly distinct ones. Among practically all Metazoa only the Siphonophora approach the upper limit of coloniality, where polymorphism and specialization of zooids provide difficulty in distinguishing them functionally from organs. In the opinion of some authors the integration of the whole colony approaches closely an ''individual of higher rank,'' something more than cormus (Beklemishev, 1950, 1964). The evolution of other colonial Metazoa (especially Bilateria) escaped such extreme results, and the Graptolithina with their evolution lasting some 200 millions of years never approached this upper boundary of colonial organization. The extreme integration and subordination of zooids modified to function as colonial organs of some sort are by no means an overwhelming trend in the evolution of all colonial organisms, nor an inescapable final result of this organization when once established, but are a rather special case. To view the biology of colonial organisms only in this aspect is an exaggeration, comparable with the analysis of all evolutionary trends in solitary organisms from the standpoint of their achievements in cerebration alone, because in several important instances the latter has been a key character and source of great evolutionary success.

Since 1960, the present author has attempted to suggest another line of reasoning, interpreting colonies (cormae) as a certain form of superorganismal integration, as a particular type of population (Urbanek, 1960, 1966). The very simply organized and slightly integrated colonies of graptolites cannot be regarded as a single organism of higher order, a superindividual, whose separate zooids function as organs. Although this point of view has been developed independently, a similar approach to the problem of coloniality was outlined earlier by Plate (1922, 153) in simple form: "Eine Kolonie ist kein Individuum dritter Ordnung, sondern etwas Ueberindividuelles, eine feste Vereinigung von Individuen, ebenso wie die Tiergesellschaft ein freier Verband derselben ist." To distinguish this line of reasoning from the haeckelian one, I suggest it be referred to as the "plateian line."

The following arguments may be quoted in favor of the "plateian line": (1) The existence of numerous intergradations in the interdependence of individuals, starting with colonial populations (swarm-like assemblages of solitary organisms) and ending with true colonies (cormae); see also Urbanek, 1960, p. 129–131; (2) the presence of highly integrated colonial populations with a degree of dependence comparable with that in primitive cormae (societies of insects, pseudocolonies); and (3) the mere presence of tissue connection, degree of dependence apart, cannot be regarded as a sharp delimitation between cormae and populations. Tissue interconnection may be temporary, long lasting but deciduous, or permanent. Tissue interconnection is by no means an indispensable prerequisite of a strong centralization of assemblages.

In the opinion of Beklemishev (1950, 1964), clone-like assemblages of Synascidiae, in which particular zooids are integrated only by a tunica (comparable with cuticula or epidermal secretions), but sometimes lacking contact between their vascular and stolonal systems, are nevertheless colonies.

On the other hand, the presence of tissue interconnections should not be undervalued. The physical continuity between zooids in a colony is an important feature by which it differs from the physically discontinuous organization of a population. It approaches fundamentally the single individual of solitary organisms. But it is also true that the mere appearance of physical continuity may change but little the degree of integration of zooids within the colony.

Although colonies are colonies, and cannot be reduced in all aspects simply to populations, much more may be gained heuristically through their interpretation as a supraorganismal form of integration. This is best thought of by comparison with highly integrated populations rather than by operations with "individuality transfer" or stating that a colony is peculiar and unique in all aspects of organization. This approach agrees better with the modern concepts of levels of biological organization and integration than does the haeckelian line. The present author is deeply convinced that particular levels of biological organization cannot be arranged in one ascending order, along one line (see also Zavadskij, 1966, 1968). Such concepts are only remnants of eighteenth-century ideas on Scala Naturae.

Of particular importance is the difference between organismal and supra-organismal levels of integration. In the animal kingdom the fundamental stages of organismal level are cellular and multicellular individuals, which may function as independent organisms. The individual is a notion operative with respect to the organismal level of integration, and further extrapolation of this notion to supraorganismal levels of integration will lose a firm basis and may lead to the misunderstandings so dramatically exposed by Mackie (1963). This is another reason why the present author prefers the "plateian line" of approach to the problems of coloniality as a heuristically superior one.

The speculations presented above may have some bearing on the proper selection of research strategies in the investigation of coloniality. Two approaches may be distinguished. The colony may be considered first as a biological system, unique in all its aspects, neither comparable with a single individual or with any kind of population. This is, however, a heuristically sterile, or self-destroying viewpoint. A more realistic approach to the problems of colonies is somewhat eclectic or dualistic. Although the colony is a biological system which cannot be simply reduced to an individual or to a form of a population, it can be compared with them in certain important aspects. In some cases interpretation of a colony as a single organism may be quite adequate, and in others comparison with highly integrated populations may be more satisfactory. (Compare also the above-quoted viewpoint of Stepanianc, 1967, with respect to siphonophores). This latter viewpoint has been less explored and seems promising, especially with respect to the relatively slightly integrated colonies. In each case one should keep in mind that a colony is a collective category, which phylogenetically always appears composed of a number of zooids, homologous with individuals, and which preserves traces of this condition in its morphogenesis and physiology.

As indicated earlier, there is another reason why colonies may be considered single individuals; namely because all zooids of the colony are genetically identical. This is a widespread point of view (Mayr, 1963, p. 408; Ehrlich and Holm, 1963, p. 73–75; Van Valen, 1969, p. 207–208), but in the opinion of the present author, it is not quite appropriate. In an earlier paper (Urbanek, 1966) the validity of the genetic criterion of individuality was questioned. The genetic concept of an individual is an excessive extrapolation of this concept which is primarily morphological and physiological. If an individual is a genotype, then the individuals formed from the same zygote are only one individual (identical twins, individuals formed through polyembryony, the polyp and several medusae produced by it, a number of daughter individuals coming from a single parent individual through budding). Perhaps, if genetic criteria are decisive in distinguishing all biological units, the cells formed by cleavage from the zygote are only one cell! The extreme consequences of such an approach to the individuality problem are even more devastating than "naturephilosophic" speculations about it. Although certain genetic differences usually accompany individual characters, the present author is of the opinion that, on the whole, the genetic concept of an individual is a case of extreme "genetic reductionism" which considers all biological objects and phenomena as primarily or exclusively genetic. The decisive criteria of individuality are morphologic and physiologic, and introduction of a genetic aspect entirely changes the extension of the meaning of this term. The concept of an individual is based on distinctness or separateness rather than on difference (Urbanek, 1966), and it is not the only biological category founded on such premises. To a lesser extent the species is another example of a biological category "defined by distinctness rather than by difference" (Mayr, 1963, p. 20).

Nevertheless, even with respect to the unity of the genotype, colonies (cormae) are not unique among different forms of biological integration on the supraorganismic levels. Certain colonial-like populations, such as the pseudocolonies of *Cephalodiscus* (Pterobranchia) are probably clones, produced by budding from a single oozooid, the founder of the assemblage. Similarly, the gregarious assemblages of *Phoronis* (Phoronoidea) are clones formed by fission from a single parental zooid (Beklemishev, 1950, see also Urbanek, 1960, p. 130). Such pseudocolonies are assemblages of genetically identical organisms, being also only a special case of an ecological population. There is no difference with respect to genetic organization between such pseudocolonial clones and true cormae.

Our considerations may be summarized, as follows:

1. The interdependence of zooids in graptolite colonies, and especially their dependence on the siculozooid, was facultative and has been relatively small as compared with upper levels of coloniality.

2. The interconnection of zooids by soft tissues does not automatically deprive them of their whole autonomy (= individuality).

3. The colony (cormus) is a special form of integration at the supraorganismal level. The term "individuum" should be restricted to the organismic level and should not be used with respect to an entire colony.

Conclusions which may be drawn from these premises are the following:

1. Terms elaborated for solitary organisms may be used for the description and explanation of phenomena concerning particular zooids, especially in the case of slightly integrated cormae (i.e., in majority of colonies of Bilateria).

2. Nonintegrated or slightly integrated cormae may be treated as assemblages of individuals and adequately interpreted in many aspects as a particular case of a population. When compared with other forms of supraorganismal integration, colonies are not unique in all respects of their organization (including genetic), except for tissue interconnection (which they possess uniquely by definition!).

3. The problem of coloniality is wide open in the present state of our knowledge, and a proper selection of approaches is essential for further advances in research. It seems that much more may be gained through comparison with other forms of supraorganismal integration than through operation with such ideas as "individuality orders" and "individuality transfer," or through "genetic reduction." Skillful application of many organismic notions may also contribute to the explanation of processes going on in colonies treated as spatially organized assemblages of zooids, or in particular as highly integrated collective units.

Since a graptolite colony should be interpreted as an assemblage of zooids, each essentially comparable with the individual in solitary organisms, the application of the terms "expressivity" and "penetrance" seems to be fully justified for the following reasons: (1) Each zooid of a colony had genes responsible for the formation of a given character, (2) varying numbers of them manifested it (penetration) to a different degree (expressivity), and (3) although the genetic environment of a colony had been stable, the internal environment influenced the degree of phenotypic manifestation. All these elements are included in the classical meaning of these phenogenetic terms as defined by Timoféeff-Ressovsky (1931, 1934; see also Timofféef-Ressovsky and Ivanov, 1966). The use of other terms is poorly justified because the essential similarity of the phenomena involved is too great. The terms "incomplete" and "complete expression" as suggested by Van Valen, do not seem to be satisfactory, because they attract attention only to one aspect of the changes. In graptolite colonies there are two aspects: the number of zooids affected and the severity of the effect. For many technical reasons the use of two terms is required, and Bulman (1963, 1970) coined the very useful terms "expressivity intermediates" and "penetrance intermediates" to denote certain morphological forms which differ in the first or second aspect.

However, populations of colonial organisms have a stage structure being composed of colonies, i.e., groups of zooids with identical genotype formed through seriation (particular clone), and group of clones, i.e., groups of series of zooids with different genotypes. Urbanek (1966) distinguished "a colonial penetrance" or, perhaps better, "an intracolonial penetrance" and "a population penetrance" or, perhaps better, "an intercolonial penetrance." In fossil organisms we can speak safely only about the *intra*colonial penetrance, since an incomplete spreading of a particular character among different colonies in a given population may also be ascribed to the absence of a corresponding gene from the different genotypes. This latter point has not been emphasized by Urbanek (1966) when describing certain phenomena which could also be interpreted as changes in intercolonial penetrance.

SOME OPEN PROBLEMS

One of the greatest obstacles hindering further progress in research on the mechanisms of evolutionary change in graptoloid colonies is our ignorance of their reproductive biology and the recognition of how this may affect the genetic structure of their populations.

According to a well-founded concept of Kozlowski (1949), graptoloid zooids were hermaphrodites. In the light of his concept, the origin of Graptoloidea has been connected with a shift from gonochorism to hermaphroditism. Such a profound transformation of sexuality was involved with the radical changes at the boundary of dendroids and graptoloids. The transition between both of these groups has been gradual in many respects, in keeping with a persistent trend controlled by natural selection. It seems probable that the change in sexual conditions was (per se) advantageous. Since the disadvantage of "being a hermaphrodite" is commonly emphasized, the demonstation that, at least in the origin of the Graptoloidea, a shift from gonochorism to hermaphroditism was probably involved in macroevolutionary changes which were almost certainly adaptively advantageous would be highly instructive.

The occurrence of hermaphroditism in zooids of Graptoloidea, considered here as highly probable, does not imply, however, a particular mode of reproduction. In spite of the widespread occurrence of hermaphroditism in the animal kingdon, very few hermaphrodites have been found to be truly bisexual and self-fertilizing. The vast majority of hermaphrodites are consecutive, showing sex reversal with only a short bisexual stage in their life history. In this way, they escape certain extreme consequences of self-fertilization, which may result in great homozygotization due to frequent inbreeding, splitting of the population into numerous pure lines, reduction of hidden variation and an extremely high rate of evolution as a result of the rapid spread of mutations, or extinction. This last result is the most probable fate of organisms with such a genetic system. On the other hand, unisexual hermaphrodites (protandric or protogynic) do not

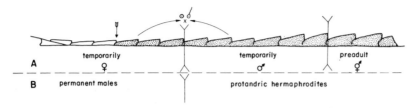

Figure 35

Ideogram to illustrate the hypothetical sexual conditions within graptoloid colony and their consequences.

A (upper row). Presuming that zooids are protandric hermaphrodites, sex reversal into female being correlated with age of each zooid, all younger being males and all mature females with a category of unripe hermaphroditic zooids at the growing tip.

B (lower row). Alternative interpretation, presuming that smaller proximal zooids are permanent males and all remaining are protandric hemaphrodites.

Perpendicular arrow indicates point of appearance of somatic mutation changing the genotype of derivative blastozooid tissues (stippled part of rhabdosome).

Mutated somatic gene may be introduced into gene pool of population as result of somatogenic embryogeny, and gametogeny based on mutated cells (arcuate arrow indicates inbreeding possible under suggested conditions).

differ essentially in the genetic structure of their populations from that in gonochoristic organisms (Mayr, 1963, 1970).

As the present author's faith in the wisdom of nature stands unshaken, he is convinced that graptoloids escaped any extreme consequences of their breeding mechanisms and were benefited by the advantages of cross fertilization. This belief cannot be proved directly from the paleontological record, but the following arguments may be quoted in support of it:

1. There is little doubt that graptoloids showed a gregarious habit and lived in large populations as typically planktonic organisms. Under these conditions there was easy access to gametes produced by zooids of other colonies, and reciprocal fertilization among zooids of the same colony was not a *malum necessarium*.

2. A large number of graptoloid species were very widely distributed geographically showing over great areas a limited range of variation. This implies great cohesion of their genotypes—much greater than that expected in self-fertilizing organisms, which should quickly split into a number of lines or stocks differing in a variety of characters.

3. Although the rate of evolution is higher and the average duration of taxa shorter in graptoloids as compared with dendroids, quite a number of species were long-lasting and typically bradytelic. Such excessive survival is less probable for self-fertilizing organisms, with their rapid spread of mutation due to frequent inbreeding. Given the alternatives of "change or extinction," their persistence under these conditions seems to be less probable.

Although the problem is widely open to discussion, the present writer is inclined to think on the basis of these arguments that graptoloids probably were essentially cross fertilizers. The elaboration of reproductive mechanisms ensuring cross fertilization among hermaphroditic colonial organisms, however, is most probably not an easy task. As all zooids of a colony are bearers of the same genotype, any reciprocal fertilization between them would result in selfing, i.e., would not differ in its genetic consequences from ordinary self-fertilization. Mechanisms based on consecutive sex reversal of zooids linked with their aging would not be effective enough as the age of zooids is different, and any simultaneous presence of male and female individuals in a colony might result in selfing (Fig. 35A). On the other hand, a rapid shift of sexuality at the same time for the whole colony, although very effective, seems unrealistic as a physiological mechanism in graptoloid colonies. In contrast to Bryozoa, there is no evidence for any physiological and morphogenetic cyclic periodicity in the life history of a graptoloid colony. In the former group a common colonial control of sexuality, resulting in shift from male to female phase, is rather suggestive, as a number of other facts (astogenetic repetitions; see Boardman, Cheetham, and Cook, 1969; brown-body formation, degeneration, and rejuvenation of polypides) indicate repeating, cyclic phenomena. A fundamental difference in the organizational patterns of graptoloid and bryozoan colonies may lie in the fact that in the former group colonial control is expressed generally only in the spatial order of the zooid distribution (e.g., distribution of sexes along the rhabdosome) but probably was not expressed in the timing of their sexual activity.

Another theoretical possibility is the existence of permanent sex differences among zooids of a colony. For instance, proximal zooids, which are in connection with the size gradient relatively small, may be considered as permanently arrested in the male (protandric) phase. The larger distal zooids may be considered as hermaphroditic, perhaps with predominance of the female phase in their life cycle. Such conditions may be compared with andromonoecious conditions as found in some plants which show a number of staminate flowers in the lower part of the stem and perfect flowers in its upper part (Bacci, 1965). In the animal world evidence for such a suggestion may be seen in the fact that the majority of consecutive hermaphrodites are in the young stages males since the production of female gametes needs greater physiological efficiency. In colonies of some tunicates (*Pyrosoma*) the oozooid is sterile, the next four blastozooids are permanent males, and all remaining are hermaphrodites (Beklemishev, 1964). This approaches to some extent the conditions presumed above for graptoloid colonies (Fig. 35B). The previously quoted example of *"Monograptus" limatulus* Törnq. (compare Fig. 33) is rather suggestive of such conditions. Its thread-like, diminutive proximal thecae housed zooids of restricted size, being in this connection probably sexually immature (neuter or sterile) or being permanent males. Their fully grown distal thecae may correspond to sexual, and probably protandric hermaphroditic, zooids. According to the important contribution of Stebbing (1970), colonies of Recent *Rhabdopleura* may contain, in addition to neuter individuals, both male and female zooids. Such colonies are therefore also "monoecious" (however, see critical remarks of Mayr, 1963, against use of this term in relation to animals), true hermaphroditism being impossible because the zooids have only one gonad. According to Stebbing, the possibility cannot be excluded, however, that even *Rhabdopleura* is a consecutive (protandric?) hermaphrodite. In *Cephalodiscus* buds of both sexes may be found attached to the stalk of the same parent zooid, and occasionally some zooids were found which possess both an ovary and a testis. (*Cephalodiscus* has paired gonads.)

In this connection, size differences among zooids in a graptoloid colony could be correlated also with sexual differences, extreme dimorphs being connected by a graded series of sexually intermediate zooids (intersexes, or permanent hermaphrodites). Size gradient, a general feature of the Graptoloidea, in the light of such a consideration, may be explained as a sexual gradient, and the role of morphogenetic substances produced by the siculozooid would be comparable with that of a sex hormone with a masculinizing effect. Further morphogenetic functions of this sex hormone have been secondary in respect to sex determination, as the structural gradients are secondary as compared with the size gradient.

This suggested organization of a graptoloid colony would result in a permanent presence within a colony of male and female zooids which were probably simultaneously reproductively active. Under such conditions, however, any effective mechanism protecting against selfing, except perhaps the most general and intrinsic mechanism of cross incompatibility, is hardly imaginable. Recently Longwell and Stiles (1970) have found that incompatibility of genes may have played a significant role in the genetic system of hermaphroditic oysters, which develop and spawn both eggs and sperm simultaneously. Cross incompatibility ensuring preferential fertilization of certain eggs by certain sperm types prevents inbreeding and promotes outcrossing and is a rather widespread genetic mechanism. Longwell and Stiles (1970) showed, moreover, the general significance of incompatibility for the evolution of organisms living in a marine environment which allows broad interconnections and, as compared with inland biotopes, offers few isolating barriers. This seems especially true with respect to planktonic graptoloids, where colonial organization (seriation of genetically identical individuals), hermaphroditism, and presumed differences in sexual phases in zooids placed near one another, all act to promote inbreeding and to reduce the share of outcrossing in their genetic systems.

In the present author's opinion, the best solution which will agree satisfactorily with the scanty data available on the biology of Graptoloidea (and with an almost equally imperfect knowledge of the biology of Recent hermaphroditic organisms) is the supposition that the former were unisexual hermaphrodites with consecutive and heterochronic sex reversal among all zooids within a colony, and consequently with a rather prolonged bisexual stage in the life history of a given colony (Fig. 35A). Such a bisexual stage in the life cycle of each colony would result in comparatively frequent reciprocal fertilization (selfing), but on the whole graptoloids were essentially cross fertilizers, because the share of inbreeding was probably restricted by cross incompatibility, as a most general and effective biological means to promote outcrossing. The duration of the supposed bisexual stage within a colony could be controlled by natural selection and the share of self-fertilization in the breeding system of a population could be kept within reasonable limits. Similar control of breeding mechanisms has been suggested for recent hermaphroditic planktonic organisms by van der Spoel (1971) and termed the "telescoping" of the sexual cycle, through shortening of the male phase, so that succession of the stages occurs more rapidly, which may enable fertilization among genetically closely related organisms. Telescoping, or absence of it, might result respectively in decrease or increase of variability, both being advantageous in certain specific ecological situations. Certain harmful effects of inbreeding limited in this way (e.g., reduced reproductive fitness as a result of homozygotization) may be imagined as compensated for by an unusually high proliferation of graptoloids, as indicated by the masses of siculae usually accompanying adult colonies. On the other hand, limited inbreeding opens certain possibilities for rapid mobilization of hidden variation, which in part could be adaptively useful, a possible source of high rate of evolution in graptoloids.

The speculative picture of the reproductive system in graptoloids suggested above, if ever realized, could be at least very convenient, combining little risk of failure with certain chance of success. Hermaphroditism enables even a single colony to be the founder of a new population and, combined with a high breeding potential of the population, avoids blind alleys created by obligatory inbreeding.

It is widely accepted that somatic mutations, while producing considerable alteration of the phenotype, are of rather minor evolutionary value, and especially do not result in instantaneous speciation "because the gametes carrying the new mutation are not reproductively isolated from those carrying the 'parental' growth gene" (Mayr, 1963; 1970, p. 251).

Nevertheless, some evolutionary significance of somatogenic inheritance in organisms capable of both asexual and sexual reproduction cannot be denied, at least theoretically. A somatic mutation affecting the tissues in the zone of bud formation may subsequently be transmitted to a series of daughter individuals. Moreover, it is theoretically possible that such a mutated gene may be introduced into a gene pool in a case of gametogeny based on mutated tissue material. Hermaphroditism of individuals formed through the budding of mutated somatic cells may contribute by selfing to a sudden homozygotization of such a gene, which in turn may accelerate the whole effect. However, formation of a bud in asexually reproducing animals involves usually a cluster of different cells inside a growth field, and there is little probability that the whole group may frequently be a cell population (a cell clone) derived from a single parental somatic cell affected by mutation. One would rather expect that a bud and resulting zooids would be in this case chimaeras formed by tissues with different genotypes. Transmission of a somatically mutated gene into the genotypes of the next generation would be possible only in those rare cases when gonads and subsequent gametes are produced from mutated somatic cells in the course of blastogeny (= somatic embryogeny). Presuming the same in a greater number of hermaphroditic zooids, we could expect through selfing a sudden homozygotization of a newly introduced somatic mutation and its quick elimination or spreading in the gene pool. Such an event, suggested by Fig. 35, is, however, rather improbable and it seems that somatic mutations could be generally neglected as a source of genetic variation in graptolites.

ACKNOWLEDGMENTS

This work was supported in part by a Smithsonian Institution fellowship, and was stimulated by a preliminary announcement of the symposium of Development and Functions of Animal Colonies to be held at the annual meeting of the Geological Society of America in the autumn of 1971.

I express my warm gratitude to Roman Kozlowski (Warsaw), for reading the manuscript and stimulating discussions. My colleagues in Washington—K. M. Towe, Richard S.

Boardman, and Alan H. Cheetham, Smithsonian Institution, and William A. Oliver, Jr., U.S. Geological Survey—were kind enough to go through the rough draft of the manuscript of this paper, and their suggestions, comments, and advice contributed much to improvement of its contents. I express also sincerest gratitude to V. Jaanusson (Swedish Museum of Natural History, Stockholm, and Visiting Professor for 1971/72 at The Ohio State University, Columbus, Ohio) for reading the manuscript and sending important comments.

Finally, I feel indebted to members of the technical staff at the Institute of Paleozoology of the Polish Academy of Sciences in Warsaw, Poland; Mrs. K. Budzyńska, Mrs. D. Sawik, and Miss M. Czarnocka for their help in preparation of illustrations.

REFERENCES

Bacci, G. 1965. Sex Determination. Oxford. 306 p.

Barth, L. G. 1938. Quantitative studies of the factors governing the rate of regeneration in Tubularia. Biol. Bull., *74*(2): 155–177.

———. 1940. The process of regeneration in hydroids. Biol. Rev., *15*: 405–420.

Beer, G. de. 1958. Embryos and Ancestors. Oxford. ix + 197 p.

Beklemishev, V. N. 1950. K problem individualnosti v. biologii. Uspekhi Sovr. Biol., *29*(1): 91–120.

———. 1964. Osnovy Sravnitelnoi Anatomii Bezpozvonochnykh. Moskva. 3rd ed., I/II, p. 3–432, p. 3–446.

Berrill, N. J . 1961. Growth, Development, and Pattern. San Francisco. p. 1–155.

Berry, W. N. B., and R. S. Takagi. 1970. Electron microscope investigation of Orthograptus quadrimucronatus from the Maquoketa Formation (Late Ordovician) in Iowa. Jour. Paleont. *44:* 117–124.

———, and R. S. Takagi. 1971. Electron microscope study of a diplograptus species. Lethaia, *4*(1): 1–13.

Bertalanffy, L. von. 1952. Problems of Life. London. 11 + 216 p.

Boardman, R. S., A. H. Cheetham, and P. Cook. 1969. Intracolony variation and the genus concept in Bryozoa. N. Am. Paleont. Conv., Proc., pt. C, p 294–320.

Bouček, B. 1933. Monographie der obersilurischen Graptolithen aus der Familie Cyrtograptidae. Trav. Inst. Geol.-Paläont. Univ. Charles, *1*: 1–84.

———, and A. Přibyl. 1953. O rodu Diversograptus z ceskeho siluru-Sborn. Ústr. Úst. Geol., *20*: 485–576.

Braverman, M. H., and R. G. Schrandt. 1966. Colony development of a polymorphic hydroid as a problem in pattern formation, *in* W. T. Rees, ed., The Cnidaria and Their Evolution. London. p. 169–198.

Bulman, O. M. B. 1941. Some dichograptids of the Tremadocian and Lower Ordovician. Ann. Mag. Nat. Hist., *11*(7): 100–121.

———. 1944, 1947. A monograph of the Caradoc (Balclatchie) graptolites from limestones in Laggan Burn, Ayrshire. Palaeontogr. Soc., *98*, pt. I: 1–42; pt. III: 59–78.

———. 1950. Graptolites from the Dictyonema Shales of Quebec. Quart. Jour. Geol. Soc., *106*(1): 63–99.

———. 1951. Notes on thecal variation in Monograptus. Geol. Mag., *88*: 306–328.

———. 1955. Graptolithina, 1st ed., *in* R. C. Moore, ed., Treatise on Invertebrate Paleontology, Kansas. pt. V, p. 1–101.

———. 1958. Patterns of colonial development in graptolites. Jour. Linn. Soc. London, Zool., *44*(295): 24–32.

———. 1963. The evolution and classification of the Graptoloidea. Quart. Jour. Geol. Soc. London, *119*: 401–418.

———. 1970. Graptolithina, 2nd ed., *in* R. C. Moore. ed., Treatise on Invertebrate Paleontology, Kansas. pt. V, p. 32 + 163.

———, and R. B. Rickards. 1966. A revision of Wiman's dendroid and tuboid graptolites. Bull. Geol. Inst., Univ. Uppsala, *43*: 1–72.

Child, Ch. M. 1915, Individuality in Organisms. Chicago. p. 1–213.

———. 1941. Patterns and Problems of Development. Chicago. p. 1–811.

Driesch, H. 1890. Tektonische Studien an Hydroidpolypen. Jenaische Ztschr. Naturwiss., n.f., *17*: 189–226.

Ebert, J. D. 1965. Interacting Systems in Development. New York (Polish transl., Warszawa, 1970).

Ehrlich, P. R., and R. W. Holm. 1963. The Process of Evolution. New York—London. 9 + 347 p.

Elles, G. L. 1922. The graptolite faunas of the British Isles. Geol. Assoc., Proc., *33*: 168–200.

———, and E. M. R. Wood. 1901–1918. A Monograph of British Graptolites, pts. I-XI. Palaeontogr. Soc., London, *21*: 171–539.

Foucart, M.-F., and Ch. Jeuniaux. 1965. Paléobiochémie et position systématique des Graptolithes. Ann. Soc. Roy. Zool. Belg., *95*(2): 39–45.

Gast, R., and E. Godlewski. 1903. Die Regenerationserscheinungen bei Pennaria carolini. Arch. Entw.-mech. Org., *16*(1): 76–116.

Haeckel, E. 1866. Generelle Morphologie der Organismen. v. I/II, p. 1–606, p. 1–662.

———. 1906. Prinzipien der Generellen Morphologie der Organismen. Berlin (Polish transl. Warszawa, 1960).

Ivanov, A. V. 1968. Proiskhozdenie mnogokletochnykh zivotnych. Leningrad. p. 5–287.

Ivanova-Kazas, O. M., and A. V. Ivanov. 1967. O proiskhozdenii Metazoa i ikh ontogeneza. Trudy Zool. Inst. AN SSSR, *44*: 5–25.

Jaanusson, V. 1960. Graptoloids from Ontikan and Viruan (Ordov.) Limestones of Estonia and Sweden. Bull. Geol. Inst. Uppsala, *38*: 289–366.

Jaeger, H. 1959. Graptolithen und Stratigraphie des jüngsten Thüringer Silurs. Abh. deutsch. Akad. Wiss., *2*: 1–197.

———. 1960. Uber Diversograptus. Paläont. Ztschr., *34*(1): 13–14.

———. 1969, *in* H. Jaeger, V. Stein, and R. Wolfart, Fauna (Graptolithen, Brachiopoden) der unterdevonischen Schwarzschiefer Nord-Thailands. N.Jb. Geol. Paläont. Abh., *133*(2): 171–190.

Kozłowski, R. 1949. Les Graptolithes et quelques nouveaux groupes d'animaux du Tremadoc de la Pologne. Palaeont. Polonica, *3*: 1–235.

———. 1953. Etude d'une nouvelle espèce du genre Corynoides. Acta Geol. Polon., *3*(2): Consp., 68–81.

———. 1956. Nouvelles observations sur Corynoididae (Graptolithina). Acta Palaeont. Polon., *1*(4): 259–269.

———. 1963. Le développement d'un Graptolite tuboide. Acta Palaeont. Polon., *8*(2): 103–134.

———. 1966. On the structure and relationships of graptolites. Jour. Paleont., *40*(3): 489–501.

Kraatz, R. 1958. Stratigraphische und paläontologische Untersuchungen (besonders in Gotlandium) im Gebiet zwischen Wiedan und Sorge (Südl. Westharz). Ztschr. deutsch. geol. Ges., *110*(1): 22–70.

Kühn, A. 1909. Studien zur Ontogenese und Phylogenese der Hydroiden. I Sprosswachstum und Polypenknospung bei den Thecaphoren. Zool. Jahrb., *28*(2): 387–476.

Leopold, A. C. 1964. Plant Growth and Development. New York (Russian transl., Moskva, 1968).

Lesh, G. E., and G. E. Laurie. 1971. Observations on pseudocolonial growth in hydra. Biol. Bull. *141*(2): 278–298.

Longwell, A. C., and S. S. Stiles. 1970. The genetic system and breeding potential of the commercial American Oyster. Endeavour, *29*(107): 94–99.

Mackie, G. O. 1963. Siphonophores, Bud Colonies and Superorganisms, *in* E. C. Dougherty et al., eds., The Lower Metazoa. Berkeley. p. 329–337.

Mayr, E. 1963. Animal Species and Evolution. Cambridge, Mass. 14 + 797 p.

———. 1970. Populations, Species and Evolution. Cambridge, Mass. 15 + 453 p.

Monsen, A. 1925. Über eine neue ordovicische Graptolithen Fauna. Norsk Geol. Tidsskr., *8*: 147–187.

Naumov, D. V. 1960. Gidroidy i gidromeduzy morskich, solonovatych i presnovodnych basseinov SSSR, *in* Opredel. po faune SSSR, Moskva-Leningrad, *70*: 1–585.

Palmer, D. 1971. The Ludlow graptolites *Neodiversograptus nilssoni* and *Cucullograptus (Lobograptus) progenitor*. Lethaia, *4*(4): 357–384.

Plate, L. 1922. Allgemeine Zoologie und Abstammungslehre. Jena. I, 6 + 629 p.

Regnéll, G. 1949. On large-sized rhabdosomes in monograptids. Kungl. Fysiograf. Sällskap. Förh., *19*(12): 1–11.

Ruedemann, R. 1902. Growth and development of Goniograptus thureaui M'Coy. N.Y. State Mus., Bull., *52*: 576–605.

———. 1904. Graptolites of New York. N.Y. State Mus. Mem., *7*.

———. 1947. Graptolites of North America. Geol. Soc. America Mem., *19*: 1–652.

Schellhorn, M. 1969. Probleme der Struktur, Organisation und Evolution biologischer Systeme. Jena. p. 5–134.

Schenck, D. A. von. 1965. Die Kormentektonik der Plumulariiden (Coelent., Hydrozoa). Revue Suisse de Zool., 72(4): 885–1021.

Skvington, D. 1963. Graptolites from the Ontikan Limestones (Ordov.) of Oland, Sweden, 1, Dendroidea, Tuboidea, Camaroidea and Stolonoidea. Univ. Uppsala, Bull. Geol. Inst., 42: 1–81.

Smith, J. M. 1968. Mathematical Ideas in Biology. Cambridge.

Snow, R. 1925. The correlative inhibition of the growth of auxillary buds. Ann. Bot., 39: 841–859.

———. 1929. The young leaf as the inhibiting organ. New Phytol., 28: 345–358.

Spiegelman, S. 1945. Physiological competition as a regulatory mechanism in morphogenesis. Quart. Rev. Biol., 20(2): 121–146.

Spjeldnaes, N. 1963. Some Upper Tremadocian graptolites from Norway. Palaeontology, 6(1): 121–131.

Spoel, S. van der. 1971. Some problems in infraspecific classification of holoplanktonic animals. Zeitschr. zool. System. Evolutionsforschung, 9(2): 107–138.

Stebbing, A. R. D. 1970. Aspects of the reproduction and life cycle of Rhabdopleura compacta (Hemichordata), Mar. Biol., 5(3): 205–212.

Stepanianc, S. D. 1967. Sifanofory morei SSSR i severnoi chasti Tikhogo okeana, in Opredeliteli po faune SSSR. Leningrad. p. 1–216.

Sudbury, M. 1958. Triangulate monograptids from the Monograptus gregarius zone (Lower Llandovery) of the Rheidol Gorge (Cardiganshire). Phil. Trans. Roy. Soc. London, Biol. Sci., 685 (241): 485–555.

Tardent, P. 1956. Propf-Experimente zur Untersuchung des regenerationshemmenden Stoffes von Tubularia. Rev. Suisse Zool, 63(2): 229–236.

———. 1963. Regeneration in the Hydrozoa. Biol. Rev., 38(3): 293–333.

———. 1965. Developmental aspects of regeneration in coelenterates in V. Kiortis, and H. A. L. Trampusch, eds., Regeneration in Animals and Related Problems. Amsterdam. p. 71–88.

———, and H. Eymann. 1959. Some physical and chemical properties of the regeneration-inhibitor in Tubularia. Arch. Entw.-Mech. Org., 151: 1–37.

———, and R. Tardent. 1956. Wiederholte Regeneration bei Tubularia. Publ. Staz. Zool. Napoli, 28: 367–396.

Thimann, K. V. 1937. On the nature of inhibitions caused by auxins. Am. Jour. Bot., 24: 407–412.

———. 1939. Auxins and the inhibition of plant growth. Biol. Rev., 14(3): 314–337.

———, and F. Skoog. 1933. Studies on the growth of plants, III, The inhibiting action of the growth substance on bud development. Natl. Acad. Sci. Wash., Proc., 19: 714–716.

———. and F. Skoog. 1934. On the inhibition of bud development and other functions of growth substance in Vicia Faba. Roy. Soc. London. Proc., B, 114: 307–339.

Thorsteinsson, R. 1955. The mode of cladial generation in Cyrtograptus. Geol. Mag., 92: 37–49.

Timoféeff-Ressovsky, N. W. 1931. Gerichtetes Variieren in der phänotypischen Manifestierung einiger Generationen von Drosophila funebris. Naturwiss., 19(23-25): 493–497.

———. 1934. Über den Einfluss des genotypischen Milieus und der Aussenbedingungen auf die Realisation des Genotyps. Genmutation vti (venae transversae incompletae) bei Drosophila funebris. Nachr. Ges. Wiss. Goettingen, Math.-Phys. Kl., Biol., n.f., 1(6): 53–106.

———, and V. I. Ivanov. 1966. Nekotorye voprosy fenogenetiki, in C. I. Alichanian, ed., Aktualnye voprosy sovremennoi genetiki. Moskva. p. 114–130.

Tokin, B. P. 1970. Obschaya embryologia Moskva. p. 3–507.

Törnquist, S. L., 1899. Researches into the Monograptidae of the Scanian Rastrites Beds. Lunds Univ. Arskr., 35(2, 1): 1–25.

Towe, K. M., and A. Urbanek. 1972. Collagen-like structures in Ordovician graptolite periderm. Nature, 237(5356): 443–445.

Tweedel, K. S. 1958. Inhibitors of regeneration in Tubularia. Biol. Bull., 114(2): 255–269.

Urbanek, A. 1953. Sur deux espèces de Monograptidae. Acta Geol. Polon., 3(2): 277–298; Consp., 100–107.

———. 1958. Monograptidae from erratic boulders of Poland. Palaeont. Polon., 9: IV + 105.

———. 1960. An attempt at biological interpretation of evolutionary changes in graptolite colonies. Acta Palaeont. Polon., 5(2): 127–234.

———. 1963. On generation and regeneration of cladia in some Upper Silurian monograptids. Acta Palaeont. Polon., 8(2): 135–254.

————. 1966. On the morphology and evolution of the Cucullograptinae (Monograptidae, Graptolithina). Acta Palaeont. Polon., *11*(3/4): 291–544.

————. 1967, *in* L. Kuźnicki and A. Urbanek, Zasady nauki o ewolucji. Warszawa. I, p. 5–617.

————. 1970. Neocucullograptinae n. subfam. (Graptolithina)—their evolutionary and stratigraphic bearing. Acta Palaeont. Polon., *15*(2/3): 163–388.

Van Valen, L. 1969. Variation genetics of extinct animals. Am. Naturalist, *103*(931): 193–224.

Waddington, C. H. 1940. Organizers and Genes. Cambridge. p. 1–160.

————. 1956. Principles of Embryology. London-New York. p. 1–495.

Walker, M. 1953. The development of Monograptus dubius and Monograptus chimaera. Geol. Mag., *90*(5): 362–373.

Went, F. W., and K. V. Thimann, 1945. Phytohormones. New York. 11 + 283. p.

Westoll, T. S. 1950. Some aspects of growth studies in fossils, *in* S. Zuckermann, A. discussion on the measurement of growth and form. Roy. Soc., Proc., ser. B, *137*(889): 490–509.

Whittington, H. B., and R. B. Rickards. 1969. Development of Glossograptus and Skiagraptus, Ordovician graptoloids from Newfoundland. Jour. Paleont., *43*(3): 800–817.

Willefert, S. 1962. Quelques Graptolites du Silurien supérieur du Sahara septentrional. Bull. Soc. Géol. France, *7*(4, 1): 24–39.

Wiman, C. 1901. Über die Borkholmer Schicht im Mittelbaltischen Silurgebiet. Bull. Geol. Inst. Uppsala, *5*(10): 149–222.

Zavadskij, K. M. 1966. Osnovnye formy organizacii zizni i ikh podrazdelenia *in* Filosofskie Problemy Sovremennoi Biologii. Leningrad.

————. 1968. Vid i vidoobrazovanie. Leningrad. p. 3–404.

Morphological Discontinuities in the Evolution of Graptolite Colonies

Valdar Jaanusson   Naturhistoriska Riksmuseet

ABSTRACT

The evolution of graptolite colonies progressed in many instances over morphological discontinuities. This process was involved for example when the number of stipes increased or decreased; when budding changed from linear to alternate or vice versa; when biserial scandent rhabdosomes changed into uniserial rhabdosomes; and when porusresorption was replaced by the formation of the lacuna. The morphological discontinuity had to be bridged by a succession of populations in which the morphological types on either side of the discontinuity coexisted as genetic polymorphs. This seems to be an important evolutionary process, some implications of which are discussed.

INTRODUCTION

Phylogenetic studies of some groups of invertebrates (Jaanusson, 1971, and unpublished data) have shown that during evolution functional and morphological changes could not always take place continuously by a gradual process through a series of microevolutionary changes. In many instances evolution progressed over functional and morphological discontinuities where no intermediate stages were possible. Such discontinuities must have been surmounted in one step.

Author's address: Naturhistoriska Riksmuseet, Paleozoologiska sektionen, Stockholm 50, Sweden.

An analysis of the possible ways to bridge a morphological discontinuity shows that this process must take place over a succession of populations in which phenotypic variability includes morphologies on either side of the discontinuity. In other words, such populations must be genetically polymorphic with regard to these morphologies and functions. If the new morph on the other side of the discontinuity becomes fixed, the discontinuity is bridged.

Genetic polymorphism within a species is a widespread phenomenon (for summaries, see Huxley, 1955, and Ford, 1965). The particular polymorphism associated with bridging a functional and morphological discontinuity during evolution is distinguished mainly by its evolutionary effect. The phylogenetic significance of the effect varies widely. It depends upon the degree of difficulty in bridging the discontinuity within the limits of the particular genetic and developmental systems, as well as upon the functional importance of the difference between the two morphs and the subsequent history of the group.

Experience from discussions in which these ideas have been presented shows that, for brevity and clarity, it is useful to have a special term for populations in which a major evolutionary discontinuity has been bridged. The term used by me on such occasions has been *dithyrial populations,* that is, "populations with two doors." The term designates populations that include a morph on either side of a functional and morphological discontinuity of clear phylogenetic significance. The population is said to be dithyrial with respect to this particular pair of morphs.

The evolution of graptolites included a number of morphological discontinuities that were successfully bridged. However, graptolites are not the best group for illustrating the process of bridging of such discontinuities for the following reasons: (1) knowledge of functional morphology and reproductive biology of graptolites is limited and in part rather speculative; (2) several pertinent morphological transitions during the evolution of this group have not yet been analyzed, mainly because suitably preserved material is lacking; and (3) graptolites were colonial animals, and this imposes particular problems. For these reasons the discussion of morphological discontinuities in the evolution of graptolites is confined here to a few selected examples.

GENETIC ASPECTS OF GRAPTOLITE POPULATIONS

In a graptolite colony, as in animal colonies in general, individuals were produced by asexual budding. Thus all zooids within a colony had the same genotype (Urbanek, 1960), with the exception of differences caused by occasional mutations during astogenetic development. Within a colony individual zooids were integrated by and subordinated to a coordinating organization that served the functions of the colony as a whole. This implies that, in addition to having been able to duplicate the zooids, information embodied in the genotype also had to duplicate the organization of the colony. This is valid irrespective of how the genetic code was decoded during the astogenetic development of the colony. That is, the implication holds if the siculozooid acted as organizer and inductor (Urbanek, 1960) or if genetic information was transmitted in a more complicated way.

The integrated effect of zooidal and colonial genetic information in the genotype of a zooid makes it difficult to apply to individuals within an advanced colonial organism some of the genetic concepts used for individual organisms. In graptolites the principal morphological and functional differences between zooids within a colony are not phenogenetic phenomena since the differentiation is caused by the internal genetic environment, the instructions of the genetic code. Superimposed on the internal, "cormotypic" differentiation is phenotypic variation, which is due to the interaction between zooids during their development and the external environment. In graptoloids phenotypic variability is mostly slight compared to "cormotypic" differentiation of thecae, and it ordinarily affects the particular astogenetic stage of the colony as a whole.

Our knowledge of the reproductive biology of graptolites rests on deductive evidence and is in part fairly speculative. Kozlowski (1949) suggested that dendroid graptolites were gonochoristic and that the loss of bithecae at the transition from dendroids to graptoloids involved a change from gonochoristic to hermaphroditic reproduction. Other interpretations are possible. For instance, at a later stage of evolution at least some of the graptoloid groups may have returned to a gonochoristic mode of reproduction without changes in thecal morphology. However, this can neither be proved nor disproved. Hermaphroditism (for a recent survey, see Ghiselin, 1969) is widespread among animals. Self-fertilization seems to be rare since various mechanisms are known that reduce or completely eliminate this process. Many hermaphrodites are cross-fertilizing. For the purpose of this paper, it is assumed that graptolites had cross-breeding populations.

From the point of view of population genetics, colonies (rhabdosomes) and not zooids constituted populations. In graptolites selection acted on the colony as a whole and on individual zooids only in their capacity as integrated constituents of the colony. In graptolites with a "free sicula" the rhabdosomes were the functional carriers of the gene flow since migrations of the floating colonies certainly exceeded the distance traveled by their gametes before fertilization. Mutated gametes were produced by each fertile zooid separately, but in this respect the colony can be treated as an organism with several reproductive organs.

SOME MORPHOLOGICAL DISCONTINUITIES
IN THE EVOLUTION OF GRAPTOLITES

A simple type of morphological discontinuity is that associated with meristic characters in which only a whole number (a, b, etc.) of a feature x is possible (such as x, $2x$, $3x$, etc.) but not intermediate stages. In such cases an evolutionary change from ax to bx must take place over a succession of consecutive populations that include both ax and bx as morphs. Epigenetic meristic polymorphic features due to the action of a phenotypic threshold on underlying continuous variation are not considered here.

Meristic differences are seldom important functionally unless a certain number of an organ is important functionally or if some number of a feature has become tied up with important parts of the organization of an organism by pleiotropic or other correlative effects. Many meristic discontinuities have been relatively easy to bridge during evolution.

In graptolites the development of pauciramous forms from early multiramous forms has generally been attributed to the principle of stipe reduction (Nicholson and Marr, 1895). The number of stipes in the rhabdosome has decreased successively along different phylogenetic lineages, leading in some lineages ultimately to biramous forms such as

Kiaerograptus or *Didymograptus*. Current opinion on this matter has been summarized by Bulman (1970, p. 104–105). Also the reverse process, increase in the number of stipes, is a common phenomenon, particularly in nemagraptids and the suborder Monograptina.

A graptoloid rhabdosome is a well-balanced structure, particularly in forms with relatively few stipes. A hydrodynamically balanced construction of the rhabdosome was obviously important to the colony, not only for efficient floating but also for feeding, particularly in forms that may have had the ability to rotate the rhabdosome (Bulman, 1964). Decrease in number of stipes during evolution can hardly have taken place through a gradual reduction of one or more of the stipes because the intermediate stages would result in geometrically unbalanced constructions of the rhabdosomes. No undoubted example of reduction of stipes by this process is known. Therefore, it is very likely that the process of stipe reduction involved sudden elimination of one or more of the stipes (Skevington, 1967).

The stratigraphic succession of different forms of *Tetragraptus fruticosus* (Harris and Thomas, 1938; Berry, 1960) may constitute an example of the process of stipe reduction. The oldest form referred to this species has four stipes. Higher up in the sequence forms with four and three stipes coexist. In still higher beds only a form with three stipes is found. The different forms of *T. fruticosus* have never been studied in isolated specimens, so there is some uncertainty as to the taxonomic homogeneity of the species as currently defined. However, if the taxonomy is correct, *T. fruticosus* had a long succession of populations that were polymorphic with respect to the number of stipes. Polymorphism seems to have been balanced during a fairly long time span until the four-stiped morph finally became eliminated.

Many other stipe reductions must have followed the pattern illustrated by *Tetragraptus fruticosus* except that the succession of populations that were polymorphic with respect to the number of stipes was mostly short, and chances of finding remains of such populations are limited. In many cases polymorphism may have been transient (Ford, 1940). Skevington (1966, 1967) discussed the possibility that the two-stiped *Holmograptus lentus* and the one-stiped *Nicholsonograptus fasciculatus* might have been polymorphs within one species.

The genetic variability or mutation pressure for increased or reduced number of stipes probably existed in the gene pool of many uniserial graptoloid populations, but the unusual number of stipes presumably was selectively disadvantageous. Jaanusson (1960) described as *Didymograptus?* sp. *A* a specimen with three stipes that was otherwise identical with an associated two-stiped *D. pakrianus*. Further studies confirmed the suspicion that the specimen represents the latter species. The three-stiped rhabdosome is otherwise quite normal and fully grown. From only a single specimen it cannot be decided whether the specimen represents a rare polymorph or a mutant.

Increase in the number of stipes during evolution took place by the same process as decrease. Linograptids differ from monograptids by the presence of one or more sicular cladia developed by supraapertural budding (for details, see Urbanek, 1963). The difference involves a morphological discontinuity since a sicular cladium is either present or absent and intermediate stages are hardly possible. Specimens with a diversograptid type of sicular cladium occur first as rare specimens associated with apparently conspecific normal monograptids. (*Neodiversograptus nilssoni,* Urbanek, 1963: "*Monograptus*" *runcinatus,* Strachan, 1952, Bulman, 1970, p. V109). These populations were obviously polymorphic with respect to the presence or absence of a sicular cladium (Urbanek, 1970). The polymorphism was probably balanced for a fairly long time and affected several phylogenetic lineages.

Most graptoloids show a conspicuous gradational change in the size and apertural morphology of thecae along the stipe. This phenomenon has been discussed by a number of writers, and several different explanations for the thecal differentiation have been suggested (for a comprehensive review, see Urbanek, 1960). The explanations have dealt almost exclusively with intrinsic causal factors during astogenetic and phylogenetic development of the rhabdosomes. Differences in thecal morphology and other gradational changes along the stipe can also be explained as responses to differential selection pressures at consecutive astogenetic stages of the rhabdosomes. Hydrodynamic properties of young, small rhabdosomes were probably somewhat different from those of full-grown rhabdosomes, and thus these properties presumably changed with increase in size of the rhabdosome. Size-dependent differences in floating properties of the rhabdosome very likely affected the feeding mechanism to various degrees. This, in turn, may in part be reflected in the morphology of thecal apertures. In graptoloids thecae completed their growth very early during the astogenetic development of a stipe (see also Urbanek, 1966) with the exception of some dichograptids with simple tubular thecae that continued to grow at least until the rhabdosome had attained its full size (Jaanusson, 1960). For this reason the morphology of thecae at different astogenetic stages is mostly preserved along a stipe and may reflect differential adaptation of zooids during the growth of the rhabdosome. It is possible that in extreme cases (biform rhabdosomes), young rhabdosomes may even have occupied an ecologic niche different from that of subsequent astogenetic stages.

A phylastogenetic gradient is shown also by the proximal change from alternate to linear budding in diplograptids (Davies, 1929; Waern, 1948; Urbanek, 1960, p. 184–187) in that the dicalycal theca shifts progressively distally in discontinuous steps and the median septum shortens accordingly. Shift in position of the dicalycal theca is associated with a discontinuity, which must have been bridged by populations that were polymorphic in this respect. The presence of such populations, polymorphic as to the position of the dicalycal theca, seems actually to be demonstrated in the material described by Waern (1948; *Climacograptus scalaris* var. *transgrediens* β and γ-δ).

Boundaries between high taxonomic categories seem frequently to coincide with functional and morphological discontinuities, and several key innovations (Miller, 1950) are associated with discontinuous change in function or morphology or both. This suggests that during evolution functional and morphological changes may have been more difficult and more profound when major discontinuities had to be overcome than when the change could take place by a gradual process.

The suborder Monograptina differs from the other graptoloid suborders principally through two characters (Bulman, 1970): (1) the scandent, secondarily uniserial rhabdosome and (2) the monograptid type of budding from sicula by way of the sinus and lacuna stages. With respect to both characters monograptids are separated from their probable ancestors by a morphological discontinuity. The first character is of meristic nature.

The change from a biserial to a uniserial rhabdosome may have taken place through successive discontinuous steps by eliminating theca after theca in one row in a distal direction until a uniserial stipe was produced. Something similar to this process may have taken place if derivation of monograptids from dimorphograptids is postulated. The change can also have taken place in one step (Bulman 1970, Fig. 76:3) before or after the first theca (th1) attained its initially upward direction of growth. In either case an intermediate morphological stage with respect to this character simply could not have existed. The discontinuity had to be bridged by a succession of dithyrial populations, which included both a secondarily uniserial morph and a morph that was biserial at least distally. The change from a porus, formed secondarily by resorption of sicular periderm, to a sinus and lacuna formed during the growth of the sicula is also associated with a functional and morphological discontinuity and had to pass through a succession of populations that were dithyrial in this respect.

Experience from a study of some other groups suggests that, during the evolution of a group, bridging of several major functional and morphological discontinuities seldom, if ever, took place simultaneously through one, "polythyrial" succession of populations. Ordinarily one discontinuity after another was bridged through separate dithyrial populations, which even belonged to different taxa of various categories. This implies that a definition of a group based on two or more characters associated with functional and morphological discontinuities is usually not exact. During evolution these characters very likely appeared successively and not simultaneously. Hence intermediate taxa that possess some but not all of the diagnostic characters may exist. The practical value of a definition associated with bridging several major functional and morphological discontinuities depends on how fast the complex of discontinuities as a whole was passed through the consecutive dithyrial populations.

With regard to the suborder Monograptina, it is likely that the change from partially biserial to secondarily uniserial rhabdosomes did not take place simultaneously with the change from porus resorption to the formation of the lacuna. It is probable that forms existed in which partially biserial rhabdosomes were associated with a lacuna type of development, or secondarily uniserial scandent rhabdosomes had a resorption of porus.

CONCLUSIONS

As in other groups of organisms, the evolution of graptolites did not always proceed by a gradual process through a series of continuous or microdiscontinuous changes in function and morphology. Several morphological changes during evolution were discontinuous and these evolutionary steps had to be passed by way of a succession of populations in which morphological types on either side of the discontinuity coexisted as genetic polymorphs. In graptolites the simplest and most obvious discontinuities were meristic and associated with changes in the number of stipes and similar phenomena. A discontinuity was involved also in the ability to secrete microfusellar tissue (Urbanek, 1970) and in a change from the porus type of development in diplograptids to the lacuna type of development in monograptids. The nature of several other morphological transitions has yet to be analyzed. A functional discontinuity or instability is not always clearly reflected in morphological change of the skeleton. This makes it difficult to evaluate, without a thorough functional analysis, some morphological changes that deductively may have been continuous but in which the change may have been controlled by bridging a functional discontinuity.

ACKNOWLEDGMENTS

The final manuscript of this paper was prepared while the writer was at the Department of Geology, The Ohio State University, Columbus, Ohio. W. C. Sweet kindly revised the manuscript linguistically and provided valuable suggestions.

REFERENCES

Berry, W. B. N. 1960. Graptolite faunas of the Marathon region, West Texas. Univ. Texas Publ., *6005*: 1–179.

Bulman, O. M. B. 1964. Lower Palaeozoic plankton. Geol. Soc. London, Quart. Jour., *120*: 455–476.

———. 1970. Graptolithina with sections on Enteropneusta and Pterobranchia *in* R. C. Moore, ed., Treatise on Invertebrate Paleontology, Kansas pt. V, second ed., 163 p.

Davies, K. A. 1929. Notes on the graptolite faunas of the Upper Ordovician and Lower Silurian. Geol. Mag., *66*: 1–27.

Ford, E. B. 1940. Polymorphism and Taxonomy, *in* J. Huxley, ed., The New Systematics. 493–513.

———. 1965. Genetic Polymorphism. M. I. T. Press, Cambridge, Mass. 101 p.

Ghiselin, M. T. 1969. The evolution of hemaphroditism among animals. Quart. Rev. Biol., *44*: 189–208.

Harris, W. J., and D. E. Thomas. 1938. A revised classification and correlation of the Ordovician graptolite beds of Victoria. Mining Geol. Jour., *1*: 62–73.

Huxley, J. S. 1955. Morphism and evolution. Heredity, *9*: 1–52.

Jaanusson, V. 1960. Graptoloids from the Ontikan and Viruan (Ordov.) limestones of Estonia and Sweden. Bull. Geol. Inst. Univ. Uppsala, *38*: 289–366.

———. 1971. Evolution of the brachiopod hinge. Smithsonian Contr. Paleobiol., *3*: 33–46.

Kozlowski, R. 1949. Les graptolithes et quelques nouveaux groupes d'animaux du Tremadoc de la Pologne. Palaeont. Polon., *3*: 1–235.

Miller, A. H. 1950. Some ecological and morphological considerations in the evolution of higher taxonomic categories, *in* E. Mayr and E. Schüz, eds., Ornithologie als biologische Wissenschaft. Carl Winter, Heidelberg. p. 84–88.

Nicholson, H. A., and J. E. Marr. 1895. Notes on the phylogeny of the graptolites. Geol. Mag., *42*: 529–539.

Skevington, D. 1966. The morphology and systematics of "*Didymograptus*" *fasciculatus* Nicholson, 1869. Geol. Mag., *103*: 487–497.

———. 1967. Probable instance of genetic polymorphism in the graptolites. Nature, *213*: 810–812.

Strachan, I. 1952. On the development of *Diversograptus* Manck. Geol. Mag., *89*: 365–368.

Urbanek, A. 1960. An attempt at biological interpretation of evolutionary changes in graptolite colonies. Acta Palaeont. Polon., *5*: 127–234.

———. 1963. On generation and regeneration of cladia in some Upper Silurian monograptids. Acta Palaeont. Polon., *8*: 135–254.

———. 1966. On the morphology and evolution of the Cucullograptinae (Monograptidae, Graptolithina). Acta Palaeont. Polon., *11*: 291–544.

———. 1970. Neocucullograptinae n. subfam. (Graptolithina)—their evolutionary and stratigraphic bearing. Acta Palaeont. Polon., *15*: 163–388.

Waern, B. 1948. The Silurian strata of the Kullatorp core. Bull. Geol. Inst. Univ. Uppsala, *32*: 433–473.

Life Forms and Feeding Habits of Graptolites

Bernd-Dietrich Erdtmann   Indiana University at Fort Wayne

ABSTRACT

Studies of Recent living pterobranch colonies demonstrate the ability of integrated zooids to produce multidirectional inhalent and unidirectional exhalent feeding currents through ciliated lophophores. In Recent attached benthic pterobranchs irregular growth patterns of zooidal tubes neutralize the potentially directional effect of coordinated lophophore currents, but analogy of pterobranch soft-part structures with the more integrated hypothetical organs of attached Dendroidea and unattached Graptoloidea permits reconstruction of coordinated feeding currents which could have facilitated self-propellant mobility for some graptolites. Rhabdosome growth patterns, certain proximal and distal end structures, and thecal apertural morphologies aid in interpretation of life modes. Ecological observations of cool-water pterobranchs lend support to geological biofacies analyses of graptolites with regard to geographical distribution and oceanic habitat stratification.

INTRODUCTION

During the past few years scores of studies have been published dealing with physiological, embryological, and ecological aspects of living Pterobranchs and others dealing with functional morphology and paleofacies interpretation of extinct Paleozoic

Author's address: Department of Earth and Space Sciences, Indiana University, Fort Wayne, Indiana 46805.

graptolites. With the phylogenetic affinities of pterobranchs and graptolites being securely established as individual classes of the Hemichordata (Kozlowski, 1938, 1949, 1966; Bulman, 1970), functional analogies may become more meaningful for the interpretation of graptolite life forms and feeding habits.

The fundamental knowledge about pterobranch biology was established through the investigations of Allman (1869), Lankester (1884), and Schepotieff (1907). During more recent years data on ecology, physiology, reproduction, and life cycles of some ptero-branchs were added through studies by Kozlowski (1949), Burdon-Jones (1954), and Stebbing (1970a, 1970b). The literature on graptolite ecology and biology is equally extensive, although few analyses have succeeded in contributing so fully to our present knowledge about graptolite and pterobranch biology as that of Kozlowski through his now classic work on "Les graptolithes et quelques nouveaux groupes d'animaux du Tremadoc de la Pologne" (1949). New impulses for research and imaginative ideas of graptolite paleobiology resulted from Kozlowski's fundamental work, and the radiations of his impulses are still continuing today. It would be too onerous here to list all contributions to the interpretation of graptolite ecology and life habits which have appeared lately, but Kirk's (1969) refreshingly unconventional and imaginative thoughts should be mentioned at this point for having stimulated this author to reconsider some of his rusty concepts on this topic.

FEEDING HABITS OF LIVING PTEROBRANCHIA IN RELATION TO THE LIFE HABITS OF GRAPTOLITES

Among the living pterobranchs the genus *Rhabdopleura* with its roughly half-dozen described species is better suited for ecological correlations with graptolites than *Cephalodiscus* because of morphological criteria, as discussed by Kozlowski (1949, 1966). *Rhabdopleura* grows in microcolonies (see Fig. 1) with individual zooids housed in characteristic creeping or repent cylindrical tubes which are mostly compounded in clumps due to their "thigmophilic" (touch-loving) mode of growth (Stebbing, 1970a). The erect portions of the zooidal tubes have a fusellar periderm and show occasional zigzag sutures, such as those characteristic of the graptolites.

Inside the tubes, however, three types of heteromorphous zooids are observed: A majority of "normal" or neuter zooids which engage in the feeding and food digestion within the colony, a variable number of "dormant" zooids, and two equally small groups of male and female (temporary nonfeeding) zooids. Stebbing (1970b) described the morphologies and functions of these heteromorphous zooids for *Rhabdopleura compacta* from the English Channel. It is important to note that the heteromorphous nature of the pterobranch zooids, particularly of the neuter and dormant individuals, never becomes reflected in the skeletal morphology. This may have significance for the interpretation of the possible existence of functionally heteromorphous zooids in the graptolites.

Little has so far been published on the feeding modes of pterobranchs, but Gilchrist (1915) stated that *Cephalodiscus gilchristae* possesses numerous cilia on multiple-paired lophophores which capture phytoplankton by inducing multidirectional inhalent feeding currents with these cilia. The food particles then travel via a mucilaginous film along

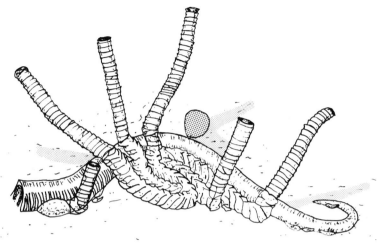

Figure 1

Colony of Recent *Rhabdopleura compacta* showing repent and erect tubes in association with *Serpula vermicularis*. (After Stebbing, 1970a.)

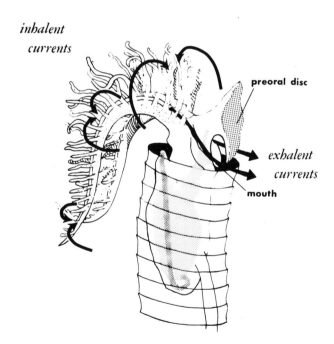

Figure 2

Extended feeding zooid of *Rhabdopleura normani* with arrows indicating directions of ciliary induced inhalent currents for left lophophore and exhalent currents for both lophophores. (After Bulman, 1970.) (Redrawn from *Treatise on Invertebrate Paleontology,* courtesy of the Geological Society of America and The University of Kansas Press.)

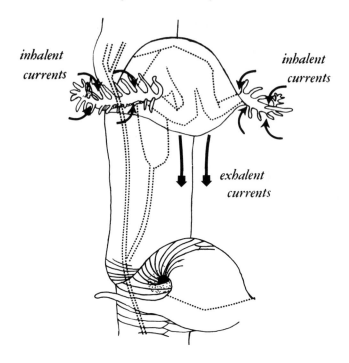

Figure 3

Reconstruction of possible current regimen and soft parts of *Monograptus exiguus*. Downward direction of coordinated exhalent currents would propel the rhabdosome upward. (After Bulman, 1970.) (Redrawn from *Treatise on Invertebrate Paleontology,* courtesy of the Geological Society of America and The University of Kansas Press.)

small grooves toward the precephalic mouth opening, but most food matter finally bypasses the cephalic disk and actually misses the oral opening (see Fig. 2). A weak unidirectional exhalent current develops from the coordinated flow of water along the central grooves of the lophophores. In the randomly organized colonies of *Cephalodiscus* or *Rhabdopleura* the combined effect of these currents, however, becomes neutralized as should be expected, since a colony has no "need" for directional currents which would introduce undue strain on the anchoring structures. But the capability of inducing directional currents could have developed as an adaptive advantage to the Graptoloidea during their Early Ordovician anagenesis to unattached modes of life (see Fig. 3).

Kirk (1969) reviewed a number of both dendroid and graptoloid species to investigate in which ways their major rhabdosomal morphologies might best fit a hydrodynamic model that would facilitate a self-propelled upward mobility and a resistance to gravitational subsidence of different rhabdosomal structures. Lophophore-induced feeding currents were called upon for generating the necessary propulsion, but there was only little discussion of the types of currents which the hypothetical lophophores of graptolites could have induced or of the highly integrated functional organization that would be required of a graptolite colony to produce differential movements, such as spinning or torsion or even maintaining a straight course in agitated environments. Furthermore, obvious problems arise where zooids were differently oriented through successive growth stages during graptolite astogeny, such as in *Phyllograptus,* reclined *Tetragraptus, Isograptus,* most scandent biserial forms, and in monograptids. In these and other graptolites, at least the sicular individual would have to invert its life orientation during astogeny unless it either lost its function as a feeding individual, or the other zooids possessed highly flexible and protractable lophophores which would allow for differential movements. In all graptolites with simple open apertures, the lophophores may have been able to extend reasonably far outside the thecae to facilitate differential movements of the rhabdosome and possibly optional life positions either in conventional form (proximal end down) or in inverted position. Graptolites with complex thecal apertures such as *Dicranograptus, Dicellograptus, Dicaulograptus,* some monograptids, and some other forms may be notable exceptions and may have been confined to a particular life position and rotational movements only. The possession of either proximal or distal buoyancy structures in some biserial graptolites could be interpreted as supporting the possibility of both types of life positions. These may have been chosen in response to functional adaptation and developed as evolutionary grades. While a self-propelled vertical mobility for most unattached graptolites is generally acceptable many questions remain unanswered, mainly concerning the possible role and extent of extrathecal organic matter and devices to control differential movements and depth adjustments. These problematic points may never become known unless traces of graptolite soft parts are found.

At the present stage of knowledge it may be inconclusive to correlate general physical habitat requirements, such as temperature, salinity, and photic and bathymetric tolerances of living pterobranchs with those of graptolites. Although a number of hemichordates thrive in warm waters of varying salinities, *Rhabdopleura* and other living pterobranchs are quite stenothermic to cold water (5–10°C) and relatively euryhaline (3.3–3.5 percent) according to Burdon-Jones (1954) and Stebbing (1970a). If an analogy with graptolites is at all tenable, the exclusion of warm-water habitats for living pterobranchs lends support to the hypothesis that graptolites preferred similar cool temperatures for their biotopes.

The bathymetric distribution of living pterobranch species, such as *Rhabdopleura normani* and *R. compacta,* extends from near sea level to 800 m, suggesting little dependence of these species upon light and pressure variation. Laboratory investigations have also demonstrated (Burdon-Jones, 1954) that *Rhabdopleura normani* is completely insensitive to light.

Autoplanktonic graptolites probably were able to inhabit a wide range of bathymetric zones as suggested by investigations of paleofacies distribution of Silurian brachiopods and graptolites (Ziegler, 1965; Berry and Boucot, 1971), but certain assemblages show distinct selectivity for stratified and even for nonoverlapping habitat zones. Such habitat selectivity of many species suggests that graptolites must have possessed hydrostatic control mechanisms or other pressure-sensitivity devices and remained relatively independent of photic and warm temperature requirements.

At this point, however, the question of nutrient sources for graptolites may arise. Phytoplankton might not have been the direct source of food for deeper-water assemblages because phytoplankton proliferation is dependent on the limited range of the photic zone (maximum of 200 m in pelagic areas, less in neritic environments).

The general morphological features of graptolite rhabdosomes may give some indication of energy levels in their habitat. For example, widely extended multiramous Dichograptidae usually do not occur together with the smaller more compact forms throughout the Lower Ordovician. Berry (1962) assumed that the larger rhabdosomes reflect quieter, deeper habitats and that more compact forms lived in near-surface strata. This habitat stratification also could have resulted in differential paleogeographic distribution because deeper living graptolites had reduced chances of current dispersal. The geological studies on this important relation of form to paleoecofacies and stratigraphy of graptolites are yet to be undertaken. Study of nongraptolitic biota and their life habits could possibly help to elucidate some of the current confusion about graptolite provinciality and noncorrelating graptolite zones in various parts of the world.

Unusual "infant" mortalities are often recorded on some graptolitiferous bedding planes which are crowded with siculae or young rhabdosomes of various sympatric communities. Broken or fragmented rhabdosomes are rarely found, however, unless they show evidence of postmortem redistribution and size sorting. This demonstrates that mass mortalities occurred without damage to the highly flexible rhabdosomes prior to disintegration of the soft parts. There is, however, little conclusive evidence as to whether these mortalities were caused by high mechanical energy (e.g., by storm-induced wave movement) or by inversion of toxic water masses. In low-diversity neritic graptolite communities such as those of the Late Ordovician Eden Shale or Maquoketa Limestone of North America, high-energy destruction may have occurred rather than death by toxic environments as demonstrated for the higher diversified taphocoenoses* of deep-water shales in orogenic areas. One of the causes of catastrophic death of planktonic organisms may be sudden sharp fluctuations in water temperature beyond the limits of endurance for the species. This may be more common than toxic conditions.

Besides taphonomic* indicators for graptolite habitat analysis there are, of course, numerous intrinsic characteristics. Most of the observable morphological features of graptolites are characteristic of species, genera, and families, and these features have been interpreted and reinterpreted in various ways. Kirk related the common trend of the Graptoloidea toward a high degree of symmetry to their planktonic life habit and considered symmetry to be additional evidence for more efficient exploitation of that water column through which a rhabdosome may have traveled during its diurnal feeding movements. The spiral morphologies found in various unrelated forms such as *Nemagraptus gracilis, Dicellograptus caduceus, Dicranograptus furcatus, Monograptus turriculatus,* and *Cyrtograptus solaris* (see Fig. 4) are obvious examples which demonstrate the feasibility of active gyratory upward movement. Other forms, such as biserial graptolites and monograptids, may simply have rotated around their axes or moved straight up. Species that possessed vesicular structures, either on their proximal or distal ends, probably were unable to alter their vertical position unless the fluid or gas pressure in these vesicles was variable, as in the living *Nautilus.* Some of these forms may have lived in epipelagic or near-surface habitats with little active mobility necessary because of wave dispersal and nutrient resources constantly being available.

---

*"Taphocoenosis" or "taphonomic" defines the actual fossil burial community; this being distinct from "thanatocoenosis" which includes postmortem transportation and selectivity in preservation or other post-mortem distortions of the "death community."

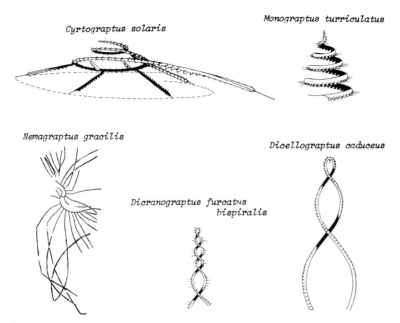

Figure 4

Assumed life positions of some spiral graptolites. *N. gracilis* is shown in obverse view.

Certain multiramous horizontally oriented graptolites, such as *Stellatograptus stellatus, Loganograptus logani, Dichograptus octobrachiatus,* and *Tetragraptus headi* (see also Bulman, 1964, fig. 4) possess interramous web structures connecting the proximal stipe portions to form a functionally solid disc. Similar flange-like extensions of cortical tissues parallel to the stipes, but tapering distally, are observed in *Clonograptus callavei* and *Stellatograptus stellatus* (Erdtmann, 1967a). The function of these structures is not clear, but they are suggested by Bulman (1964) to have served either as buoyancy devices or to retard sinking after active upward migration of the rhabdosome (Kirk, 1969). Interramous structures are also one of the most convincing evidences for extrathecal tissues which must have constructed these layers, since almost all thecal apertures within these cortical discs had been overgrown by cortex and thus lost their function of housing feeding zooids. Other graptolites which have been interpreted as horizontally oriented, such as *Clonograptus flexilis, Etagraptus, Tetragraptus quadribrachiatus, Goniograptus* and *Trochograptus,* among others, probably possessed a life orientation more resembling that of an inverted cone with the sicula at the apex and all thecal apertures pointing toward the sea bottom. Such a life orientation has previously been demonstrated for the quadriramous *Allograptus canadensis* (Erdtmann, 1967b, fig. 6). It seems reasonable enough to assume that most ''horizontal'' graptolites with or without funiculate primary stipes had a life orientation analogous to that of *Allograptus canadensis,* with the laterally extended thecal lophophores generating propulsion for upward and possible rotational motion of the entire rhabdosome. During nonfeeding periods the rhabdosome would have begun to subside, but the flexible stipes may have reclined as a result of gravity, and this could have aided in retarding the downward sinking in a manner analogous to the extended and upwardly inclined fins of a stingray. Physical strain must have been most concentrated near the bifurcation points of the secondary stipes, where cortical thickening is usually more extensive than in other positions of the rhabdosome, in order to strengthen such points of hinge movement.

Life orientations of bottom-attached dendroid graptolites have been described by previous authors (Ruedemann, 1925, 1947; Bulman, 1970). As stated above, attached dendroids are more comparable to the attached living pterobranchs in feeding mode and life form. In situ preservation of benthonic graptolites, however, is relatively rare. So far the best known instances of in situ dendroids are from the Gasport interreef channel (Ruedemann, 1925) and similar Niagaran (Middle Silurian) environments of the Lockport Limestone of New York or Hamilton, Ontario. Coeval interreef lenses containing well-preserved dendroids also occur in the Mississinewa Siltstone of Indiana (Cumings and Shrock, 1928) of which a new exposure is currently under investigation by the author. At this new location near Huntington, Indiana, some of the cone-shaped colonies of *Dictyonema filiramum* are found in their life position (apex down) without major postmortem compression of the rhabdosomes. This and other species indicate exposure to directional bottom currents as their shapes are assymetrically extended in response to downcurrent growth (average orientation is E 110°S). Thecal apertures of these rhabdosomes point upward or concentrically toward the axes of cone-shaped forms; this would indicate better feeding efficiency in this apparently badly agitated shallow habitat.

SUMMARY

The purpose of this paper is to point out possible morphological analogies related to feeding between living pterobranchs and extinct graptolites, and to suggest that many graptolites may have been able to move or to remain up in the epipelagic zone by means of lophophore engendered water movements that originated as feeding currents. The author also took this opportunity to comment on other aspects of graptolite functional morphology and paleoecology, some of which were discussed in earlier publications.

It is here suggested that benthonically attached dendroids lived mostly in offshore mudbanks, often below photic influence or more rarely in shallow interreef zones, passively feeding from current- or wave-agitated water or actively abstracting food particles through lophophore-induced feeding currents. The feeding currents were multidirectional to the extent that the combined effect became neutralized in regard to the total rhabdosome.

Nonbenthonically living graptolites, most of them belonging to the order Graptoloidea, followed one or (during stages of their astogeny) a sequence of several of the following modes of life:

1. As inactive plankton ("epiplankton").
2. As plankton attached to other plankton ("pseudoplankton").
3. As actively mobile plankton ("autoplankton").

From geological evidence it appears that larval siculae and juvenile rhabdosomes of many graptolite species initially lived either as inactive epiplankton or pseudoplankton attached to Sargasso-type algae within or near to the photic pelagic zone. Adult graptolites may have remained pseudoplanktonic, but this may have been dependent on rhabdosome bulk and size, since there is no geological evidence for the existence of large epiplanktonic algae which would have been capable of carrying relatively bulky rhabdosomes. Further, any rotational movement of the rhabdosome, self-propelled or current-induced, might break the nema and separate the rhabdosome from its attachment to a host.

It can be hypothesized, using functional interpretation of morphological features, that unattached graptolites were able to rotate around their vertical axes and to propel themselves upward for more efficient nutrient exploitation of the water column or, at least in a few cases, to gyrate their rhabdosomes for the same purpose. It seems highly improbable that bulky inactive graptolite rhabdosomes would have remained in suspended equilibrium with the surrounding water masses, although such quasi-suspension may have existed during short intervals of nonfeeding. Inadvertently, the rhabdosome would have subsided during such periods of rest, but some types of rhabdosomal structures, such as interramous webs or thecal spines which are observed in many graptolites, may be interpreted as having aided in retarding the sinking of the rhabdosomes during periods of inactivity.

The geologically well-documented diversity in shape and form of graptolites, however, is coupled with the repetition of certain functional structures which must be regarded as having developed in natural response to diverse modes of life in diverse habitats with diverse nutrient resources.

ACKNOWLEDGMENTS

This paper was presented at the symposium of Development and Function of Animal Colonies which was held during the Annual Convention of the Geological Society of American in Washington, D.C. in 1971. The author wishes to express his thanks to the organizers of this symposium and to numerous anonymous contributors to lively discussions about graptolite functional morphology. This manuscript was reviewed by V. Jaanusson, Stockholm, Sweden, whose suggestions and comments were incorporated and are gratefully acknowledged. All drafting of illustrations should be credited to Larry Wahlie, Indiana University, Fort Wayne.

REFERENCES

Allman, G. J. 1869. On *Rhabdopleura*, a new form of polyzoa from deepsea dredgings in Shetland. Quart. Jour. Micros. Sci., n.s. *IX*: 57–63.

Berry, W. B. N. 1962. Graptolite occurrence and ecology. Jour. Paleont., *36*: 285–293.

———, and A. J. Boucot. 1971. Depth distribution of Silurian graptolites. Geol. Soc. America, Abstracts 1971 Annual Meet., *3*: 505.

Bulman, O. M. B. 1964. Lower Paleozoic plankton. Geol. Soc. London, Quart. Jour., *120*: 455–476.

———. 1970. Graptolithina, *in* R. C. Moore, ed., Treatise on Invertebrate Paleontology. Geol. Soc. America, and Univ. Kansas Press, *V* (rev.): VI–V163.

Burdon-Jones, C. 1954. The habitat and distribution of *Rhabdopleura normani* Allman. Univ. Bergen Àrbok, 1954, Naturv. rekke, *11*: 1–17.

Cumings, E. R., and R. R. Shrock. 1928. The geology of the Silurian rocks of northern Indiana. Indiana Dept. Conserv. Pub., *75*: 1–226.

Erdtmann, B. D. 1967a. A new fauna of Early Ordovician graptolites from St. Michel, Quebec. Canadian Jour. Earth Sci., *4*: 335–355, 1 pl.

————. 1967b. *Allograptus canadensis,* n. sp.; its significance to Lower Ordovician correlation. Canadian Jour. Earth Sci., *4*: 609–617, 2 pls.

Gilchrist, G. D. F. 1915. Observations of the Cape *Cephalodiscus (C. gilchristae,* Ridewood) and some of its early stages. Ann. Mag. Nat. Hist., ser. 8, *16*: 233–243.

Kirk, N. H. 1969. Some thoughts on the ecology, mode of life and evolution of the Graptolithina. Geol. Soc. London, Proc., *1959*: 273–292.

Kozlowski, R. 1938. Informations préliminaires sur les Graptolithes du Tremadoc de la Pologne et sur leur portée théorique. Ann. Mus. Zool. Polon., *13*: 183–196.

————. 1949. Les Graptolithes et quelques nouveaux groupes d'animaux du Tremadoc de la Pologne. Palaeont. Polon., *3*: 1–235, 6 pls.

————. 1966. On the structure and relationships of graptolites. Jour. Paleont., *40*: 489–501.

Lankester, E. R. 1884. A contribution to the knowledge of *Rhabdopleura* Quart. Jour. Micros. Sci., n.s., *XIV*: 622–647.

Ruedemann, R. 1925. Some Silurian (Ontarian) faunas of New York. N.Y. State Mus., Bull., *265*: 5–83, 24 pls.

————. 1947. Graptolites of North America Geol. Soc. America, Mem., *19*: 1–652, 92 pls.

Schepotieff, A. 1907. Die Pterobranchier, Zool. Jahrb. Abt. Anatomie, *23*: 463–534.

Stebbing, A. R. D. 1970a. The status and ecology of *Rhabdopleura compacta* Hincks (Hemichordata) from Plymouth. Jour. Mar. Biol. Ass. U.K., *50*: 209–221.

————. 1970b. Aspects of the reproduction and life cycle of *Rhabdopleura compacta* (Hemichordata). Mar. Biol., *5*: 205–212.

Ziegler, A. M. 1965. Silurian marine communities and their environmental significance. Nature, *207*: 270–272.

Diplograptid and Monograptid Fine Structure and Colonial Development

William B. N. Berry and Robert S. Takagi   University of California, Berkeley

ABSTRACT

Electron microscope study of *Diplograptus, Orthograptus,* and *Monograptus* species indicates that the periderm fine structure in these graptolite colonies may be divided into two principal types: that of the prosicula and that of the metasicula and thecae. Prosicula structures include a wall that is a loosely woven spongy mesh of fibers; a band, also formed from fibers, that spirals down the prosicula; and longitudinal structures. The metasicula and thecae have a basically two-layered structure which varies little throughout the colony. The inner or fusellar layer is formed by adjacent sets of half rings. The individual half-rings are composed of a spongy fibrous mesh bounded by a membrane-like structure. Each half-ring presumably was a growth increment. The outer or cortical layer is formed by alternations of mesh similar to that in the fusellar layer and long fibers that may be interwoven to have an almost fabric-like aspect. Fine structures of the metasicula and thecae differ little among the graptolites examined, even from genus to genus. Colony fine structure appears to have been little influenced by the environment, and it appears scarcely to have changed throughout the history of the graptoloid graptolites. The prosicula probably housed a sexually produced individual. The structural differences between the pro- and metasicula suggest that a metamorphosis took place in the development of the initial individual in the colony that heralded asexual budding. Each zooid was, in turn, the growing tip of

Author's address: Department of Paleontology, University of California, Berkeley, California 94720.

the colony, and each was connected to the others by a stolonal system. Cortical layer structure and appearance suggests that the zooids were connected by extrathecal tissue as well.

INTRODUCTION

Graptolites are an extinct group of colonial animals that lived in marine environments during the Paleozoic. The hard substance, herein referred to as periderm, secreted by individual zooids that formed the colonies is the only preserved part of graptolites. Any information concerning form and function of graptolite colonies must be sought through analysis of the periderm.

The fossil record of graptolite colonies reveals a diversity in form that reflects different modes of life. Many different graptolites were sessile benthonic, some having an encrusting habit whereas others were essentially rooted to the substrate and had a form of stem from which the colony developed. One group of graptolites, the graptoloids, were planktonic.

Associations of Silurian graptoloids with shelly (dominantly brachiopod) faunas indicative of marine benthic communities that were arrayed in order of increasing water depth away from Silurian shorelines suggests that the graptoloids were depth-distributed in the oceanic water column (Berry and Boucot, 1971). A few graptoloids apparently lived in relatively near-surface waters, whereas others were distributed down the oceanic water column at successively deeper levels. Greatest graptoloid diversity is found in rocks and faunal associations suggestive of life in water depth of about those over the edge of the continental shelf and over the continental slope (Ziegler, 1965; Berry and Boucot, 1972).

The evidence bearing on the phyletic relationships of the graptolites has been summarized by Kozlowski (1949, 1966). The closest living relatives of the graptolites appear to be the Pterobranchiates (Hemichordates), *Rhabdopleura* and *Cephalodiscus*, both of which are colonial and have encrusting modes of life in relatively cold waters.

Although all graptolites are similar in many morphological aspects, only a few of them have been examined by the authors with the electron microscope. Those examined form the basis for this paper.

All graptoloids developed from an initial individual that secreted an essentially conical test, the sicula. Siculae have two morphologically distinct parts, an apical part or prosicula and an apertural part or metasicula (see Kraft, 1926). The prosicula is formed from a thin membrane over which runs a thin band that spirals downward from the apex to the aperture. The metasicula, in contrast, is formed from two basic layers. Kozlowski (1966, p. 490) pointed out that "because no growth stages occur in it, the prosicula was surely secreted as a whole" by the individual it housed. Kozlowski (1966, p. 490) suggested that that individual could have been some larval form.

The boundary between the pro- and metasicula is commonly thickened. That thickening as well as the structural differences between the two parts of the sicula led Kozlowski (1966, p. 491) to state: "The contrast between the structure of the wall of the metasicula and prosicula is so great that an inescapable conclusion is drawn that the oozoid which produced the prosicula was subject to a thorough metamorphosis perhaps even before it started to secrete the metasicula part."

A pore in the metasicula part of graptoloid siculae indicates that budding took place from the individual housed in the sicula when the metasicula had formed or at least had begun to form (see Kraft, 1926). Eisenack (1942) noted that in certain of the youngest graptoloids, representatives of the genus *Monograptus,* a sinus developed along the apertural margin of a growth band in the initially formed part of the metasicula and that that sinus was closed subsequently by added growth bands in the metasicula. Some monograptids apparently commenced budding from the metasicula relatively soon after the individual that secreted the metasicula phase became established. In other graptoloids, the metasicula individual apparently was fully or nearly fully developed before budding commenced.

The periderm from selected graptoloid species has been examined using transmission electron microscopy to ascertain what the fine structure of the periderm might be and if aspects of the fine structure might shed some light on the nature and function of the graptoloid colony. A species in each of the diplograptid genera *Diplograptus* and *Orthograptus* from Late Ordovician strata in the United States, species in the monograptid genera *Lobograptus* and *Pristiograptus* from Early Ludlow (Late Silurian) strata found as glacial boulders in North Germany, and *Monograptus* in a boulder from the Arctic Canadian Silurian succession have been examined using transmission electron microscopy. The observations of the diplograptid peridermal construction have been recorded by the authors (Berry and Takagi, 1970, 1971). The data concerning monograptid peridermal structure as well as information given us by Adam Urbanek (1971), personal communication) from his transmission electron microscope investigation of the peridermal structure of certain graptoloids indicates close similarities with the peridermal structures described among the diplograptids.

METHODS OF STUDY

The periderm structures have been examined using microtome serial sections under both light and transmission electron microscopes. In addition, small sections of the periderm from all parts of the colonies examined have been studied directly under the transmission electron microscope without resorting to serial sections. A combination of the observations of periderm structure by viewing unsectioned specimens directly as well as by study of serial sections has been used in our interpretations of the structures seen. The specimens have been examined with a Zeiss EM 9A transmission electron microscope operated at a fixed voltage of 60 kV and with a Hitachi 650-kV transmission electron microscope. Details of specimen preparation for study by transmission electron microscopy have been discussed by Berry and Takagi (1970, 1971).

FINE-STRUCTURE OBSERVATIONS

Background

The frame of reference for any observations of the graptolite perderm was established by Kozlowski (1949) in his classic study of graptolites freed from Ordovician cherts. Kozlowski noted that the periderm of the graptolite metasicula and all thecae is comprised

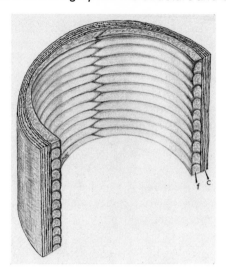

Figure 1

Schematic sketch of a part of a diplograptid theca showing aspects of the fusellar (f) and cortical (c) layer structure deduced from analysis of electron microscope examination of diplograptid thecal walls.

of two layers, an inner or fusellar and an outer or cortical. He indicated that the fusellar layer developed in growth increments that had the form of half-rings, each set of half-rings joining an adjacent set along the ventral and dorsal sides of the thecae and the metasicula in every graptoloid graptolite rhabdosome or colony. The half-ring boundaries form a sort of envelope for the fusellar material (Fig. 1 and Kozlowski, 1949, fig. 5). Only the prosicula has a different peridermal structure. The authors' electron microscope examination of the graptolite periderm has been directed at understanding the fine structure of the periderm in both the cortical and fusellar layers. The nature of the prosicula periderm has not been so thoroughly investigated as that of the metasicula and thecae.

Fusellar Layer

The fusellar layer in all graptoloids examined is essentially a spongy material, composed of fibers woven together, that has the general aspect of a mesh (Figs. 2–4). Urbanek (1971, personal communication) suggested that the fusellar layer of Kozlowski (1949) be termed the fusellar tissue. The spongy material that forms the fusellar tissue

Figure 2

Orthograptus quadrimucronatus metasicula.

A. Unsectioned fragment viewed from the exterior showing fibrous sublayer of the cortical layer (note "fabric-like" aspect) underneath the meshy or spongy sublayer of the cortical layer; ×20,000.
B. Section of the wall cut at a moderate angle to the sicula axis, showing both fusellar and cortical layers. The f is situated close to the closure in the membrane-like substance that forms

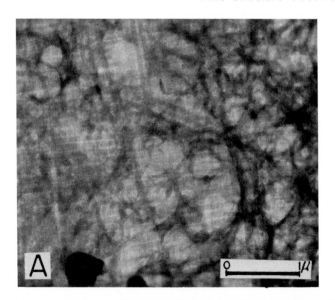

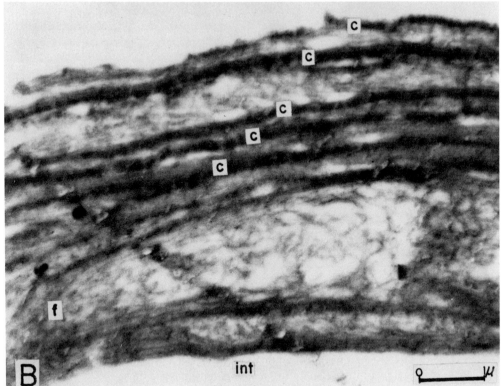

the boundary of one fusellar layer half-ring and encloses the fusellar spongy mesh material. The c indicates cut fibers in the fibrous sublayer of the cortical layer. The int indicates the interior of the metasicula, ×18,000.

Figure 3

Orthograptus quadrimucronatus, section of metasicula cut parallel to the long axis of the sicula showing three fusellar layer half-rings; the three U-shaped structures (marked with an f close to the closure of each) are the half-rings and the dark borders are the membrane-like substance; int indicates interior of metasicula; ×4950.

is bounded by relatively dense material that has the aspect of a thin membrane or sheet. It forms the boundaries of the half-ring growth increments (see Figs. 1–4).

Fusellar tissue is identical throughout the colony, from the metasicula through all of the thecae, in every colony examined. It is remarkably similar among all species examined: its general aspect differs little from species to species or genus to genus.

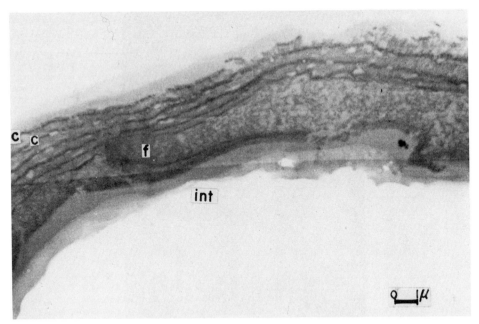

Figure 4

Orthograptus quadrimucronatus section of metasicula cut at an angle to the axis of the sicula showing both fusellar and cortical layers. (f is located in the nose of the U-shaped membrane-like substance that encloses the fusellar layer spongy mesh material; c is situated on the cut ends of fibers that comprise the fibrous sublayer of the cortical layer; int denotes the interior of the metasicula); ×6500.

Cortical Layer

The cortical layer of Kozlowski (1949) appears to be composed of alternations of spongy mesh which is identical in appearance with the spongy mesh in the fusellar tissue, with broad bands formed of rows of fibers (Figs. 1, 2, 4, 5). The sublayer formed from rows of fibers may include at least two sets of fibers that cross each other at high angles or are normal to each other and so have an almost "fabric-like" aspect (Berry and Takagi, 1970, pl. 27, fig. 2; and see Fig. 2A). The spongy, mesh-like material that forms the fusellar layer or tissue and one of the two sublayers in the cortical layer or tissue appears to be the primary peridermal substance secreted by the zooids that comprise the colony. The rows of fibers that form the second type of sublayer in the cortical layer appear to have been deposited upon a substrate formed from the spongy material.

The cortical layer appears to thicken and thin around any metasicula or theca. The thicker parts appear to be formed from several sets of fibers deposited one on top of another. The rows of fibers are built up most markedly over the interval of the zigzag sutures that is the area where adjacent sets of fusellar layer half-rings join. Bands of

Figure 5

Lobograptus sp., showing ventral wall of theca in the vicinity of the zigzag overlap between adjacent sets of fusellar layer half-rings. Two half-rings and the join between them (marked "growth line") may be seen beneath long fibers in the cortical layer. Unsectioned wall fragment viewed from the exterior; ×9750.

cortical fibers appear to cross many half-ring growth increments (Fig. 5 and Berry and Takagi, 1971, fig. 8). Certain of the bands of fibers appear to cross broad sections of the whole colony diagonally.

The cortical layer is markedly thicker in the proximal or older part of all colonies examined than it is in the distal part. The evidence suggests that cortical layer materials were being formed throughout the life of the colony. The cortical layer appears to be somewhat thinner in the monograptid colonies studied than in the diplograptid. It may not be present at all in the younger parts of some monograptid colonies.

Virgula

The virgula in the diplograptids and monograptids studied is hollow and tubular (Fig. 6). It extends upward from the prosicula but appears to be separated from the prosicula

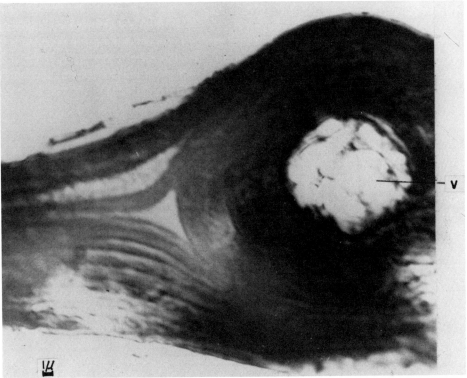

Figure 6

Monograptus priodon, cross section of virgula showing concentric structure of virgula wall; v indicates hollow interior of virgula; ×3000.

apex by a thin membrane. Virgula interiors are relatively narrow, having approximately 1/15–1/50 the diameter of the thecae in the specimens examined. The virgula appears to have been built up from a number of layers of spongy fusellar material which is dense at intervals. The dense intervals appear as thin membranes that separate spongy material. The dense, membranous-appearing intervals give the wall of the virgula tube a concentric appearance and suggest that the tube was formed by periodic secretions of spongy material that was bounded by a membrane. The virgulae commonly extend a distance above the thecate part of the colony, thus indicating the presence of tissues capable of secreting spongy material above that part of the colony where thecae were being formed.

Prosicula

The prosicula wall is a spongy mesh (Fig. 7A). The weave of the mesh is commonly more open than that in the fusellar tissue. The pattern of the weave differs somewhat from species to species and is the primary difference in fine structure among the graptoloid species studied. The spiral thread or band appears to be formed from mesh the weave of which is somewhat more close knit than that of the wall (Fig. 7B, C). The longitudinal

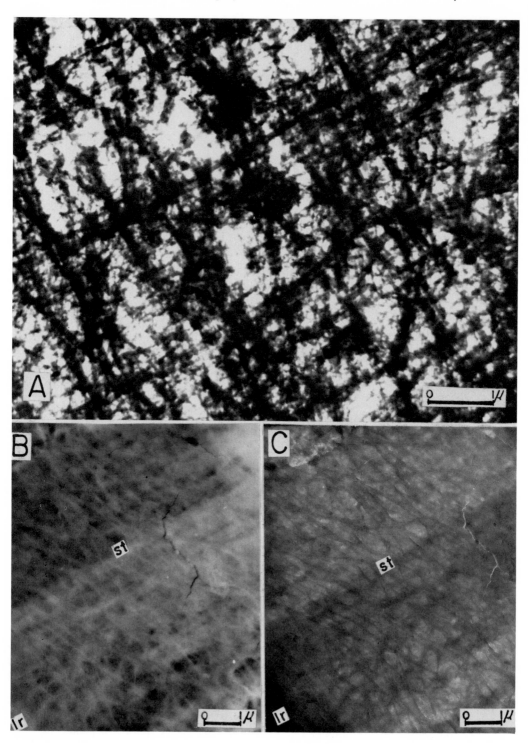

strengthening rods appear to be built from long fibers or from layers of sheet-like material. The longitudinal rods are situated outside the main wall. The longitudinal rods (or rod-like structures) are present in most graptoloid prosiculae, extending down from close to the apex to the boundary with the metasicula.

Composition

Dark field–bright field illumination and diffraction-ring studies of the fibers in the prosicula (Fig. 7B, C) and in the fusellar layer in the thecae suggest that the fibers in the specimens examined are preserved as single crystals. Biochemical studies of graptolite peridermal material by Florkin (1965) suggest that the periderm had a proteic nature. If so, and if the fibers were single crystals initially as they are now preserved, then the fibers may have been formed from long protein molecules.

GRAPTOLOID GRAPTOLITES AS COLONIES

Graptoloid graptolites developed as colonies from an individual housed within the sicula. In the colonies examined, the metasicula began to develop before budding took place, thus suggesting that the apparent metamorphosis from the individual in the prosicula to that living in the metasicula phase of the sicula may have been necessary to the onset of a condition related to asexual budding.

The marked uniformity of the peridermal fine structure of both cortical and fusellar layers within colonies as well as the regularity seen in budding among members of the colony suggests that genetic control of phenotypic expression was relatively strong and that the effect of the environment, even that of microenvironments, was slight.

Each individual zooid in the colony apparently developed from its predecessor as budding proceeded. Each individual was, for a time at least, a growing tip of the colony. Each individual was connected with its neighbors, and the connections among zooids collectively formed a sort of stolonal system. Individuals developed toward maturity secreting a housing (the theca) as they grew. Every theca has the same basic peridermal structure as the metasicula. The appearance and construction of the cortical layer, particularly the pattern in which bands of fibers of cortical tissue are formed on the fusellar tissue, suggests that the individual zooids were connected by some form of extrazooidal tissue as well as by the stolonal system.

The size of the asexually budded individuals in the colonies examined is progressively greater away from the sicula until a maximum size was attained. That size was then maintained throughout the remainder of the colony. Urbanek (1966) discussed the development of graptolite colonies as being controlled by morphogenetic (perhaps hormonal)

Figure 7

Unsectioned fragments of diplograptid prosiculae viewed from the exterior.

A. *Orthograptus quadrimucronatus* wall fragment, showing mesh; ×18,000.

B, C. *Diplograptus* sp. fragments, showing wall, spiral thread or band (marked st), and the edge of a longitudinal rod (lr). B is in dark field illumination; ×10,000. C is in bright field illumination; ×10,000.

substances. Those substances could have acted as inhibitors to maximum thecal development until a certain threshold level had been achieved when the thecae with maximum size began to develop.

The individual graptoloid zooids were presumably suspension feeders. No individuals appear to have been specialized for particular activities, and there was no apparent action in concert among individuals in the colony although some or all of the zooids might have acted together to create small water currents about the colony or even perhaps to turn the colony.

SUMMARY

Graptoloid graptolite colonies appear to have been formed from structurally connected individuals each of which had its turn as the growing tip of the colony and each of which was connected to the others by a form of stolonal system and perhaps by extrathecal zooidal tissue as well. Additions to the cortical peridermal layer appear to have continued, at least periodically, throughout the life of the whole colony.

Although the graptoloids studied include specimens of Late Ordovician, Middle Silurian, and Late Silurian age, no discernible change in mode of colony integration was noted. Indeed, colony integration appears to have remained relatively constant throughout the history of the graptoloid graptolites, from early in the Ordovician through the early part of the Devonian. A general trend in colony form through the early history of the graptoloid graptolites appears to have been from many individuals to few and from an orientation in which most of the zooids were oriented downward toward the sea bottom to one in which the individuals were directed upward toward the sea surface. The types of colonies that persisted for the longest period of time were those in which the thecae were arranged uniserially and the individuals in them were apparently oriented generally upward.

Graptoloid graptolites appear to have had a periderm suited to planktonic modes of life. The periderm was basically spongy, yet strong and perhaps flexible. No obvious change in this basic fine-structural plan took place in the history of the graptoloids. Interestingly, therefore, the basic structural plan as well as colonial integration appear to have remained relatively constant throughout most if not all of the history of the graptoloids. Graptoloid extinction did not follow after any obvious changes in coloniality.

ACKNOWLEDGMENTS

The authors are indebted to Hermann Jaeger for supplying the graptolite-bearing block from Germany and to Raymond Thorsteinsson for the monograptid-bearing material from Arctic Canada. They thank Adam Urbanek for his helpful suggestions and discussions of graptoloid peridermal structure.

REFERENCES

Berry, W. B. N., and A. J. Boucot. 1971. Depth distribution of Silurian graptolites. Geol. Soc. America, Abstracts with Programs, *3*: 505.

————, and A. J. Boucot. 1972. Silurian Graptolite depth zonation. 24th Internat. Geol. Congress, Proc., section 7, p. 59–65, Montreal.

————, and R. S. Takagi. 1970. Electron microscope investigation of *Orthograptus quadrimucronatus* from the Maquoketa Formation (Late Ordovician) in Iowa. Jour. Paleont., *44*: 117–124.

————, and R. S. Takagi. 1971. Electron microscope study of a *Diplograptus* species. Lethaia, *4*: 1–13.

Eisenack, Alfred. 1942. Uber einige Funde von Graptolithen aus ostpreussischen Silurgeschieben. Zeitschr. Geschiebeforschung, *18*: 29–42, 2 pls.

Florkin, Marcel. 1965. Paleoproteins. Acad. Roy. Belgique (Sciences), Bull., ser. 5, *51*: 156–169.

Kozlowski, Roman. 1949. Les Graptolithes et quelques nouveaux groupes d'animaux du Tremadoc de la Pologne: Palaeont. Polon., *3*: 1–235, 6 pls.

————. 1966. On the structure and relationships of graptolites. Jour. Paleont., *40*: 489–501.

Kraft, Paul. 1926. Ontogenetische Entwicklung und Biologie von *Diplograptus* und *Monograptus*. Palaont. Zeitschr., *7*: 207–249, pls. 3–17.

Urbanek, Adam. 1966. On the morphology and evolution of the Cucullograptinae (Monograptidae, Graptolithina). Acta Palaeont. Polon., *11*: 291–544, pls. 1–47.

Ziegler, A. M. 1965. Silurian marine communities and their environmental significance. Nature, *207*: 270–272.

Introduction to Sponges

Convention places the sponges "below" the coelenterates in general grade of development, and zoological texts normally treat them first among multicelled animals because there is a logic to moving from simple to more complex. However, it is clear that "individuality" and "coloniality" have different meanings (or no meaning) at this parazoan level and the significance of the following discussions can better be appreciated after a discussion of the other groups; thus the present sequence.

Authors in this volume do not agree on the definition of "individual" and "colony" in sponges. **Simpson** considers sponges to be colonies by definition but that there is no qualitative difference between colonies and individuals; he considers other "colonial" metazoans to be individuals also. **Hartman and Reiswig** add historical perspective by reviewing the development of thought on the "coloniality" or "individuality" of sponges; through morphological, developmental, and physiological arguments they conclude that the level of integration within sponges reflects a primitive individuality of the whole rather than a loss of zooid individuality. **Finks** examines an extinct group of sponges and demonstrates that skeletal chambers and functional (exhalent) units do not necessarily correspond, thus adding another argument against the interpretation of sponges as colonies.

Coloniality Among the Porifera*

Tracy L. Simpson   University of Hartford

ABSTRACT

Sponges consist of choanocytes and pinacocytes; the former line a water-filled cavity and the latter form an outer covering and line canals which lead to the choanocytes. A single functional unit in sponges consists of a number of incurrent openings, formed by pinacocytes, which lead to a layer or groups of choanocytes from which excurrent canals lead to a single oscular (excurrent) opening. Functional units can be arranged in a number of ways, the most common of which is leuconoid structure. Most sponges possess multiple functional units and have therefore been referred to as colonial. Colonialism in sponges and other groups is the capacity and realization of asexual duplication of adult functional structure which maintains continuity with older functional units. However, colonial animals reproduce sexually in the same way as noncolonial animals and there is thus no qualitative difference between them. In sponges a new animal begins with fertilization and the formation of a free swimming larva; this larva initially produces a sponge which possesses a single functional unit. The continuity of the mesenchyme between functional units and physiological data dealing with the functioning of the canal system support the conclusion that adult sponges should be considered individuals. This conclusion is also applicable to other colonial animal groups. Colonialism is thus a descriptive term referring to the asexual duplication of adult units. Immunological individuality in sponges appears not to be well developed.

Author's address: Department of Biology, University of Hartford, West Hartford, Connecticut 06117.
*This work was supported by NSF Research Grants GB-192, GB-4094, GB-7012, and GB-18613.

Sponges are ancient animals with a minimum of cellular organization. Considering their primitive level of organization it is amazing that they possess the capacity for the formation of an organized and specific inorganic skeleton. Related extinct groups, including the Archeocyathida, Sphinctozoa, and Stromatoporoidea, also possessed this capacity (the latter two groups are discussed in this volume). The inorganic skeleton of these animals may be in the form of separate spicules; or it may consist of a fusion of spicules or of a crystalline mass and thus is continuous and relatively solid. Both types of skeletal elements (spicules and a solid skeleton) occur in fossil sponges and in present-day forms; in some groups, such as the pharetronids (Calcarea), both a spicular and solid inorganic skeleton is present in the same animal. While the deposition of an inorganic skeleton is an outstanding feature of the Porifera, not all sponges possess the capacity to secrete this type of skeleton. In particular, the keratosan demosponges (bath sponges) exclusively secrete an organic skeleton of collagenous fibers (spongin) (Gross, Sokal, and Rougvie, 1956) and a few sponges (*Oscarella, Halisarca*) develop no obvious skeleton at all (namely, one that can be recognized on the light microscope level). The composition of the inorganic skeleton in sponges is either calcium carbonate or silicon dioxide. Among living sponges, the Calcarea produce crystalline spicules of calcite (Jones, 1970) and one group, the pharetronids, produces a solid skeleton of the same composition in addition to spicules (Vacelet, 1964). The Demospongiae deposit siliceous spicules consisting of hydrated silica, as do the Hexactinellida (Vinogradov, 1953). In the latter two groups there are species in which the spicular silica is fused into a continuous skeleton. The Sclerospongiae are most unusual in that

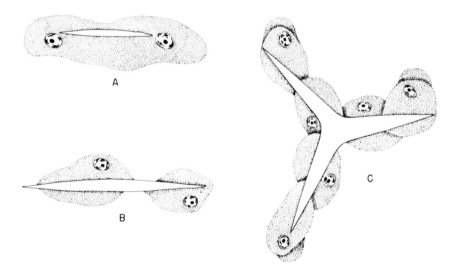

Figure 1

Schematic representation of spicule secretion in calcareous sponges. *A* and *B* are stages in the deposition of a monaxon. *C* represents an early stage in the secretion of a triaxon. Note that each ray is laid down by two cells.

they deposit a solid aragonitic (calcareous) skeleton in addition to siliceous spicules (Hartman and Goreau, 1970).

The biochemical basis of mineral deposition in sponges is totally uninvestigated. On the cellular level we have relatively few reports describing the mechanism of spicule deposition. In the Calcarea spicule secretion is initiated by two or more cells (Fig. 1), depending upon the type of spicule produced (Jones, 1970). Among the Demospongiae a single cell (Fig. 2) initiates spicule secretion (Lévi, 1963; Simpson, 1968a). In both cases present evidence indicates that mineral deposition is intracellular; that is, the new spicule is surrounded by a limiting membranous sheath. It is to be emphasized that considerably more work on spicule secretion must be undertaken before we can formulate more definite generalizations.

Although collagen (Fig. 3) appears to be of universal occurrence only in the Demospongiae [Lévi and Porte (1962) report the presence of banded fibrils on an ultrastructural level in *Oscarella*, a genus which lacks any obvious skeleton], this supporting material may be present in all sponges, at least in the form of fibrils. Tuzet and Connes (1964) have reported fibril occurrence in a calcareous sponge, as has Jones (1967). Further

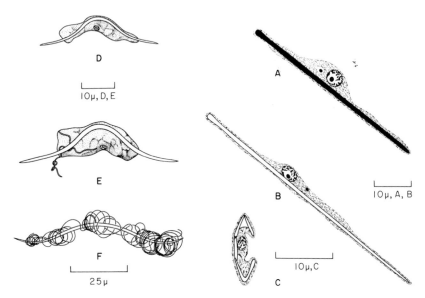

Figure 2

Spicule secretion in demosponges. In *A* an initial stage of secretion of a megasclere is pictured in which an organic core of material is visible. Not all demosponges lay down a visible organic thread. In *B* a newly secreted spicule is present within a single cell. The secretion of one type of microsclere is pictured in *C*. This spicule also is clearly intracellular. *D, E,* and *F* picture various means by which a second type of microsclere is deposited. In *F*, no apparent cell is associated with the new spicule; however, membranous material of cellular origin is clearly visible. This figure displays spicule secretion in a small group of demosponges. The secretion of other types of spicules in other demosponges is not well documented and awaits further investigation.

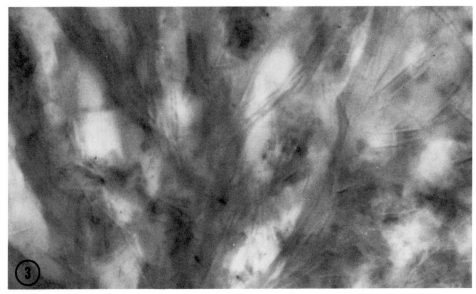

Figure 3

Collagenous fibers (spongin) which are present in a marine demosponge. Note the considerable amount of material which is present and the occurrence of spicules embedded in the collagen.

electron microscopic analysis is necessary before this generalization can be accepted. It is of extreme importance that fossil sponges and related groups be examined for the presence of collagen, for, if present and identifiable, its recognition would be a new criterion for the identification of fossil sponges.

CHARACTERISTICS OF SPONGES

General Features (See Fig. 4)

Sponges can be described as aquatic, sessile animals which possess (1) special flagellated cells (choanocytes); at the base of the flagellum of each cell is a collar made up of microvillus extensions of the cell membrane (Fjerdingstad, 1961); and (2) pinacocytes, which form an outer epithelial layer (Lévi, 1970). In the vast majority of sponges which have been studied a third type of cell, the archeocyte or nucleolate cell, is present in the mesenchyme. Choanocytes line a water-filled cavity which is connected to the environment by canals. The pinacocytes, in addition to forming an outer covering, form openings and canals which convey water to and away from the choanocytes. Between the choanocyte–canal layer and the external pinacocytes is mesenchyme, which contains ground substance, collagen fibers, motile cells, reproductive cells, and an inorganic skeleton when present.

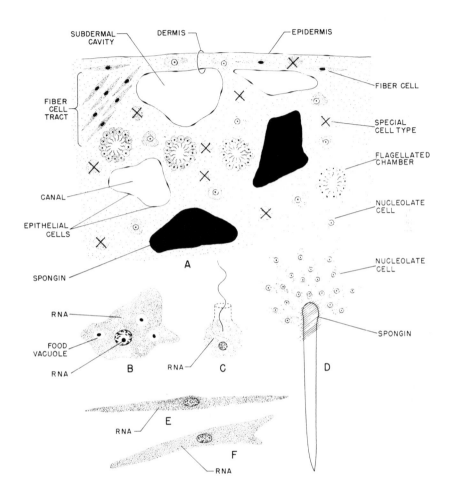

Figure 4

A. Schematic representation of a section of a leuconoid sponge. The epidermis consists of a layer of pinacocytes; the subdermal cavity could be part of either the excurrent or incurrent canal system. Groups of choanocytes form flagellated chambers (connections with the canal system are not shown; see Fig. 5). The mesenchyme (stippled) contains nucleolate cells (archeocytes) and special mesenchymal cells as well as spongin. Fiber cells are frequently present in the mesenchyme also. Spicules are not pictured. *B*. Nucleolate cell showing the distribution of ribonucleic acid and the presence of food vacuoles. *C*. Choanocyte showing the flagellum and surrounding collar, which consists of extensions of the cell membrane. *D*. Group of nucleolate cells is shown initiating spongin (collagen) secretion around a spicule. *E*. Pinacocyte. *F*. Fiber cell. RNA, ribonucleic acid. The features shown in this figure are present in most spnges. See text and refer to Fig. 5.

Functional Structure

The simplest structure found in living sponges is exemplified by relatively few species, most of which belong to the genus of calcareous sponges *Leucosolenia*. In these animals the choanocytes line a single cavity which opens to the environment by a single excurrent pore, the osculum (Fig. 5A). Water enters the cavity through numerous small pores (ostia) formed by pinacocytes (Jones, 1966). In this type of sponge there are no canals which connect the ostia with the choanocytes or the choanocyte layer with the osculum; consequently, the mesenchyme is very thin. A single functional unit thus consists of ostia, choanocytes, and a single osculum. The beating of choanocyte flagella results in a flow of water from ostia to osculum. This structure is referred to as asconoid. A more complex type of organization, syconoid, involves the outpocketing of the choanocyte layer and the production of an internal pinacocyte layer. In other respects syconoid structure is comparable to asconoid structure (Fig. 5B).

In most sponges the choanocytes form spherical masses (flagellated chambers) which lie at some distance from the surface pinacocytes. Ostia lead into incurrent canals rather than directly to the choanocytes; the incurrent canals then lead to the flagellated chambers from which excurrent canals lead to an oscular opening. In sponges with this type of functional system (leuconoid) the mesenchyme is very highly developed, occupies a large volume, and is permeated with excurrent and incurrent canals. Each oscular opening has a number of exhalant canals which lead into it (Fig. 5C). In some leuconoid sponges (mostly in encrusting species) the exhalant canals emerge from the interior of the sponge and at the surface form radiating patterns (astrorhizae) which end in an osculum. In other species the exhalant canals merge below the surface (Fig. 6) and only the osculum is visible. In leuconoid sponges a single functional unit consists of a single osculum and the canal structures connected to it (ostia, incurrent canals, flagellated chambers, and excurrent canals).

Some sponges have evolved a growth form in which upright oscular chimneys rise from a basal region (Fig. 7). In these species we are dealing with a special type of compound leuconoid structure. Ostial openings permeate the outer wall and lead by incurrent canals to flagellated chambers, which in turn empty into excurrent canals which open through pores within the chimney. These internal pores are functionally equivalent to oscular openings. At the tip of the chimney is a single opening (hereafter referred to as a supraoscular opening). If each of the internal pores is considered a single oscular opening, then each chimney consists of multiple functional units. On the other hand, if the supraoscular opening at the tip of each chimney is considered homologous to oscular openings in other sponges, then the whole chimney represents a single functional unit. Because of the absence of detailed studies of these animals, it is not possible to decide which of these formulations is more meaningful. Because of the arrangement of pores, the structure of compound leuconoid sponges is of much interest in considering, in particular, the structure of the archeocyathids as well as other fossil groups in which there is a double layer of pores.

In each of the types of functional structure discussed above (asconoid, syconoid, leuconoid, and compound leuconoid) there are theoretically two possible kinds of sponges: those in which only a single functional unit is present, and those in which multiple functional units are present in the same sponge. The vast majority of sponges fall into

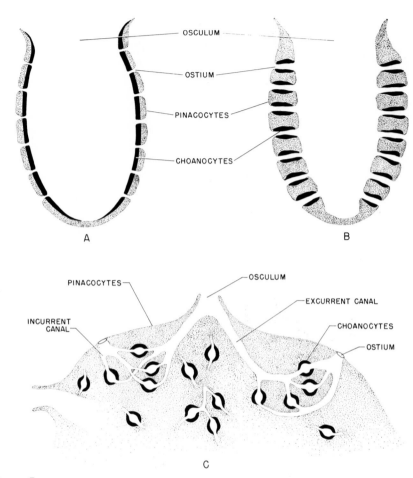

Figure 5

Schematic drawings of the organization of the canal system in sponges. The choanocyte layer is represented by a solid thick line. *A*. Asconoid structure as is found in *Leucosolenia*. Canals which connect the choanocytes to the environment are absent and the mesenchyme, occurring between the pinacocytes and choanocytes, is very sparse. *B*. Syconoid structure, which is present in *Sycon*. The mesenchyme in these sponges is much more substantial than in asconoid sponges. *C*. Leuconoid structure, which is present in most sponges. The mesenchyme in these animals occupies a considerable volume; in most leuconoid sponges both spicules and collagen fibers occur (neither of which are pictured). Numerous functional units such as that pictured occur throughout the animal. Compound leuconoid structure is like leuconoid structure except the functional units are vertically oriented on an upright cylinder with ostia on the outer surface and oscula opening within the chimney (see Fig. 7).

the latter category. For example, in *Leucosolenia* (asconoid structure) numerous tubes, each consisting of a single functional unit, are asexually produced by the same animal. In leuconoid sponges (most genera of sponges) adults asexually duplicate functional units; in some cases asexual duplication occurs throughout the life of the animal. In

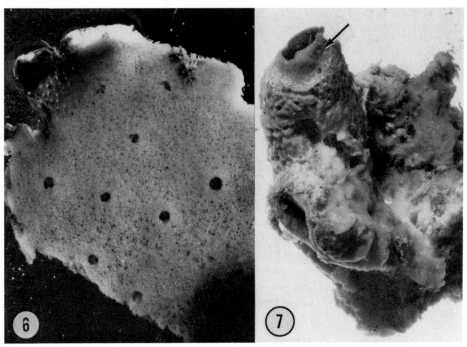

Figure 6
Surface view of a marine demosponge *(Microciona seriata)* in which numerous functional units are present. Each osculum is clearly visible and is formed by the union of a number of large exhalant canals. See Fig. 5C.

Figure 7
Marine demosponge *(Tedania ignis)* which possesses compound leuconid structure. The oscular chimney is obvious; at its tip (arrow) is the supraoscular opening. The mesenchymal cells of the supraosculum are contractile and the opening can be closed.

compound leuconoid sponges, such as some species of *Callyospongia,* numerous chimneys are asexually produced by the same animal. In this case, regardless of whether we consider a single chimney as consisting of a single functional unit or of multiple units (see above), the whole animal possesses multiple units. There are relatively few sponges in which there is a single functional unit present throughout the adult life span, an example being calcareous sponges in the genus *Sycon* (syconoid structure). Some pharetronids, archeocyathids, and sphinctozoa which have a single chimney-like structure can, if we consider a single chimney as a single functional unit, be given as additional examples (Ziegler and Rietschel, 1970; Zhuravleva, 1970), as can species of hexactinellid sponges in the genera *Rosella* and *Staurocalyptus.* Among the demosponges it is rare to find species which grow as single chimneys. On occasion members of *Verongia* occur in this form. The sclerosponges and stromatoporoids apparently all possess multiple functional units. A thorough review of the literature will have to be undertaken before a complete summary can be given of the distribution of sponges which possess single functional units and those which possess multiple functional units.

COLONIALISM

Because adult sponges asexually duplicate functional units, they have been referred to as colonial. In essence this is what is meant by the term "colonial" when it is applied to any animal group—the asexual duplication of adult functional structure, with each new unit maintaining continuity with the older units. The capacity to asexually duplicate adult structure is doubtless an attribute of the genetic systems present in these groups. The evolution of this capacity is most certainly related to the efficiency of sexual reproduction in colonial animals as well as to their ability to survive; the duplicative process increases sexual reproductive potential and the ability to obtain necessary nutrients. As special as this capacity is, its existence does not demand that we view colonial animals as being qualitatively different from noncolonial animals. Sexual reproduction in colonial animals is like that in other animals; the initiation of new individuals begins with a single egg, and consequently the evolution of colonial animals does not differ from other animals.

SEXUAL REPRODUCTION

In sponges, sexual reproduction takes place through the production of egg and sperm cells, fertilization, and the formation of a free-swimming larva (Simpson, 1968b); most sponges are hermaphroditic (Figs. 8–10). The structure of the larva is variable, but in all cases it is ciliated (Lévi, 1956; Simpson, 1968a, 1968b; Berquist, Sinclair, and Hogg, 1970; Borojevic, 1970). Some sponges are oviparous and release eggs into the environment prior to fertilization (Lévi, 1956; Borojevic, 1967). The process of fertilization is very poorly documented among the Porifera. The single larva which results from fertilization eventually, after spending a number of days as a free-swimming entity, attaches to a substratum and undergoes metamorphosis into a young sponge. This initial sponge which is produced always consists of a single functional unit (Simpson, 1968b) (Fig. 11).

NONCOLONIAL ASEXUAL REPRODUCTION

Many animals are capable of asexually reproducing a new adult which is autonomous, that is, not attached to the original parent. In a genetic sense this new adult is not different from a new functional unit in a colonial animal, it is simply spatially separate from the original animal. The asexual production of the new animal increases sexual reproductive potential and survival; more gametes are produced and new geographic environments are available for securing nutrients. Asexual reproduction is uncommon in sponges. Species of the demosponge *Tethya* produce buds which break away from the sponge and initiate a new animal (Connes, 1967). Burton (1948) has observed asexual reproduction in a number of encrusting marine demosponges. This involves the pinching

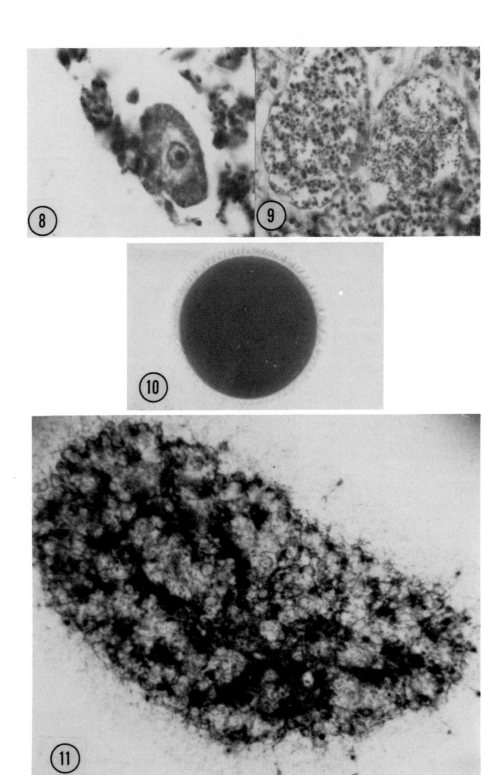

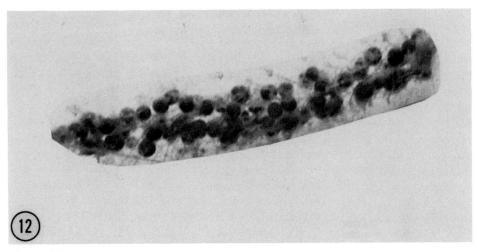

Figure 12

Part of a branch of the freshwater demosponge *Spongilla lacustris,* showing numerous gemmules which are embedded in the skeleton of the animal.

off of part of an adult sponge to produce two sponges. The production of gemmules in freshwater sponges could be considered as asexual reproduction because we know in the laboratory that each gemmule which is produced is capable of initiating a separate new animal (Rasmont, 1962). Furthermore, when gemmules are produced a single sponge will form hundreds or thousands of them (Fig. 12). However, there is no indication from field work that gemmules normally function by breaking away from the parent sponge and forming a new animal. On the contrary, studies on gemmules indicate that these structures remain in situ and when active they reform the original parent sponge (Simpson and Gilbert, unpublished observations). Asexual reproduction in sponges and other animals can be viewed as a special case of colonialism in which the new, duplicated adult loses attachment with the parent.

Figure 8

Mature egg present in the mesenchyme of a demosponge *(Tubella pennsylvanica).*

Figure 9

Two spermatogenic masses present in the mesenchyme of a demosponge *(Spongilla lacustris).* The cells in the right-hand mass are entering the spermatid stage.

Figure 10

Mature free-swimming larva of *Microciona prolifera* in which the surface ciliation is visible. This larva after attachment will form a new sponge.

Figure 11

Surface view of a newly formed sponge. This sponge consists of a single functional unit and was formed from a single larva. Flagellated chambers are visible in the tissue.

INDIVIDUALITY AND SPONGES

Since sponges duplicate functional units the question can be raised as to whether an adult sponge can be referred to as an individual. Because adult sponges are derived from fertilized eggs and begin as genetic systems comparable to noncolonial animals, there is no question that each adult sponge is an individual in a genetic sense. They are, as are other colonial animals, special in the sense that each individual possesses *and realizes* the capacity to asexually duplicate adult structure. Additional data which support the individuality of sponges follow. The mesenchyme in adult sponges is continuous and mesenchymal cells can migrate throughout the sponge as well as into new growth areas (Simpson, 1963). Tuzet, Pavans de Ceccaty, and Paris (1963) have reported the migration of egg cells in the mesenchyme of *Tethya* and *Hippospongia* from one functional unit to another. When spermatogenesis and oogenesis occur in sponges both processes occur simultaneously throughout the mesenchyme (Simpson, 1968a; Lévi, 1956; Tuzet, Garrone, and Pavans de Ceccaty, 1970). The adult thus acts as an individual in terms of gametogenesis. When freshwater sponges produce gemmules, these overwintering structures are formed throughout the sponge simultaneously (Simpson and Gilbert, unpublished data). Again, each sponge functions as a single gemmule-producing unit. Furthermore, the cell types which are present in the mesenchyme of sponges are the same throughout the animal. When adult sponges grow the new growth area is initially an integral part of the adult system (Fig. 13) (Simpson, 1963, 1968a). It is only after new growth areas become extensive that they develop into new, additional functional units. These data suggest that new functional units are produced in new growth areas because it is inefficient to maintain continuity with the old excurrent system, owing to the large increase in volume of the canal system. Detailed physiological investigations are needed to support this view. Finally, the recent work of Hartman and Reiswig (reported in this volume) strongly supports the individuality of sponges on the basis of the functioning of the canal system: the whole sponge acts as a single physiological unit. The results of their work are the best physiological data we have dealing with the canal system. Based upon the data reviewed above it is clear that each adult sponge should be considered an individual.

Two additional problems are still outstanding. First, when sponges or other colonial animals asexually reproduce new, separate adults, these animals are not genetically new adults. By referring to them as individuals, the implication is that they are genetic individuals in the sense that they possess genetic variability which is the result of sexual reproduction. In practice the result is that upon occasion we are implying a genetic individuality which does not exist. There is no satisfactory way to avoid this dilemma in any group of animals short of genetically analyzing all members of a species. The second problem is: How do we now employ the term colonial; what do we mean by it if colonial animals are individuals? As previously stated, colonialism is the asexual duplication of a new adult unit which maintains continuity with the older unit or units. Sponges, ectoprocts, and other taxa in which adults possess multiple units are animal groups consisting of colonial individuals. Colonialism is thus a descriptive term which refers to the genetic capacity and realization of a multiple-unit state. Among the ectoprocts and cnidaria, in particular, colonialism has become specialized in that the multiple units

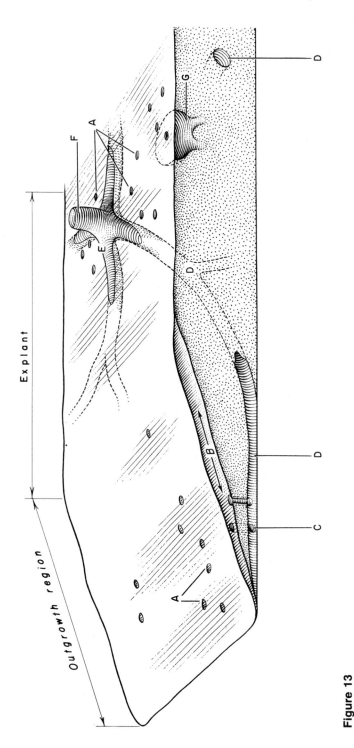

Figure 13

Schematic drawing of an explant of a marine demosponge and the adjoining new outgrowth region. The canal system of the outgrowth region at this stage of growth is part of the adult (explant) system. The mesenchyme is common. *A*, ostia; *B*, incurrent canal; *C*, flagellated chamber; *D*, excurrent canal which emerges from the sponge at *E*; *F*, osculum; *G*, incurrent canal. Mesenchyme is stippled; flagellated chambers are not pictured in the explant and in most areas of the outgrowth region.

produced may not be identical; many of these animals are polymorphic. Polymorphic colonialism represents a special case of colonialism.

A SPECIAL CASE OF INDIVIDUALITY IN THE PORIFERA

Individuality also implies that transplants of tissues or organs are usually not possible, owing to intraspecific and interspecific genetic differences which occur among individuals (Ebert, 1959). In sponges, however, there are indications that this kind of immunological individuality is not well developed. Of particular importance are reports of the fusion of whole sponge larvae prior to metamorphosis. Van De Vyver (1970) has observed this in free-swimming larvae of the freshwater sponge *Ephydatia fluviatilis* and in the marine sponge *Crambe crambe*. In the latter species not all larvae were capable of fusion. In both cases fusion was observed in larvae which were produced by the same sponge. Lévi (1956) has also recorded cases of larval fusion, as has Borojevic (1967). These reports are of interest as they relate to immunological individuality; in the cited cases whole organisms (larvae) undergo fusion. Although transplants between siblings in vertebrates have a higher probability of "take," in none of these are we dealing with a mixture of cells from whole organisms (Ebert, 1959) as is the case with sponge larvae. Warburton (1958) has reported fusion between adult individuals in the boring sponge *Cliona celata* as well as in *Cliona lobata*. In addition, Van De Vyver (1970) has shown that gemmules from different individuals of *E. fluviatilis* fuse upon hatching to produce a single individual. This situation, however, is complicated by the fact that in this sponge there are a number of different strains which occur and gemmular fusion occurs only between individuals of the same strain.

There are also well-documented cases of nonfusion in sponges, which is comparable to transplant rejection in higher animals. Van De Vyver (1970) has found that gemmules derived from different strains of *E. fluviatilis* do not fuse upon hatching and larvae derived from individuals of different strains also do not fuse. Paris (1960) has shown that intra- and interspecific grafts of two marine demosponges are rejected, a result comparable to that found in vertebrates (Ebert, 1959). In *Microciona prolifera* nonfusion between individuals can be observed occasionally in the field. The new growth areas of explants of each individual also do not fuse (Figs. 14 and 15). Nonfusion in this case is thus not dependent upon an aging process but is inherent.

This field of research in sponges is relatively new and it is too early to generalize. It is probably to be expected that sponges and other lower invertebrates do not develop

Figure 14

Two explants of the demosponge *Microciona prolifera* which were taken from the same sponge have produced new growth areas which have completely fused. An excurrent canal (arrow) common to both explants is visible.

Figure 15

Two explants of *Microciona prolifera*, taken from two different sponges. The adult sponges displayed a line of nonfusion. The outgrowth areas of each explant also show (arrow) a line of nonfusion. Incompatability of tissue in this instance is not related to an aging process.

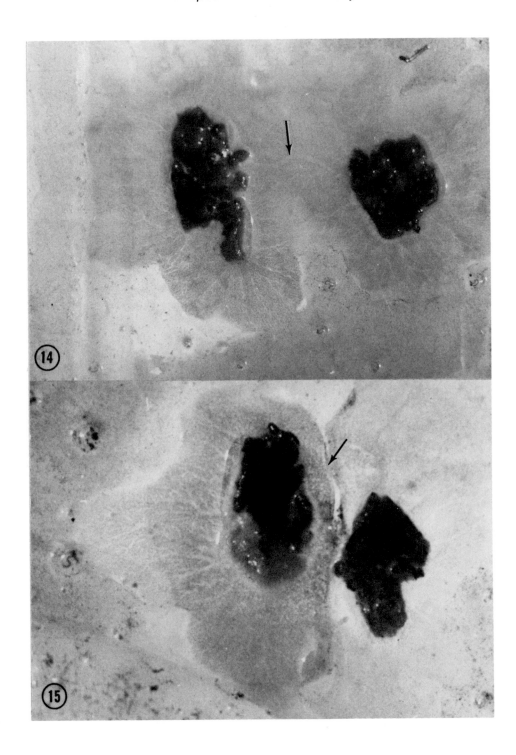

individuality on an immunological level to the extent that it is manifest in higher groups. For example, Tardent (1968) has shown that grafts between individuals of *Hydra attenuata* are accepted and indeed male grafts induce masculinity in female recipients.

PERSPECTIVE

Some interesting questions can now be raised about colonial animals. Prime among these is the question of whether colonialism, the asexual duplicative process, has a common genetic basis in sponges, ectoprocts, cnidarians, tunicates, and other colonial animals. Furthermore, one can ask whether all animals possess the potential to be colonial; that is, whether colonialism is a latent capacity of genetic systems in noncolonial animals. The answers to these questions lie to a large extent in the field of developmental biology, in that they focus attention on the ability of adult cells to act, in the absence of fertilization, in an embryonic fashion through developing a new adult. In colonial animals, adult cells display this capacity. Finally, we can ask if there is a direct relationship between colonialism and the level of development of immunological individuality. Do colonial animals develop less immunological individuality than noncolonial animals and, if so, is this related to their capacity for and realization of colonialism?

ACKNOWLEDGMENT

The author expresses his gratitude to John H. Felber and Ernest Weinberger for their encouragement and help.

REFERENCES

Berquist, P. R., M. E. Sinclair, and J. J. Hogg. 1970. Adaptation to intertidal existence: reproductive cycles and larval behavior in Demospongiae, *in* The Biology of the Porifera, W. G. Fry, ed., Acad. Press, London. Zool. Soc. London, Symp., *25*: 247–271.

Borojevic, Radovan. 1967. La ponte et le développement de *Polymastia robusta* (Demosponges). Cah. Biol. Mar., *8*: 1–6.

———. 1970. Différenciation cellulaire dans l'embryogenèse et la morphologenèse chez les spongiaires, *in* The Biology of the Porifera, W. G. Fry, ed., Acad. Press, London. Zool. Soc. London, Symp., *25*: 467–490.

Burton, Maurice. 1948. Observations on littoral sponges, including the supposed swarming of larvae, movement and coalescence in mature individuals, longevity and death. Zool. Soc. London, Proc., *118*: 893–915.

Connes, Robert. 1967. Structure et développement des bourgeons chez l'éponge siliceuse *Tethya lyncurium* Lamarck. Recherches expérimentales et cytologiques. Arch. Zool. Exp. Gén., *108*: 157–195.

Ebert, James. 1959. The aquisition of biological specificity, *in* J. Brachet and A. E. Mirsky, eds., The Cell. Acad. Press, New York, *1*: 619–693.

Fjerdingstad, E. J. 1961. The ultrastructure of choanocyte collars in *Spongilla lacustris* (L.). Zeit. f. Zellforsch., *53*: 645–657.

Gross, Jerome, Z. Sokal, and M. Rougvie. 1956. Structural and chemical studies on the connective tissue of marine sponges. Jour. Histochem. Cytochem., *4*: 227–246.

Hartman, W. D., and T. F. Goreau. 1970. Jamaican coralline sponges: their morphology, ecology, and fossil relatives, *in* W. G. Fry, ed., The Biology of the Porifera. Acad. Press, London. Zool. Soc. London, Symp., *25*: 205–243.

Jones, Clifford. 1966. The structure of the porocytes in the calcareous sponge *Leucosolenia complicata* (Montagu). Jour. Roy. Micros. Soc., *85*: 53–62.

———. 1967. Sheath and axial filament of calcareous sponge spicules. Nature, *214*: 365–368.

———. 1970. The composition, development, form, and orientation of calcareous sponge spicules, *in* W. G. Fry, ed., The Biology of the Porifera. Acad. Press, London. Zool. Soc. London, Symp., *25*: 91–123.

Lévi, Claude. 1956. Étude des *Halisarca* de Roscoff. Embryologie et systématique des Demosponges. Arch. Zool. Exp. Gén., *93*: 1–181.

———. 1963. Scléroblastes et spiculogenèse chez une éponge siliceuse. C. R. Acad. Sci., *256*: 497–498.

———. 1970. Les cellules des éponges, *in* W. G. Fry, ed., The Biology of the Porifera. Acad. Press, London. Zool. Soc. London, Symp., *25*: 353–364.

———, and A. Porte. 1962. Etude au microscope électronique de l'éponge *Oscarella lobularis* Schmidt. et da sa larve amphiblastula. Cah. Biol. Mar., *3*: 307–315.

Paris, Jean. 1960. Contribution à la biologie des éponges siliceuses *Tethya lyncurium* Lmck. et *Suberites domuncula* O.: Histologie des greffes et sérologie. Theses: Cansse, Graille, Castelnau, Imprimeur, Montpellier, 74 p.

Rasmont, R. 1962. The phsiology of gemmulation in fresh-water sponges. Soc. Study Dev. Growth, *20*: 3–25.

Simpson, T. L. 1963. The biology of the marine sponge *Microciona prolifera* (Ellis and Solander). I. A study of cellular function and differentiation. Jour. Exp. Zool., *154*: 135–152.

———. 1968a. The structure and function of sponge cells: new criteria for the taxonomy of Poecilosclerid sponges (Demospongiae). Peabody Mus. Nat. Hist., Bull., Yale Univ., *25*: 1–141.

———. 1968b. The biology of the marine sponge *Microciona prolifera* (Ellis and Solander). II. Temperature-related, annual changes in functional and reproductive elements with a description of larval metamorphosis. Jour. Exp. Mar. Biol. Ecol., *2*: 252–277.

Tardent, Pierre. 1968. Experiments about sex determination in *Hydra attenuata* Pall. Dev. Biol., *17*: 483–511.

Tuzet, O., and R. Connes. 1964. Sur la présence de lophocytes chez le *Sycon*. C. R. Acad. Sci., Paris, *258*: 4142–4143.

———, M. Pavans de Ceccaty, and J. Paris. 1963. Les éponges sont-elles des colonies? Arch. Zool. Exp. Gén., *102*: 14–19.

———, R. Garrone, and M. Pavans de Ceccaty. 1970. Observations ultrastructurales sur la spermatogenèse chez la Demosponge *Aplysilla rosea* Schulze (Dendroceratide): une métaplasie exemplaire. Ann. Sci. Nat. Zool., ser. 12, *12*: 27–50.

Vacelet, Jean. 1964. Étude monographique de l'éponge calcaire Pharetronide de Méditerranée, *Petrobiona massiliana* Vacelet et Lévi. Les Pharetronides actuelles et fossiles. Thèses. Louis-Jean-Gap, Imp., 125 p.

Van De Vyver, G. 1970. La non-confluence intraspécifique chez les spongiaires et la notion d'individu. Ann. Embryol. Morphogen., *3*: 251–262.

Vinogradov, A. P. 1953. The elementary chemical composition of marine organisms. Mem. Sears Foundation. Mar. Res., *2*: 647 p.

Warburton, F. E. 1958. Boring sponges, *Cliona* species, of eastern Canada, with a note on the validity of *C. lobata*. Can. Jour. Zool., *36*: 123–125.

Zhuravleva, T. 1970. Porifera, Sphinctozoa, Archeocyathi—their connections, *in* W. G. Fry, ed., The Biology of the Porifera. Acad. Press, London. Zool. Soc. London, Symp., *25*: 41–59.

Ziegler, B., and S. Rietschel. 1970. Phylogenetic relationships of fossil calcisponges, *in* W. G. Fry, ed., The Biology of the Porifera. Acad. Press, London. Zool. Soc. London, Symp., *25*: 23–40.

The Individuality of Sponges

Willard D. Hartman and Henry M. Reiswig　　　　　　Yale University

ABSTRACT

Historically the cell, the choanocyte chamber, the oscule with its associated canal systems, or the entire mass of sponge substance bounded by pinacocytes have been regarded as the individual in sponges. Most recent authors have favored one of the last two hypotheses. The plasticity of sponges manifest in the random distribution and ease of transformation of some cell types, the freedom of cell migration throughout the sponge, and the continual remodeling of the canal systems and oscules fails to support the notion that sponges are composed of discrete individual zooids that reproduce by asexual processes of budding. Phenomena such as the synchronous contraction of oscules and melanization that occur simultaneously throughout a sponge without regard to subunits imply mechanisms of coordination and favor consideration of the entire sponge as an individual. The calicular units of the calcareous skeleton of the sclerosponge *Ceratoporella,* while resembling superficially the zooidal skeletal units of such colonial animals as Cnidaria and Ectoprocta, house small units of tissue supplied with one or two dermal pores. The canal systems of many such calicular units drain into each oscule, indicating that the skeletal units are not homologous to the zooidal compartments of colonial animals.

Authors' present addresses: W. D. H., Department of Biology and Peabody Museum of Natural History, Yale University, New Haven, Connecticut 06520; H. M. R., Redpath Museum, McGill University, Montreal 101, Quebec, Canada.

INTRODUCTION

The concept of the individual among the Porifera has engaged the fancy of naturalists since the animality of these organisms was early suggested by Ellis (1766) and demonstrated more conclusively by Grant (1826). Early naturalists such as Dujardin (1841) and Lieber-kühn (1856) regarded the entire sponge body as an individual. These authors recognized the existence of protozoa-like cells that make up a sponge but emphasized the fundamental unity of the structure and function of the sponge as a whole. Lieberkühn attached particular importance to the fact that every sponge develops from a fertilized egg by way of a free-swimming larva. Contemporary and subsequent nineteenth-century authors tended to look upon the amoeboid or flagellated cells or aggregations of these as the sponge individual. Carter (1848) regarded the freshwater sponge as a congeries of amoebae and in this view was joined by Perty (1852). Carter (1857, 1869, 1871) noted pseudopodial processes extending from the cell bodies of choanocytes and hence regarded them as compound amoeboid-flagellated cells; Perty argued that the flagellated cells of sponges are foreign elements in the colony. In contrast to the views of these authors, who emphasized the amoeboid nature of the sponge individual, James-Clark (1866, 1867) regarded the sponge *Leucosolenia* as a compound, flagellated protozoan. Kent (1878a, 1878b) supported him in this view, and Carter (1872) proposed the name ''spongozoon'' for the flagellated collar cells. Because of the general random distribution of cell types, the freedom of cell migration, and the relative independence of cellular function, this view of the single cell as the basic unit of individuality has persisted as a minor but recurring interpretation of sponge organization (Bidder, 1937; Simons, 1963; Stempien, 1970).

Haeckel (1872), in developing his logical but fanciful system of tectology, defined the sponge individual in terms of the ''gastral cavity.'' Sponges such as *Leucosolenia* or *Clathrina*, of tubular form, with an outer layer of flattened cells or pinacocytes and an inner layer of choanocytes that line the ''gastral cavity'' in Haeckel's view, constitute the simplest sponge persons. Species of *Clathrina* with an anastomosing mass of tubules surmounted by a single oscule were regarded by Haeckel as colonies composed of numerous asconoid tubules that had arisen from the original postlarval individual through a process of budding.

A more complicated grade of construction among the Calcarea, the syconoid, is achieved theoretically by a folding of the wall of an asconoid tubule so that the choanocytes are now restricted to elongate finger-like outpushings of the wall, called radial canals. In Haeckel's view the syconoid sponge is a colony with each radial canal representing an individual that has arisen by strobiloid bud formation from a simple asconoid sponge.

The leuconoid grade of construction is achieved by a thickening of the wall of the sponge and the development of a network of branching canals. Haeckel at first (1872) accepted Schmidt's (1864) suggestion that the individual in a leuconoid sponge is defined by each oscule and the system of canals that drain into it. In branching sponges the colonies are composed of as many persons as there are longitudinal axes or branches. Later (1889, 1896), however, Haeckel accepted the choanocyte chambers as the true individuals of the leuconoid sponge as well, just as Carter (1857) had done during one period of his life.

The notion that the choanocyte chamber is the true sponge person, although exhaustively developed by Haeckel, is all but forgotten today. The series of grades of construction among the Calcarea, from asconoid through syconoid to leuconoid, may be considered as phenomena related to growth in size. As size increases, considerations of hydrodynamic efficiency demand a multiplication of the propelling units, the choanocyte chambers, and an increase in absorptive area also results.

A more acceptable theory, first suggested by Schmidt (1864) and cited initially by Haeckel (1872) in his postulate about individuality among leuconoid sponges, equates the sponge person to that unit of sponge tissue centered about a single oscule. Each oscule with the excurrent channels that flow into it as well as that portion of the incurrent system and choanocyte chambers that are drained by it, constitutes an individual. This theory was stated succinctly by Minchin (1900), who pointed out that regardless of the type of canal system, the metamorphosis of a larva or development of a gemmule results in the formation of a small sponge with a single oscule that represents the physiological and morphological center of the organism. In many sponges increase in size is accompanied by an increase in number of oscules, each of which may be considered the center of a new individual and the resultant sponge a colony. Hyman (1941) accepted this hypothesis with some reservation, primarily because of the lack of evidence of functional integration and hence "individuality" in entire multioscular sponges. More recently Beklemishev (1964), Hadzi (1966); and Brien (1967, 1968) have regarded the oscule as the sign of individuality among sponges. In the view of these authors an encrusting sponge with, say, six oscules would represent a colony composed of six physiological units each defined as an individual. In tall tubular sponges (Fig. 1) with a large central atrial cavity, many exhalant apertures open into the atrium, and their effluent currents unite to produce a powerful flow of water through an apical opening. It is more difficult to conceive of such sponges as being composed of numerous individuals. Indeed, as both Brien (1967) and Beklemishev (1964) point out, a gradual process of integration may occur during the course of evolution in colonial animals, with colonies expressing increasing degrees of individuality. In *Leucosolenia variabilis* Haeckel a series of upright tubes each surmounted by an oscule arises from a basal stolon in a fashion comparable to that of hydroid polyps budding from hydrorhizae; it is possible to consider each upright tube as an individual forming part of a colony. On the other hand, in the remarkable and highly specialized demosponge *Disyringa dissimilis* (Ridley), with incurrent and excurrent tubes at opposite ends of the spherical body, it stretches the imagination to consider the four to sixteen component exhalant canals running along the excurrent tube as signs of individuals forming a highly unified colony.

A source of difficulty in discussions of individuality versus coloniality among the Porifera is that in most sponges "le bourgeonnement se confond avec la croissance," to quote Brien (1967). As a sponge increases in size, new water current systems arise in such a way that it is difficult to define their origin as a result of a process of budding. Thus it is only by inferred homology with forms like the asconoid calcareous sponges that an oscule can be considered the sign of an individual in most leuconoid sponges.

Tuzet, Pavans de Ceccatty, and Paris (1963) made a strong case for regarding each anatomically isolated sponge as an individual on the basis of morphological, embryological, and physiological evidence. With their work, hypotheses about the

Figure 1

Verongia gigantea (Hyatt) at 48 m, Discovery Bay, Jamaica. Specimen about 55 cm in height. Photo by H. M. Reiswig.

individuality of sponges had come a full circle and returned to that of the naturalist Dujardin, albeit on a more sophisticated basis. A group of sponge biologists gathered together at the Station Biologique at Roscoff, France, during the summer of 1967 addressed themselves to problems of sponge terminology and defined an individual sponge as "a mass of sponge substance bounded by a continuous pinacoderm" (Borojević et al., 1968). In so doing, they agreed in essence with the views of Tuzet et al. (1963). Borojević (1970) has emphasized the importance of the pinacoderm in isolating the internal milieu of a sponge from its environment and thus providing a physiological unity to the whole organism. Thus there is a consensus among modern sponge biologists that sponges are individuals and not colonies, although Beklemishev (1964), Hadzi (1966), and Brien (1967) dissent from this view.

BEHAVIOR OF SPONGES

If the last three authors are correct we should expect evidence that each oscular unit of a sponge behaves independently of its neighbors with respect to such phenomena as contraction and water flow. Oscules vary greatly in ability to contract from species to species (Topsent, 1888). Among intertidal and shallow-water species asynchronous

Figure 2

Encrusting demosponge at 63 m, Discovery Bay, Jamaica. Note morphological continuity of adjoining excurrent canal systems. Photo by H. M. Reiswig.

contraction of oscules has been reported (Bowerbank, 1857; Parker, 1910; Pavans de Ceccatty, 1969). At the same time it is apparent that the excurrent water systems of adjacent oscules have a morphological continuity, a feature that is especially clearly seen in encrusting sponges (Fig. 2). The subdermal cavities commonly form an intercommunicating space throughout the sponge as well (Parker, 1910). Thus, under conditions in which one or more oscules of a sponge are closed while others remain open, as reported for *Spongia officinalis* Linn. and *Hippospongia communis* Lmk by Pavans de Ceccatty (1969), water currents may nonetheless be able to pass through the entire canal system by bypassing closed exhalant openings. The adaptive significance of oscular closure in intertidal sponges is suggested by Parker's (1910) experiments, in which he showed that the cessation of water currents over the tip of an oscule or its exposure to air causes closure. Thus as water recedes from a sponge during periods of ebbing tide, oscules exposed to air may close to prevent air from entering the sponge while those still submerged may remain open and allow the sponge to continue to pump water. It is more difficult to understand the asynchronous closure of oscules in subtidal habitats. It seems clear, however, that when some oscules of a sponge are closed, water currents may continue to flow through the entire sponge, albeit at a reduced rate.

Water being pumped from a specific underlying portion of tissue may exit through one of several oscules. In freshwater sponges newly developed from gemmules and growing between cover glass and microscope slide Kilian (1952) observed a reciprocal relationship in the behavior of oscules. Depending upon which one is functioning at the time, the direction of water in the exhalant canal system reverses to flow toward

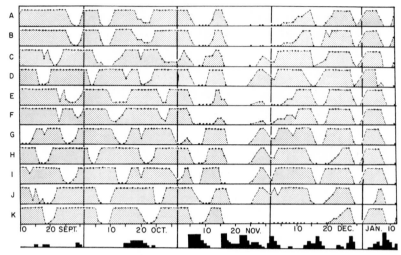

Figure 3

Variations in areas of eleven oscules of *Tethya crypta* (de L.) based on almost daily observations over a 4-month period. States of expansion are shown as shaded areas. Oscules of the same individual are represented by pairs A, B and E, F: note their synchrony of behavior. Relative index of wave strength shown in lowermost graph. (From Reiswig, 1971.)

the open oscule. This common arrangement in both marine and freshwater Demospongiae is adaptive in that it enables pumping and filtration to continue in tissues below a contracted oscule; it reinforces the independence of the active choanosomal tissues from oscular behavior.

Furthermore, in long-term studies of the behavior patterns of sponges in situ on coral reefs in Jamaica, Reiswig (1971) has shown that all oscules of a single sponge of some

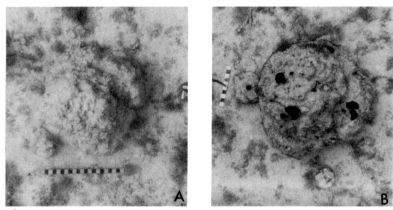

Figure 4

Tethya crypta (de L.) photographed at 2.4 m, Discovery Bay, Jamaica, showing same specimen with all oscules closed (A) and with all dilated (B). Scale: each black or white unit = 1 cm. Photos by H. M. Reiswig.

Figure 5

Verongia gigantea (Hyatt) photographed at 48 m, Discovery Bay, Jamaica, showing specimen (A) with all apertures of the exhalant canals open and another (B) with all apertures closed. Scale: black bar in B = 10 cm; same scale for A. Photos by H. M. Reiswig.

species behave in concert and not as independent entities. Through the use of continuously recording battery-powered thermistor sensors and hand-held current meters, the exhalant currents of several species of sponges were monitored, and species-specific activity rhythms were demonstrated. In *Tethya crypta* (de Laubenfels), for example, in which one to five oscules occur in the central region of the upper surface of the cake-shaped sponge, a basic diurnal rhythm occurs as well as a long-term cycle characterized by 1 to 5 days of complete contraction at intervals of 9 to 21 days. In a graph of the long-term cycle (Fig. 3), oscules of the same sponge (A, B and E, F) show identical activity patterns although different individuals within a population behave asynchronously, suggesting that the cycles are intrinsic and not environmentally generated. When the sponge is active all oscules are open, while all are closed during periods of inactivity (Fig. 4).

In *Verongia gigantea* (Hyatt), Reiswig found that water pumping activities cease every 19 hours (6.5–38.2 hr) on the average for periods of about 40 minutes. Again contraction of the apertures of the exhalant canals is simultaneous in the entire sponge (Fig. 5). The rhythm is again assumed to be intrinsic, since the periodic cessations are asynchronous in local populations.

Monitoring of individual oscules of subtidal encrusting, branching, or lobate sponges that possess many exhalant apertures distributed over the surface has not yet been carried out, and it is uncertain whether individual oscules in these sponges show a rhythmic behavior independent of their neighbors on the same sponge. In *Ectyoplasia ferox* (Duch. and Mich.) and *Anthosigmella varians* (Duch. and Mich.), Demospongiae with scattered oscules, synchrony of behavior of the oscules of any one sponge has been observed by Reiswig (1971), although the time characteristics of the behavior are not yet known.

ORGANIZATION OF SPONGE TISSUES

A further critique of the views of Beklemishev, Hadzi, and Brien necessitates a brief description of the tissue organization of the Porifera with inferences about the processes determining this organization. In the higher Demospongiae, cell types, except for pinacocytes and choanocytes, tend to be distributed in an apparent random fashion throughout the tissues without repeatable patterning, although peripheral zones of collencytes, lophocytes and some other cells may occur in some species. All sponges, however, play the role of suspension feeders with unidirectional flow through choanocyte chambers, the basic pumping and filtation units. In all sponges inhalant and exhalant flows are isolated at the chamber level and this separation is maintained throughout entire specimens, resulting in minimized recycling of expelled water (Bidder, 1923; Kilian, 1952). In very thin-walled sponges composed of tubes or simple large choanocyte chambers, it is sufficient for water to be taken into and expelled directly from the tubes or chambers. In order to permit thicker tissue layers with chambers more and more removed from the ambient environment, formation of discrete canals and canal systems are prerequisites. In a sponge with chambers randomly distributed through a tissue layer of uniform thickness, simple factors of canal system optimization (minimization of canal length and frictional debts) would result in a more-or-less regular distribution of inhalant and exhalant canals; spacing and diameter would depend upon specific metabolic rate, layer thickness, and proportion of active to passive tissue volumes. All evidence available indicates that the relationships between the tissue masses of sponges and the distribution and placement of inhalant and exhalant canal systems is a direct result of optimization of the following physical factors: reduction of friction, minimization of canal length, and maximization of separation of inhalant and exhalant systems. Time-lapse studies (Wintermann, 1951; Kilian, 1952, 1964) have shown that placement and distribution of oscules and/or canal systems vary with time and appear to exist in a state of dynamic equilibrium. Mergner (1964, 1970) has demonstrated that spheres of oscular tissue formed after excision of oscular tubes from spongillids may induce the formation of new oscules when placed on the pinacoderm of young specimens of the same species. Long-term observations of field populations of Demospongiae (Reiswig, 1971) and the cursory examination of almost any ramose or plate-shaped sponge clearly shows that oscules are transitory structures, constantly being closed, obliterated, or fused over while new ones arise on previously inhalant surfaces, apparently as canal systems are internally reorganized to maintain optimal conditions in the face of varying external and internal factors (Fig. 6). Time-lapse photography (Wintermann, 1951; Kilian, 1952, 1964; Efremova, 1967) has also shown that individual cells, like canal systems, are

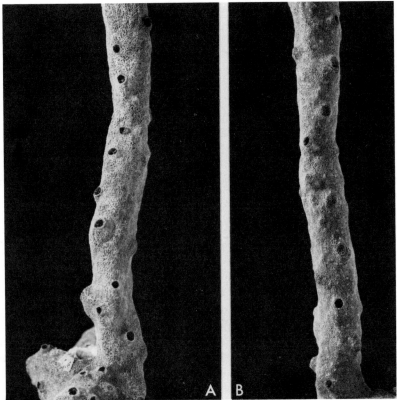

Figure 6

Portion of a branch of the West Indian demosponge *Haliclona rubens* (Pallas), showing many functional oscules on one side (A) and most oscules overgrown by ectosomal tissue on the opposite side (B) of the same specimen, probably resulting from a reorientation of the sponge in its habitat. ×7/8.

in continual motion within the sponge. Cells such as collencytes and lophocytes appear to be transient states of amoebocytes, depending upon their position within the sponge (Efremova, 1967). Recent studies have shown that histological samples taken during natural periods of pumping cessation reported by Reiswig (1971) in the Demospongiae *Tethya crypta* (de L.) and *Verongia gigantea* (Hyatt) exhibit the disappearance of both inhalant and exhalant fine canal systems within the choanosome, perhaps resulting from the general contraction of the sponge at these times comparable to the observations of Wintermann (1951) in spongillids. Peripheral canal reorganization is apparently a normal daily phenomenon in these species in natural situations.

This evidence leads us to the inescapable conclusion that an oscular unit cannot be interpreted as a structural unit of individuality; it represents an organizational *pattern* which is dictated by selective forces optimizing the relationships among a basically structureless aggregation of cells and flagellated chambers, the prerequisites for inhalant and exhalant water supplies to each chamber, and the hydrodynamic factors governing water movement to and from these chambers. All elements of this system are in more-or-less constant flux, without long-term integrity (Wintermann, 1951; Kilian, 1952, 1964, Borojević, 1970).

Both Hadzi and Brien interpret formation of new oscular units in sponges to be the result of asexual reproductive processes rather than simple growth. As indicated above, the thickness of the tissue layer in any given species would be expected to dictate the spacing of inhalant and exhalant canals, and thereby the spacing of oscules as well. Processes of size increase in most sponges (encrusting, tubular, and plate-shaped forms) effect an increase in areal extent of a tissue layer of relatively constant thickness. Cell division occurs randomly and more or less continually throughout the tissue mass, and edge extension is accomplished by cell migration and displacement (Simpson, 1963). Since tissue thickness remains static in the "parental" regions, new canal systems originate *de novo* marginally, not from subdivisions of the systems of the parental regions. In other species (cake-shaped and some encrusting forms) tissue thickness increases without substantial increase in area or density of surface canals and oscules. Instead, both inhalant and exhalant canals increase in diameter to handle larger fluid loads. In some cases, new oscular systems do arise by subdivision of preexistent parental systems, but these are once again interpreted to represent hydrodynamic optimization of the aquiferous system by minimizing the canal length to the exterior. In view of these simple methods of canal reorganization and addition of new aquiferous systems, the interpretation of increases in size and complexity of sponges as processes of asexual reproduction is fully dependent upon accepting oscular units as valid subindividuals, an untenable position as discussed above. Clearly cell division is widespread and is not localized at the extending surfaces. A newly forming oscule is not related to neighboring extant oscules by way of tissue anlagen in the same sense as the bud of a cnidarian carries with it an extension of the gastrovascular canal of its parental polyp; rather does it arise in response to hydrodynamic needs of the sponge as a whole. The entire process of size extension in the Demospongiae must be regarded simply as a true growth process, and only in the most obscure and theoretical sense can it be construed as being similar to a true budding process. The situation in the asconoid Calcarea, from which Hadzi and Brien obtained their basic data for regarding the oscule as a sign of the individual, is simply an extreme case of the process of surface extension. Tissue layers are generally restricted to two cells in thickness and the radial tubular form grows by layer extension; new oscules open as required to maintain hydrodynamic optimization and minimal recycling. The case of longitudinal fission in *Clathrina blanca* (Miklucho-Maclay) must be considered a special adaptation toward these ends. It is not known whether or not the oscules of specimens of asconoid calcareous sponges behave in unison, but it is of interest to note that transitory melanization of cells (Pavans de Ceccatty, 1958) occurs throughout specimens of *Leucosolenia botryoides* Ellis and Solander and *Clathrina coriacea* Montagu, indicating that component tubes do not behave individually in this respect.

The deep-water Indonesian demosponge *Esperiopsis challengeri* (Ridley) has a unique form (Fig. 7) comprising six or seven mushroom-shaped lamellae each attached by a stalk to a peduncle that is in turn attached to the substratum. Each lamella has numerous oscules on the upper surface and pores below, a common arrangement among fan-shaped sponges. Although it is possible to consider each of the component lamellae as originating

Figure 7

Esperiopsis challengeri (Ridley). Off Celebes, 1485 m. × 2/3. Individual or colony? (From Ridley and Dendy, 1887.)

by budding with each bud further subdividing to form numerous oscules and their associated canal systems, even in this curious sponge the entire organism is enclosed by pinacoderm that delimits an internal milieu endowed with a physiological unity. Mesenchymal strands undoubtedly run through the spicule-laden stalks and provide a morphological continuity between the subunits.

When two young spongillids of the same species, newly developed from gemmules, meet, they may fuse (Wintermann, 1951; Brien, 1967) and acquire a new individuality. Neighboring specimens of intertidal marine sponges may also coalesce when they meet (Burton, 1949). These observations again emphasize the regulatory capacities of sponges. Mature sponges may fail to fuse upon coming in contact (Brien, 1967), however, and among certain spongillids physiological strains occur wherein two specimens of the same strain fuse upon meeting while specimens of different strains fail to coalesce (van de Vyver, 1970). In the marine sponge *Crambe crambe* (Schmidt) fusion does not take place when either adults or metamorphosing larvae come together provided the latter have a different parentage (van de Vyver, 1970). It is possible that the specific differences in coalescence among these sponges relate to differing times at which biochemical specificities come into' action during the life cycle. Of special concern here are those instances in which fusion occurs, and the coalesced sponges reorganize their canal systems to form a unified whole.

WHAT IS AN OSCULE?

An additional but important criticism of the theory relating an oscule to an individual in sponges is the fact that throughout the phylum designations of oscules are not universally agreed upon. A generally accepted definition of an oscule is "an aperture through which water leaves a sponge" (Borojević et al., 1968), but what is considered to be an oscule by one worker may simply be the aperture of an exhalant canal leading by way of an atrial cavity to an oscule by another. Although in most cases structures interpretable as oscules can be agreed upon, a not inconsiderable number of cases represent intermediate and transitory conditions. For example, in a vase-shaped species of *Mycale* studied by Reiswig (1971) and in tubular specimens of *Callyspongia vaginalis* (Lamarck) what may be considered exhalant canal openings expel water into an atrial cavity that opens out apically by way of an oscule. In fan-shaped specimens of the same species, however, water flows out directly to the external medium by way of exhalant canal openings that are now called oscules in the absence of an atrium with its large terminal opening. Similarly, in young encrusting specimens of the same species of *Mycale* exhalant apertures identifiable as oscules become exhalant canal openings incorporated into the inner wall of the vase-like form of an adult; a new oscule that allows passage of water from all the new internal exhalant canal openings develops. Thus the application of the term "oscule" throughout the phylum is by necessity inconsistent since structural and functional homology of exhalant apertures does not exist over the ontogeny of a given specimen, between species, and certainly not across the entire phylum. If the term "oscule" cannot be applied with a high degree of certainty, the recognition of the individual implied by an oscule must also remain a relatively uncertain concept. This accumulation of structural, physiological, developmental, and semantic objections to the one oscule–one individual interpretation revived by Beklemishev, Hadzi, and Brien clearly leads us back to the concept of individuality expressed by Tuzet et al. and Borojević et al. Individuality within the Porifera is recognizable at the cellular and entire specimen levels.

Members of the phylum do not exhibit intermediate levels of subindividuality and, as such, cannot be considered as colonial animals.

DEVELOPMENT AND COORDINATION

The homology between cnidarian polyps and sponge olynthus, rhagon, and oscular units presumed by Brien (1967) is open to question. The regional specialization within tissue layers of the cnidarian polyp, the constancy of organization throughout the phylum, and the permanent relationship between apertures and internal structures such as septa are without analogy in the Porifera. The similarity of the two-layered larval stage of primitive cnidarians and the few sponges with olynthus stages represent the best evidence for the relationship, but here the homology of the two layers is in doubt (Borojević, 1966, 1970). The developmental processes of sponges remain unique among multicellular animals. The free-swimming larvae are blastulae (Lèvi, 1963); gastrulation occurs after settlement and in the case of most Demospongiae and Hexactinellida metamorphosis involves a reversal or reorganization of the larval cell layers. The outer flagellated cells of the larva migrate into the interior and form the flagellated chambers; the internal cells of the stereoblastula differentiate into two lines of cells, the pinacocyte–collencyte line and the archeocyte line, the pinacocytes taking up a position as the external boundary layer of the sponge. Among the calcaronean calcareous sponges the amphiblastula gastrulates as the anterior flagellated cells invaginate into the interior, while in the calcinean Calcarea the larva is a coeloblastula, the cavity of which later becomes filled with cells that migrate into it from the external larval layer. Differentiation of the cells of the now solid larva depends upon their position in the cell mass (Borojević, 1966, 1970). Borojević (1970), following Willmer (1960), recognizes two cell lines in developing sponges, an amoeboid and a flagellated line. The former differentiates early, especially in parenchymella larvae, into a pinacocyte–collencyte line and an amoeboid line. The common origin of the last-mentioned cell lineages sets sponges apart from the Eumetazoa, where mechanocytes and amoeocytes differentiate from the animal and vegetal halves of the embryo. Details of the developmental process thus fail to support a close homology between cnidarian polyps and sponges.

In commenting upon the reorganization of dissociated sponge cells Huxley (1912) says, "the image of a general directing his army, even of an architect arranging his materials, springs to mind: but again, where is the general, where the architect?" The search for coordinating mechanisms in sponges continues. Nerve cells comparable to those of other multicellular animals clearly do not occur in sponges, for their reactions are slow and action potentials have not been demonstrated (Jones, 1962; Prosser, Nagai, and Nystrom, 1962). The occurrence of periodic cessations of pumping and the simultaneous closure of oscules of entire sponges as observed by Reiswig (1971) strongly indicates some mechanism of coordination, however. General contractions might result from responses of contractile cells to stretch occasioned by the contraction of nearby cells or might be mediated by way of intercellular junctions (Pavans de Ceccatty, Thiney, and Garonne, 1970) or by way of neurohormones and neurosecretory products (Lentz,

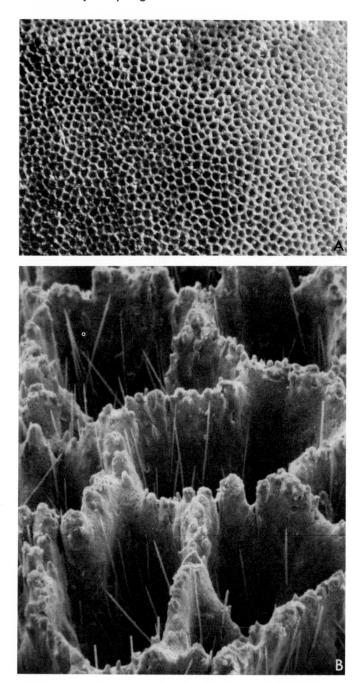

Figure 8

Surface of basal aragonitic skeleton of *Ceratoporella nicholsoni* (Hickson) showing calicular units. A. ×10. (From Hartman and Goreau, 1970.) B. ×100. (From Hartman and Goreau, 1972.)

1966). The existence of conducting cells is strongly indicated by the work of the last two authors, and such cells may play an important role in the coordinated behavior that supports the concept of individuality in sponges.

SCLEROSPONGES

The West Indian sclerosponge *Ceratoporella nicholsoni* (Hickson) secretes a basal aragonitic skeleton divided into calicular units (Fig. 8) that show a striking resemblance to the skeletal units that house the zooids of certain colonial cnidarians and ectoprocts. In *Ceratoporella*, however, each calicle bears a portion of the sponge tissue that is supplied by one or occasionally two inhalant pores (Hartman and Goreau, 1970). Water so entering the canal system of the sponge passes by way of choanocyte chambers to tiny exhalant canals that join to larger canal systems on the surface of the sponge. Each stellate exhalant system of canals opens by way of an oscule that therefore services the tissue of numerous calicles. Neighboring exhalant canal systems are often confluent as in encrusting Demospongiae (Fig. 9). The flow of water from individual calicle masses is here again governed by the path of least resistance, and, if the nearest oscule is

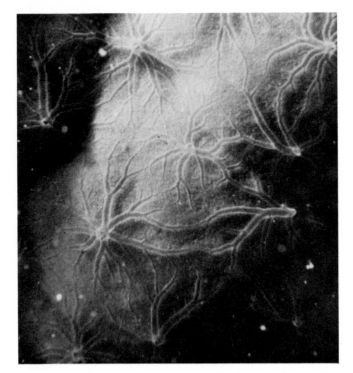

Figure 9

Surface of living specimen of the sclerosponge *Ceratoporella nicholsoni* (Hickson), showing dilated exhalant canal systems. Note morphological continuity of neighboring systems. ×2.5. (From Hartman and Goreau, 1970.)

contracted, the pumping and filtration activity can be maintained by diversion of the exhalant stream to more distant but connected oscules. In cases where the extent of dilation and contraction have been resolvable in nature, all oscules of entire specimens have been found to be in essentially identical states, implying integration at the entire specimen level once again. In spite of the expected coloniality implied by skeletal structures, all investigations of microscopic anatomy and field behavior of the organisms indicate a basic similarity to encrusting Demospongiae and, as such, they cannot be regarded as colonial organisms.

CONCLUSION

We have discussed interpretations of the organization of the Porifera on morphological, developmental, and physiological grounds and have presented new evidence of specimen integration. We find that individuality is recognizable at only two levels, in that of single cells and of entire confluent specimens and are unable to substantiate claims of the existence of intermediate units of subindividuality. It is possible to interpret the lack of recognizable subunits in one of two ways: either the individuality of previously existing zooids has been reduced to the point that they are unrecognizable and the colony has become highly individualized, or such zooids have never existed. We believe that our discussion supports the second conclusion. ''It is better to believe in the historical individuality of the cells and to wonder at the idea of the whole's form that can thus penetrate the substance and absorb the individualities of its parts, robbing them of all ancestral freedoms'' (Huxley, 1912).

REFERENCES

Beklemishev, W. N. 1964. Osnovy Sravnitelnoy Anatomii Bespozvonochnykh, Vol. I. Nauka, Moscow. (Engl. transl. by J. M. MacLennan, 1969, Principles of Comparative Anatomy of Invertebrates, v. I, Promorphology. Oliver & Boyd, Edinburgh, 490 p.)

Bidder, G. P. 1923. The relation of the form of a sponge to its currents. Quart. Jour. Micros. Sci., *67*: 293–323.

———. 1937. The perfection of sponges. Linn. Soc. London, Proc., *149*: 119–146.

Borojević, Radovan. 1966. Etude expérimentale de la différenciation des cellules de l'éponge au cours de son développement. Dev. Biol., *14*: 130–153.

———. 1970. Différenciation cellulaire dans l'embryogénèse et la morphogénèse chez les spongiaires, *in* W. G. Fry, ed., The Biology of the Porifera. Zool. Soc. London, Symp., *25*: 467–490.

———, W. G. Fry, W. C. Jones, Claude Lévi, Raymond Rasmont, Michele Sarà, and Jean Vacelet. 1968. Mise au point actuelle de la terminologie des éponges. (A reassessment of the terminology for sponges.) Mus. Nat. Hist., Paris, Bull., sér. 2, *39*: 1224–1235.

Bowerbank, J. S. 1857. On the vital powers of the Spongiadae. Brit. Ass. Advance. Sci., Rept., *1856*: 438–451.

Brien, Paul. 1967. Les éponges: leur nature métazoaire; leur gastrulation; leur état colonial. Ann., Soc. Roy. Zool. Belg., *97*: 197–235.

———. 1968. The sponges, or Porifera, p. 1–30 *in* Marcel Florkin and B. T. Scheer, eds., Chemical Zoology, v. 2. Academic Press, New York.

Burton, Maurice. 1949. Observations on littoral sponges, including the supposed swarming of larvae, movement and coalesence in mature individuals, longevity and death. Zool. Soc. London, Proc., *118*: 893–915.

Carter, H. J. 1848. Notes on the species, structure and animality of the freshwater sponges in the tanks of Bombay. Ann. Mag. Nat. Hist., ser. 2, *1*: 303–311.

———. 1857. On the ultimate structure of *Spongilla* and additional notes on freshwater Infusoria. Ann. Mag. Nat. Hist., ser. 2, *20*: 21–41, 1 pl.

———. 1869. On *Grayella cyathophora*, a new genus and species of sponges. Ann. Mag. Nat. Hist., ser. 4, *4*: 189–197, pl. 7.

———. 1871. A description of two new Calcispongiae, to which is added confirmation of Prof. James-Clark's discovery of the true form of the sponge-cell (animal) and an account of the polype-like pore-area of *Cliona corallinoides* contrasted with Prof. E. Haeckel's view on the relationship of the sponges to the corals. Ann. Mag. Nat. Hist., ser. 4, *8*: 1–27, pl. 1, 2.

———. 1872. Proposed name for the sponge-animal, viz. "Spongozoon"; also on the origin of thread-cells in the Spongiadae. Ann. Mag. Nat. Hist., ser. 4, *10*: 45–51.

Dujardin, Félix. 1841. Histoire naturelle des zoophytes. Infusoires, comprenant la physiologie et la classification de ces animaux et la manière de les étudier à l'aide du microscope. Librairie Encyclopédique de Roret, Paris. 684 p., 22 pl.

Efremova, S. M. 1967. The cell behavior of the fresh-water sponge *Ephydatia fluviatilis*: a time-lapse microcinematography study Acta Biol. Hung., *18*: 37–46.

Ellis, John. 1766. On the nature and formation of sponges: in a letter from John Ellis, Esquire, F. R. S. to Dr. Solander, F. R. S. Roy. Soc. London, Phil. Trans., *55*: 280–289, pls. 10, 11.

Grant, R. E. 1826. Observations and experiments on the structure and functions of the sponge. Edinb. Phil. Jour. *14*: 113–124.

Hadzi, Jovan. 1966. Vprašanje individualitete pri spužvah. [Le problème de l'individualité chez les éponges.] Slov. Akad. Znan. Umet., Razpr. 9 (Razred 4): 167–204.

Haeckel, Ernst. 1872. Die Kalkschwämme, v. 1 (Genereller Theil.) Biologie der Kalkschwämme. Georg Reimer, Berlin. 484 p.

———. 1889. Natürlicheschöpfungs-Geschichte. George Reimer, Berlin. 832 p., 20 pl.

———. 1896. Systematische Phylogenie, pt. 2, Systematische Phylogenie der wibellosen Thiere (Invertebrata). Georg Reimer, Berlin. 720 p.

Hartman, W. D., and T. F. Goreau. 1970. Jamaican coralline sponges: their morphology, ecology and fossil relatives, *in* W. G. Fry, ed., The Biology of the Porifera. Zool. Soc. London, Symp., *25*: 205–243.

———, and T. F. Goreau. 1972. *Ceratoporella* (Porifera: Sclerospongiae) and the chaetetid "corals". Conn. Acad. Arts Sci. Trans., *44*: 133–148.

Huxley, J. S. 1912. The Individual in the Animal Kingdom. Cambridge Univ. Press, Cambridge. 167 p.

Hyman, L. H. 1940. The Invertebrates: Protozoa Through Cetenophora. McGraw-Hill, New York. 726 p.

James-Clark, H. 1866. Conclusive proofs of the animality of the ciliate sponges and of their affinities with the Infusoria Flagellata. Am. Jour. Sci., ser. 2, *42*: 320–324.

———. 1867. On the Spongiae Ciliatae as Infusoria Flagellata: or, observations on the structure, animality and relationship of *Leucosolenia botryoides* Bowerbank. Boston Soc. Nat. Hist. Mem., *1* (pt. 3): 1–36, pl. 9, 10.

Jones, W. C. 1962. Is there a nervous system in sponges? Biol. Rev., *37*: 1–50.

Kent, W. S. 1878a. Observations upon Professor Ernst Haeckel's group of the "Physemaria" and on the affinity of the sponges. Ann. Mag. Nat. Hist., ser. 5, *1*: 1–17.

———. 1878b. Notes on the embryology of sponges. Ann. Mag. Nat. Hist., ser. 5, *2*: 139–156, pl. 6, 7.

Kilian, E. F. 1952. Wasserströmung und Nahrungsaufnahme beim Süsswasserschwamm *Ephydatia fluviatilis*. Z. vergl. Physiol., 34: 407–447.

———. 1964. Zur Biologie der einheimischen Spongilliden: Ergebnisse und Probleme unter besonderer Berücksichtigung eigener Untersuchungen. Zool. Beitr., ser. 2, *10*: 85–159.

Lentz, T. L. 1966. Histochemical localization of neurohumors in a sponge. Jour. Exp. Zool., *162*: 171–180.

Lévi, Claude. 1963. Gastrulation and larval phylogeny in sponges, p. 375–382, *in* E. C. Dougherty, Z. N. Brown, E. D. Hanson, and W. D. Hartman, eds., The Lower Metazoa: Comparative Biology and Phylogeny. University of Calif. Press, Berkeley.

Lieberkühn, N. 1856. Zusätze zur Entwickelungsgeschichte der Spongilliden. Arch. Anat. Physiol., *1856*: 496–514, pl. 18.

Mergner, Hans. 1964. Uber die Induktion neuer Oscularrohre bei *Ephydatia fluviatilis*. Roux Arch. Entwick-lungsmech., *155*: 9–128.

———. 1970. Ergebnisse der Entwicklungsphysiologie bei Spongilliden, *in* W. G. Fry, ed., The Biology of the Porifera. Zool. Soc. London, Symp., *25*: 365–397.

Minchin, E. A. 1900. Sponges, *in* E. R. Lankester, ed., A Treatise on Zoology, pt. 2. Adam and Charles Black, London. p. 1–178.

Parker, G. H. 1910. The reactions of sponges, with a consideration of the origin of the nervous system. Jour. Exp. Zool., *8*: 1–41.

Pavans de Ceccatty, Max. 1958. La mélanisation chez quelques éponges calcaires et siliceuses: ses rapports avec le système réticulohistiocytaire. Arch. Zool. Exp. Gén., *96*: 1–51.

———. 1969. Les systèmes des activités motrices, spontanées et provoquées des éponges: *Euspongia officinalis* L. et *Hippospongia communis* LMK. Acad. Sci. Paris, sér. D, Compt. Rend. *269*: 596–599.

———, Y. Thiney, and R. Garonne. 1970. Les bases ultrastructurales des communications intercellulaires dans les oscules de quelques éponges, *in* W. G. Fry, ed., The Biology of the Porifera. Zool. Soc. London, Symp., *25*: 449–466.

Perty, J. A. M. 1852. Zur Kenntnis kleinster Lebensformen nach Bau, Funktionen, Systematik, mit Spezialver-zeichniss der in der Schweiz beobachteten. Bern. 228 p.

Prosser, C. L., T. Nagai, and R. A. Nystrom. 1962. Oscular contractions in sponges. Comp. Biochem. Physiol., *6*: 69–74.

Reiswig, H. M. 1971. *In situ* pumping activities of tropical Demospongiae. Mar. Biol., *9*: 38–50.

Ridley, S. O., and Arthur Dendy. 1887. Report on the Monaxonida collected by H.M.S. Challenger during the years 1873–76. Voyage of H.M.S. Challenger, Rep. Sci. Res., Zool., *20*: 1–275, 51 pl.

Schmidt, Oscar. 1864. Supplement der Spongien des adriatischen Meeres, enthaltend die Histiologie und systematische Ergänzungen. Wilhelm Engelmann, Leipzig, 48 p., 4 pl.

Simons, J. R. 1963. Sponges—the first republicans. Australian Nat. Hist., *14*: 194–197.

Simpson, T. L. 1963. The biology of the marine sponge *Microciona prolifera* (Ellis and Solander), I, A study of cellular function and differentiation. Jour. Exp. Zool., *154*: 135–152, 2 pl.

Stempien, M. F., Jr. 1970. Sponges. Animal Kingdom, 7(4): 2–7.

Topsent, Emile. 1888. Contribution à l'étude des clionides. Arch. Zool. Exp. Gén., sér. 2, *5* bis, suppl.: 1–165, pl. 1–7.

Tuzet, Odette, Max Pavans de Ceccatty, and Jean Paris. 1963. Les éponges sont-elles des colonies? Arch. Zool. Exp. Gén., *102*: 14–19.

van de Vyver, Gysèle. 1970. La non-confluence intraspécifique chez les spongiaires et la notion d'individu. Ann. Embryol. Morphogen., *3*: 251–262.

Willmer, E. N. 1960. Cytology and Evolution. Academic Press, New York. 403 p.

Wintermann, Gertrud. 1951. Entwicklungsphysiologische Untersuchungen an Süsswasserschwämmen. Zool. Jahrb. Abt. Anat., *71*: 427–486.

Modular Structure and Organic Integration in Sphinctozoa*

Robert M. Finks   Queens College (CUNY)

ABSTRACT

The Sphinctozoa are an extinct group of calcareous sponges whose skeleton is organized into a series of hollow chambers. These chambers form a level of structural organization intermediate between that of cells of an organism and that of individual members of a colony. The size of these modular units,[1] which is relatively constant in a given individual and is species-characteristic, seems to be determined by internal factors. These are operative at the time of growth, for the organism increases in size by adding chambers peripherally. Subsequently the flesh may be withdrawn from a chamber, moving distally as shown by a succession of imperforate internal partitions as in *Girtycoelia*. The modules are partly independent of larger functional units such as the exhalent system (osculum plus tributary canals). In *Guadalupia* a monolayered sheet of chambers is surmounted by the skeletal traces of the exhalent systems, permitting direct analysis of the relations between the two organizational units. In two species the units are independent with respect to size; the larger exhalent systems

Author's address: Department of Geology, Queens College, The City University of New York, Flushing, New York 11367.

*Reprinted from Geological Society of America, Abstracts with Programs (1971 Annual Meeting, Washington), v. 3, no. 7, p. 764–765.

[1]A *modular unit* is a structural unit which makes up, in the aggregate, a given individual or other entity. A *colony* would be a structural entity made up of modular units having the form of individual organisms. (*Note added in proof.*)

serve more chambers than the smaller. In another species, however, chamber volume varies directly with the total volume served by the exhalent system, the number of chambers per exhalent system being relatively constant. This, as well as the fact that oscular diameter varies directly with chamber height, suggests some coupling between factors determining chamber size and those determining exhalent system size. The exhalent systems are not equivalent to individuals nor are skeletal chambers equivalent to choanocyte chambers.

INDEX TO BIOLOGICAL NAMES

[Only subgenera and higher taxa are indexed.]

I. Coelenterata

Acanella, 82
Acanthogorgiidae, 82
Acropora, 33
Acroporidae, 33
Actinacidae, 20, 24
Agaricia, 33
Agaricidae, 33
Alcyonacea, 71, 75–81, 91
Alcyonaria, 46, 47, *see also* Octocorallia
Alectorurus, 90
Anemones, pelagic, 109
Antennella, 484
Antennelopsis, 484
Anthomedusae, 110–111, 113
Anthoplexaura, 70
Anthozoa, 1–93
Antipatharia, 85
Astreopora, 44
Astrocoeniida, 33
Auloporida, 13

Bathyalcyon, 81
Bougainvillia, 111

Calyptrophora, 83
Campanulariidae, 111
Caryophyllida, 33
Caryophyllidae, 33
Cavernularia, 88
Chaetetida, 18
Charnia, 90
Chelophyes, 101

Chondrophora
  as colonies, 108ff.
  not as colonies, 99
Chrysaora, 113, 115
Chrysogorgia, 85
Chrysogorgiidae, 84, 86, 91
Chrysomitra, 108
Chunella, 88
Cladocora, 33, 46, 49
Clathrodictyon, 52
Clava, 97
Clavidae, 111
Clavularia, 73–75
Clavulariidae, 73
Cnidaria, 396, 560, 564, 576, 579, 581
  holopelagic, 2, 107–117
  pelagic, 108
Coelenterata, vii, 1–118, 230, 396, 547, 560,
    564, 576, 579, 581
Coelogorgia, 91
Coenothecalia, 70, 90
Colpophyllia, 33
Coralliidae, 81
Corallium, 82
Cordylophora, 97–98, 104, 105
Corhiza, 484
Cornularia, 72
Cornulariidae, 73
Cystihalysites, 15, 21
Cystiphylloides, 62
Cytaeis, 111, 112

II. Bryozoa (p. 119–437)

III. Graptolithina (p. 439–545)

SUBJECT INDEX